CROP–WATER RELATIONS

CROP–WATER RELATIONS

Edited by

I. D. Teare

Agricultural Research and Education Center, University of Florida, Quincy, Florida

M. M. Peet

North Carolina State University, Raleigh, North Carolina

A Wiley-Interscience Publication

JOHN WILEY & SONS

New York Chichester Brisbane Toronto Singapore

Library of Congress Cataloging in Publication Data:

Main entry under title:

Crop-water relations.

"A Wiley Interscience publication."
Includes index.
1. Crops and water. 2. Field crops—Water requirements. I. Teare, I. D. (Iwan Dale), 1931–
II. Peet, M. M. (Mary Monnig), 1947–

SB185.6.C76 1982 633 82-8671
ISBN 0-471-04630-2 AACR2

Printed in the United States of America

10 9 8 7 6 5 4 3 2 1

CONTRIBUTORS

D. A. Ashley, Department of Agronomy, University of Georgia, Athens, Georgia

B. L. Blad, Department of Agricultural Engineering, University of Nebraska, Lincoln, Nebraska

K. J. Boote, Department of Agronomy, University of Florida, Gainesville, Florida

G. J. Burch, Division of Plant Industry, CSIRO, Canberra, ACT, Australia

G. J. Gascho, Department of Agronomy, Georgia Coastal Plains Experiment Station, Tifton, Georgia

A. J. Halterlein, Delta Branch Experiment Station, Stoneville, Mississippi

G. H. Heichel, Department of Agronomy and Plant Genetics, University of Minnesota, St. Paul, Minnesota

W. R. Jordan, Temple-Blackland Conservation Research Center, Texas A&M University, Temple, Texas

E. T. Kanemasu, Evapotranspiration Laboratory, Kansas State University, Manhattan, Kansas

M. B. Kirkham, Evapotranspiration Laboratory, Kansas State University, Manhattan, Kansas

G. E. Kleinkopf, Plant and Soil Science Department, University of Idaho, Kimberley, Idaho

D. R. Krieg, Department of Agronomy, Texas Technical University, Lubbock, Texas

G. N. McCauley, Agricultural Research Center, Texas A&M University, Beaumont, Texas

S. S. Martin, USDA Crop Research Laboratory, Colorado State University, Fort Collins, Colorado

S. F. Shih, Belle Glade Research and Education Center, University of Florida, Belle Glade, Florida

S. J. Thien, Department of Agronomy, Kansas State University, Manhattan, Kansas

F. T. Turner, Agricultural Research Center, Texas A&M University, Beaumont, Texas

N. C. Turner, Division of Plant Industry, CSIRO, Canberra, ACT, Australia

R. P. Waldren, Department of Agronomy, University of Nebraska, Lincoln, Nebraska

PREFACE

Water relations in crops is the domain not only of the plant physiologist and soil physicist, but also of the micrometeorologist, ecologist, agronomist, and horticulturist. In consequence, we have prepared this book with two main objectives. The first of these is to present a review of current research on the physical and physiological aspects of water transfer in the biosphere. We have included research concerning the atmospheric factors that affect energy exchanges in the biosphere, particularly evaporation; the soil factors that affect soil water status and water supply to crop roots; the distribution, movement, and function of water in plant cells, tissues, and organs; and the development of internal water deficits in relation to physiological processes.

The second objective is to present a review of the water relations of the twelve most important food and fiber crops grown today. In dealing with individual crops, we have asked each contributor to emphasize root growth, the canopy structure, the interaction of the crop with its environment, and the applied aspects of irrigation. We have also asked them to identify physiological stages of development for each crop and to relate them to critical water demand periods.

Technical terms from all fifteen chapters are included in a glossary at the end of the book. Symbols, however, are listed separately at the end of each chapter in their order of usage.

The subject matter coverage was developed from a course of lectures given to advanced graduate students at Kansas State University entitled "Crop Water Relations" during the years 1975 through 1978. At that time the senior author was a member of the Evapotranspiration Laboratory at Kansas State University and a Professor in the Department of Agronomy. We would like to acknowledge Dr. P. J. Kramer, Dr. Nasser P. Sionit, and other members of the Phytotron staff at Duke University for their generous assistance during the period this book was being prepared.

I. D. TEARE
M. M. PEET

Quincy, Florida
Raleigh, North Carolina
July 1982

CONTENTS

CROP–WATER RELATIONS

1

ATMOSPHERIC DEMAND FOR WATER

Blaine L. Blad

INTRODUCTION

Water has many unique properties including its ability to exist in the solid, liquid, or vapor phase within the range of air temperatures occurring on earth. It is possible for all three phases to exist at the same time, but it is the change from one phase to another that is of interest to us, particularly the change from liquid to vapor. The phase change from ice to water requires 80 cal g^{-1} (334 kJ kg^{-1}). The change from liquid to vapor, the latent heat of vaporization (L), requires approximately 585 cal g^{-1} (2.44 MJ kg^{-1}) at 20°C.

The physical process for the change of liquid water to gaseous water vapor is called evaporation (E). Evaporation occurs from oceans, lakes, ponds, rivers, soil, or other wet surfaces. Transpiration (T_p) is the evaporation of water that has been absorbed through plant roots, transported through the plant, and then removed from leaf surfaces. It is controlled by the stomatal aperture and by the vapor pressure gradient from the leaf to the air. The term evapotranspiration (E_t) is commonly used to describe the total process of water vapor transfer into the atmosphere from vegetated land surfaces.

ENERGY EXCHANGE AMONG LEAF, CANOPY, AND ENVIRONMENT

General

The vast quantities of energy consumed in E_t are supplied almost entirely by sources traceable to the sun. This energy may come from direct, diffuse, or reflected sunlight; it may come from solar radiation absorbed by molecules and particles in the air or from objects on the surface of the earth where it is reradiated as long-wave radiation; or it may come through energy transferred to cool plants from air heated by solar radiation.

Solar radiation received at the earth's surface is concentrated primarily in the 0.30–3.0 μm waveband portion of the electromagnetic spectrum (short-wave radiation), and is composed of a direct beam and a diffuse component. Long-wave radiation, radiation concentrated primarily in the 3.0–40.0 μm waveband, also contributes energy for E_t. The difference between the incoming and the outgoing radiation streams is termed net radiation (R_n) and may be stated in the form of an equation as follows:

$$R_n = R_s - \alpha R_s + R_a - R_g \tag{1.1}$$

where R_s and αR_s are incoming and outgoing short-wave radiation and R_a and R_g are incoming and outgoing long-wave radiation, respectively. The sign convention adopted in this chapter is such that fluxes toward the surface are positive and those away from the surface are negative. Maximum midday values of R_n over actively growing vegetation will seldom exceed about 1 cal cm^{-2} min^{-1} (697 W m^{-2}).

Net radiation provides energy for E_t, for heating the air, soil, and plants, and for photosynthesis. It is partitioned into the energy consuming processes:

$$R_n = -(LE + H + S + P_s + M) \tag{1.2}$$

where LE is latent heat flux (energy used in E_t), H is the sensible heat exchange with the air, S is the soil heat flux, P_s is the energy for photosynthesis, and M is the term for any miscellaneous energy exchanges including energy for metabolic activity or energy storage in the plant canopy. Most of the energy supplied by R_n is consumed as LE, lesser but significant amounts go into H and S, while the amount utilized in P_s and M is generally less than 1–2% of the total. Thus, Eq. (1.2) can be simplified to:

$$R_n = -(LE + H + S) \tag{1.3}$$

Leaf

Both sides of a leaf exchange radiation with all objects "seen" by the leaf (Fig. 1.1). For a horizontal leaf above the soil surface the net amount of radiant energy absorbed by the leaf (R_l) is:

$$R_l = R_s - \alpha_l R_s + \alpha_g R_s - \alpha_l \alpha_g R_s + R_a + \epsilon_g \sigma T_g^4 - \tau_l R_s - \tau_l \alpha_g R_s$$
$$- (1 - \epsilon_l) R_a - (1 - \epsilon_l)\epsilon_g \sigma T_g^4 (1 - \epsilon_g) R_a - (1 - \epsilon_g)(1 - \epsilon_l) R_a \tag{1.4}$$

where σ is the Boltzmann constant; ϵ_g and ϵ_l are the emissivity of the soil surface and leaf, respectively; α_l and α_g are the reflection coefficients for the leaf and ground, respectively; T_g is the temperature (°K) of the ground; and τ_l is the transmission coefficient for the leaf.

Certain of the radiation streams, i.e., $(1 - \epsilon_l)R_a$, $(1 - \epsilon_g)R_a$, and $(1 - \epsilon_l)\epsilon_g \sigma T_g^4$, are very small and can be neglected. The equation can, therefore, be rearranged and simplified to give:

$$R_l = R_s(1 - \alpha_l)(1 + \alpha_g) + R_a + \epsilon_g \sigma T_g^4 - \tau_l(R_s + \alpha_g R_s) \tag{1.5}$$

The absorbed energy, R_l, is dissipated by reradiation, convection, and transpiration. The reradiated energy is equal to $2\epsilon_l \sigma T_l^4$ where T_l is the temperature (°K) of the leaf. This term may be combined with those in Eq. (1.5) to represent the net radiation of the leaf (R_{nl}) as follows:

$$R_{nl} = R_s(1 - \alpha_l)(1 + \alpha_g) + R_a + \epsilon_g \sigma T_g^4$$
$$- 2\epsilon_l \sigma T_l^4 - \tau_l(R_s + \alpha_g R_s) \tag{1.6}$$

Additional discussion of factors affecting net radiation of a leaf is contained in Monteith (1).

The following example is illustrative of typical values that might be encountered on a sunlit leaf during a clear, summer, midday period. $R_s = 1.4$ cal cm^{-2} min^{-1} (976 W m^{-2}), $R_a = 0.40$ cal cm^{-2} min^{-1} (279 W m^{-2}), $T_l = 300$°K, $T_g = 310$°K, $\alpha_l = 0.25$, $\tau_l = 0.25$, $\alpha_g = 0.15$, $\epsilon_l = 0.97$, $\epsilon_g = 0.95$. Using these values we have, by Eq. (1.6):

$$R_{nl} = 1.20 + 0.40 + 0.71 - 1.58 - 0.40$$
$$= 0.33 \text{ cal cm}^{-2} \text{ min}^{-1} \text{ (231 W m}^{-2}\text{)}$$

If this leaf were shaded by other leaves in a canopy, then R_{nl} could be drastically different. The solar radiation impinging on the leaf would be reduced to about one-fourth of full R_s so that R_s in Eq. (1.6) would

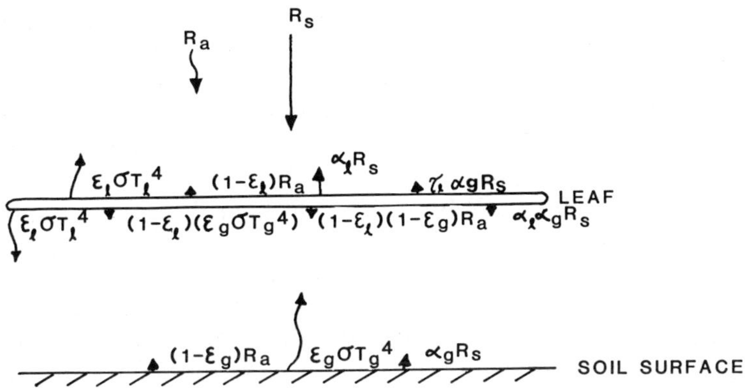

Figure 1.1. Diagram showing all radiation streams impinging on and leaving a leaf surface over the ground. Straight arrows indicate short-wave fluxes and wavy arrows indicate long-wave fluxes of radiation. Length of the arrows indicates relative flux densities that might be expected during a clear, summer midday period. R_s is incoming solar radiation, R_a is incoming long-wave radiation, ϵ_l and ϵ_g are emissivities, α_l and α_g are reflextivities, τ_l is the transmissivity of the leaf, and T_l and T_g are temperatures (°K) of the leaf and ground, respectively.

be replaced by $0.25R_s$ and the long-wave radiation from the sky, R_a, would be replaced by the radiation emitted by leaves "seen" by the shaded leaf. In this case a reasonable estimate of R_{nl}, if T_l remains at about 300°K, would be:

$$R_{nl} = 0.30 + 0.79 + 0.71 - 1.58 - 0.10$$

$$= 0.12 \text{ cal cm}^{-2} \text{ min}^{-1} \text{ (84 W m}^{-2})$$

The difference in R_{nl} between sunlit and shaded leaves has important implications for leaf temperature, leaf transpiration, and the exchange of sensible heat with the surrounding air. Because R_{nl} of shaded leaves is low such leaves are generally at or near the temperature of the surrounding air, whereas the sunlit leaves may be a few degrees warmer than the air. Sunlit leaves can transpire large amounts of water but much less water is lost from shaded leaves. Shaded leaves exchange almost no sensible heat with the surrounding air but sunlit leaves may exchange a significant amount.

The sensible heat exchange (H_l) of a leaf with the surrounding air may be described by:

$$H_l = \frac{\rho_a C_p (T_a - T_l)}{r_a} \tag{1.7}$$

where T_a is the temperature of the air (°K), ρ_a is the air density, C_p is the specific heat of air at constant pressure, and r_a is the resistance to the flow of heat through the boundary layer of air surrounding the leaf. Heat is transferred either by natural convection (due to buoyancy effects) or by forced convection (due to windspeed effects). Except for very low windspeeds the amount of heat transferred by forced convection is much greater than that due to natural convection. Parlange et al. (2) show that, even for velocities as low as 60 cm/sec, forced convection carries 10 times more heat than natural convection. If the leaf is warmer than the air some fraction of R_{nl} will be utilized to heat the surrounding air. If $T_a > T_l$, heat will be transferred from the air to the leaf. H_l will, therefore, be an additional source of energy for transpiration or for increasing the leaf temperature.

The latent heat flux from a leaf is the third major mechanism for the exchange of energy between a leaf and its environment. This flux may be described by:

$$LE = \frac{M_w/M_a}{P} L\rho_a \frac{(e_l - e_a)}{r_a + r_s} = \frac{\rho_a C_p}{\gamma} \frac{(e_l - e_a)}{r_a + r_s} \tag{1.8}$$

where M_w/M_a is the ratio (0.622) of the mole weights of water vapor (M_w) and air (M_a), P is the atmospheric pressure, e_l is the vapor pressure at the leaf surface, e_a is the vapor pressure of the air, γ is the psychrometric constant ($\gamma = PC_p/LM_w/M_a = 0.66$ mb °C^{-1} at 20°C and $P =$ 1000 mb), and r_s is the diffusive resistance to the flow of water vapor from the internal to the external surface of a leaf. This latter term is principally the resistance to the diffusion of water vapor through the plant stomata. Only minor amounts of water vapor may, in some plants, diffuse through the leaf cuticle. r_s is usually on the order of 1 to 10 or 20 sec cm^{-1} and r_a is generally 0.1 to 1 or 2 sec cm^{-1} (2). More detailed discussion of energy exchanges between a leaf and the surrounding environment can be found in Gates (3) and Montieth (1).

Plant Canopy

A plant canopy is composed of plants and their individual parts. The exchange of energy between the aerial environment and the canopy is more complicated than that just described for a single leaf. Leaves within a canopy exchange energy with neighboring leaves and other plant parts. The irradiance of leaves at various levels within a plant canopy is influenced by the amount of plant cover and by the manner in which plant, parts, particularly leaves, are arranged, distributed, oriented, and inclined within the canopy. A review describing procedures for estimating the penetration of short-wave radiation into plant canopies is provided by Lemeur and Blad (4). Net radiation generally decreases with increasing depth into the canopy and most of the radiation is captured in the upper canopy levels. Thus, for example, Denmead et al. (5) found that almost three fourths of the net radiation absorbed by a corn canopy was captured in the upper half of the canopy.

In the exchange of energy with the atmosphere, various levels within a plant canopy may serve either as sinks or sources for sensible heat, water vapor, CO_2, and other materials. Because of the complexities and difficulties associated with describing energy and mass exchanges within plant canopies there are few studies that have been made and few models developed which accurately simulate these processes. Begg et al. (6) conducted one of the few such studies of mass and energy exchanges using a crop of bulrush millet. One of the best simulation models of these processes was recently developed by Norman (7).

In some cases it is only necessary to understand what the entire canopy is doing and not what is happening at different levels within the canopy. In such cases, the radiation and energy exchanges between the canopy and the atmosphere can be determined by evaluating differences between the radiative and energy streams passing through some defined plane above the canopy and another plane just above the soil surface. The net radiation above the canopy (R_n) may be written as:

$$R_n = R_s - \alpha_c R_s + R_a - R_a(1 - \epsilon_c) - \epsilon_c \sigma T_c^4$$
$$= R_s(1 - \alpha_c) + \epsilon_c R_a - \epsilon_c \sigma T_c^4 \quad (1.9)$$

where α_c is the albedo of the crop, ϵ_c is the crop emissivity, and T_c is the crop temperature. Similarly, the net radiation below the canopy (R_{nb}) can be described. Each radiation stream will be modified significantly by the plant canopy. We are interested in the energy absorbed by the canopy, i.e., $R_n - R_{nb}$. This difference is related to the canopy energy balance by:

$$R_n - R_{nb} = T_p + H_c \quad (1.10)$$

where H_c is the sensible heat exchange between the canopy and the surrounding air. H_c may be calculated by:

$$H_c = \rho_a C_p \frac{T_a - T_c}{r_a} \quad (1.11)$$

The energy balance at the soil surface can be expressed as:

$$R_{nb} = E + H_s + S \quad (1.12)$$

where H_s is the sensible heat exchange between the soil surface and the air above.

Kanemasu and Arkin (8) show the differences in canopy energy balance between sorghum grown in wide (0.91 m) and narrow (0.46 m) rows. These data were obtained under conditions of a dry soil surface so that E is assumed to be negligibly small. Therefore, E_t is essentially the canopy transpiration in this case. Components of the energy balance in wide and narrow sorghum are shown in Fig. 1.2.

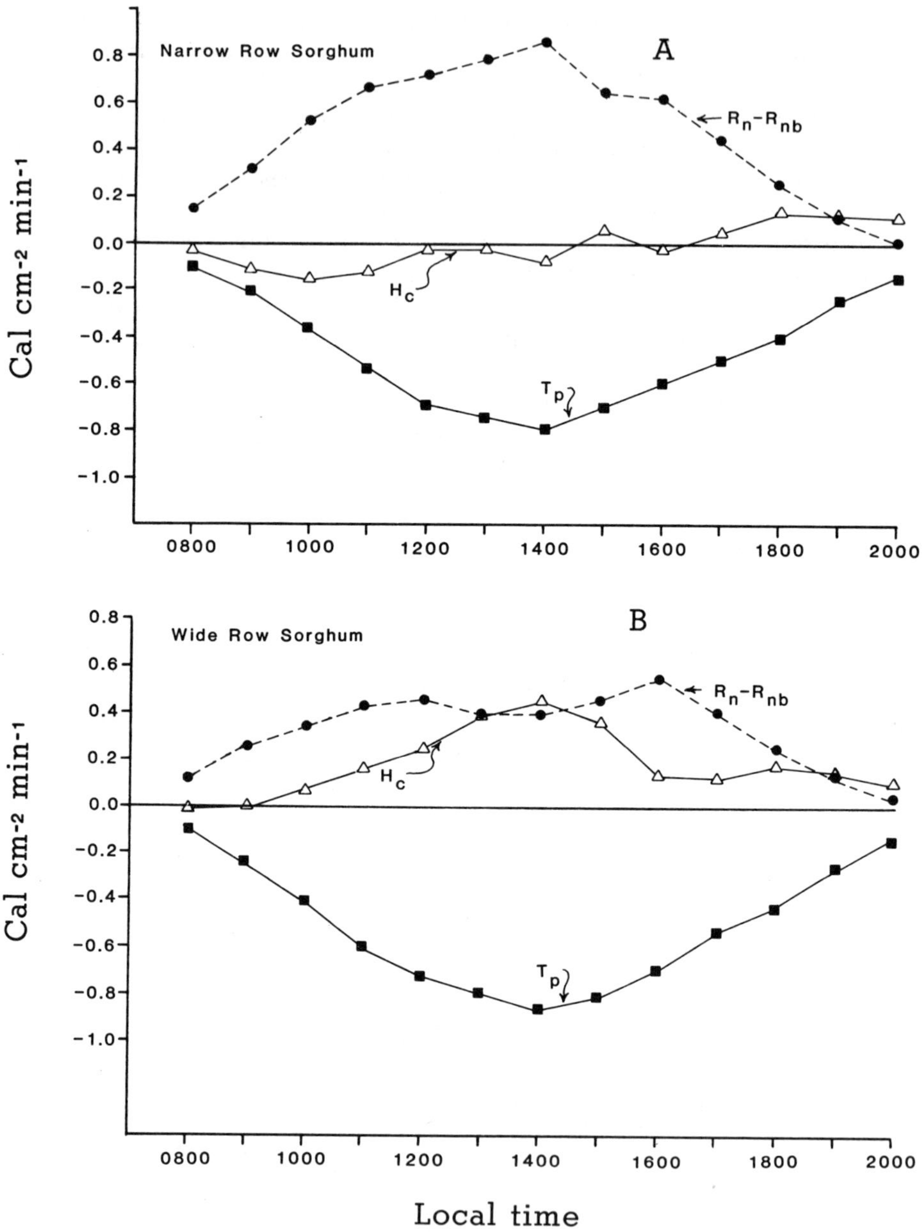

Figure 1.2. Above canopy minus below canopy net radiation ($R_n - R_{nb}$), transpiration (T_p), and sensible heat flux between the canopy and air (H_c) for narrow-row sorghum (A) and wide-row sorghum (B). [Adapted from Kanemasu and Arkin (8).]

It is evident that the wide-row sorghum in this study transpired more water than did the narrow-row sorghum. Over the entire season the wide-row sorghum was observed to use about 10% more water. A significant proportion of the energy used throughout the day in E_t in the wide-row sorghum was supplied by sensible heat (H_c has a positive sign). The narrow-row sorghum, on the other hand, absorbed sensible heat only after about 1400 hr. The major source of sensible heat consumed by the plants in the wide rows came from the air which was heated at the soil surface. This was especially true in the morning. In the afternoon sensible heat was also drawn from the warm air above the crop to the plants. Chin Choy et al. (9, 10) and McCauley et al. (11) also reported reduced water loss from peanuts and sorghum planted in narrow as compared to wide rows. However, in very wide rows (perhaps 2 m or wider) the total water use of the field will be decreased because evaporation from the soil, especially when the soil surface is not saturated, will be less than transpiration from the plants. In these very wide rows, much of the heat generated at the soil surface may not be captured by the plants that occupy only a relatively small portion of the total air–plant volume.

In regions where the soil surface is well supplied with water, E_t from wide- and narrow-row vegetation will be about the same. In humid regions R_n generally sets the upper limit on the amount of energy consumed in LE (12). Lemon et al. (13), for example, reported that R_n in the eastern United States in summertime was proportioned as follows: LE (40–90%), H (10–60%), S (5–10%), and P_s (1–5%). In contrast LE in arid regions may often exceed R_n because of the advection of sensible heat (14, 15).

Hanks et al. (16) identified various sources of advected energy. These include between row advection, energy advected from nearby fields, and energy transported from regions remote from a given area. Between row advection occurs when sensible heat is generated at the soil surface between the rows as illustrated in the previous discussion of wide- and narrow-row sorghum.

Sensible heat is advected to a crop from nearby fields when wind blows across a surface that is discontinuous in temperature and/or humidity. This so-called local advection is common in areas where irrigation is practiced because irrigated fields tend to be cooler and wetter than adjacent nonirrigated areas. It also occurs where fallowed fields are adjacent to crop covered fields. As the surface of a fallowed field dries it becomes a potential source of sensible heat for nearby crops. Local advection causes E_t rates to be greatest near the field border and to decrease with increasing distance into the field (16–19). Generally, the effects of local advection are minimal at downwind distances of about 80–100 m (18).

Regional advection occurs when sensible heat is transported from regions remote from the field. McIlroy and Angus (20) reported that regional advection of sensible heat caused E_t rates double that when net radiation was the only source of energy. In research conducted in the central Great Plains, Brakke et al. (19) found that local advection supplied 1–14% and regional advection supplied 7–40% of the total energy consumed in E_t. The contribution of regional advection was greatest on days of strong wind.

Thus, the atmospheric demand for water from vegetative surfaces depends on two sources of energy. The primary source of energy is R_n but especially in arid to subhumid climates advected sensible heat may contribute as much as 50% of the energy consumed in E_t.

The Soil–Plant–Atmosphere Continuum

The transpiration process involves the absorption of water by the plant roots, the transport of water through the conducting tissues of the plant, and the passage of evaporated water through the leaves and into the air, primarily through the stomatal aperture.

The hydraulic system within intact plants acts as a true continuum. This was demonstrated by research that showed that pressure changes at one end of the plant system, the roots, are faithfully manifest at the other end of the system, the leaves (21). The movement of water through the soil–plant–atmosphere system can be explained using laws and principles of thermodynamics. A detailed discussion of these laws and principles of thermodynamics will not be attempted here. Additional information on the thermodynamics of water movement in biological systems may be obtained from Leyton (22).

For the purposes of this chapter it is necessary only to know that water will move through the soil and into the plant, through the plant and into the atmosphere in response to a water potential gradient, that is, water

Table 1.1. Approximate magnitudes of water potential for a transpiring corn plant with soil moisture at field capacity and relative humidity of 50% and air temperature of 30°C

Component	Water Potential	
	$N\ m^{-2}$	Bars
Soil	-1×10^4	-0.1
Roots	-1.5×10^5	-1.5
Leaves	-8×10^5	-8
Air	-8×10^7	-800

moves along a gradient of decreasing water potential (ψ). Data in Table 1.1, from Neuman et al. (23) and Leyton (22), give typical values of ψ for an actively transpiring corn plant.

According to Baars (24) the movement of water through plants obeys an Ohm's law relationship, i.e., current = driving force ÷ resistance. Thus, the water transfer process is more clearly understood if the flow of water is regarded as being driven by a difference in water potential, $\Delta\psi$, against a resistance, r (van den Honert, 25). Any increase in transpiration rate must be matched by a proportionate increase in the water potential gradient.

Under steady-state conditions, flow through each segment of the continuum is equal and the potential gradients and resistances are related as follows:

$$\text{water flux} = \frac{\psi_s - \psi_r}{r_m} = \frac{\psi_r - \psi_l}{r_r + r_x} = \frac{\psi_l - \psi_a}{r_s + r_a} \tag{1.13}$$

where r_m is the resistance due to the soil matrix, r_r is the root resistance, r_x is the resistance through the xylem in plant stems, r_s is the stomatal resistance, and r_a is the aerial resistance. ψ_s, ψ_r, ψ_l, and ψ_a are the water potential of the soil, root, leaf, and air, respectively. Those who attempt to model this process make the basic assumption of constant flow of water in the liquid phase from roots through the stems to the evaporating sites.

The greatest resistance to water flow in the soil–plant–atmosphere continuum is the resistance to flow in the vapor phase (26). The resistances r_a and r_s are influenced by wind velocity and stomatal aperture, respectively. To overcome these resistances requires a large drop in water potential between leaf and air. This, in fact, does occur (Table 1.1).

Although the flux of water through a plant system is dynamic rather than steady-state, the flux is typically much greater than any change in the net amount of water stored within the plant. Consequently, diurnal changes in flux may be described by a sequence of steady-state conditions even under variable environmental conditions (26). According to Kaufman and Hall (27), the flow of water through the system is primarily controlled by climatic and plant factors influencing gaseous flow from the leaf to the atmosphere, that is, water flow through the plant is driven by water potential gradients resulting from transpiration. They further state that other soil and plant factors indirectly influence the flow of water through feedback effects on stomatal aperature.

There are limitations to the application of the Ohm's law approach to describing water flow in the soil–plant–atmosphere system. It has been demonstrated that, at least for some species, transpiration rates may increase severalfold with no increase in the driving force (leaf water potential) (24, 27, 28). For this to occur there must be a reduction in resistance to liquid flow through the plant. In spite of these limitations, the Ohm's law approach still provides an acceptable approximation to promote understanding of water flow through the soil and plant and into the atmosphere (29).

The flow of water within soils and through plants is examined in detail in Chapters 2 and 3. The transfer of water vapor from the plant to the atmosphere will be emphasized in greater detail in the following pages. This transfer process is controlled by both climatic and plant factors. As explained previously, R_n and H_c provide energy to change liquid water into water vapor. The transpiration process is also dependent on the water potential of the surrounding air. The water potential of air, ψ_a, is given by:

$$\psi_a = \frac{RT}{V_w} \ln \frac{e_a}{e_s} = \frac{RT}{V_w} \ln RH \tag{1.14}$$

where R is the gas constant, T is the temperature (°K), V_w is the volume occupied by 1 mole of water vapor, e_a is the vapor pressure of the air, e_s is the saturation vapor pressure of the air at the given air temperature, and RH is the relative humidity (e_a/e_s) of the air.

From Eq. 1.14 it is seen that ψ_a increases as the air becomes more dry resulting in a concomitant increase in the atmospheric demand for water. Air is seldom at 100% relative humidity, especially during daytime periods, so ψ_a will exceed ψ_l, often by several hundred bars. Thus, water potential conditions of the atmosphere are generally favorable for transpiration; however, the rate at which transpiration proceeds will be controlled, in large measure by the magnitudes of r_s and r_a. r_a is a function of the wind speed and aerodynamic crop roughness. The crop roughness varies among other things with plant height, leaf size, stem, and leaf pubescence. r_a decreases with increasing wind velocity (30) and is generally an order of magnitude smaller than r_s, although in still air r_a and r_s may be of similar magnitudes. If adequate water is available r_s will be relatively small and the transport of water vapor from plant surfaces will proceed at rates that are restricted almost exclusively by the amount of energy available to evaporate water. If water becomes limiting, stomata will close, at least partially, r_s will increase, and vapor flow will be restricted. When stomata are completely closed some water vapor may still diffuse through the leaf cuticle but the amount evaporated in this manner will be very small. For a more detailed discussion of water transport in the soil–plant–atmosphere continuum, the reader is referred to Plaut and Moreshet (31).

WATER BALANCE

Essentially all water on the earth is held in the oceans, lakes, and ice or snow fields. Only about 0.5% of the total water on earth is involved at any given time in the hydrologic cycle, that is, the cycle that includes the evaporation of water from the surface of the earth, the transport of water vapor in the atmosphere, and the eventual return of the water to the land surfaces and water bodies. Of the water that falls on the United States it is estimated that about 30% goes into drainage systems or percolates deep into groundwater storage. The remaining 70% of the water returns to the atmosphere by evaporation or transpiration through vegetation (32). In some areas 90% or more of the water that falls will be lost through E_t.

All water balance models attempt to determine what happens to water that is applied to or falls on a given area. The water balance of a system is the difference between the inputs to the system and the flow of water out of the system or the storage of water within the system. The inputs are generally precipitation or, in some cases, irrigation although, depending on the boundaries of the system, it could also include water brought into the system through runoff. An equation describing the water balance may be written as:

$$P_t + \Delta W \pm O - U - E_t = 0 \qquad (1.15)$$

where P_t is the precipitation and/or irrigation, O is the runoff, U is the percolation or deep drainage, and ΔW is the water stored in soil. Inputs into the system are positive and outputs from the system are negative. Percolation is the amount of water passing beyond the root zone.

According to Slayter (33), Eq. (1.15) can be written more specifically as:

$$\int_{t_1}^{t_2} [(P_t \pm O) - E_t - U_z]\, dt = \int_{t_1}^{t_2} \int_{z_0}^{z} \frac{\partial \theta}{\partial t} \qquad (1.16)$$

where $t_2 - t_1$ is the time interval over which measurements are made, z_0 is the soil surface (depth $= 0$), z is the depth to the lowest point of measurement, U_z is the net downward flux of water at depth z, and θ is the volumetric soil water content.

These equations can be used on any scale, ranging from continental land masses (34) and hydrologic catchments (35–37) down to small fields or even individual plants. In most water balance studies all elements but E_t are measured or otherwise estimated. E_t is obtained by difference.

The measurement of precipitation or irrigation is generally more straightforward than for other components of the water balance equation. But even in the case of precipitation, depending on the scale, there may be marked differences in the quantities of water reaching the ground. This unevenness may result on the larger scale from a lack of spacial uniformity in precipitation. The interception of precipitation by vegetation and channelization of water flow down plant stems or into soil cracks may explain a lack of uniformity on a microscale. Some water is evaporated from the plant surface before it reaches the soil. Some of the water which reaches the ground runs off when the rate of precipitation exceeds the rate of infiltration. The amount of runoff affects the final pattern of soil water recharge. The slope of the land generally plays an important role in determining runoff. Soil water recharge is reduced at the top of a slope and increased at the bottom.

E_t is often determined on the assumption that gain or loss of water to or from the zone below maximum root depth is negligible. This assumption can lead to very serious errors in many cases. Rose and Stern (38) observed that as much as 40% of the total change in water content in the surface meter of soil was due to drainage. Goltz et al. (39) observed that drainage from irrigated onions growing on a sandy soil exceeded E_t. Other evidence suggests that the thermally or hydraulically induced flux of water into or out of the soil root zone can be of sufficient magnitude to introduce serious errors if it is neglected (40, 41).

Rose and Stern (38) have suggested a means for calculating the vertical flow of water from the hydraulic conductivity and the soil water potential profile of the soil. The equations are:

$$U_z = K + K\frac{\partial h}{\partial z} \tag{1.17}$$

and

$$U = \int_{t_1}^{t_2} U_z \, dt \tag{1.18}$$

where K is the hydraulic conductivity and $\partial h/\partial z$ is the rate of change of soil water tension with depth. h may be obtained from ψ_s values by the relationship:

$$h = -\psi_s \rho_w g \tag{1.19}$$

where ρ_w is the density of water and g is the acceleration due to gravity. Unless h is very small $\partial h/\partial z$ is usually greater than unity and K in Eq. 1.17 is negligible. Another approach to estimating U is suggested by Davidson et al. (42).

When U is properly accounted for, then E_t can be estimated from changes in ΔW that occur during periods between rains or irrigations. In other instances, primarily for water balance studies over entire drainage systems, ΔW is assumed to be zero, P_t and O are measured, and U is neglected or otherwise accounted for. In this case $E_t = P_t - O$. This approach is generally used over long time periods or over shorter periods when the soil water content at the beginning and ending of the measurement period is essentially the same.

According to Slayter (33), the advantages of the soil water balance approach, compared to aboveground measurements of water vapor flux, are the ease of data processing and the integration, by the soil water reservoir, of extraction rates between observations. The disadvantages are the low level of measurement accuracy and the difficulties of assessing E_t during rainy periods. Even with careful measurement, it is difficult to detect soil water changes with an accuracy of greater than ± 2 mm of water. Errors associated with the water balance approach are likely to invalidate the use of this approach for quantification of normal daily E_t. Determinations at 2 or 3 day intervals, however, should provide an acceptably quantitative picture of E_t provided that U can be adequately accounted for. The water balance approach is used rather widely, especially for studies of E_t from large areas such as watersheds or in research where instrumentation to measure the flux of water vapor from soil to atmosphere is unavailable. Perhaps, the major problem associated with the water balance method, especially for large areas, is not in the method itself, but in the lack of good spatial averaging of inputs and outputs due to the variation in rainfall over extended areas and to a lack of homogeneity in land surfaces and the soils that underlie them.

POTENTIAL EVAPOTRANSPIRATION

Evaporation from the soil and vegetation of a region depends on the availability of both water and energy. Potential evapotranspiration (E_0) is a concept introduced independently in 1948 by both Thornthwaite and Penman to describe E_t from a surface which is continuously wet. E_0 is a function of the climate alone, that is, the magnitude of E_0 is energy-limited. Morton (43) defines E_0 as the "evaporation from a continuously moist surface with regional insolation absorption, vapor transfer, and heat transfer characteristics and with an area so small that convected heat and vapor have no significant effect on the passing air." Rosenberg (32) defines potential evapotranspiration as "the evaporation from an extended surface of a short green crop which fully shades the ground, exerts negligible resistance to the flow of water and is always well supplied with water. E_0 cannot exceed free water evaporation under the same weather conditions."

E_0 is generally estimated from models such as that of Penman (44) which use only meteorological inputs. The concept of E_0 is useful for many purposes and has gained widespread acceptance. The truth is, however, that actual E_t will almost always differ from E_0. This is due to the effects of soil and plant factors on E_t.

Plants influence E_t by the amount of leaf area and plant cover they provide, by their height, and by the resistance they offer to the flow of water through the plant and into the atmosphere. Soils influence E_t by the amount of moisture they contain that is evaporated from the surface and by the amount of water available in the soil reservoir for transpiration.

Plant Cover

The partitioning of E_t into its component parts (transpiration from plants, or evaporation from the soil, E_s) is dependent on the amount of plant cover, especially when the soil is wet. Evaporation from the soil surface decreases with increased plant shading of the soil. As plant cover increases, the relative importance of transpiration likewise increases.

Luebs et al. (45) observed that with adequate soil water E_t by a barley crop was related to the amount of plant cover. Brun et al. (46) showed that the fraction of the total E_t accounted for by transpiration was closely correlated to the amount of leaf area covering a given soil area (the leaf area index, *LAI*, as it is commonly called). At an *LAI* value of 2, T_p was about 50% of the total E_t but at an *LAI* of 4, it was about 95%. Denmead and McIlroy (47) reported that E_s could be large relative to T_p with *LAI* between 1 and 3 and a wet soil surface.

Studies of several different crops suggest that E_t rates do not increase significantly with *LAI* greater than about 3. For example, Stewart et al. (48) found for corn that the ratio of crop water use to pan evaporation did not change as *LAI* increased from 3 to 6. Ritchie (49) reported data showing that T_p increased with increasing *LAI* to equal E_0 at an *LAI* value of about 3 (Fig. 1.3). Kristensen (50) found that an *LAI* of 3 was required for E_t to equal E_0 in barley, sugar beets, and grass.

Plant Height

El Nadi and Hudson (51) reported on the effect of alfalfa and wheat crop height on E_t. They found that when complete plant cover had been established, an increase in plant height was accompanied by an increase in E_t rate. They attributed this increase to an increase in surface roughness, resulting in increased turbulent mixing and transfer of water vapor to the atmosphere, and to an increase in leaf area exposed to the wind. In areas of strong sensible heat advection, taller crops are able to extract more sensible heat from the air than do shorter crops, thereby increasing the E_t rates of tall plants relative to short ones. Slabbers (52) stressed the importance of taking crop roughness, especially for tall crops, into account when calculating E_t.

Plant Resistance to Water Flow

In the section on the soil–plant–atmosphere continuum, the resistance to the flow of water vapor from the plant to the air was discussed. The concept of E_0 requires that plants behave passively as wicks and that plants exercise no control over transpiration. The amount of control that plants exert appears to depend on several factors. These factors include the plant species and the previous history of the crop, that is, whether or not the plants have always had adequate water or if they have, instead, had periods when they were subjected to water stress.

Research has shown that alfalfa, for example, exerts little or no control of E_t as long as adequate water is available (53). Thus, alfalfa may be

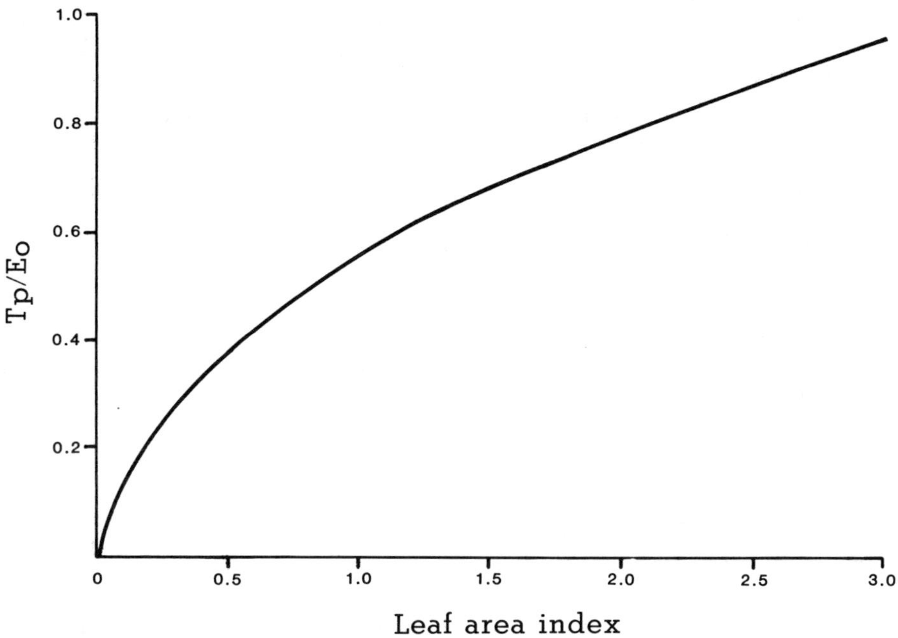

Figure 1.3. Relative transpiration rate (T_p/E_0) in well-watered cotton and sorghum as a function of leaf area index. [After Ritchie (49).]

considered to be a potential evaporator. Other crops such as corn, sorghum, and many grasses apparently offer some resistance to vapor flow and have E_t rates that are, consequently, lower than the potential evaporators such as alfalfa. This is true even when adequate water is available (54, 55).

The resistance of plants to the diffusion of water vapor may be determined from measurements made with diffusion porometers. These are instruments with small chambers that contain a humidity sensor. The chamber is clamped onto a leaf. The initial vapor concentration in the chamber is low and increases as the leaf transpires through the leaf stomata or cuticle. The time needed to decrease the electrical resistance of the humidity sensor between two fixed values is determined. The diffusion resistance can be directly related to the time difference when the evaporating surface and chamber temperatures are known, assuming, of course, that the instruments are properly calibrated. For descriptions and use of diffusion porometers see Kanemasu et al. (56) and Stigter et al. (57). Recent, as yet unpublished, research by John Norman at the University of Nebraska (personal communication) has shown some serious errors in estimation of r_s with porometers of the type described in Kanemasu et al. (56). Porometers of a new design which eliminate these errors have been built, tested, and are now commercially available.

To determine the average resistance of a leaf, r_s, the stomatal resistance of the top (adaxial), r_t, and bottom (abaxial), r_b, sides of leaves are determined with a diffusion porometer. Considering r_t and r_b as resistances connected in parallel the average resistance of a leaf, is obtained as:

$$r_s = r_b \text{ or } r_t \tag{1.20a}$$

if the leaf is hypostomatous (stomata on one side only),

$$r_s = \frac{r_b}{2} = \frac{r_t}{2} \tag{1.20b}$$

if the leaf is amphistomatous (equal number of stomata on both sides of the leaf), or

$$r_s = \frac{r_b r_t}{r_b + r_t} \tag{1.20c}$$

if there are unequal number of stomata on the top and bottom of the leaf. The latter case is most common.

There have been some procedures given in the literature to calculate the canopy resistance (r_c) term (58–60). These procedures include considering the canopy resistance to be the harmonic mean of r_s values for all leaves measured in the canopy or to consider r_c to be the mean of resistances for various layers of leaves in a canopy. The layer resistances are treated as parallel resistances in this procedure. A better estimate of r_c, however, should be obtained by calculating the average stomatal resistance for the sunlit leaf area and the average stomatal resistance for the shaded leaf area. The mean of those two values should give the most reliable estimate of r_c. In considering the overall canopy resistance to water vapor flux from a canopy, the boundary layer resistance, r_a, must also be included.

Soil Water Availability

Water will evaporate from the soil surface at a potential rate as long as the surface remains wet. As the surface dries, E_s begins to fall and E_t becomes less than E_0 (61, 62). The length of time that soils will evaporate at the potential rate was shown by van Bavel and Hillel (63) to be influenced by the climate and type of soil. Water movement to the evaporating sites near the surface is controlled by the hydraulic properties of the soil. Even though E_t rates from soils with a dry surface are generally much lower than E_0 rates, evaporation from the soil can continue for several days at measurable rates. As plant cover increases, the contribution of E_s to total E_t decreases in importance. With increasing crop cover, the ratio E_t/E_0 depends on the soil water status and on the ability of the plant roots to extract the available water. Thus, as Ritchie has shown (61, 64), plants are able to extract water from the soil at rates near E_0 until a critical or lower limit of soil water is reached. After this limit is reached, the soil water availability to the plants decreases until, eventually, transpiration ceases entirely.

It is apparent that, while the concept of E_0 is often useful for estimating the quantities of water evaporated from plant, soil and water surfaces, the regulating influences of plants and soils will generally cause actual E_t to differ from E_0.

Because plants and soils are directly involved in the evaporation, it is possible, through the manipulation of various soil and plant factors, to exercise some control of the process. Stone (65) provides a review of ways to manipulate plant, soil, and climatic factors to either reduce or increase E_t depending on the desired outcome. Among the methods cited by Stone are:

- Reflectorizing or shading of plants to reduce energy absorption
- Mulches on soils to provide vapor barriers
- Irrigation methods to control the amount and placement of water to either increase or decrease E_t
- Windbreaks to modify the plant environment
- Planting and spacing of crops to alter absorption of energy
- Manipulation of root structure of plants
- Increasing the stomatal resistance of crops through the use of antitranspirants or through plant preconditioning practices

MEASUREMENT OF EVAPOTRANSPIRATION

The need for expanded food production to feed a hungry world has provided impetus for dramatic increases in the amount of land being brought under irrigation in the subhumid to semiarid regions of the world. Coupled with increased demands for water for agriculture is increased competition from industries, municipalities, and recreational and environmental groups. This swelling demand for water, combined with increased energy costs to move the water, make it imperative to improve the management of our valuable water resources. The accurate assessment of E_t is a vital part of proper water management schemes.

Since Dalton first introduced the mass transport equation, there have been numerous models and methods to evaluate evaporation from plant, soil, and water surfaces. Some methods are accurate and reliable, others are only capable of providing a first approximation of E_t rates. A review of several methods for measurement or estimation of E_t fluxes follows.

Lysimeters

Lysimeters are essentially large blocks of soil, isolated from, but as identical as possible with surrounding soils, in which the changing water content of the soil is measured. According to Pelton (66), lysimeters were first used to study the percolation of water through soils. It was not until the twentieth century that modifications in lysimeter design were made to permit the study of E_t. There are several different types of lysimeters ranging widely in the accuracy with which changes in the soil water content are detected. The most accurate lysimeters can detect losses as small as 0.01 mm of water and can be used to accurately detect E_t rates over time periods shorter than 1 hr (measurements over 5 to 10 min periods have been reported). Other kinds are only sensitive enough to detect changes that occur over daily or longer periods.

Lysimeters provide the only direct measure of water flux from a vegetative surface. As such, they provide a standard against which other methods can be tested and calibrated. To provide reliable measurement of E_t, lysimeters should meet the following criteria (66):

1. They should be constructed so that their moisture relationships correspond closely to those of soils under natural conditions.
2. The lysimeters should be sufficiently deep to extend well below the plant root zone or should use a tensioning device at the bottom of the lysimeter to maintain soil moisture at or near the same level as surrounding areas.
3. Lysimeters should be managed in exactly the same manner as the surrounding area.
4. The ratio of the wall surface area to the enclosed lysimeter area should be small to avoid small-scale advection from the uncropped surface.

Reviews of various kinds of lysimeters are given by Harrold (67) and Rosenberg (32). Some of the more common types of lysimeters will be described below.

Potential Evapotranspirometers

Potential evapotranspirometers are lysimeters in which the water table is held constant in a mass of soil so that evaporation may proceed at a potential rate. By continually monitoring the water added to the system and the amount of drainage, it is possible to estimate the rate of E_t. Because of inaccuracies involved, potential evapotranspirometers should not be used to provide estimates of E_t for periods shorter than 1 day.

Potential evapotranspirometers were initially developed by Thornthwaite (68). They have been used in studies to evaluate the performance of the Thornthwaite and Penman methods (69), to determine the effect of water table depth on yields of timothy (70), and to measure water uptake by orchard grass from shallow water tables (71).

Floating Lysimeters

Floating lysimeters are instruments that float on liquids such as water, oil, or heavy liquids such as solutions of zinc chloride. In floating lysimeters, the displacement of the fluid by the lysimeter bin is measured, generally with a manometer, and the weight determined using the Archimedean principle.

King, et al. (72) developed one of the first floating type lysimeters. Their lysimeter floated on water and used a modified water stage recorder to detect water level changes. This lysimeter required large air spaces in the soil-filled bin to maintain buoyancy. McMillan and Paul (73) described a lysimeter which consisted of a large soil container floating in a solution of zinc chloride. The use of heavy liquids such as the solution of zinc chloride eliminates the need for maintaining large air spaces in the lysimeter bins.

In other cases, liquids are placed in bags. The pressure exerted on the bags, which is a function of the total lysimeter weight, is then measured with manometers (74, 75).

Floating-type lysimeters equipped with manometers should not be used to obtain estimates of E_t for periods shorter than 1 day and, even then, care must be exercised to account for errors due to temperature fluctuations. For periods as long as 1 week, however, such lysimeters may be expected to provide precise measurements of E_t (76). Floating lysimeters have been equipped with linear variable differential transducers to continuously record weight changes of the floating tank (77). This system has a reported sensitivity of 0.025 mm of water and is not sensitive to diurnal temperature changes. With this system, E_t changes for periods of less than 1 day can be reliably measured.

Weighing Lysimeters

A simple weighing lysimeter can be made by filling a small container with soil and burying it in the ground. The container can be removed from the soil periodically and weighed on a scale. For very large lysimeters, such as those described by Harrold and Dreibelbis (78) and Pruitt and Angus (79), mechanical scales are installed below the ground. Recently, Young et al. (80) described the use of three electronic strain gauge transducers to measure E_t from a monolith lysimeter.

Very accurate estimates of E_t can be obtained from precision weighing lysimeters. Ritchie and Burnett (81) developed a precision weighing lysimeter with dimensions of 1.8 × 1.8 × 1.2 m deep. Weight changes are detected with a strain gauge load cell with a sensitivity equivalent to ±0.025 mm of water change. An even more precise, although somewhat more expensive, precision weighing lysimeter is described by van Bavel and Meyers (82). Improvement in the design of this lysimeter was suggested by Rosenberg and Brown (83). This lysimeter is 1 $m^2 \times 1.6$ m deep. The soil bin, when filled, weighs about 3 tons. Weight changes, also detected with a strain gauge load cell, permit the measurement of E_t to an accuracy approaching 0.01 mm of water.

Chamber Techniques

It is possible to estimate the amount of water evaporated from plants by enclosing them in a chamber and measuring the air flow rate and the concentration of water vapor entering and exiting the chamber. The change in water vapor concentration multiplied by the volume of air passing through the chamber is equal to the amount of water lost by the vegetation and soil inside the chamber. Because the placement of a chamber can drastically disturb the plant environment, data should be collected soon after the chamber is placed over the plants. Even then care should be exercised in interpreting chamber data.

Sebenik and Thomas (84) measured E_t from a tree by enclosing it in a large plastic tent. Peters et al. (85) describe a chamber that travels along a track for measuring evaporation from crops. The chamber can be stopped at various locations and a set of measurements made within about 20 sec. An inexpensive portable chamber, which can be mounted on the front of a tractor for making rapid measurements of E_t in the field, was designed by Reicosky and Peters (86). Reicosky and Peters report that errors in E_t estimates made with the chamber are on the order of 11–19% as long as dry and wet bulb temperatures of the air can be measured with ±0.1°C accuracy before and 1 min after the chamber is put in place. They concluded that accuracy and rapidity of measurement combined with portability and relatively low cost make the chamber technique useful for evaluating the influence of soil and water management practices on plant water use.

Sap Flow Techniques

The transpiration rate of a plant can be monitored by evaluating the rate at which sap flows through the conducting tissue. There are several techniques for describing the length of time it takes for sap to move from one point in a plant to another point. These methods include the heat pulse velocity method (87–89), the movement of injected tritiated water through a plant (90), the monitoring of radioactive phosphorous movement up a stem (91), and a magnetohydrodynamic method for measuring sap flow in small stems (92). The transpiration rate is calculated from the volume of water that passes between the two points during a measured amount of time. These are obviously somewhat specialized methods, applicable primarily to measuring water flow through individual plants, especially larger plants such as trees, and are not likely to be used for measuring E_t from an agricultural field.

Atmometers

Various types of atmometers are described by Livingston (93), Carder (94), Pelton and Korven (95), and Dilley and Helmond (96). Atmometers are generally constructed of a porous ceramic or paper evaporating surface, generally colored black or white. These surfaces are continuously supplied with water. Atmometers are, thus, passive devices that expose a moist porous surface to the atmosphere. They respond relatively quickly to environmental influences, particularly radiation, wind speed, and the dryness of the air. According to Rosenberg (32) they appear to be especially sensitive to wind so that care should be exercised in the choice of elevation and exposure.

Atmometers are relatively inexpensive instruments that have been successfully used for obtaining estimates of potential evapotranspiration, E_0. Stanhill (97) found that weekly and monthly evaporation rates from a Piche atmometer were linearly related with measured E_t. However, correlation coefficients were only about 0.69. Dilley and Helmond (96) found that estimates of E_0 from atmometers were in good agreement with E_t from weighing lysimeters. Pelton and Korven (95) reported that a black Bellani plate and a black porous disc could be used to estimate daily E_t of alfalfa to an accuracy of about 1.5 mm day^{-1} using empirically derived actual E_t to atmometer evaporation ratios to adjust the atmometer data. They observed, however, that atmometers overestimated actual E_t. Wilcox and Sly (40) found a second-degree polynomial equation was required to accurately estimate E_0 from atmometer measurements. It would thus seem that properly cared for and calibrated atmometers can be used to obtain reasonably accurate estimates of E_0.

Pan Evaporimeters

Pans of many shapes and sizes have been used to measure free water evaporation. Pans are inexpensive, easily maintained, and simple to operate. However, especially in arid climates care must be taken in relating evaporation from pans to actual E_t. According to Rosenberg (32), because of the very small roughness of a water surface relative to a vegetative surface, less sensible heat energy will be extracted from the passing air by the water surface. Thus, pans may at times give unrealistically low values of evaporation. Rosenberg and Powers (98) observed, for example, that during a 5-day period in May 1967 the lysimetrically measured E_t from alfalfa was 51.91 mm of water while the evaporation from three evaporation pans (W. B. Class A type) averaged 45.66 mm during the same period. In contrast, Pruitt and Lourence (99) observed that the ratio of E_t from fescue and rye grass to pan evaporation (E_p) decreased with strong hot and dry winds. This was probably due to increased stomatal resistance of the vegetation under these stressful conditions.

In more humid regions pans may give realistic estimates of potential evapotranspiration. Usually crop water use in such climates is some 60–90% of pan evaporation (32). Pruitt and Lourence (99) reported that E_t from grasses was about 80% of that from evaporation pans except when winds were strong and the air was dry and hot. Stewart et al. (48) found that the ratio of E_t from well-watered corn to that of E_p varied as a function of the stage of growth although for much of the year the ratio of E_t/E_p was about 0.9.

The exposure of the evaporation pan is very important. Depending on the pan exposure, Fuchs (100) reported that E_0/E_p varied from 0.6 to 2.0. The larger values generally occurred when the pan was partially shaded or was inside an irrigated field.

MODELS FOR ESTIMATING EVAPOTRANSPIRATION

Precision lysimeters provide the only direct way to accurately measure E_t. However, due to the expense, time, and expertise required to install and use lysimeters, it is necessary to have other means of estimating E_t. Engineers, agronomists, hydrologists, and others need reliable estimates of E_t for numerous and different purposes. In response to this need, several E_t models, of varying levels of accuracy and complexity, have been developed. These models range from almost purely empirical models to those having an entirely physical basis. Data for some models may be readily available from weather stations. For others data may be obtained only with expensive and sophisticated equipment. Some of the more common E_t models, grouped into several somewhat arbitrary categories, will be reviewed.

Mass Transport Models

One of the earliest models for estimating E_t was developed by Dalton about 1800. This model relates the evaporation rate, E, to the difference in vapor pressure at the evaporating surface, e_s, and the vapor pressure in the air, e_a, at some point above the surface as follows:

$$E = K(e_s - e_a) \tag{1.21}$$

where K is a theoretically or empirically derived constant usually involving wind speed terms. Modifications of this form have been made by Rohwer (101), Penman (44), Harbeck (102), Blad and Rosenberg (103), and others.

The mass transport model is especially useful for estimating E from lakes or reservoirs (102, 104). The model has also been used to estimate E from bare soils (105, 106) and E_t from vegetation (103, 107). The model has some inherent difficulties associated with it, especially in evaluating the empirical constants and in determining e_s.

The ability to measure surface temperatures, with infrared thermometers and thermal scanners has simplified the measurement of e_s, as is explained below: To obtain e_s for vegetative surfaces from measurements of canopy temperatures (T_c), it must be assumed that air spaces (substomatal cavities) within the leaf are at the temperature of the leaf and at or very near a relative humidity of 100%. (Based on this assumption tables and equations can be used to obtain e_s values from T_c.)

The mass transport model will provide less accurate estimates of E_t from vegetative surfaces as crops experience water stress because with water stress the temperature of the vegetative surface (T_s) becomes elevated. e_s estimated from these elevated temperatures may be higher than actual e_s values and produce excessively high E_t estimates. In spite of several difficulties, the model does work rather well under many conditions, including advective and nonadvective cases, when water is not limiting (103). The mass transport model does offer the advantage of simplicity. With remote sensing techniques to provide measurements of surface temperature over large areas, the model will continue to be an important one for estimating evaporation, especially from bodies of water.

Resistance Models

Monteith (108) proposed a model for estimating E_t similar to the mass transport equation. In his model K in Eq. (1.21) is replaced by a term that includes the aerial resistance, r_a, and the plant resistance, r_c. Equation (1.21) now becomes:

$$LE = -\frac{M_w/M_a}{P} L\rho_a \frac{(e_s - e_a)}{(r_a + r_c)} = -\frac{\rho_a C_p}{\gamma} \frac{(e_s - e_a)}{r_a + r_c} \tag{1.22}$$

To use this model one may estimate e_s from T_s data, measure e_a, estimate r_a from wind speed measurements, and obtain r_c with methods previously described.

Brown and Rosenberg (109) proposed a somewhat more complicated resistance model which can be used to evaluate the influence of changing weather and plant factors on *LE*. Their model, which requires iterative solutions, is:

$$LE = -\frac{f(R_n + S + LE)r_a/C_p\,\rho_a + T_a - e_a\,M_w/M_a}{r_a + r_c}\,\frac{M_w/M_a}{P}\,L\rho_a \quad (1.23)$$

Verma and Rosenberg (110) evaluated the performance of this model. They also simplified the model by the use of functional expressions for r_a and r_c. They reported that the model predicted hourly and daily values of E_t that generally agreed within 10–15% of E_t measured by lysimeters.

Aerodynamic Models

The vertical flux of water vapor due to turbulent diffusion is given by:

$$LE = L\rho_a K_w \frac{\partial q}{\partial z} = \frac{\rho_a C_p}{\gamma} K_w \frac{\partial e}{\partial z} \quad (1.24)$$

where K_w is the exchange coefficient for the transfer of water vapor, and $\partial q/\partial z$ and $\partial e/\partial z$ are the changes in specific humidity and vapor pressure with height, respectively. K_w is not generally measured directly. Rather the assumption is made that $K_w = K_h$ (exchange coefficient for sensible heat transfer) $= K_m$ (exchange coefficient for momentum transfer). Techniques have been developed to estimate E_t using modified forms of Eq. 1.24. Thornthwaite and Holzman (111) proposed one such technique. Their method, based on the assumption of a logarithmic wind speed profile and $K_h = K_m$, is:

$$LE = \rho_a k^2 \frac{(q_2 - q_1)(u_2 - u_1)}{[\ln(z_2/z_1)]^2} \quad (1.25)$$

where q is specific humidity, u is wind speed, k is the von Karman constant, and z is the height of measurement.

The log-linear wind profile holds only for lapse rates between isothermal (no change in temperature with height) and dry adiabatic (a decrease in temperature with increasing height at the rate of 9.8°C km^{-1}). To achieve accurate estimates of E_t using aerodynamic models, it is necessary to make corrections depending on the stability of the air. The most widely used corrections are based on either the Richardson number (112) or the Monin–Obukhov (113) mixing length. Pasquill (114), Deacon and Swinbank (115), Munn (116), Pruitt (117), Pierson and Jackson (118), and others have developed different approaches to aerodynamic techniques for estimating E_t.

Especially with recent refinements, aerodynamic methods seem to yield sufficiently accurate results for most applications. However, due to the complexity of the models and the accuracy required for measurement of the parameters required in these models, it is unlikely that they will gain widespread use except for research applications. In addition, these models have stringent requirements for adequate fetch which is the downwind distance between the leading edge of the field and the instruments. It is generally recommended that minimum fetch distance should be 100 times the height of the instruments.

Eddy Correlation Model

Water vapor, along with heat, CO_2, and other atmospheric substances, or entities, are transported in the vertical direction by small parcels of air called eddies. The estimation of E_t with the correlation technique is based on the measurement of the upward and downward motion of these eddies along with observation of the concentration of water vapor within the discrete eddies. Swinbank (119) proposed the eddy correlation method to estimate the vertical fluxes of heat and vapor. The mean upward flux of vapor in fully turbulent flow is given by:

$$E = \frac{M_w/M_a}{P} \overline{\rho_a w e_a} \tag{1.26}$$

where the overbar indicates the average condition over a given time period and w is the vertical wind velocity. The terms on the right-hand side of Eq. (1.26) can be expressed as:

$$\rho_a = \bar{\rho}_a + \rho_a', \quad w = \bar{w} + w', \quad \text{and} \quad e_a = \bar{e}_a + e_a'$$

where the overbar indicates the mean and the prime signifies as instantaneous deviation from the mean.

Expansion of Eq. 1.26 using these relationships and removal of those terms which either equal zero or are insignificantly small yields:

$$E = \frac{M_w/M_a}{P} (\rho_a \bar{w} \bar{e}_a + \rho_a \overline{w' e_a'}) \tag{1.27}$$

with adequate upwind fetch and a horizontal surface $\rho_a \bar{w} \bar{e}_a$ can be neglected so that:

$$E = \frac{M_w/M_a}{P} \rho_a \overline{w' e_a'} \tag{1.28}$$

Theoretically, the eddy correlation technique should provide very accurate estimates of E_t. The major difficulties lie in the instrumentation and data collection. The instruments must be sensitive enough to detect simultaneous changes in e_a and w in the rapid passage of different eddies. Frequencies of eddy changes are very rapid near the surface and decrease with height. Unless instruments are sensitive enough to detect these eddy changes near the surface, they may need to be placed so high above the field that the eddies they sample may not be representative of the field being studied. Great quantities of data need to be collected so integrating circuits or analog computers must be coupled to sensors to provide water vapor fluxes over time periods longer than just a few minutes.

Energy Balance Models

Bowen Ratio–Energy Balance Model

Bowen (120) introduced a relationship between *LE* and *H* known as the Bowen ratio, β, which is defined as:

$$\beta = H/LE = \frac{\rho_a C_p}{LM_w/M_a} \frac{K_h}{K_w} \frac{\partial T/\partial z}{\partial e/\partial z} = \gamma \frac{K_h}{K_w} \frac{\partial T/\partial z}{\partial e/\partial z} \quad (1.29)$$

This relationship is generally simplified by assuming that $K_h = K_w$ and that $(\partial T/\partial z)/(\partial e/\partial z) \approx \Delta T/\Delta e$. Equation (1.29) then becomes:

$$\beta \approx \gamma \frac{\Delta T}{\Delta e} \quad (1.30)$$

Combining Eq. 1.30 with Eq. 1.3 gives:

$$LE = \frac{-(R_n + S)}{1 + \gamma \Delta T/\Delta e} \quad (1.31)$$

the so-called Bowen ratio–energy balance method of estimating E_t.

Tanner (121), Pruitt and Lourence (99), and Denmead and McIlroy (122) observed good agreement between estimates of *LE* made with Eq. 1.31 and lysimeter measurements in nonadvective conditions. Blad and Rosenberg (123) reported that the model underestimated *LE* under advective conditions. They suggested that the underestimation is due to an inequality in K_h and K_w. Later studies by Verma et al. (124) have shown that $K_h > K_w$ when advective conditions prevail. Estimates of *LE* with the Bowen ratio method in regions of sensible heat advection will be improved by using actual K_h/K_w ratios rather than assuming that $K_h/K_w = 1$.

Fuchs (100) suggests that daily estimates of E_t cannot be made directly from daily average $(R_n + S)$ and daily average β. Integration of 30 to 60 min estimates of *LE* over the day will yield the best results. The availability of good instrumentation for accurate measurements of R_n and temperature and vapor pressure gradients permit reliable measurements of E_t to be made over short or long periods with the Bowen ratio–energy balance method if proper adjustments are made for advective conditions.

Energy Balance–Resistance Model

The energy balance–resistance model is derived by substituting for *H* in Eq. 1.3 with the expression for H_c given in Eq. 1.11. Thus, the model becomes:

$$LE = -\left(R_n + S + \rho_a C_p \frac{T_a - T_c}{r_a}\right) \tag{1.32}$$

This model has been shown to provide reliable estimates of *LE* fluxes under advective and nonadvective conditions and on both a short period and a daily basis by Verma et al. (30) and Blad and Rosenberg (103). E_t estimates are sensitive to errors in the measurement of T_c, especially in nonadvective cases, but are less strongly affected by errors in estimation of r_a. Until recently all terms in this equation except T_c could be routinely measured or estimated from readily available meteorological data (see discussion of Penman method that follows). The availability of surface temperature data from remote sensing measurements provides the means to apply this model to estimate E_t over large areas.

Combination Models

Penman Model

Penman (44) combined energy balance and aerodynamic equations for evaporation to give an expression for E_0 from extensive wet surfaces in terms of easily measureable parameters. The original Penman equation was designed to estimate *E* from open water surfaces and may be written as:

$$E_0 = \frac{mR_n + \gamma E_a}{m + \gamma} \tag{1.33}$$

where m is the slope of the curve for water saturation vapor pressure versus temperature and E_a is given by a modification of Eq. (1.21) or:

$$E_a = (a + bu_z)(e_{sa} - e_a) \tag{1.34}$$

where a and b are empirically derived constants, e_{sa} is the saturation vapor pressure at mean air temperature, and u_z is the wind speed at height z. Penman used values of 0.27 mm Hg °F^{-1} for γ, 0.35 for a, and 3.43×10^{-3} for b when u_z was in miles per day measured at 2 m. Using these values with R_n expressed in equivalent mm water day^{-1}, Eq. (1.33) gives E_0 in mm day^{-1}.

The model requires knowledge of vapor pressure, air temperature, wind speed, and net radiation. All but the last term are routinely measured at weather observatories. Net radiation is not generally measured but may be estimated with procedures such as those described by Penman (44) or Linacre (125, 126). Methods to estimate R_n require the estimation of net short-wave radiation and net long-wave fluxes. The net short-wave radiative flux can be estimated from extraterrestrial radiation (solar radiation at the top of the atmosphere) or from solar radiation at the earth's surface or from sunshine duration combined with information about surface albedo. Net long-wave fluxes can be estimated from formulas such as those of Brunt (127), Swinbank (128), or Idso and Jackson (129).

Although the Penman model does have a sound physical basis the empirical nature of Eq. 1.34, the empiricism involved in estimating R_n, when it cannot be directly measured, and the lack of a soil heat flux term cause Penman's model to often be classified as an empirical model.

To estimate E_t from vegetative surfaces and under conditions where optimum water supply is not maintained, Penman (130) modified his original model and included a term that is the ratio of E_t from well-watered turf to E from open water. He also developed a drying curve to account for the influence of limited soil water on the actual E_t from the turf. Penman's model has been widely used to estimate E_0, often with good success (131). With conditions of strong sensible heat advection Rosenberg (14), however, found that the Penman method consistently underestimated E_t by alfalfa.

Penman Model Modified by Monteith

Monteith (108, 132) introduced resistance terms into the method of Penman and arrived at the following equation for LE from surfaces with either optimal or limited water supply:

$$LE = -\frac{mR_n + \rho_a C_p(e_{sa} - e_a)/r_a}{m + \gamma\,(r_a + r_c)/r_a} \tag{1.35}$$

This model has been successfully used to estimate E_t from crops (52) and from forests (133) but, of course, this model requires r_a and r_c data which are not readily available and somewhat difficult to measure. Thus, its use will be limited primarily to research applications, at least until such time that r_a and r_c can be estimated more readily than is presently possible.

Van Bavel Model

Following the method of Penman, van Bavel (134) derived the following expression:

$$LE = -\frac{m(R_n + S) + \gamma L B_v(e_{sa} - e_a)}{m + \gamma} \tag{1.36}$$

where L is the latent heat of vaporization and B_v is defined as:

$$B_v = \frac{\rho_a M_w/M_a k^2}{P}\,\frac{u_z}{[\ln(z/z_0)]^2} \tag{1.37}$$

where z_0 is a surface roughness parameter and other terms are as previously defined. Van Bavel stated that this model is an improvement over the Penman model because it contains no empirical constants or functions. He obtained excellent agreement between measured *LE* rates and those calculated with Eq. 1.36 for open water, wet bare soil, and well-watered alfalfa. However, Rosenberg (14) found the van Bavel method to be very sensitive to windiness and very dependent on the value assigned to z_0. The model underestimated in calm conditions and overestimated in strong winds.

Priestly–Taylor Model

Priestly and Taylor (135) introduced a semiempirical model for estimating *LE* fluxes over land surfaces. Their model may be expressed as:

$$LE = -\alpha \frac{m}{m + \gamma}(R_n + S) \tag{1.38}$$

where α is an empirically determined constant. The model may be considered to be a simplified form of the Penman equation in which the additive γR_a term in Eq. 1.33 is replaced by a multiplicative constant, α.

The constant, α, is also related to the Bowen ratio, β, by:

$$\alpha = \frac{m + \gamma}{m(\beta + 1)} \tag{1.39}$$

Using sets of reliable data from diverse well-watered surfaces, Priestly and Taylor obtained values for α between 1.08 and 1.34 with an overall mean of 1.26. Davies and Allen (136) showed that the value for α varied slightly, depending on the air temperature, but was about 1.26 for temperatures between 15 and 30°C. α is also a function of the soil moisture at the surface and Eq. 1.38 can be made more general if α is modified to account for the change due to soil moisture influences (136).

Thompson (137) verified the value of 1.26 for wet surfaces and at the same time showed that as LE/R_n decreased so likewise did α. Tanner and Jury (138) stated that α ranges from 1.35 ± 0.10 depending somewhat on the crop and local climate if daily values of R_n are used. If R_n values during daytime periods only are used, α will be smaller than when 24 hr R_n values are used. Jury and Tanner (139) showed that α will increase with advection and suggested a procedure for adapting the Priestly–Taylor model to advective conditions.

The Priestly–Taylor model has been shown to be most reliable in humid areas. It has not yet been adequately tested in arid regions. Because the model utilizes meteorological data that are readily available, it is likely to become an increasingly important model for estimating E_0. It may, with modification, be reliable for estimating actual E_t.

Air Temperature Based Formulas

Thornthwaite Model

Thornthwaite (140) developed an equation for estimating E_0 utilizing the monthly mean temperature, t, and day length. The Thornthwaite equation may be written as:

$$E_0 = 1.6 \frac{L}{12} \frac{N}{30} \frac{10t}{I} a \tag{1.40}$$

where L is actual daylength in hours, N is the number of days in a month, t is the mean monthly air temperature (°C) and a is defined as:

$$a = 6.75 \times 10^{-7} I^3 - 7.71 \times 10^5 I^2 + 1.79 \times 10^{-2} I + 0.49 \tag{1.41}$$

where I is a heat index derived from the sum of 12 monthly index values, i, obtained from:

$$i = \left(\frac{t}{5}\right)^{1.514} \tag{1.42}$$

Nomograms and tables have been prepared which greatly simplify the procedure for estimating E_0 with this method (141).

There are several shortcomings in the method, especially if E_t estimates are made for periods less than a month in length. The method has given some misleading results and is no longer widely used except to obtain some estimate of E_t for a growing season or for annual E_t estimates. It may be useful, however, for estimating E_t in regions where weather data are limited to temperature measurements.

Blaney–Criddle Model

Blaney and Criddle (142) developed a model for estimating actual E_t or as they have termed it "consumptive use." The consumptive use, c_u, on a monthly basis is:

$$c_u = K_m f \tag{1.43}$$

where c_u is in inches, K_m is in empirically derived monthly consumptive use coefficient (a function of the type of crop), and f is a monthly consumptive use factor equal to $tp/100$ where t is the monthly mean temperature (°F) and p is the monthly percentage of total annual daylight hours. The total consumptive use for the seasonal period, u, is:

$$u = K_s F = \sum c_u = \sum K_m f \tag{1.44}$$

where K_s is the seasonal consumptive use coefficient and $F = \sum f$ for the total period.

The method is relatively easy to use and necessary data are readily accessible. The model has been used extensively, especially in the western United States, with results accurate enough for several practical applications.

Linacre Model

Linacre (143) suggested a simplification of the Penman model to a form which requires only values for the elevation, latitude, and daily maximum and minimum temperatures and mean dew point temperature. The formula is:

$$E_0 = \frac{700T_m/(100 - A) + 15(T - T_d)}{80 - t} \tag{1.45}$$

where $T_m = T + 0.006h$, h is the elevation (m), T is the mean temperature (°C), A is the latitude (degrees), and T_d is the mean dew point temperature (°C). Linacre found that values of E_t given by Eq. (1.45) typically differed from measured values by 0.3–1.7 mm day^{-1} depending on whether the estimate was for an annual mean or for a single day.

Solar Radiation Methods

Experimental evidence (144) suggests that the energy for E_t, even in arid regions, is derived from solar radiation. Thus E_t is correlated with solar radiation. The relationship between E_t and solar radiation generally changes with the season as climate and surface conditions change. To account for these changes some solar radiation models also include a temperature term.

Makkink Model

Makkink (145) proposed the following formula for estimating E_0 from solar radiation measurements:

$$E_0 = R_s \left(\frac{m}{m + \gamma}\right) + 0.12 \tag{1.46}$$

where E_0 and R_s are in units equivalent to millimeters of water. Makkink's formula has given good results in cold wet climates but has not been very satisfactory in arid regions.

Regression Models

As shown by Aslyng (146) evapotranspiration is linearly and strongly related to global radiation. Thus, empirical relationships between E_0 and R_s have been given by simple linear regression such as:

$$E_0 = aR_s + b \tag{1.47}$$

where a and b are empirical constants which change with location and season. Stanhill (97) and Tanner (147) give several proposed values for a and b. These regression models are simple to use, but because of their highly empirical nature, they have only very limited applicability. The lack of adequate networks providing routine measurement of R_s is also a drawback to the use of R_s-based models.

Jensen–Haise Model

Jensen and Haise (148) summarized data collected in arid regions of the Western United States and developed the following formula:

$$E_0 = R_s(0.025T + 0.08) \tag{1.48}$$

where T is the mean daily temperature (°C), R_s is the daily total solar radiation in units equivalent to mm of water, and E_0 is in mm day^{-1}. This model has been tested against lysimetric measurements and as reported by Rosenberg (32) the model tended to seriously underestimate E_t under advective conditions but gave good results under nonadvective conditions. Jensen et al. (149) propose a modification of Eq. 1.48 but according to Rosenberg (32) this modified method tended to underestimate E_t under advection even more seriously. The Jensen–Haise model and others based on similar principles are described and discussed by Linacre (150).

Solar Thermal Unit Model

Caprio (151) proposed a model to estimate E_0 based on the solar thermal unit (STU) concept. STU is defined as the product of the mean daily temperature (°F), minus a threshold temperature of 31°F times the daily total solar radiation in cal cm^{-2}. Caprio states that E_0 in inches can be computed by multiplying the STU by 10^{-5}. The model has not been widely tested yet but certainly it should be subject to the same limitations of the other solar radiation and temperature based empirical models.

Models for Calculating Evaporation and Transpiration Separately

Evaporation from the soil, E_s, and transpiration from vegetation, T_p, are both complex processes driven by the atmospheric demand for water. These processes are limited by the amount of water available in the soil and by soil and plant characteristics. When plants cover only a small portion of the soil surface, E_t is dominated by the soil evaporation rate. Transpiration becomes increasingly important as the amount of leaf area increases. As crop cover becomes complete, transpiration accounts for essentially all of the water lost by E_t. It is sometimes desirable to model the two processes separately. Approaches for doing this are described below.

Ritchie Model

In 1972 Ritchie introduced a model for predicting evaporation from a row crop with incomplete cover. He calculated E_s in two stages: the constant rate and the falling rate stages. In the constant rate stage the soil is sufficiently wet to permit the transport of water to the surface at a rate adequate to maintain E_0. In this stage, the supply of energy reaching the soil limits E_s. The water supply to the soil surface is limited primarily by the hydraulic properties of the soil. Thus, they appear to be the main cause for determining when the constant rate stage ceases and the falling rate stage begins.

When plants begin to cover the soil, E_s will not be the same as evaporation from a bare soil because of shading, lowered wind speed, and reduced vapor pressure deficits ($e_{sa} - e_a$). The equation to estimate E_s from the soil in the constant rate stage is:

$$E_s = \left(\frac{m}{m + \gamma}\right) R_{ns} = \left(\frac{m}{m + \gamma}\right) R_n e^{-0.4LAI} \qquad (1.49)$$

where R_{ns} is the net radiation at the soil surface and other terms are as previously defined. When Eq. (1.49) is applied to soil surfaces partially covered by mulch, the exponent becomes $-0.4(LAI + 2.5M)$ where M is the fraction of the soil surface covered by the mulch. Ritchie (152) suggests that E_s estimates made with Eq. 1.49 can be improved by multiplying the right hand sides of that equation by the Priestly–Taylor α term [Eq. 1.38]. He gives the following empirical expression for calculating α:

$$\alpha = 0.92 + 0.4 \frac{R_{ns}}{R_n} \qquad (1.50)$$

During the falling rate state, E_s is more dependent on the hydraulic properties of the soil and less dependent on the available energy. Black et al. (153) have shown that the cumulative evaporation from the soil can be expressed as:

$$\sum E_s = \alpha_s t^{\frac{1}{2}} \qquad (1.51)$$

where α_s is dependent on soil hydraulic properties, α_s is approximately proportional to the hydraulic conductivity at 0.1 bar soil matric potential and can be evaluated for individual soils from data on cumulative evaporation for a single drying cycle, t is the time (in days) from the onset of the falling rate stage, and E_s for a given day is obtained by subtracting the $\sum E_s$ for all previous days from $\sum E_s$ through the current day.

The equation developed by Ritchie and Burnett (154) for estimating T_p when soil water is not limited is:

$$T_p = E_0(-0.21 + 0.70\, LAI^{\frac{1}{2}}) \qquad 0.1 \leq LAI \leq 2.7 \qquad (1.52)$$

Ritchie suggests using either the Penman model [Eq. (1.33)] or the Priestly–Taylor model [Eq. (1.38)] to calculate E_0. For crop canopies with *LAI* greater than about 2.7 T_p is obtained from:

$$T_p = E_0 - E_s \qquad LAI > 2.7 \tag{1.53}$$

Ritchie (152) and Ritchie et al. (155) suggest that soil water reaches a critical threshold value when the amount of extractable water, W_e, remaining in the root zone of the soil is approximately 25% of the total amount of extractable water, W_u. When soil water is limiting the following equations for calculating T_p are suggested:

$$T_p = 4(-0.21 + 0.70\, LAI^{\frac{1}{2}})\, E_0\left(\frac{W_e}{W_u}\right) \qquad LAI < 2.7 \tag{1.54}$$

and

$$T_p = 4E_0\left(\frac{W_e}{W_u}\right) \qquad LAI > 2.7 \tag{1.55}$$

The models have been used by Ritchie and co-workers to estimate E_t with reasonably good success from agricultural crops and small watersheds. The major problem with their models has been an apparent overestimate of T_p with wet soils and large *LAI* values. Other problems may arise when the model is applied in areas where sensible heat advection is common because of the difficulty of obtaining reliable estimates of E_0 under advective conditions.

Tanner–Jury Model

Tanner and Jury (138) proposed an approach similar to that of Ritchie but one in which the procedures for estimating potential T_p and potential E_s have been modified to prevent the serious overestimates of T_p encountered under certain conditions with the Ritchie approach.

For the calculation of E_s under nonlimiting soil water conditions Eq. 1.49 is modified to give:

$$E_s = \alpha_E \tau R_n\left(\frac{m}{m + \gamma}\right) \tag{1.56}$$

where α_E for a wet bare soil is the same as α of the Priestly–Taylor model for full cover vegetation, α_E decreases under increasing canopy cover to a value approaching one for full crop cover, and τ is the fraction of the total R_n exchanged at the soil surface and can be described by:

$$\tau = R_{ns}/R_n = e^{-bLAI} \tag{1.57}$$

where b is an empirical coefficient. Methods for calculating α_E are given by Tanner and Jury (138).

During the falling rate stage, Tanner and Jury utilize two methods for calculating E_s. Method A is essentially the same as that used by Ritchie. It is assumed in this method that E_s proceeds according to Eq. 1.51 and that it is uninfluenced by T_p. Method B considers the entire root zone to have constant properties with a water content obeying the equation:

$$D \frac{d\theta}{dt} = P_t - E_t - u \tag{1.58}$$

where D is the depth of the root zone, u is the drainage rate, and θ is the volumetric water content in the root zone. When θ drops below some threshold value, θ_c, then t_0, the beginning of the falling rate stage is set, and E_s is described by Eq. 1.51 regardless of any water added to the soil until θ is recharged above θ_c. According to Jury and Tanner the primary differences in the two methods are (1) the falling rate stage for E_s is determined by a water balance in the root zone which accounts for drainage and transpiration in Method B but not in Method A, (2) small precipitation additions always raise the soil-limited E_s rate in Method A but not in Method B, and (3) Method A accounts for drainage with a field capacity concept and Method B employs a drainage function. Method B applies when the root system is confined mainly to the plow layer but probably will not work well for deep rooted zones.

When water is not limiting, Tanner and Jury use the following equation for calculating T_p:

$$T_p = \alpha_T(1 - \tau)\left(\frac{m}{m + \gamma}\right) R_n \tag{1.59}$$

and

$$\alpha_T = \frac{\alpha - \alpha_E \tau}{1 - \tau} \tag{1.60}$$

Tanner and Jury do not suggest a procedure for calculating T_p when soil moisture becomes limiting. Tanner and Jury found that estimates of E_t made with both methods are very similar and compared fairly well with lysimetric measurements.

LIST OF SYMBOLS

R_n net radiation
R_s solar radiation (short wave)
R_2 atmospheric radiation (long wave)
R_g outgoing radiation (long wave)
R_{ns} net radiation at the soil surface

α	reflection coefficient (albedo)
τ	transmissivity
ϵ	emissivity
T	temperature
T_a	air temperature
T_c	canopy temperature
T_g	ground temperature
T_s	temperature of the vegetative surface
E	evaporation
E_t	evapotranspiration
E_s	evaporation from the soil
E_o	potential evapotranspiration
E_p	pan evaporation
T_p	transpiration
LE	latent heat flux
H	sensible heat flux
H_c	sensible heat exchange between the soil surface and the air above
S	soil heat flux
ρ_a	air density
C_p	specific heat of air at constant pressure
r_a	aerial resistance
r_s	stomatal resistance
r_c	canopy resistance
ψ	water potential
ψ_a	atmospheric water potential
e_s	saturation vapor pressure of surface
e_{sa}	saturation vapor pressure of air
e_a	air vapor pressure
M_w	molecular weight of water
M_a	molecular weight of air
L	latent heat of evaporation
P	atmospheric pressure
γ	psychrometric constant
K_w	exchange coefficient for water vapor
K_h	exchange coefficient for heat
u	wind speed
z	height or depth
β	Bowen ratio
m	slope of curve for a plot of saturation vapor pressure vs. temperature

LAI leaf area index
k von Karman constant
σ Boltzmann constant

REFERENCES

1. J. L. Monteith, *Principles of Environmental Physics,* Edward Arnold, London, 1973, 241 pp.
2. J. Y. Parlange, P. E. Waggoner, and G. H. Heichel, *Plant. Physiol.* **48,** 437–442 (1974).
3. D. M. Gates, *Energy Exchange in the Biosphere,* Harper & Row, New York, 1962.
4. R. Lemeur and B. L. Blad, *Agric. Meteorol.* **14,** 255–286 (1974).
5. O. T. Denmead, L. J. Fritschen, and R. H. Shaw, *Agron. J.* **54,** 505–510 (1962).
6. J. E. Begg, J. F. Bierhuizen, E. R. Lemon, O. K. Misra, R. O. Slayter, and W. R. Stern, *Agric. Meteorol.* **1,** 294–312 (1964).
7. J. M. Norman, Modeling the Complete Crop Canopy, in B. J. Barfield and J. F. Gerber, Eds., *Modification of the Aerial Environment of Crops.* American Society of Agricultural Engineers, St. Joseph, Mich., 1979, pp. 249–277.
8. E. T. Kanemasu and G. F. Arkin, *Agric. Meteorol.* **14,** 211–225 (1974).
9. E. W. Chin Choy and E. T. Kanemasu, *Agron. J.* **66,** 98–99 (1974).
10. E. W. Chin Choy, J. F. Stone, and J. E. Garton, *Soil Sci. Soc. Am. J.* **41,** 428–432 (1977).
11. G. N. McCauley, J. F. Stone, and E. W. Chin Choy, *Agric. Meteorol.* **19,** 295–304 (1978).
12. L. H. Parmele and J. L. McGuinness, *J. Hydrol.* **22,** 239–251 (1974).
13. E. Lemon, D. W. Stewart, and R. W. Shawcroft, Science **174,** 371–378 (1971).
14. N. J. Rosenberg, *Agron. J.* **61,** 879–886 (1969).
15. N. J. Rosenberg, *Agric. Meteorol.* **6,** 179–184 (1969).
16. R. J. Hanks, L. H. Allen, and H. R. Gardner, *Agron. J.* **63,** 520–527 (1971).
17. N. E. Rider, J. R. Philip, and E. F. Bradley, *Quart. J. R. Meteorol. Soc.* **89,** 507–530 (1963).
18. A. J. Dyer and T. V. Crawford, *Quart. J. R. Meteorol. Soc.* **91,** 345–348 (1965).
19. T. W. Brakke, S. B. Verma, and N. J. Rosenberg, *J. Appl. Meteorol.* **17,** 955–963 (1978).
20. I. C. McIlroy and D. E. Angus, *Agric. Meteorol.* **1,** 201–224 (1964).
21. R. A. Nulsen, G. W. Thurtell, and K. R. Stevenson, *Agron. J.* **69,** 951–954 (1977).
22. L. Leyton, Some Thermodynamic Concepts of Water Movement, in *Fluid Behavior in Biological Systems,* Oxford University Press, New York, 1978, Chap. 11.
23. H. H. Neumann, G. W. Thurtell, and K. R. Stevenson, *Can. J. Plant Sci.* **54,** 175–184 (1974).
24. H. D. Barrs, Controlled Environment Studies of the Effects of Variable Atmospheric Water Stress on Photosynthesis, Transpiration and Water Status of *Zea mays L.* and Other Species, in *Plant Response to Climatic Factors, Proceedings of the Uppsala Symposium* UNESCO, 1973, pp. 249–258.
25. van den Honert, *Discuss. Faraday Soc.* **3,** 146–153 (1948).
26. I. R. Cowan and F. L. Milthorpe, Plant Factors Influencing the Water Status of Plant

Tissues, in T. T. Kozlowski, Ed., *Water Deficits and Plant Growth,* Vol. 1, Academic Press, New York, 1968, pp. 137–193.

27. M. R. Kaufmann and A. E. Hall, *Agric. Meteorol.* **14,** 85–98 (1974).
28. R. Stoker and P. E. Weatherley, *New Phytol.* **70,** 547–554 (1971).
29. G. K. Hansen, *Acta Agric. Scand.* **24,** 84–92 (1974).
30. S. B. Verma, N. J. Rosenberg, B. L. Blad, and M. W. Baradas, *Agron. J.* **68,** 776–782 (1976).
31. Z. Plaut and S. Moreshet, Transport of Water in the Plant-Atmosphere System, in B. Yaron, E. Danfors, and Y. Vaadia, Eds., *Arid Zone Irrigation,* Springer-Verlag, New York, 1973, pp. 123–141.
32. N. J. Rosenberg, *Microclimate: the Biological Environment,* Wiley, New York, 1974, 315 pp.
33. R. O. Slayter, *Agroclimatological Methods, UNESCO Natl. Resources Res.* **7,** 73–87 (1968).
34. G. P. Malhotra and P. Brock, Hydrologic Budget of North America and Subregions Formulated Using Atmospheric Vapor Flux Data, in *World Water Balance, Proc. Reading Symp. Int. Assoc. Sci. Hydrol.,* 1970, pp. 501–524.
35. A. Baumgartner, Water and Energy Balances of Different Vegetation Covers, in *World Water Balance, Proc. Reading Symp. Int. Assoc. Sci. Hydrol.,* 1970, pp. 57–65.
36. D. A. Woolhiser, R. E. Smith, and C. L. Hanson, Evaporation Components of Watershed Models for the Great Plains, in *Evapotranspiration in the Great Plains,* Great Plains Agric. Council Publ. No. 50, Kansas State University, Manhattan, 1970, pp. 111–136.
37. H. H. Lettau and M. W. Baradas, *Monthly Weather Rev.* **101,** 636–649 (1973).
38. C. W. Rose and W. R. Stern, *Aust. J. Soil Res.* **3,** 95–100 (1965).
39. S. M. Goltz, C. B. Tanner, A. A. Miller, and A. R. G. Lang, *Agron. J.* **63,** 762–765 (1971).
40. J. C. Wilcox and W. K. Sly, *Can. J. Plant Sci.* **54,** 559–564 (1974).
41. C. H. M. van Bavel, K. J. Brust, and G. B. Stirk, *Soil Sci. Soc. Am. Proc.* **32,** 317–321 (1968).
42. J. M. Davidson, L. R. Stone, D. R. Nielsen, and M. E. Larue, *Water Resources Res.* **5,** 1312–1321 (1969).
43. F. I. Morton, Evaporation and Climate—A Study in Cause and Effect, Scientific Series No. 4, Dept. Energy, Mines and Resources, Ottawa, 1968, 32 pp.
44. H. L. Penman, *Proc. R. Soc. London, Ser. A* **193,** 120–145 (1948).
45. R. E. Luebs, A. E. Laag, and P. A. Nash, *Agron. J.* **67,** 339–343 (1975).
46. L. S. Brun, E. T. Kanemasu, and W. L. Powers, *Agron. J.* **64,** 145–148 (1972).
47. O. T. Denmead and I. C. McIlroy, *Agric. Meteorol.* **7,** 285–302 (1970).
48. J. I. Stewart, R. M. Hogan, W. O. Pruitt, R. E. Danielsen, W. T. Franklin, R. J. Hands, J. P. Riley, and E. B. Jackson, Optimizing Crop Production through Control of Water and Salinity Levels in the Soil, PRWG 151-1, Utah Water Research Laboratory, Utah State University, Logan, 1977, 191 pp.
49. J. T. Ritchie, The Contribution of the Surface Energy Balance in Climatic Stress for Agriculture, Paper presented at Climate-Technology Seminar, Oct. 25–26, Columbia, Missouri, 1977.
50. K. J. Kristensen, *Nordic. Hydrol.* **5,** 173–182 (1974).
51. A. H. El Nadi and J. P. Hudson, *Exp. Agric.* **1,** 289–298 (1965).
52. P. J. Slabbers, *J. Hydrol.* **34,** 181–191 (1977).
53. C. H. M. van Bavel, *Agric. Meteorol.* **4,** 165–176 (1967).

54. L. J. Fritschen, *Agron. J.* **58,** 339–342 (1965).
55. B. L. Blad and N. J. Rosenberg, *Agron. J.* **66,** 248–252 (1974).
56. E. T. Kanemasu, G. W. Thurtell, and C. B. Tanner, *Plant Physiol.* **44,** 881–885 (1969).
57. C. J. Stigter and B. Lammers, *Meded. LandbHogesch. Wageningen* **21,** 1–76 (1974).
58. T. A. Black, C. B. Tanner, and W. R. Gardner, *Agron. J.* **62,** 66–69 (1970).
59. L. S. Brun, E. T. Kanemasu, and W. L. Powers, *Agron. J.* **65,** 326–328 (1973).
60. G. Szeicz, C. H. M. van Bavel, and S. Takami, *Agric. Meteorol.* **12,** 361–389 (1973).
61. J. T. Ritchie, *Agron. J.* **65,** 893–897 (1973).
62. S. B. Idso, R. J. Reginato, R. D. Jackson, B. A. Kimball, and F. S. Nakayama, *Soil Sci. Soc. Am. Proc.* **38,** 831–837, 1974.
63. C. H. M. van Bavel and D. I. Hillel, *Agric. Meteorol.* **17,** 453–476 (1976).
64. J. T. Ritchie, *Water Resources Res.* **8,** 1204–1213 (1972).
65. J. F. Stone, *Soil Crop Soc. Fla. Proc.* **37,** 1–11 (1977).
66. W. L. Pelton, The Use of Lysimetric Methods to Measure Evapotranspiration, *Proc. 2nd Hydrol. Symp.*, Toronto, Cat. R 32-36112, Queen's Printer, Ottawa, 1961, pp. 106–134.
67. L. L. Harrold, Measuring Evapotranspiration by Lysimetry, in *Evapotranspiration and Its Role in Water Resources Management*, Proc. Am. Soc. Agric. Eng., 1966, pp. 28–33.
68. H. G. Wilm, *Trans. Am. Geophys. Union* **27,** 720–725 (1946).
69. M. J. Gilbert and C. H. M. van Bavel, *Trans. Am. Geophys. Union* **35,** 937–942 (1954).
70. F. M. Wilhite, A. R. Grable, and H. K. Rouse, *Agron. J.* **5,** 479–481 (1965).
71. J. C. van Schaik and K. K. Krogman, *Can. J. Soil Sci.* **48,** 281–288 (1968).
72. K. M. King, C. B. Tanner, and V. E. Suomi, *Trans. Am. Geophys. Union* **37,** 738–742 (1956).
73. W. D. McMillan and H. A. Paul, *Agric. Engr.* **42,** 498–499 (1961).
74. J. Glover and S. A. Forsgate, *Nature (London).* **195,** 1330 (1962).
75. R. J. Hanks and R. W. Shawcroft, *Agron. J.* **57,** 634–636 (1965).
76. J. M. Fulton and W. I. Findlay, *Can. J. Plant Sci.* **46,** 685–686 (1966).
77. F. J. Lourence and W. B. Goddard, *J. Appl. Meteorol.* **6,** 489–492 (1967).
78. L. L. Harrold and F. R. Dreibelbis, Evaluation of Agricultural Hydrology by Monolith Lysimeters (1944–1955), *U.S. Dep. Agric. Tech. Bull. No. 1179*, 1958, 166 pp.
79. W. O. Pruitt and D. E. Angus, *Trans. Am. Soc. Agric. Eng.* **3,** 13–18 (1960).
80. D. W. Young, D. D. Evans, T. W. Sammis, and C. Constant, Measurement of Evapotranspiration with a Monolith Lysimeter, *US/IBP Desert Biome Res. Memo 76-34*, Utah State University, Logan, 1976, 5 pp.
81. J. T. Ritchie and E. Burnett, *Agron. J.* **60,** 545–549 (1968).
82. C. H. M. van Bavel and L. E. Meyers, *Agric. Eng.* **43,** 580–583, 587–588 (1962).
83. N. J. Rosenberg and K. W. Brown, *Water Resources Res.* **6,** 1227–1229 (1970).
84. P. G. Sebenik and J. L. Thomas, *Prog. Agric. Ariz.* **19,** 10–11 (1967).
85. D. B. Peters, B. F. Clough, R. A. Graves, and G. R. Stahl, *Agron. J.* **66,** 460–462 (1974).
86. D. C. Reicosky and D. B. Peters, *Agron. J.* **69,** 729–732 (1977).
87. J. E. Bloodworth, J. B. Page and W. R. Cowley, *Soil Sci. Soc. Am. Proc.* **19,** 411–414 (1955).
88. R. H. Swanson, *Agric. Meteorol.* **10,** 277–281 (1972).

89. J. Balek and O. Pavlik, *J. Hydrol.* **34,** 193–200 (1977).
90. J. R. Kline, J. R. Martin, C. F. Jordan, and J. J. Kovanda, *Ecology* **51,** 1068–1073 (1970).
91. P. W. Owston, J. L. Smith, and H. G. Halverson, *For. Sci.* **18,** 266–272 (1972).
92. D. W. Sheriff, *J. Exp. Bot.* **23,** 1086–1095 (1972).
93. B. E. Livingston, *Ecology* **16,** 438–472 (1935).
94. A. C. Carder, *Can. J. Plant Sci.* **40,** 700–706 (1960).
95. W. L. Pelton and H. C. Korven, *Can. J. Plant Sci.* **49,** 615–621 (1969).
96. A. C. Dilley and I. Helmond, *Agric. Meteorol.* **12,** 1–11 (1973).
97. G. Stanhill, *Isr. J. Agric. Res.* **11,** 159–171 (1961).
98. N. J. Rosenberg and W. Powers, Potential for Evapotranspiration and Its Manipulation in the Plains Region, *Proc. Symp. Evapotranspiration Great Plains,* Great Plains Agric. Council Publ. No. 50, 1970, pp. 275–300.
99. W. O. Pruitt and F. J. Lourence, Correlation of Climatological Data with Water Requirement of Crops, Dept. of Water Sci. and Engr. Paper No. 9001, University of California, Davis, 1968.
100. M. Fuchs, Water Transfer from the Soil and the Vegetation to the Atmosphere, in B. Yaron, E. Danfors, and Y. Vaadia, Eds., *Arid Zone Irrigation,* Springer-Verlag, New York, 1968, pp. 143–152.
101. C. Rohwer, Evaporation from Free Water Surfaces, *USDA Tech. Bull. 271,* 1931, 96 pp.
102. G. E. Harbeck, Jr., *U.S. Geol. Surv. Paper* **272-E,** 101–105 (1962).
103. B. L. Blad and N. J. Rosenberg, *Agron. J.* **68,** 764–769 (1976).
104. T. L. Richards and J. G. Irbe, Estimates of Monthly Evaporation Losses from the Great Lakes 1950 to 1968 Based on the Mass Transfer Technique, *Proc. 12th Conf. Great Lakes Res.,* 1969, pp. 469–487.
105. J. Conaway and C. H. M. van Bavel, *J. Appl. Meteorol.* **6,** 650–655 (1967).
106. C. D. Ripple, J. Rubin and T. E. A. van Hylckama, Estimating Steady-State Evaporation Rates from Bare Soils under Conditions of High Water Table, *U.S. Geol. Sur. Open-file Rep. Water Res. Div.,* Menlo Park, Calif., 1970, 62 pp.
107. W. O. Pruitt and M. J. Aston, Atmospheric and Surface Factors Affecting Evapotranspiration, *Final Report Task 3A99-27-005-08,* 1963.
108. J. L. Monteith, Gas Exchange in Plant Communities, in L. T. Evans, Ed., *Environmental Control of Plant Growth,* Academic Press, New York, 1963, pp. 95–112.
109. K. W. Brown and N. J. Rosenberg, *Agron. J.* **65,** 314–347 (1973).
110. S. B. Verma and N. J. Rosenberg, *Agron. J.* **69,** 332–335 (1977).
111. C. W. Thornthwaite and B. Holzman, Measurement of Evaporation from Land and Water Surfaces, *USDA Tech. Bull 817,* 1942, 75 pp.
112. L. F. Richardson, Proc. R. Soc. **A97,** 354–373 (1929).
113. A. S. Monin and A. M. Obukhov, *Akad. Nauk USSR Acad. Sci.* **24,** 163–187 (1954).
114. F. Pasquill, *Quart. J. R. Meteorol. Soc.* **76,** 287–301 (1950).
115. E. L. Deacon and W. C. Swinbank, Comparison between Momentum and Water Vapor Transfer, *Proc. Canberra Symp. Climatol. and Microclimatol. Arid Zone Res. UNESCO,* 1958, pp. 38–41.
116. R. E. Munn, Energy Budget and Mass Transfer Theories of Evaporation, *Proc. 2nd Hydrol. Symp.,* Toronto, Cat. R32-361/2, Queen's Printer, Ottawa, 1961, pp. 8–30.
117. W. O. Pruitt, Application of Several Energy Balance and Aerodynamic Evaporation

Equations under a Wide Range of Stability, *Final Report to USAEPG on Contract No. DA-36-039-SC-80334*, University of California, Davis, 1963, pp. 107–124.

118. F. W. Pierson and A. P. Jackman, *J. Appl. Meteorol.* **14**, 477–487 (1975).
119. W. C. Swinbank, *J. Meteorol.* **8**, 135–145 (1951).
120. I. S. Bowen, *Phys. Rev.* **27**, 779–787 (1926).
121. C. B. Tanner, *Soil Sci. Soc. Am. Proc.* **24**, 1–9 (1960).
122. O. T. Denmead and I. C. McIlroy, *Agric. Meteorol.* **7**, 285–302 (1970).
123. B. L. Blad and N. J. Rosenberg, *J. Appl. Meteorol.* **13**, 227–236 (1974).
124. S. B. Verma, N. J. Rosenberg, and B. L. Blad, *J. Appl. Meteorol.* **17**, 330–338 (1978).
125. E. T. Linacre, *Agric. Meteorol.* **5**, 49–63 (1968).
126. E. T. Linacre, *J. Appl. Ecol.* **6**, 61–75, 1969.
127. D. Brunt, *Physical and Dynamical Meteorology*, Cambridge Univ. Press, London, 1934.
128. W. C. Swinbank, *Quart. J. R. Meteorol. Soc.* **89**, 339–348 (1963).
129. S. B. Idso and R. J. Jackson, *J. Geophys. Res.* **74**, 5397–5403 (1969).
130. H. L. Penman, *J. Soil Sci.* **1**, 74–89 (1949).
131. H. L. Penman, *Quart. J. R. Meteorol. Soc.* **76**, 372–383 (1950).
132. J. L. Monteith, *Proc. Symp. Exp. Biol.* **19**, 205–234 (1965).
133. I. R. Calder, *J. Hydrol.* **33**, 247–265 (1977).
134. C. H. M. van Bavel, *Water Resources Res.* **2**, 455–467 (1966).
135. C. H. B. Priestly and R. J. Taylor, *Monthly Weather Rev.* **100**, 81–92, (1972).
136. J. A. Davies and C. D. Allen, *J. Appl. Meteorol.* **12**, 649–657 (1973).
137. J. R. Thompson, *J. Appl. Meteorol.* **14**, 1399–1401 (1975).
138. C. B. Tanner and W. A. Jury, *Agron. J.* **68**, 239–243 (1976).
139. W. A. Jury and C. B. Tanner, *Agron. J.* **67**, 840–842 (1975).
140. C. W. Thornthwaite, *Geogr. Rev.* **38**, 55–94 (1948).
141. C. W. Thornthwaite and J. R. Mather, *Climatology* **8**, 1–104 (1955).
142. H. F. Blaney and W. D. Criddle, *USDA Soil Conservation Service Tech. Paper No. 96*, 1950, 48 pp.
143. E. T. Linacre, *Agric. Meteorol.* **18**, 409–424 (1977).
144. C. B. Tanner and E. R. Lemon, *Agron. J.* **54**, 207–212 (1962).
145. G. F. Makkink, *Neth. J. Agric. Sci.* **5**, 290–305 (1957).
146. H. C. Aslyng, *Nordic Hydrol.* **5**, 247–256 (1974).
147. C. B. Tanner, Measurement of Evapotranspiration, in R. M. Hagen, M. R. Horse, and T. W. Edminster, Eds., *Irrigation of Agricultural Lands*, Am. Soc. Agron., Madison, Wisc., 1967, pp. 534–574.
148. M. E. Jensen and H. R. Haise, *J. Irrig. Drain. Div. ASCE* **89**, 15–41 (1963).
149. M. E. Jensen, D. C. N. Robb, and C. E. Franzoy, *J. Irrig. Drain. Div. ASCE* **96**, 25–38 (1970).
150. E. T. Linacre, *J. Irrig. Drain. Div. ASCE* **93**, 61–79 (1967).
151. J. M. Caprio, The Solar Thermal Unit Concept in Problems Related to Plant Development and Potential Evapotranspiration, in H. Leith, Ed., *Phenology and Seasonality Modeling*, Springer-Verlag, New York, 1974, pp. 353–364.
152. J. T. Ritchie, Evaluating Irrigation Needs for Southeastern U.S.A., in *Contribution of Irrigation and Drainage to World Food Supply*, Proc. Irrigation and Drainage

Div., Specialty Conf., Biloxi, Miss., Aug. 14–16, 1974, Am. Soc. Civil Engr., New York, 1975, pp. 262–279.

153. T. A. Black, W. R. Gardner, and G. W. Thurtell, *Soil Sci. Soc. Am. Proc.* **33**, 655–660 (1969).

154. J. T. Ritchie and E. Burnett, *Agron. J.* **63**, 56–62 (1971).

155. J. T. Ritchie and E. D. Rhoades, and C. W. Richardson, *Trans. ASAE* **19**, 1098–1103 (1976).

2

THE SOIL AS A WATER RESERVOIR

Steve J. Thien

INTRODUCTION

Soil is a growth medium providing plants with support, nutrients, and water from a dynamic, three-phase, exceedingly complex system. Even though a separable area of study, investigation of only the liquid soil phase has not been regarded by scientists as particularly meaningful. The dynamics of soil–water relations results from the chemical and physical interaction of free water and the soil in which it resides. A true interaction exists, for when the soil and water are in contact with each other, the properties of both change.

The edaphic functions of soil may be subdivided into the collection, storage, and release of water for plant use, although this simplistic organization belies the comlexity involved. The activity of water in soils reflects the net effect of many forces. However, before analyzing the effects of soil components on water behavior, this chapter first establishes a foundation of basic terminology for soil properties and then discusses the role of the soil as a water reservoir in crop–water relations.

SOIL PROPERTIES THAT AFFECT WATER RELATIONS

The primary soil properties important in water relations include texture, structure, porosity, and colloidal content.

Soil Texture

Soil texture describes the relative proportions of different-sized mineral particles in soil. The principal feature of soil components when considering only their size is the surface area they present. When expressed as the amount of surface area per unit of weight (i.e., $cm^2\ g^{-1}$), this property is termed *specific surface area*. Particle size and specific surface area vary inversely; for every 10-fold decrease in average particle radius, a corresponding 10-fold increase occurs in surface area for the same weight of material. The surfaces attributable to clay-sized particles so dominate most soils that the contribution from sand- and silt-sized particles is of little significance (Table 2.1) (1). On these surfaces, soil reactions take place, including the one of principal interest in this book, the attraction and binding of water.

Soil texture is a property inherited from parent material and becomes a rather permanent feature of a soil. Erosional deposition and removal, however, represent ways that surface textures can be rapidly changed. Within the soil profile, soil development processes tend to increase the clay content of the subsoil (B horizon) at the expense of the surface soil (A horizon). The more advanced a soil is in its development, the more pronounced will be the textural differences between these horizons. Up to a certain point, increases in the amount of clay in the subsoil are desirable because the clay will increase the water and nutrient storage capacity and reduce leaching losses by slowing percolation. If, however, the accumulation of clay becomes too great, a claypan layer forms that is dense enough to severely restrict the movement of air and water and the penetration of roots. When this happens, the subsoil below the pan remains unexplored by roots and the nutrients or water stored there remain unused. Also, the pan restricts deep percolation of water, and profiles fill quickly. Thus claypans occurring near the surface on sloping soils contribute to increased erosion.

Soil Structure

Soil structure describes the arrangement of primary soil particles (sand, silt, and clay) into secondary particles (peds). Soil scientists have many

Table 2.1. Distribution of surface area[a] in a 100 g soil sample of three different textures

	Surface Area (cm^2)		
	Texture and Percentage Composition (Sand–Silt–Clay)		
	Sandy Loam (65–25–10)	Silt Loam (20–60–20)	Silty Clay Loam (15–55–30)
Sand	1,950	600	450
Silt	37,500	90,000	82,500
Clay	30,000,000	60,000,000	90,000,000

[a] Computed from data by Kohnke (1).

ways of categorizing soil structure, e.g., shape, hardness, genesis, size, but porosity and stability classifications are probably the most meaningful when considering their effect on plant–soil–water relations.

Soil structure creates a unique three-dimensional framework of solids and voids. The configuration of particles *within* peds forms numerous crevices and cavities (pores) that trap and store water. Likewise, irregular void (pore) spaces created *between* peds play major roles in water dynamics. Sometimes peds become so dense or coated with restrictive clay skins that air, water, and roots penetrate the soil only through interped voids. The abundance and size of the soil pores are probably the most important aspects of soil structure.

Preservation of soil structure is difficult. Rather weak forces bind particles into peds, making them very susceptible to natural and man-caused physical and/or chemical disruption. Raindrop impact, machinery compaction, dispersion, and oxidative loss of organic binding agents through excessive tillage are probably the most prevalent aggregate destroying factors. Once the peds are broken, the finer particles fill open pores and seal the surface of the soil. This process is called *puddling*. Relatively impermeable crusts form upon drying and hinder infiltration, air exchange, and seedling emergence. Improper tillage can also create a compacted, structureless zone at the bottom of the tilled zone, i.e., a tillage pan, which is similarly restrictive. The ability of peds to remain aggregated in the presence of disruptive forces is known as *aggregate stability*. The only stability factor currently measured quantitatively is water stability, as proper procedures for correlating aggregate strength with other disruptive forces are lacking.

Soil Porosity

Soil porosity describes the percentage of the soil's volume that is void space. The porosity of a soil is primarily determined by soil texture but is modified according to the degree of aggregation or compaction. Sandy soils appear very porous because of the prominence of large pores but actually are less porous than finer-textured soils. Porosity values range from 35% for the most severely compacted soils to 65% for well-aggregated, high colloid, high organic matter soils. A soil composed of half solids and half pores has a bulk density of approximately 1.3 g cm^{-3}.

The term porosity has limited use, however, because there is no reference to pore size, the feature that actually determines the function of the pore in soil–water relations. Micropores (<0.01 mm) are considered most useful for water storage, medium-sized pores (0.06–0.01 mm) facilitate capillary conductivity, and macropores (>0.06 mm) allow rapid infiltration and drainage and promote adequate aeration. Obviously soils should contain some of each type and the most desirable pore size distribution will depend on such factors as climate, irrigation, species to be grown, and groundwater occurrences. While most tillage practices alter the pore size distribution in the surface soil on a short-term basis, little can be done to change soil pore characteristics in the subsoil.

Soil pores are not uniform diameter, smooth-sided, interconnected channels that completely obey the theories of capillary flow. Rather, as typified in Fig. 2.1, soil pores may contain irregular constrictions and innumerable crevices that greatly complicate the prediction of water behavior.

Soil Colloids

Soil colloids are matter in a very fine state of subdivision. Although arbitrary, the upper limit of colloid size is often set at 0.0002 mm and

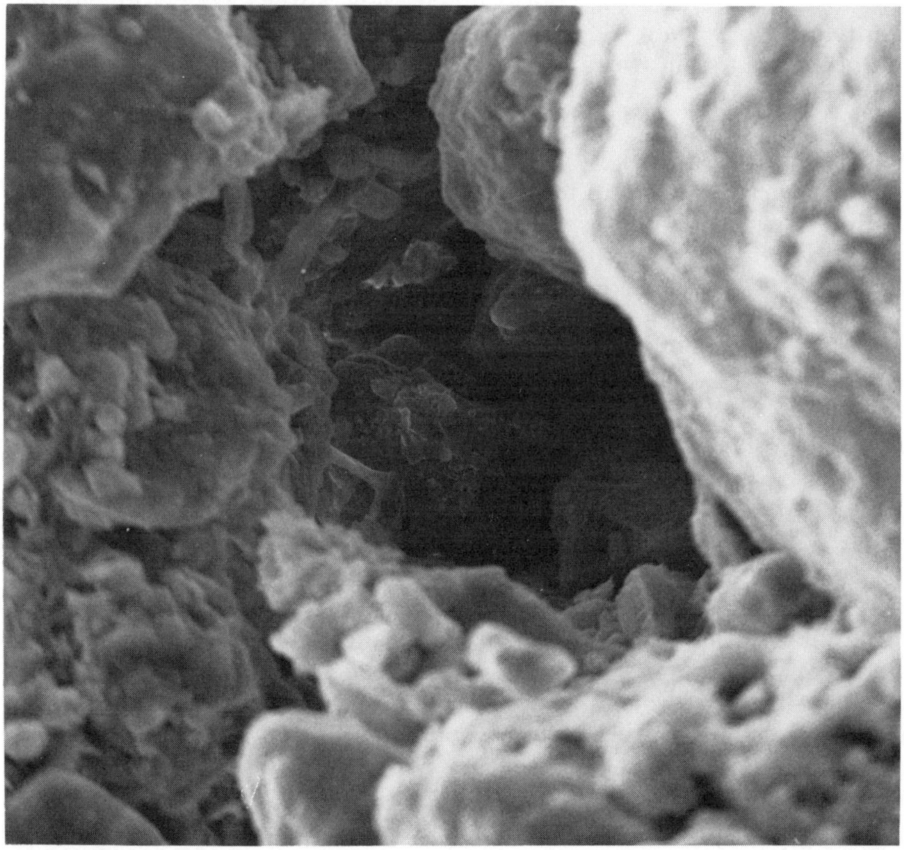

Figure 2.1. Scanning electron micrograph of a pore extending into the interior of a soil aggregate. The large particles are silt-sized and clay-sized material (<2 μm) is also evident. In the pore, strands of some organic material can be seen. [Photo by Steve J. Thien, Kansas State University.]

the lower limit is that of a single molecule. Clays are less than 0.002 mm and thus only the smaller soil mineral particles described as clay are actually colloids. Organic colloids also exist in soils, and both types gain their importance through the tremendous amounts of surface area contributed to soil systems. The specific surface area of soil colloids is large because of their small size and also in the case of clay minerals because of their platelike shape and their layered architecture, which contribute active interior surfaces. The most prevalent mineral colloids are crystalline, layered aluminosilicates. Soils contain two main types of aluminosilicates as distinguished by the structural ratio of silicon tetrahedron sheets to aluminum octahedron sheets: (1) 1 : 1 minerals such as kaolinite and (2) 2 : 1 minerals such as montmorillonite, which swell on wetting and shrink on drying, and illite, which, like kaolinite, does not possess a shrink–swell potential.

The actual contact layer between colloidal surfaces and soil solution is called an *interface*. The soil solution at this interface has properties quite different from those of the bulk solution. A higher ionic concentration in this layer (with either cationic or anionic predominance depending on the nature of the colloid) and its attraction to the colloid by electrostatic forces make it act as part of the colloid. If the colloid were to move, the water bound to it would also move. In fact, the orientation of dipolar water molecules with respect to the charge sites on the colloid gives adsorbed water a rigid structure. During adsorption to a surface, water loses energy because the adsorbed molecules have lost some freedom of motion. This energy loss can be monitored as a rise in temperature and is known as the *heat of wetting* value of a soil. Taylor and Ashcroft (2) present a comprehensive treatment of changes in interfacial energy as related to water adsorption by soils. The colloid plus its surrounding water layer (shell of hydration) is commonly called a *micelle*.

SOIL WATER POTENTIAL

The complexity of soil–water systems requires the use of thermodynamic terminology to express the energy status and physical conditions of the water involved. Energy associated with a soil–water system can be partitioned into two forms: kinetic and potential. Because water movement in soils is usually quite slow, the contribution due to kinetic energy is considered negligible. On the other hand, because water can move in response to certain forces within a soil it does possess potential energy. This potential energy is dynamic, changing from place to place within a soil and at the same place over time. In pursuit of thermodynamic equilibrium, soil water will obey the universal tendency to move from where the potential energy is higher to where it is lower. This simple relationship, plus knowledge of the components and conditions that affect potential energy, form the basis for describing intake, transport, storage, and release of water by soils.

In soils disciplines, *potential energy* frequently becomes shortened to *potential.* Numerical expression of soil water potential is in units of work on a per unit mass, unit volume, or unit weight basis. In this text, the potential energy status of water in soils will be quantitatively expressed in units of *bars,* a form of volumetric potential (see appendix for conversion to other units). Numerical values of bars are negative for water conditions where the potential is less than that of pure, free water, i.e., under normal soil conditions. This retentive force that restricts water behavior in soils is referred to as *tension* in this text. Since the term tension already implies negativity, when the term *bars of tension* is used the negative sign is omitted. Explicit wording of terms can be located in a bulletin of the International Soil Science Society (3).

Total Soil Water Potential

Total soil water potential (ψ_{soil}) expresses the potential energy status of soil water relative to conditions associated with pure, free water. Because total soil water potential is considered in relative terms, differences between values become more important than absolute units. The value for free water is arbitrarily set at zero, so the energy available in soil water to do work will always be less than zero because of the energy expended at the solid–liquid interface. Other forces at work on the water also tend to reduce its potential energy as expressed in Eq (2.1):

$$\psi_{\text{soil}} = 0 - (\psi_m + \psi_s + \psi_g + \cdots) \qquad (2.1)$$

This equation states that the total soil water potential (ψ_{soil}) equals the energy value of free water (0) minus the sum of the matric potential (ψ_m), the solute potential (ψ_s), the gravitational potential (ψ_g), and any additional terms ($\cdots$). One should not attempt to partition the total potential among its components as independence does not necessarily exist. Rather, the total relative energy state of soil water should be considered as existing along a continuum which then allows us to determine how far the water in a soil system is removed from equilibrium and in which direction change can be expected.

Matric Potential

Matric potential (ψ_m) is the force that attracts and binds the water as films onto soil or plant solids and as capillary water in small openings and crevices. The restrictive nature of this force reduces the potential energy of pure, free water. At large dimensions (dry soils) the potential is mainly brought about by adhesion, at small dimensions (moist soils) by the surface tension resulting from meniscus formation. Water in thick surface films or in larger pores is not held tightly, and there is only a slight reduction of the free energy of this water. Water in small capillaries or in thin films is held very tightly and as a consequence lowers the total soil water potential greatly (i.e., makes it more negative). Under saturated conditions, the matric potential becomes a positive pressure potential [whose value, however, would be expressed in negative units in Eq. (1)].

The concept of matric potential described here has also been described using such terminology as "tension," "suction," and "capillarity." Matric potential is the most desirable measurement, however, because it describes the net effect of water attraction whether in pores, in crevices, or onto soil or plant matric constituents.

Solute Potential

Solute potential (ψ_s) is the force due to the presence of dissolved solutes that decreases the potential energy of soil water. When a substance dissolves in water, dipolar water molecules in close proximity to either the cation or the anion orient by electrostatic attraction/repulsion. Though weak and not exerted over long distances, this attraction reduces the activity or freedom of the influenced water molecules and in turn reduces the potential energy of the water mass.

This effect is frequently called the "osmotic" potential in reference to its measurability in an osmotic membrane pressure device. As the dissolved matter content in soil water increases, osmotic pressure increases and vapor pressure decreases.

Gravitational Potential

Gravitational potential (ψ_g) is the work needed to maintain soil water at some suspended location in relation to the attraction of gravity. This potential is not measured as an absolute value. Rather, an arbitrary reference point, selected to be below the root zone under consideration, determines its magnitude.

Other Potentials

There are other potentials that have been described by soil physicists to be influential in affecting the total soil water potential under some circumstances. Where applicable, these factors would be included in Eq. (2.1) as electrical potential, temperature potential, hydraulic potential, pressure potential, potential differences attributable to specific solutes, or partitioned components of those potentials already described. Ordinarily these potentials apply only in specific instances, are very difficult to measure, or are not normally influential; therefore, they will be omitted from our consideration of total soil water potential.

WATER CONTENT VERSUS WATER POTENTIAL

As indicated previously, the soil water potential forms a continuum from 0 bars in the soil's wettest condition to −10,000 bars in its driest state. Under natural conditions soil water potentials of 0 bars follows saturation, but will seldom drop below the hygroscopic coefficient (approximately −31 bars) associated with air dryness. Water held in the soil at any potential remains in that state until a suction, or tension, force acting on it exceeds the total of the retentive forces, i.e., the total soil water potential. A slight suction on a saturated soil will remove water from the largest pores and crevices. A higher suction will empty still smaller pores. As higher and higher suction forces are applied, even the smallest pores will release their water, and finally only films of water on colloidal particles will remain. Taylor (4) predicted a water film approximately 14 molecular layers thick in a silty clay loam soil at the wilting point (−15 bars). Exposing a soil to drying temperatures over 100°C will finally result in complete removal of this hydration shell on soil particle surfaces. A graph of the complete relationship between soil water potential and water content is called a *soil moisture retention curve,* an example of which is presented in Fig. 2.2. The exact shape of such a curve depends on specific soil properties, but students of soil–water–plant relations should learn the fundamental concepts demonstrated by the soil water retention curve. For example, notice that removal of succeeding increments of water always requires logarithm-

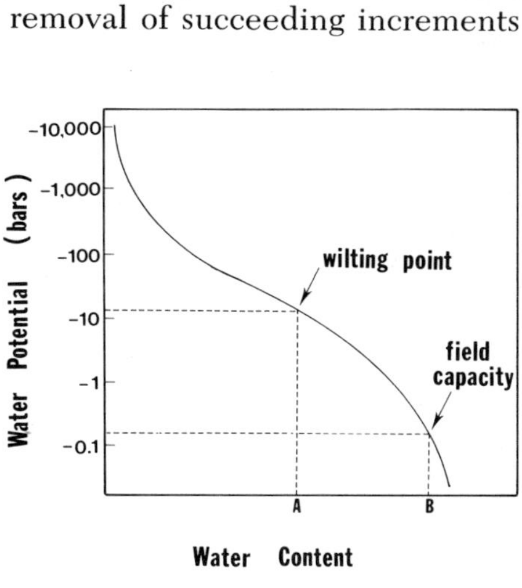

Figure 2.2. Soil moisture retention curve graphically illustrates that the amount of water in a soil is a function of soil water potential (tension). The water between points *A* and *B* represents the available water content. The shape of the curve will differ for each soil depending on specific properties that affect water relations.

ically higher amounts of energy. This explains why all water in soils is not equally accessible to plants, and, indeed, why plants wilt even when some water remains in soils.

The portion of soil water that plants can use is called *available water*. This water must be within the tension range from which plant roots are capable of exerting an extractive force greater than the opposite binding forces represented by the total soil water potential. This range is between $-\frac{1}{3}$ and -15 bars, referred to as the *field capacity* and *permanent wilting coefficients*, respectively. While water held at potentials less than -15 bars is generally considered to be unavailable in quantities needed to sustain plant growth, the exact permanent wilting point is a dynamic value determined by specific plant and soil characteristics. On the other hand, saturated soils contain some water at greater than $-\frac{1}{3}$ bar potential, but plant growth is restricted because of oxygen deficiency around the roots.

Once a soil moisture retention curve has been constructed for a soil, the amount of available water can be determined by the method demonstrated in Fig. 2.2. The available water content ($B - A$) could be expressed in gravimetric or volumetric units, or be converted to depth per unit of surface area (i.e., cm/ha).

Plant-available water always comes from the same potential range regardless of the soil condition. However, the actual amount of water contained within this range depends on specific soil properties. For example, the amount of surface area in a soil as reflected by its texture modifies the shape of the water retention curve (Fig. 2.3). Sandy soils present very little total surface area for water adsorption, and only small amounts of available water are present in such soils. Clay soils, with much greater specific surface areas, require larger amounts of water to satisfy the attractive forces of the soil matrix. This is quite evident in the more negative potentials where water retention occurs mainly as

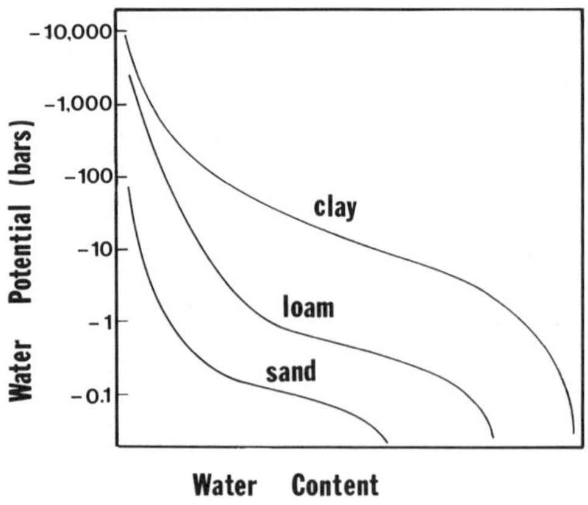

Figure 2.3. Soil texture affects water retention by governing the amount of surface area available for adsorption.

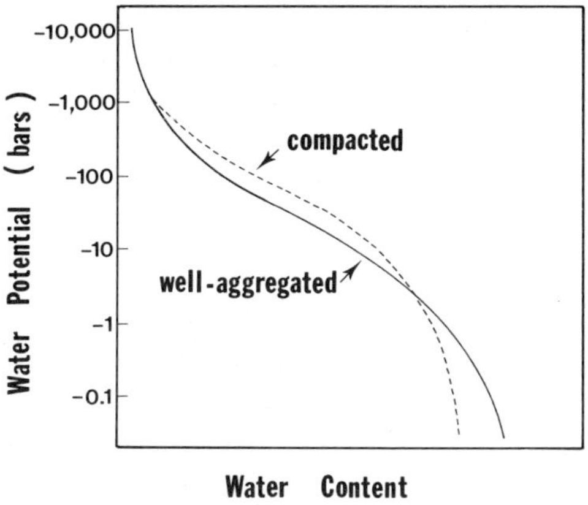

Figure 2.4. Compaction increases water retention in drier conditions by converting macropores responsible for wet-condition storage capacity to micropores.

surface films. This greater retentive capacity in high surface area soils carries over into the available water range.

Under wet soil conditions, i.e., small negative water potentials, the amount of water in a soil depends primarily on the macropore space available, which in turn is closely related to soil structure and/or compaction. The soil water retention curve for a compacted soil (Fig. 2.4) illustrates that the supply of available water is less than for a well-aggregated soil at high potential values. Under drier conditions, however, compaction can increase the water storage capacity of a soil by converting low-tension macropores to micropores capable of increased retentive capacity. Such compaction is not recommended, as the increase generally occurs only under conditions drier than the wilting point.

Another factor that affects the shape of the soil moisture retention curve is whether water is being adsorbed or desorbed, that is, whether the soil is wetting or drying. This effect (Fig. 2.5), called *hysteresis*, explains why a soil that has dried to a certain potential contains more

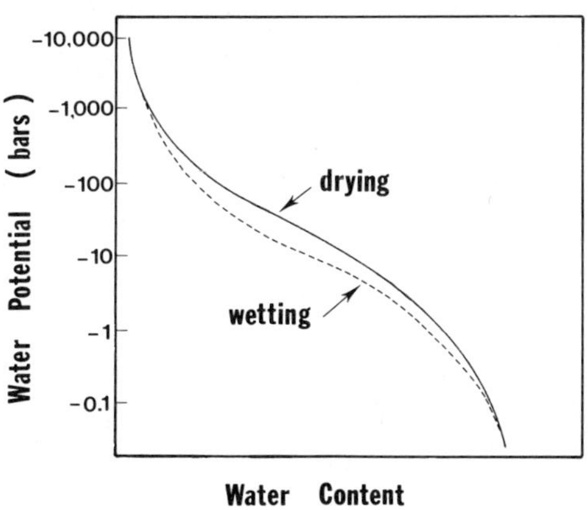

Figure 2.5. The "hysteresis effect" showing that the relationship of water content to water potential is dependent on whether water is being adsorbed or desorbed.

water than a soil that has wetted to that point. Hysteresis is commonly manifested in the capillary fringe near irrigation furrows and water tables. Several theories have been advanced to explain the hysteresis phenomenon as a failure of soil and water to reach equilibrium (2).

THE CONCEPT OF SOIL AS A WATER RESERVOIR

The soil layer on the landscape accepts, stores, and releases water in a continuous cycle and so is often likened to a reservoir. However, not all soils perform these three functions in a manner considered optimum for plant growth. Droughty or waterlogged conditions stress plant growth and are likely to occur in most soils for at least short periods of time. Most of the time, however, water relations in soils used for crop production range between field capacity and the wilting point. Specific soil properties will determine each soil's ability to optimize water relations. This section will relate soil characteristics to the infiltration, redistribution, storage, and release of water in soils.

Infiltration

Infiltration is the entry of water into a soil through pores in the surface layer. Since infiltration precedes the storage and consequent plant use of soil water, it can also be a regulating process in water use. In practically all cases it is desirable to maximize infiltration, because this significantly reduces the alternative hazards of runoff and ponding. The key to prolonging high infiltration rates is maximizing open surface pores. Hence, any condition that affects soil porosity, such as texture, structure, aggregate stability, organic matter, and tillage, can also be expected to exert some influence on infiltration. Furthermore, the initial water content of the pores will modify the infiltration potential. All except texture are properties influenced by cultural practices which make infiltration a dynamic process that is quite dependent on soil management and use.

Initial Water Content

For an initially dry soil, the rate of infiltration changes with time, as demonstrated by the typical infiltration curve in Fig. 2.6A. Water intake is greatest during the initial stages of infiltration, then slows to some final steady-state rate. Whenever the application rate exceeds the infiltration rate at any point on the curve, the excess water will pond on the surface or run off depending on the topography of the site. Either condition poses a hazard to crop production and should be avoided by a soil management goal that maximizes infiltration.

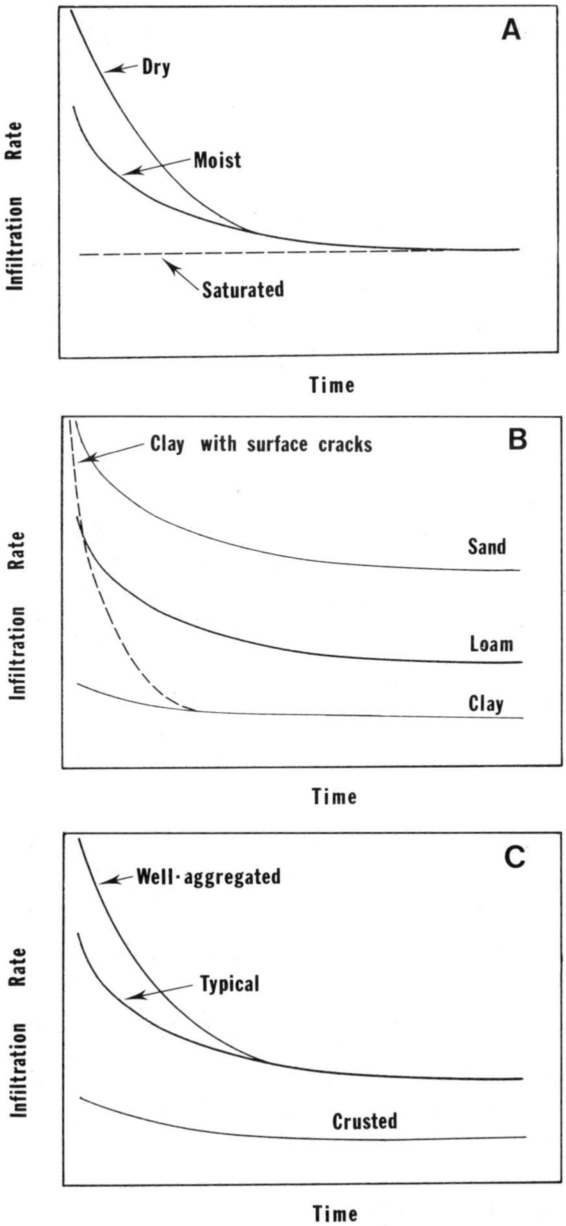

Figure 2.6. Infiltration is highly correlated with the condition of surface pores. When water is already partially or completely filling these pores (A), further infiltration is reduced. Large pores in sandy soils (B) allow faster infiltration than the small pores of finer-textured soils. When surface cracks are present, infiltration will not be limited until they swell closed. A well-aggregated soil (C) presents large, open surface pores, while a crusted condition indicates loss of structure and a sealing of surface pores. Crusting is delayed or prevented if the soil's surface is insulated from raindrop impact by either living or dead plant material (D).

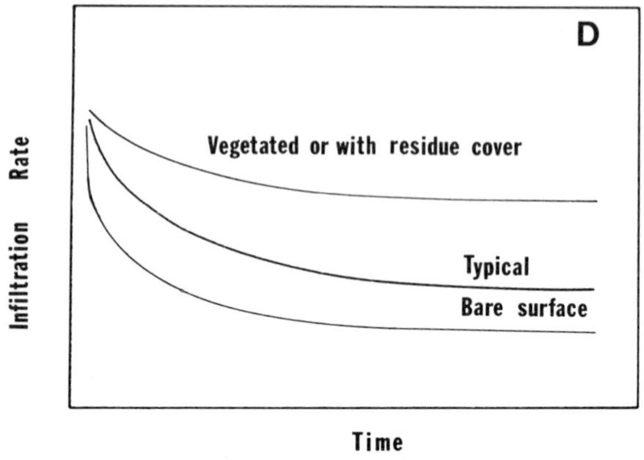

Figure 2.6. (Continued)

If the soil already contains some water in these pores, subsequent infiltration will be reduced because the tension gradient between the surface water and the soil matrix will be less than for a drier soil. The deeper in the profile a soil is wetted, the greater will be this backlogging effect and the greater will be the overall reduction in infiltration.

Soil pores not filled with water contain air. During the infiltration process, water entering the pores must displace this air, or trapped air pockets will form and seal off a pore channel to further water movement. Thus, a film of water across the soil surface may limit exhaust openings enough to impede further infiltration. [See Dixon and Peterson (5) for a more complete treatment of this concept.]

Texture

Soil texture influences infiltration through its effect on surface pore sizes (Fig. 2.6B). Sandy soil contains a large proportion of macropores, which accounts for its high infiltration rates. As the amount of silt increases, average pore size decreases and water intake is slowed. Increasing the clay content of a soil has two opposing effects on infiltration. The increased abundance of micropores associated with clay will restrict water entry and cause clayey soils to exhibit very slow *final* infiltration rates. However, clayey soil can possess some very high *initial* infiltration rates due to (1) greater aggregate stability, which can prolong the high rate, initial phase of infiltration, and (2) the extremely high initial infiltration rates that exists in soils in shrinking–swelling type clays until the cracks swell closed.

Structure, Aggregate Stability, and Tillage

The manner in which the surface soil layer is structured modifies porosity and hence influences infiltration (Fig. 2.6C). Granular, or crumb, structure is the most desirable surface structure because it tends to maximize favorable air and water relations. Sandy soils are rarely granulated, because they lack the specific surface area necessary to bind into aggregates. Silty and clayey soils are frequently granulated, except under cultivation, which tends to destroy aggregate-binding substances. Hence, most cultivated soils exhibit only weak granular structure or are structureless (massive). The degree to which aggregates resist disintegration from destructive forces is called *aggregate stability*. The most notable feature of soils with low aggregate stability in the surface layer is puddling when wet (disintegration of aggregates with resultant plugging of surface pores), followed by crusting upon drying (formation of a dense, impermeable, restrictive layer). Infiltration in puddled or crusted soils is severely restricted.

Tillage operations modify surface roughness and porosity conditions, factors that have a significant influence on infiltration. It is difficult here to do more than make general observations because there is a wide variety of types of tillage tools and sequences. The short-term effect of tillage is to increase surface roughness and create more surface macropores, both being conditions that increase infiltration. However, the stirring action of tillage tools accelerates the oxidative loss of organic aggregating agents, which contributes to reduced aggregate stability. Tilled soils are more susceptible to aggregate dispersion, puddling, and crusting. Thus, the long-term effect of tillage is to reduce infiltration. The beneficial, short-term effects of tillage on infiltration are probably eliminated by the first significant rainfall following the operation.

Organic Matter

Organic matter in either its living, dead, or decayed humus form strongly influences the infiltration process (Fig. 2.6D). A living vegetative cover insulates the fragile soil surface from the impact energy of raindrops, preserving aggregate stability. Kohnke (1) suggests that vegetated soils possess double the infiltration capacity of bare soils. Likewise, crop residue can be managed to maximize its stay on the soil surface and so protect the surface layer from puddling and crusting.

Roots and incorporated residues are converted into humus. Soils high in humus content exhibit high infiltration rates because of the favorable aggregate stability and porosity conditions that result.

WATER MOVEMENT

Following infiltration, forces of gravity, evaporation, plant uptake, and soil storage act on soil water. The inequality of thermodynamic properties associated with these forces results in movement of water throughout the soil profile. Describing this redistribution becomes quite complex because conditions vary from high tension gradients at the wetting front to saturated conditions just below the infiltration zone. In addition, changes in soil properties, ranging from gradual to sharp layering, create nonuniform matric conditions through which movement occurs. So scientists believe that, in reality, a true equilibrium of water relations within the soil profile is never achieved. At best, students should concentrate on knowing how to evaluate the current soil water status and being able to interpret the effect soil conditions will have on subsequent water behavior.

Types of Water Movement

The flow of water through soil is determined by (1) the *hydraulic conductivity** of the soil, namely, its porosity properties, and (2) the *driving force,* which reflects the steepness of the gradient between forces at work to move the water. Table 2.2 lists these considerations for three types of water movement in soils: saturated, unsaturated, and vapor flow.

Saturated Flow

When soil pores are filled with water, there is no matric tension on the water, which means that it is free to move in response to gravitational force. Thus, saturated flow (see Table 2.2) movement tends to be downward but could also be lateral along the top of an impervious soil layer.

Large, open pores increase the hydraulic conductivity of a soil for saturated flow. Thus, sandy soils exhibit higher rates of saturated flow than finer-textured soils. Likewise, soils with stable, porous structure are quite permeable, at least until the aggregates break down and clog the larger pores.

The thicker the layer of tension-free water (hydrostatic head) either on or in a soil profile, the greater will be the driving force to cause movement. The direction of movement will always be from a position of high pressure to one of low pressure (a pressure or hydrostatic head gradient).

Most soils are saturated only for short periods, but the frequency and duration that excess water resides in the profile dictate important crop culture and soil management decisions. For instance, in humid regions where the precipitation:evaporation ratio rises above 1, water accumulates and saturates lower parts of the profile. In such cases, tile drains are needed to lower the saturated zone and encourage deep rooting. Likewise, soil surface zones saturated by rainfall or irrigation can create

* Other terms frequently used to denote the same phenomenon include permeability, internal drainage, and capillary conductivity.

Table 2.2. Types and conditions of water movement in soils.

		Factor Determining		
Type of Movement	Conditions Necessary for Flow	Hydraulic Conductivity	Driving Force	Comparative Flow
Saturated	Water is at zero potential	Pore size	Hydrostatic head	Sand>loam>clay
Unsaturated	Water is between $-\frac{1}{3}$ and -31 bars potential, continuous water films	Degree of unsaturation	Potential gradient	Clay>loam>sand
Vapor	Water held tighter than -31 bars, discontinuous water films	Water-free pore volume	Vapor pressure gradient	Warm, moist>cool, dry

oxygen deficiency stress for plant roots. Supplemental surface or subsurface drainage systems raise the productivity of these poorly drained soils.

Unsaturated Flow

Most soils are normally in an unsaturated condition (see Table 2.2), meaning that both air and water occupy soil pores and the water is under tension. As matric potential values change because of wetting or drying, potential gradients form over both short and long distances. In response to these gradients, water will move from an area of high potential to one of low potential, which could be any direction in the profile. Downward flow leads to profile filling and leaching, upward flow occurs with capillary rise from a water table and evaporation, and lateral flow would be important in the study of water movement to roots, border width between irrigation study plots, and irrigation furrow spacing. The rate and distance water will move in each of these cases would depend on prevailing soil physical properties and soil water dynamics.

The hydraulic conductivity of a soil for unsaturated flow is determined by the degree of unsaturation (see Fig. 2.7). Note here that the relative conductivity of the various soils is different in the low and high potential ranges. However, in all soils as the water film on soil particles becomes thinner (i.e., potential decreases) the rate of redistribution decreases. This relationship is best explained by understanding the "least tortuous path concept" illustrated in Fig. 2.8. Sandy soils possess less specific surface area than finer-textured soils and hence in equal moisture conditions will exhibit lower hydraulic conductivity than other soils.

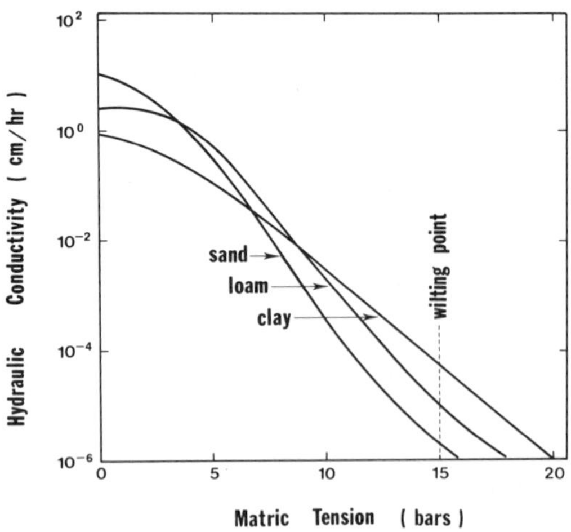

Figure 2.7. Hydraulic conductivity as a function of matric tension for three simulated soil types. [Redrawn from data by Hillel and van Bavel (6).]

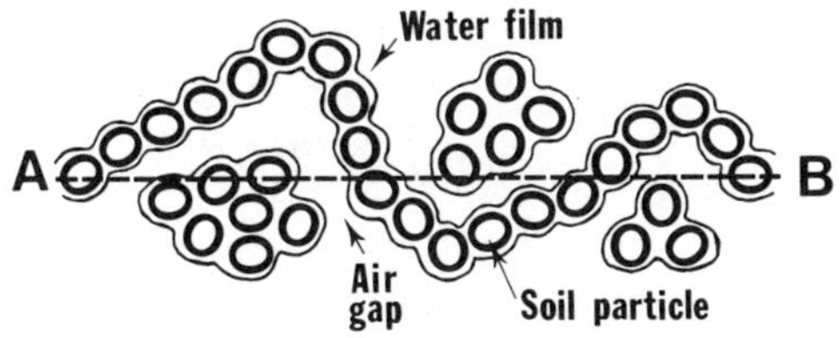

Figure 2.8. Soil particles are covered by a thin film of water under tension. The water film is not thick enough to bridge some of the larger interparticle openings. As water is removed from point *B*, a tension gradient is created that causes water to move from point *A* (area of low tension) to point *B* (area of high tension). Flow will occur as a series of water film adjustments until a tension gradient no longer exists. This type of flow can occur only along a path where the water films are continuous; i.e., the water cannot jump air gaps. The circuitousness of the route needed to circumvent the open pores, compared to the apparent straight flow path (dotted line), is termed *tortuosity*. The more unsaturated a soil, the more air gaps will be present and the more tortuous (roundabout) will be the path necessary to move from point *A* to point *B*. Hence, very dry soils present a low hydraulic conductivity for unsaturated flow.

Unsaturated flow occurs in soils when the water is held between $-\frac{1}{3}$ and -31 bars potential. At values greater than $-\frac{1}{3}$ bar saturated flow occurs, while at values less than -31 bars potential discontinuous water films preclude liquid flow.

The driving force in unsaturated flow is a function of the potential gradient between the wet and dry points involved. The steeper the gradient, the more rapid the adjustment. Water redistribution between two soils having about the same water content will be exceedingly slow.

The theoretical effects of hydraulic conductivity and the driving force suggest that unsaturated flow will occur most readily in clayey or sandy soils depending on the soil matric potential. In actual situations, however, when both time and distance of movement are considered, soils of intermediate texture appear to be the most permeable to unsaturated flow. This difference between potential and observed effect is most likely due to the slow rate of water movement in the tiny pores of clayey soils.

Vapor Movement

Under soil water conditions associated with unsaturated flow, the water-free pore spaces are maintained at nearly 100% relative humidity. But when the water is held at greater than -31 bars potential, water films become discontinuous and liquid flow ceases. Further redistribution occurs as water vapor diffusion (see Table 2.2). As the matric tension increases, water vaporizes with greater and greater difficulty and its vapor pressure becomes lower and lower. Any resulting inequality between vapor pressures creates a gradient responsible for the diffusion of water vapor from one area to another in soils. The direction of movement will always be from an area of high vapor pressure to one of low vapor pressure, which can include evaporation from the soil surface.

The hydraulic conductivity of a soil for vapor movement is directly proportional to the volume of water-free pore spaces. Pore size is not a limiting factor in vapor movement considerations because of the relatively small size of the diffusing molecules. Thus, fine-textured soils, unless compacted, are more permeable to vapor movement than coarse-textured soils.

The driving force for vapor movement is a difference in vapor pressures between two areas. This difference could be associated with relative humidity, as described above, or could be due to temperature variations. Moist and warm conditions would result in the highest vapor pressure and dry and cold conditions the lowest. Moist and cool (or dry and warm) soil conditions may create offsetting conditions and minimize vapor transfer.

Significant soil water–crop growth relations that affect soil water vapor movement include (1) maintenance of a biologically optimum environment in open soil pores, (2) loss of soil water through evaporation, and (3) dissolution of fertilizers through hygroscopic attraction.

Water Movement Through Stratified Layers

Soils are rarely uniform in properties that affect water relations. Any changes that affect soil porosity characteristics will likewise influence water redistribution. Such changes include differences due to texture, structure, compaction, tillage, and/or plant growth. Boundaries between layers may vary from gradual to abrupt. As a generalized statement, the more diverse, abrupt, and numerous the stratification, the more the impedance to water flow.

A sand layer disrupts water flow, because of the change from fine to coarse pores. In unsaturated conditions, the large pores cannot exert sufficient tension to remove water from the fine pores until the overlying zone becomes nearly saturated. A sand layer slows the downward movement of water by unsaturated flow. Conversely, a clay layer, or compacted zone, also checks downward flow under unsaturated conditions, but in a different manner. In this case, the zone with the fine pores can exert low potentials and rapidly remove water from an overlying coarse textured soil. But impedance occurs because the fine layer is slow to release water to the underlying coarse textured soil. Tillage pans are notorious for checking downward water flow and can greatly restrict the zone of optimum rooting conditions.

STORAGE AND RELEASE OF SOIL WATER

The concern about water movement in soils centers on its role in storing and releasing water for plant use. The concept of storage implies protection from losses including percolation below the root zone, evaporation, and use by undesirable growth. The release of water implies provision of usable water supply to desirable crops. The storage and release capacity peculiar to an individual soil is frequently characterized by measuring its potential *plant-available water supply*. The available water supply is defined as the water retained by a soil between *field capacity* and the *wilting point*, two terms that can best be described using the thermodynamic format introduced earlier.

First, not all water in soils is held with equal tension. On a scale that begins at 0 bars potential for very wet soils and rises to $-10{,}000$ bars potential for very dry soils, only water held between $-\frac{1}{3}$ and -15 bars potential is considered available for plant use. The $-\frac{1}{3}$ bar point, or field capacity, is defined as the amount of water in a soil after the force of gravity has removed all the water possible. Soil conditions, especially stratification, make it difficult to predict exactly when field capacity is achieved, causing some soil scientists to abandon its use. Its limited exactness notwithstanding, the concept of a field capacity soil condition is extremely useful in describing the concept of available water.

At potentials less than -15 bars, plants are generally unable to extract water from soils in sufficient quantities to prevent wilting. Hence, the water status of the soil at a potential of -15 bars is referred to as the wilting point. The remarkable water supplying capacity of single roots makes it necessary for the entire root zone to be at or above -15 bars potential before permanent wilting will occur. A common method of determining the wilting point of a soil is to use plants as bioindicators, noting the moisture content when wetting the soil failed to revive wilted plants.

Soil moisture retention curves are useful in determining the amount of available water in a soil. Figure 2.9 demonstrates the procedure for determining useful soil water information from curves illustrating conditions of both a soil of low water holding capacity and one of high water supplying capacity. Data from Fig. 2.9 show the clay loam soil to have 22 percentage units of available water and the sandy loam only 11. The clay loam soil has nearly the same amount (percentage by weight) of water remaining when plants wilt as is contained in the sandy loam at field capacity.

Since the two curves have different shapes, the release characteristics will also be different. One way to show this is to calculate the percentage of available water remaining in the soil at various potential values (see Fig. 2.9). These data show that at any selected potential value the finer-textured soil still retains a greater share of its available water. Or, stated differently, a large portion of the available water in coarse-textured soils is stored at very high potentials. More energy must be used by plants to remove half of the available water in a fine-textured soil than in a coarse-textured one. Information such as this is useful in irrigation scheduling to prevent water stress.

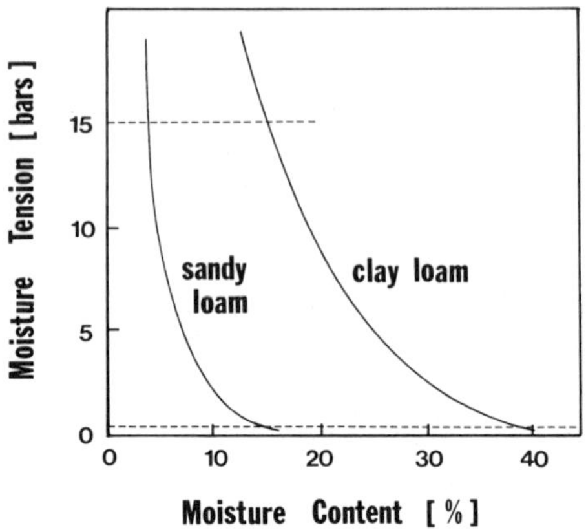

Figure 2.9. Use of soil moisture retention curves to compare the available water supply for two soils. Dotted lines indicate field capacity ($\frac{1}{3}$ bar) and the wilting point (15 bars). Note: Potential (bars) = −tension (bars).

	% Water[a]		Available water supply	Percent of available water remaining[b]				
Soil	*FC*	*WP*	(*FC* − *WP*)	$\frac{1}{3}$ bar	2	5	10	15
Clay loam	38	16	22	100	70	41	14	0
Sandy loam	15	4	11	100	50	27	5	0

[a] Data taken from Fig. 2.9.
[b] Percent available water remaining = $(IT - WP)/(FC - WP) \times 100\%$
where *IT* is water content at any indicated tension; *WP* is water content at the wilting point; and *FC* is water content at field capacity.

METHODS OF SOIL WATER MEASUREMENT

Experienced soil managers are frequently adept at discerning the water status of a soil simply by feeling and/or squeezing a soil sample. One such rule of thumb is that if a sample squeezed in one's palm retains a ball-shaped form then irrigation can wait, but if the squeezed sample readily crumbles and does not form a ball then it is time to turn on the sprinklers. While having the obvious advantage of simplicity, such a system fails to meet the exacting criteria necessary for soil water research or water conservation through proper water management.

Through the years an assorted array of methods and instruments have been developed to measure and express soil water. Before discussing these approaches to water measurement, it is necessary to examine the manner in which water content can be expressed. Basically there are three ways: (1) weight percentage, (2) volume percentage, and (3) tension. Descriptions, equations, and possibilities of conversions between these expressions are presented in Table 2.3. The choice of whether to

Table 2.3. Methods of expressing soil water

Operation	Description	Equation
Determining soil water content by weight percentage	Weight of water in 100 g of oven dry soil	$\%_w = \frac{\text{weight of water}}{\text{weight of dry soil}} \times 100\%$
Determining soil water content by volume percentage	Volume of water as a percentage of the volume of the soil sample	$\%_v = \frac{\text{volume of water}}{\text{volume of soil}} \times 100\%$
Converting between weight percentage and volume percentage	The volume percentage of water in a soil sample differs from the weight percentage by a factor equal to the bulk density (BD). (It should be noted that accurate bulk density measurements are difficult to obtain and quite subject to sampling error.)	$\%_v = (\%_w)\,(\text{BD})$
Converting from weight or volume percentage to potential	No universal conversion equation is available as many soil properties affect this correlation. Instead, it is necessary to create a conversion table for each soil type studied.	

express soil water content on a weight or a volume basis is not a critical one if the information necessary to convert from one to the other is also provided. Some researchers favor volume percentage because it gives a better concept of the water encountered by plant roots as they explore a volume of soil. Expression of water content by potentials usually signifies that the analytical method measured only potentials. Conversion between potential and either volume or weight percentage requires elaborate correlative information and is not often attempted.

Gravimetric Analysis

The traditional method of measuring the water content of a soil sample is to determine a moist and an oven dry weight (dried to a constant weight at 105°C). The weight lost in drying represents water and the percentage of water on a weight basis is calculated according to Eq. 2.2:

$$\%\ \text{water}_{\text{weight}} = \left(\frac{\text{wet weight} - \text{oven dry weight}}{\text{oven dry weight}}\right) \times 100\% \quad (2.2)$$

Gravimetric analysis can be performed on practically any weighable sample and at any water content, but requires sacrificing the sample since it would not be practical to return the sample to the field or pot after drying. Care must be exercised in collecting samples to avoid evaporative water loss before moist weights can be determined. Airtight metal or plastic containers are used for this purpose. Drying in a hot air oven to a constant weight usually requires 24–36 hr for 100–200 g samples, or longer for larger samples. This time could be considerably shortened (about 30 min for a 100 g sample containing 25% moisture) with microwave radiation drying (7), but microwave drying energy has been shown to cause erratic results in subsequent chemical analysis of the soil (8) when compared to hot air drying.

Tensiometers

The tensiometer is a practical device for the *in situ* measurement of the energy status of soil water, principally its matric potential. The tensiometer (see Fig. 2.10) consists of three parts: (1) a porous ceramic tip, connected to (2) a water-filled, airtight tube, and (3) a vacuum gauge (or mercury manometer). When the tensiometer is placed in the soil, water inside it equilibrates with soil water through the pores in the ceramic tip. As soil water is depleted and the matric potential increases, water is drawn from the tensiometer, causing a drop in its hydrostatic pressure which registers on the vacuum gauge. Replenishment of the soil water supply and the consequent lowering of the matric potential will create the opposite action. Since the water inside the tensiometer

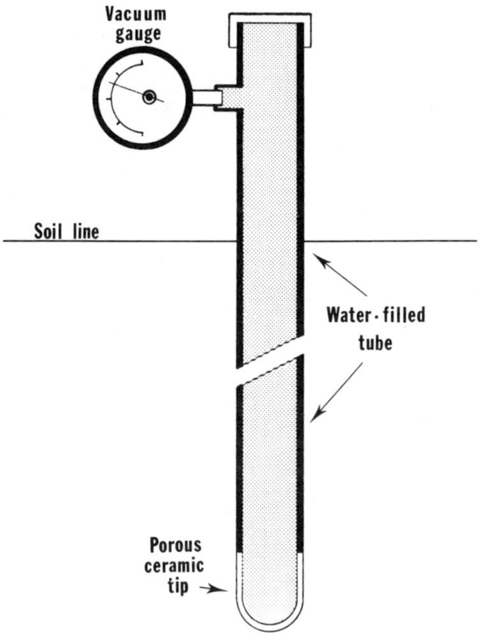

Figure 2.10. Tensiometer, a device for measuring soil water status.

equilibrates with the solute content of the soil water, the instrument does not measure the osmotic potential of the soil solution.

The tensiometer measures potential values in a range between 0 and −0.8 bars. Although this limited range represents only a small fraction of the potential values possible in a soil, it does encompass a rather large percentage of the available water (see Fig. 2.9) and is useful in determining when field capacity ($-\frac{1}{3}$ bar) is attained.

Tensiometers are inserted into openings created by properly sized core removers, so very little soil disruption occurs during installation, and repeated use creates no further disturbance. Varying the length of the tube permits tensiometers to be used to measure soil water at any depth. Frequently, irrigation managers place tensiometers at the lowest level of root development so that only the root zone will be brought to field capacity and no water will be wasted.

Electrical Resistance

The ability of a porous medium like soil to transmit an electrical current is related to its water and electrolyte content. If the influence of varying soluble salt conditions in soils can be prevented from affecting the measurement, the relationship between soil water and electrical resistance can be determined. At the heart of such a system is an *electrical resistance block*, which consists of two electrodes embedded in a gypsum block. Buried in the soil, these porous blocks equilibrate with the soil water, gaining and losing water in the same manner as the soil. Above ground, the electrodes can be connected to an ohmmeter (a device that measures electrical resistance). The block's resistance to electrical current flow is low when it is wet and increases as it dries. Such a system allows reasonably accurate correlation of resistance to either soil water content or potential over the available water range in soils ($-\frac{1}{3}$–-15 bars).

The gypsum blocks maintain a saturated solution of calcium sulfate around the electrodes and mask electrolyte changes in soils due to fertilization. However, since gypsum is slightly soluble, these block eventually need replacing, perhaps every 1–3 years. Blocks made of inert material such as nylon or fiberglass last longer but are sensitive to changes in the salt content of the soil solution.

Several blocks can be buried at different levels in the same hole, and, once they are installed, repeated readings are possible without disturbing the soil further. Hence, soil water information can be readily attained for the entire root zone.

Thermoelectronic

The rate of heat dissipation in a porous medium like soil depends on its water content. As the water films on soil particles become thinner, the path length for heat conduction increases (becomes more tortuous). The water content can therefore be measured by supplying a heat source and monitoring the temperature rise in the soil.

In practice, the heater and temperature sensor are encased in a porous ceramic block that can be buried in the soil. The water content of the block equilibrates with that of the soil, allowing the thermal response to be proportional to the water content of the soil. The sensors are calibrated to read in units of soil water matric potential.

Thermoelectronic sensors are capable of high accuracy in the 0–2 bar matric potential range. Accuracy decreases at higher potentials (for details, 9). Thermal sensors are not affected by freezing and can be permanently installed and connected to continuous recording devices.

Neutron Moderation

Atomic energy has been harnessed to measure the water content of soils in the form of a neutron probe. This instrument consists of (1) a probe containing a source of fast neutrons and a slow neutron detector and (2) a scaler which displays the number of slow neutrons detected over a measured time (see Fig. 2.11).

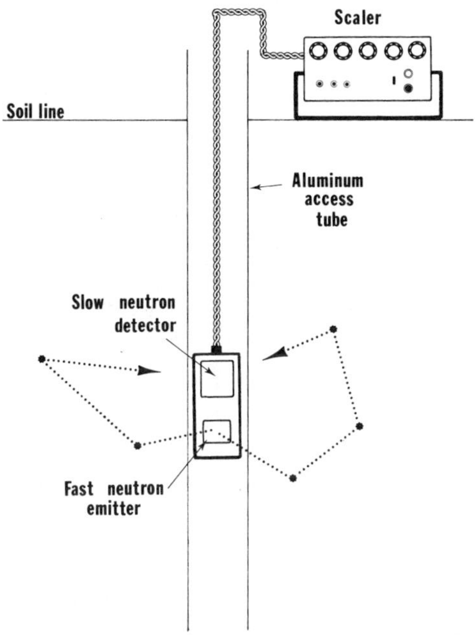

Figure 2.11. A neutron moderation device used to determine soil water content.

The probe is lowered to the desired depth in the soil via aluminum access tubes. Fast neutrons are emitted radially into the soil, where they collide with various atomic nuclei and gradually lose portions of their kinetic energy. Collisions with hydrogen nuclei, which have approximately the same mass as a neutron, maximize this loss of energy and slow the neutron. Some of these slow neutrons are deflected back to the probe, where they are counted by the slow neutron detector. Because hydrogen is chiefly responsible for slowing these fast neutrons, the rate of slow neutron detection is proportional to the hydrogen nuclei concentration in the vicinity of the probe. Because the amount of hydrogen associated with water in the soil is greater than that associated with other sources, the amount of slow neutrons detected is thus considered to be proportional to the water content in the volume of soil affected, a sphere approximately 15 cm in diameter in wet soils and up to 50 cm in diameter for dry soils. Conversion of slow neutron counts to volumetric soil water content requires calibration to specific soils (for details, see 10, or 11).

LIST OF SYMBOLS

ψ_{soil} total soil water potential
ψ_m matric potential

ψ_s solute potential

ψ_g gravitational potential

REFERENCES

1. H. Kohnke, *Soil Physics*. McGraw-Hill, New York, 1968, 224 pp.
2. S. A. Taylor and G. L. Ashcroft, *Physical Edaphology. The Physics of Irrigated and Nonirrigated Soils*, Freeman, San Francisco, 1972, 533 pp.
3. H. C. Aslyng, Soil Physics Terminology, *Int. Soc. Soil Sci. Bull.* **23**, 7 (1963).
4. S. A. Taylor, The Activity of Water in Soils, *Soil Sci.* **86**, 83–90 (1958).
5. R. M. Dixon and A. E. Peterson, Tiny Soil Channels Determine Water Infiltration, *Crops Soils* **24(5)**, 11–12 (1972).
6. D. Hillel and C. H. M. van Bavel, Simulation of Profile Water Storage as Related to Soil Hydraulic Properties, *Soil Sci. Soc. Am. J.* **40**, 807–815. (1976).
7. R. J. Miller, R. B. Smith, and J. W. Biggar, Soil Water Content: Microwave Oven Method. *Soil Sci. Soc. Am. Proc.* **38**, 535–537, (1974).
8. S. J. Thien, D. A. Whitney, and D. L. Karlen, Effect of Microwave Radiation Drying on Soil Chemical and Mineralogical Analysis, *Commun. Soil Sci. Plant Anal.* **9(3)**, 231–241 (1978).
9. C. J. Phene, S. L. Rawlins, and G. J. Hoffman, Measuring Soil Matric Potential *in Situ* by Sensing Heat Dissipation within a Porous Body, II. Experimental Results, *Soil Sci. Soc. Am. Proc.* **35**, 225–229 (1971).
10. C. H. M. van Bavel, D. R. Nielson, and J. M. Davidson, Calibration and Characteristics of Two Neutron Moisture Probes, *Soil Sci. Soc. Am. Proc.* **25**, 329–334 (1961).
11. J. W. Holmes and A. F. Jenkinson, Techniques for Using the Neutron Moisture Meter, *J. Agric. Eng. Res.* **4**, 100–109 (1959).

3

THE ROLE OF WATER IN PLANTS

Neil C. Turner and Gordon J. Burch

INTRODUCTION

Water in the plant has many functions essential to life. In most plant organs water constitutes more than 90% of the fresh weight and in few does it constitute less than 70%. Between 60 and 90% of the water is situated within cells and in part helps provide cell rigidity. The remaining 10–40% of the water permeates the cell walls where it provides a continuous medium between the specialized transport cells of the vascular bundles and the remainder of the plant. It therefore plays a role in the transport of substances both through the plant and within the cell. In addition to providing support for the plant through turgor in the cell, water also acts as a solvent for inorganic and organic solutes and for dissolved gases, and as a constituent of chemical reactions within the cell such as the conversion of carbon dioxide into carbohydrate in the process of photosynthesis.

The amount of water in the plant is, however, small compared with the amount of water that passes through the plant in transpiration. Water loss by transpiration provides cooling to the plant, but is essentially an inevitable consequence of the uptake from a dry atmosphere of carbon dioxide for photosynthesis. Thus, although in net terms six molecules of carbon dioxide and six molecules of water are required to provide one sugar molecule, the amount of water used in the production of 1 kg of wheat is approximately 1000 kg, in the production of 1 kg of rice or dried fruit is 2500 kg, and in the production of 1 kg of cotton (lint) is 5000 kg (1, 2). The 1 kg of wheat, rice, dried fruit, or cotton contains less than 0.15 kg of water, the remainder having passed through the plant into the atmosphere in transpiration.

It is appropriate, therefore, that this chapter is primarily concerned with the role of water in transpiration and the balance between the hydration of the plant and the amount of water passing through it in the transpiration stream. Since the amount of water in the plant is small relative to that passing through it, tissue water plays a minor role in buffering the daily loss of water by crop plants (3). Nevertheless, the maintenance of high water content of the tissue is essential for plant growth and development. The ability of plants to both maintain tissue hydration and maintain rapid transpiration is a function of the unique properties of water and the development of plant structures that enable plants to live in a terrestrial environment. Thus, our first undertaking is to describe the properties of water and the energy status of water in the plant. We then discuss the collection of water from the soil and the pathways of transport from the soil to the atmosphere through the plant. Previous chapters have detailed the role of the soil as a reservoir for water and the demand by the atmosphere for water; therefore, these subjects will be touched on in this chapter only when required for completeness.

The unique features that enable a plant to live in a terrestrial environment are the roots that collect water from the soil, the xylem vessels and network of capillaries in the cell walls that transport the bulk of the water through the plant, and the impermeable cuticle with its stomatal pores that operate as a variable resistance between the wet interior of a leaf and the dry exterior. In this chapter the pathway of water movement through the plant from the root surface to the stomata is therefore considered in detail, as is the growth and development of the roots and their ability to absorb water.

As the reservoir of water in the soil is depleted, plant dehydration increases such that physiological activity in the plant is affected. Thus, the development of water deficits in the plant is described together with their effect on physiological and morphological processes. Our understanding in this area has increased markedly in the past two decades due to improvements in our ability to measure the status of water in the plant: a section in this chapter is devoted to the measurement of plant water status. Not all plants suffer equally as a result of water deficits. Some are able to withstand severe aridity and maintain measurable physiological activity whereas others die. The ability of plants to withstand arid conditions has been termed drought resistance. The mechanisms that confer drought resistance on plants are therefore appropriate to this chapter and discussed in the section on Water Stress and Plant Growth.

It was pointed out earlier that a large mass or volume of water passes through the plant as an inevitable consequence of the uptake of carbon dioxide for photosynthesis and growth. The proportion of water lost in transpiration per unit of carbon dioxide fixed depends on both environmental and plant factors. The ratio of carbon gained to water lost is termed the water use efficiency and the plant and environmental factors that affect this efficiency are discussed in the final section.

Finally, it should be pointed out that as this is a chapter in a book on crop–water relations, discussion of the various topics covered is confined largely to crop species and communities. Some reference to work outside crop species is inevitable, but this is limited to those situations where it is necessary for completeness. Additionally, this chapter inevitably reflects the interests of the authors and is not a comprehensive review of the subject. For a more general treatment of the subject than is possible in the space of this chapter readers are directed to several recent books (4–9).

STATUS OF WATER IN PLANTS

Properties of Water

The dipolar configuration of a water molecule causes it to form four hydrogen bonds in a tetrahedral arrangement with adjacent water molecules. A positive charge on the hydrogen proton attracts the negative lone pair charge on an oxygen atom in another water molecule. This bonding is responsible for the phenomenon of "bound water" (10, 11), where charged surfaces and dissolved ions, especially if negatively charged, electrostatically bind water molecules to their surfaces forming an envelope of water molecules several layers thick (12). Bound water is not labile and has a freezing point of about −25°C. Water remaining in plant tissue under extreme pressure or after prolonged desiccation, which is also sometimes referred to as bound water or extracellular water (13), may include water tightly held in tissues by absorptive and capillary forces, as well as electrostatic forces.

The dipolar configuration of water also allows it to readily ionize electrolytic salts. Moreover, at any free water surface, vaporization can take place continuously, and because water has a high heat of vaporization it provides the plant with a very effective mechanism for energy dissipation. Because of the high heat of vaporization, leaves of actively transpiring plants are usually 2–3°C cooler than leaves with closed stomata.

Water has a high viscosity and the viscosity increases in close proximity to cell or membrane surfaces (14) and at low temperatures. Although high rates of laminar or capillary flow can occur in plants (11), increases in viscosity in proximity to cell surfaces and at low temperatures may restrict water movement.

Capillary Rise

In comparison with most other liquids, water has a high surface tension, 72.8 dynes cm^{-1} at 20°C, and varies little (about 8%) with temperature between 0 and 40°C, the temperature range that most crop plants experience. Crop plants must be able to maintain water uptake through xylem vessels at tensions as high as 30 bars, i.e., at water potentials as low as −30 bars, without air being drawn into the vessels and so breaking the capillary stream. The pressure (P) necessary to remove a liquid from a capillary tube of radius, x, is given by:

$$P = \frac{2\gamma}{x} \tag{3.1}$$

where γ is the surface tension (dynes cm^{-1}). This formula can be applied in an approximate way to the cell walls of a plant. At a water potential of −30 bars, air will not enter pores with a radius smaller than 50 nm. This is greater than the average size of pores in the cell wall: Gaff et al. (15) suggest that there are few pores larger than 10 or 20 nm in radius in the cell walls of plants.

The intermolecular attraction that occurs with hydrogen bonding in water is described as cohesion and the attraction between water and a solid phase as adhesion. For water in glass capillaries or in xylem vessels the water-to-wall attraction is appreciable, in which case the walls can be said to be wettable, and this causes the water to rise in the capillary or vessel. The height, h, of capillary rise depends on the contact angle, y, that the liquid makes with the capillary wall, the radius, x, of the capillary and the surface tension, γ, of the liquid (16). Any upward force due to adhesion and surface tension is balanced by the gravitational force, g, acting on a liquid of density, ρ, and volume given essentially by $\pi x^2 h$. By equating the two forces, capillary rise can be given as:

$$h = \frac{2\gamma \cos y}{x \rho g} \tag{3.2}$$

Thus, the degree of capillary rise in an open vessel is inversely proportional to the radius of the vessel. Consequently, the smaller the size of the xylem vessel or cell wall interstice, the greater will be the capillary force holding the water. For xylem cell walls with exposed polar groups, the contact angle can approach zero and $\cos y \simeq 1$. Calculations using a surface tension of 72.8 dynes cm^{-1} indicate that the water will rise to a height of 0.75 m in an open vessel with a radius of 20 μm (16). In small plants this could account for static hydration and overnight rehydration of cavitated vessels. However, it does not aid transpiration since the xylem essentially remains closed to air entry and lower potentials than those involved in capillary rise are necessary to maintain water flow.

Energy Status of Water in the Plant

Water is affected by forces arising from hydrostatic tension, the presence of solutes, and the presence of solids. Therefore, to adequately describe the energy status of water in the plant we must choose some arbitrary reference condition to account for the changing forces acting on water and the resultant changes in its energy status. In plant water relations, the energy of water at a specified location is expressed as the difference between it and the energy of pure water at standard temperature and pressure (17). This definition allows us to employ the concepts of free energy (18) to account for changes in the water status of plants. A change in free energy provides the force tending to return a system to its reference conditions. In an isothermal system, the change in free energy can be measured as a function of either a volume change or a pressure change. Pressure is force per unit area, which is dimensionally the same as energy per unit volume, e.g., 1 bar = 1×10^5 Pa = 0.1 J cm^{-3} or 0.0239 cal cm^{-3}. This then gives us the basis for describing the status of water in a plant using its free energy, or its equivalent in pressure, and since tissue water exists under tension, i.e., in a state of negative pressure, then water potential will have a negative value. For further discussion of the water potential and its derivation readers are referred to Slatyer (4), Kramer (5), or Meidner and Sheriff (19).

Components of Water Potential

Plant cells essentially consist of a semirigid cell wall and a protoplast. The protoplast is enclosed in a differentially permeable membrane, the plasmalemma, and contains cytoplasm and one or more vacuoles each enclosed in a membrane called the tonoplast (Fig. 3.1). The plasmalemma allows free movement of water into and out of the protoplast, but retards the passage of solutes. The protoplast contains solutes that lower the energy status of the water in the cytoplasm and vacuole and induce movement of free water from outside the cell into the protoplast. As the cell wall is semirigid, the increased water content increases the pressure inside the cell until the pressure exerted on the walls balances the influence of the solutes at which point no further movement of water will occur.

Thus, the *total water potential* (ψ_w) at any point in the plant can be partitioned into

(a) The *osmotic potential* (ψ_π) arising from the presence of dissolved solutes. The osmotic potential is also termed the *solute potential* or *osmotic pressure*.

(b) The *turgor potential* (ψ_p) arising from the forces exerted on the walls from the water attracted to the cell by the solutes and solids in the protoplast. The turgor potential is also termed the *turgor pressure, wall pressure, hydrostatic pressure,* or *pressure potential.*

(c) The *matric potential* (ψ_τ) arising from capillary or electrostatic forces associated with cell walls and colloidal surfaces.

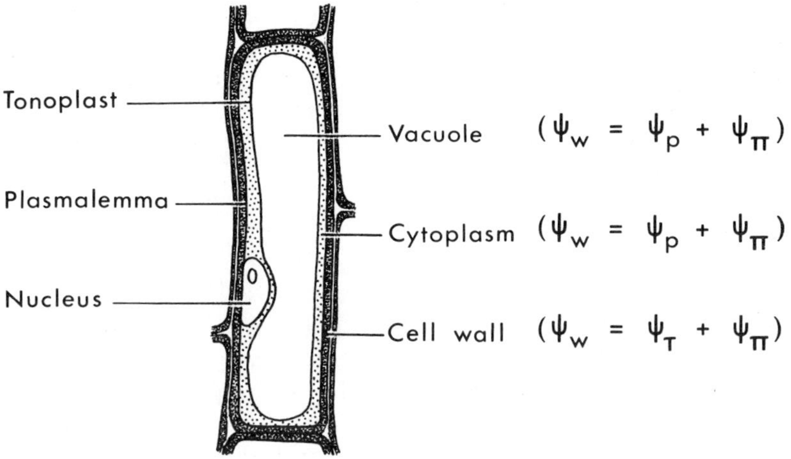

Figure 3.1. Diagrammatic representation of a plant cell showing the cell wall, cytoplasm, vacuole, plasmalemma, and tonoplast. The components of the total water potential (ψ_w) are shown for each portion of the cell. ψ_π = osmotic potential, ψ_p = turgor potential, and ψ_τ = matric potential.

(d) The *gravitational potential* arising from gravitational forces on the water in the plant. Sometimes, as in tall trees, gravitation needs to be considered (20), but in field crops the potential gradient of 0.1 bar m^{-1} required to overcome gravity is small enough to be ignored.

In cells at equilibrium with their surroundings, the total water potential will be the same in all the parts of the cell, namely, the wall, cytoplasm, organelles, and vacuole. However, the origin of the water potential may be quite different in each part of the cell. Thus as shown in Fig. 3.1, in the vacuole the total water potential arises largely from osmotic and turgor forces, in the cytoplasm it arises from osmotic and turgor forces, and in the cell wall it arises from matric and osmotic forces; as Passioura (21) has pointed out, matric potential has no meaning in the cell. Since the solute concentration of the extracellular water is low, i.e., its osmotic potential is high (22), the total water potential in the cell wall is almost equal to the matric potential.

Plant Water Balance

In crops growing in the field, transpiration responds to both diurnal and seasonal changes in atmospheric and edaphic conditions and to such factors as the development of leaf area, root growth and soil water availability (23–25). As water is lost to the atmosphere from the mesophyll and inner surfaces of the epidermal cells of the leaf, the water potential of these cells falls and a gradient in water potential is then established between plant leaves and the soil. This gradient is proportional to the rate of transpiration and to the resistances to water flow in the soil and plant. An analogy with Ohm's law has generally been adopted to describe this catenary pathway for water flow (26–30) using the equation:

$$T_p = \overset{\text{Liquid phase}}{\frac{\psi_w^s - \psi_w^l}{r_S + r_P}} = \overset{\text{Vapor phase}}{\frac{C_i - C_a}{r_a + r_l}} \qquad (3.3)$$

where T_p is the rate of flow through a plant, i.e., transpiration, ψ_w^s and ψ_w^l are the water potentials in the soil and in the plant leaves, C_i and C_a are the water vapor concentrations at the internal evaporating surfaces and in the bulk air, r_S and r_P are the resistances to liquid water flow in the soil and in the plant, and r_l and r_a are the diffusion resistances to vapor flow through the leaf epidermis and boundary layer, respectively. This analogy must be used with caution for steady-state water flow through plants (27), particularly if resistances change in response to flow rate as has been reported by several workers (31–33). Nevertheless, it provides a useful first approximation of the response in plant water status to changing rates of transpiration.

Water movement in the vapor phase, i.e., from the evaporation sites within the leaf to the bulk air, is largely through the stomata. The boundary layer resistance, r_a in Eq. (3.3), varies with wind speed, leaf shape, and leaf surface morphology (34), but is usually between 0.1 and 1.5 sec cm^{-1} for wind speeds between 10 and 0.1 m sec^{-1}. The leaf resistance, r_l in Eq. (3.3), comprises both cuticular and stomatal components in parallel:

$$\frac{1}{r_l} = \frac{1}{r_s} + \frac{1}{r_c} \qquad (3.4)$$

where r_s and r_c are the diffusive resistances of the stomatal and cuticular pathways, respectively. As the cuticular resistance is typically in the range of 25–500 sec cm^{-1} and the stomatal resistance is typically 0.5–10 sec cm^{-1} when the stomata are open (34, 35), it is clear from Eq. (3.4) that the stomatal resistance dominates the leaf resistance as long as stomata are open, and that the cuticular resistance sets the upper limit for leaf resistance. Further, provided that wind speeds are high and the boundary layer resistance is low, as is usual in the field, the stomatal resistance provides the major regulatory factor in vapor loss from plants.

The stomata respond to a range of environmental factors, notably light, humidity, carbon dioxide concentration, and temperature (36–38). They also respond to leaf turgor (35, 36) and their closure at low turgor potentials helps to regulate the water balance of the plant. Even under similar environmental conditions, stomatal resistances differ among species. Species differences are most extreme between those plants exhibiting the C_3, C_4, and CAM (crassulacean acid metabolism) carbon assimilation pathways. C_3 species tend to have lower resistances than C_4 species which have lower resistances than CAM species. Within these groups there are differences in stomatal regulation of transpiration (35, 39, 40) and even within a species there are small differences among cultivars (41, 42).

The gradient in potential between the soil and plant leaves can become appreciable even in fully turgid crops growing in moist soil (43, 44), indicating that in addition to vapor phase resistances substantial liquid phase resistances exist within the soil–plant system. Figure 3.2 shows a typical gradient at midday of water potentials through the soil–plant–atmosphere continuum for a plant growing in moist soil. Typically crop leaf water potentials can fall as low as −15 bars for plants growing in soil with a potential of around −0.1 bar. As soils dry to water potentials of around −10 bars, the leaf water potentials can reach values of −30 bars.

There is now considerable evidence that there are differences in the resistances to water flow among crops. Both field and laboratory studies indicate that resistances occur in the xylem of stems, petioles, and leaves (28, 45–47). Whether the resistances occur because water has to transfer from one cell to another within the vascular network, or arises from capillary resistances to flow, or from a breakdown in flow such as occurs with cavitation (48, 49) is not clear. Denmead and Millar (47) suggested that cavitation of the xylem stream was one cause for xylem resistances exceeding those expected from capillary flow.

The roots also have high resistances to flow (5, 50). Absorption of water by roots is subject to a large radial resistance across tissues outside the xylem vessels (30, 51–54). Additional resistances are incurred by water movement in the soil (55–57), and by poor root contact with granular soil (58) or shrinkage of the soil or root, or both, from each other (30, 59). The highest resistance in the root is most likely located at the endodermis (Fig. 3.3) where Casparian thickening of the radial walls of cells and possibly some suberization of tangential walls restricts water and ion transport (60). Suberization along the length of a maturing root (61) and in disproportionate amounts in different orders of root (62) also increases the resistance to water absorption by roots.

The resistance to water transport in the xylem is relatively unaffected by changes in plant water status when compared to the changes in resistance to root absorption of water as soils are dried and rewetted (30) and to changes in the resistance to water loss as stomata close and reopen (63). Thus, the water status of stems and leaves will remain balanced by the relative rates of root absorption and loss of water by leaves.

Jarvis (3) suggests that plants tend to balance or minimize the gradient in potential through the xylem and that this is the cause of girth increases

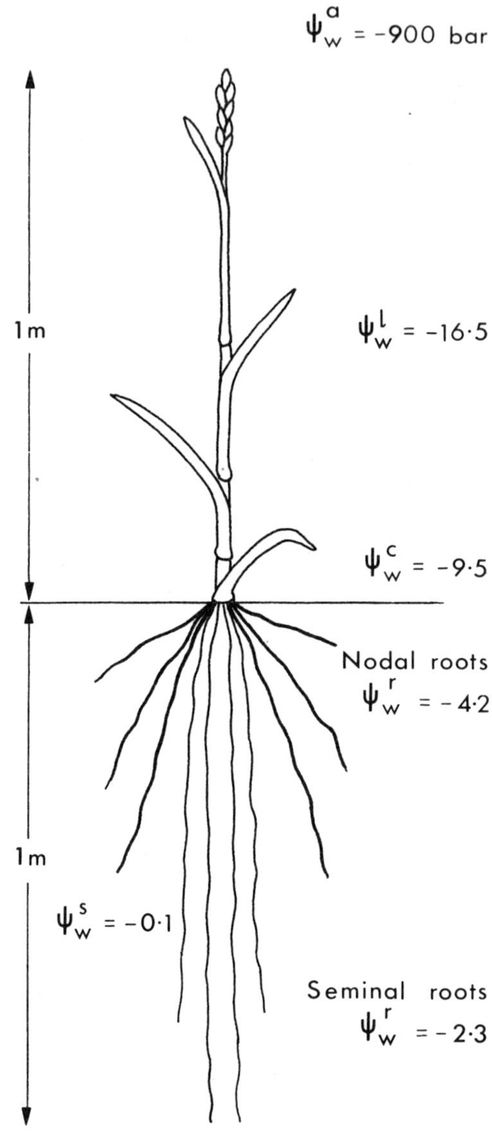

Figure 3.2. Water potentials expected at different locations throughout the pathway for water transport for a wheat plant growing in soil at a potential of -0.1 bar and in an atmosphere with a potential of -900 bar. ψ_w^s = soil water potential, ψ_w^r = root water potential, ψ_w^c = crown water potential, ψ_w^l = leaf water potential, and ψ_w^a = atmospheric potential.

by trees after thinning of dense forests. Thinning of a dense forest leads to a greater flow of water through the remaining stems and hence to steeper gradients in the xylem water potential. With the steepening of the gradient in the xylem, meristematic and expanding cells closer to the absorbing roots will have a more favorable water status than those cells further along the transpiration stream. Those cells (including immature xylem vessels) with a favored water status will then expand and

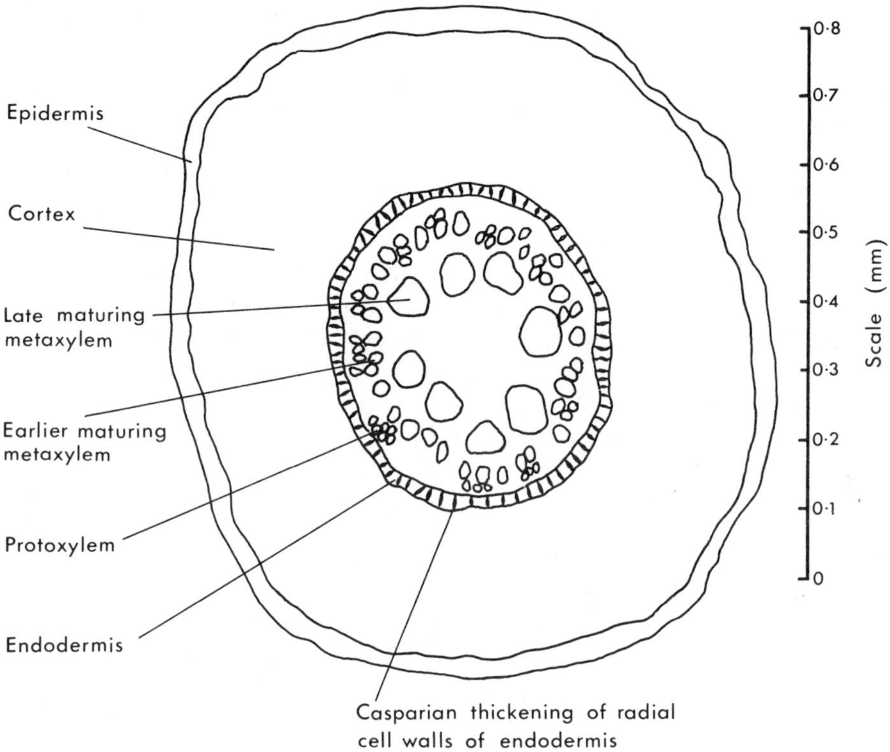

Figure 3.3. Transverse section of a maize root 10 cm from the apex. The tissue structures of the root are shown.

multiply preferentially until a balanced water status is again reached throughout the plant (64).

PATHWAY FOR WATER MOVEMENT THROUGH THE PLANT

Water moves along the potential gradient from the soil to the atmosphere through the roots, stems, and leaves of the plant. In this section we first discuss the movement of water in the leaves before moving down through the plant to the roots. Transport of water into the leaf is through the vascular network permeating the leaf tissue. This transport is in the xylem and will also be discussed in this section.

In the Leaf

Dehydration of leaf tissues is prevented by the epidermal cuticle and by active regulation of water loss by stomata (65). As mentioned earlier, the cuticle can have a diffusive resistance far in excess of 30 sec cm^{-1}, whereas with fully open stomata the diffusive resistance of the leaf can fall as low as 0.5 sec cm^{-1} (34, 35). For open stomata, diffusion can approach rates obtained from exposed inner surfaces of epidermal strips with one end immersed in water (65). This indicates that not only is vapor transfer through leaf stomata very efficient, but also that water movement along the inner surfaces of epidermal cells is usually adequate for supplying all the water for evaporation from sites within the substomatal cavities. However, the closure of stomata at high vapor pressure deficits, independent of changes in leaf water potential, could be indicative of limits to the ability of this pathway to supply water to isolated guard cells (Fig. 3.4).

The anatomical structure of a leaf (Fig. 3.4) shows that both mesophyll cells and epidermal cells surround the substomatal cavity and therefore provide alternative sites for evaporation. Meidner (65), and Sheriff and

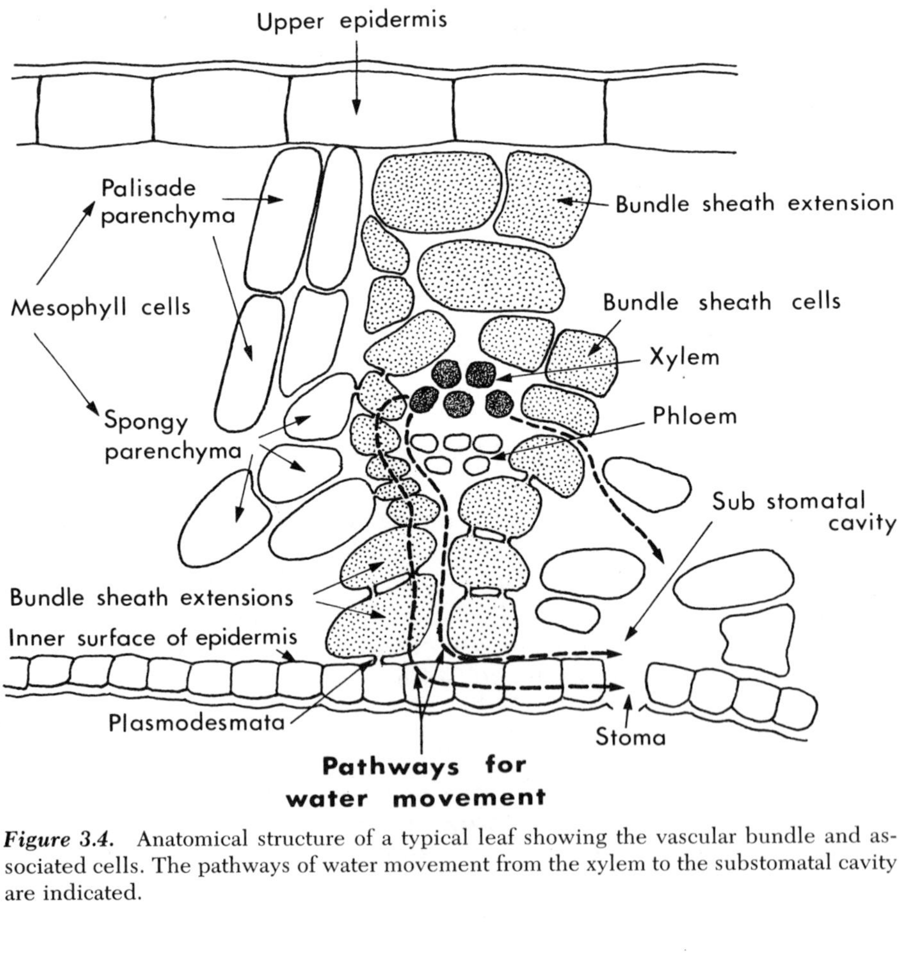

Figure 3.4. Anatomical structure of a typical leaf showing the vascular bundle and associated cells. The pathways of water movement from the xylem to the substomatal cavity are indicated.

Meidner (66) demonstrated that the inner surfaces of epidermal cells present the main sites for evaporation, and so suggested that the mesophyll cells primarily offer a moist surface for the absorprtion of carbon dioxide rather than for evaporation. Studies using the electron microscope and electron microprobe to trace the movement of agents through plant tissue confirm the importance of the epidermal pathway for water transport, but differ in the importance that they give to evaporation from mesophyll cells. Several studies suggest that appreciable quantities of water may be lost from the surface of mesophyll cells lining the substomatal cavity, particularly where they lie close to vascular tissue or abut a network of interstitial spaces containing free water (15, 67, 68). Tanton and Crowdy (69), however, found little accumulation of their marker, lead, lining the substomatal cavities of the leaves they surveyed. A recent review of water transport in plants by Boyer (70) rightly considers all cells surrounding the substomatal cavity as conducting water, but somewhat neglects the epidermal pathway in leaves.

From Fig. 3.4 it is evident that water from a vascular bundle must traverse the bundle sheath and either the spongy parenchyma or the palisade parenchyma to reach epidermal cells before passing to the stomata. Sometimes large vascular bundles will have bundle sheath extensions reaching to the epidermis on both sides of the leaf. In recent research there has been no consensus on the most likely pathway for water movement through these tissues (60, 68, 69), but it will usually be confined to one of the three alternate pathways shown in Fig. 3.5, or possibly any combination of the three pathways at different locations in the tissue. Path (A) is the vacuolar pathway in which water enters the vacuoles of the bundle sheath cells and is then passed to successive cell vacuoles by crossing two or more pairs of membranes in each step. Path (B) is through the free space (apoplast) of the cell walls or between the cell wall and the plasmalemma. Path (C) is the pathway through the symplasm in which water enters the protoplasm of the bundle sheath cells and passes from cell to cell via the plasmodesmata.

Water movement through the symplasm pathway would require a relatively high conductance for water passing through the plasmodesmata and tentative calculations by Tanton and Crowdy (71), using the geometry and dimensions of willow plasmodesmata, suggested that their conductance can be quite low. More recently, evidence of water movement through the free-space pathway in cotton leaves has been provided by Pizzolato et al. (68). As water has only to traverse the cell wall between successive cells and not cell membranes, water movement via the apoplast can proceed largely unaffected by the metabolic state of leaf cells. This contrasts with the symplastic or vacuolar pathways which are influenced by changes in cell metabolism (32).

It was mentioned in the section on plant water balance that the resistance to liquid water flow through the plant is not constant but variable

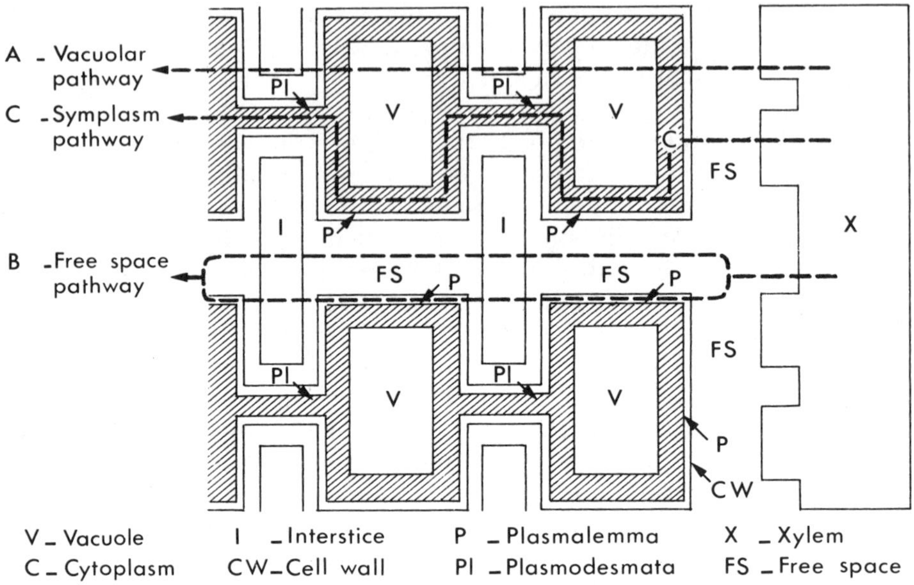

Figure 3.5. Schematic representation of cell structures showing the three alternate pathways for water movement: (A) vacuolar pathway; (B) free-space pathway; (C) symplasm pathway.

(31–33, 45, 72). This change in resistance has usually been attributed to a change in root resistance (73–76). However, Boyer (32) found that a variable resistance to water flow occurred mainly in the leaves in sunflower. In order to explain this observation, Boyer (32) proposed a model that incorporated two pathways for water transport through the leaf, viz., a high resistance pathway through the symplasm and a low resistance pathway through the free space (Fig. 3.5). He suggested that the low resistance pathway operates at high transpiration rates. Furthermore, as turgor potentials decrease in response to an increasing rate of transpiration, the leaf cells will approach zero turgor at which point there is little or no pressure from the plasmalemma on the cell wall and water flow between the wall and plasmalemma could be enhanced (68). Therefore, there is a growing consensus that the free-space pathway will carry the bulk of flow through the leaf during moderate or rapid transpiration.

In the Xylem

The passage of water through xylem vessels of the plant has generally been regarded as conforming to Poiseuille's law for capilliary flow:

$$Q = \frac{\pi x_e^4}{8\mu} \frac{d\psi}{dz} \tag{3.5}$$

where Q is the flow, $x_e = (\sum_1^n x^4)^{1/4}$ is the effective radius of n xylem elements, μ is the viscosity coefficient, and $d\psi/dz$ is the gradient in water potential. Fluid transport through the xylem is complicated by vessel geometry which is not equivalent to a simple capillary system as is often implied when using Poiseuille's law. Consequently, observed xylem flow is rarely found to agree with Poiseuille flow, although in many situations the discrepancy is small. Early observations by Emerson (77) demonstrated that xylem flow through grass roots could be half that expected from Poiseuille flow. Similar results have been reported for seminal roots in wheat (78), but larger discrepancies were found for xylem flow in the stem of wheat by Denmead and Millar (47). As mentioned in the discussion of plant water balance, this discrepancy has been variously attributed to cavitation of the xylem stream or to shrinkage of the xylem under high gradients in water potential (79, 80). An investigation by Giordano et al. (81) of xylem flow in large stem vessels showed that measured flow was less than Poiseuille flow by much as two orders of magnitude. Moreover, their study was conducted under positive pressure which eliminates the possibility that either cavitation or xylem shrinkage was responsible for the low measured flows. Therefore, great caution must be exercised in using Poiseuille's law to estimate xylem flow in plants.

In the Root

Water moves in two distinct pathway modes during its uptake by a root system. The first is the tissue mode, between the epidermal cells of the root surface and the xylem vessels within the root stele (Fig. 3.3). The second is the capillary mode, between the point of absorption into the xylem and the basal parts of the shoot. The processes of capillary flow have been discussed in the section describing xylem pathways for water movement; however, in roots the capillary flow can be influenced by water entry along the length of the xylem vessel causing flows and potentials to change independently of the effect of capillary dimension. This complicates the calculation of xylem flow in roots (82) and may result in deviations from Poiseuille flow due to the turbulent disruption of laminar flow caused by water entry through the vessel wall (83). Although these effects make any prediction of the water flux in different parts of the root system extremely difficult, Landsberg and Fowkes (84) and Taylor and Klepper (85) have attempted to simulate the essential features of capillary flow in a wheat root system.

Water movement in the tissue mode of the root can occur via the alternate pathways discussed in the section describing leaf pathways for water movement, and illustrated in Fig. 3.5. The several tissue structures of the root through which water must pass before it enters the xylem will each affect water movement as a result of the differences in their cell morphology. The greatest effect is found with differential changes in cell morphology due to maturation of the various tissues causing suberization to occur, particularly in the endodermis. The endodermal cells of plant roots are found to develop an impermeable band around the radial walls of cells called the Casparian strip (Fig. 3.3), which prevents water movement across the cell walls at this point. Maturation is found to result in a thickening of this Casparian strip, followed by suberization of the tangential walls, which increasingly inhibits apoplastic water movement across the endodermis. To traverse the endodermis, water must enter the symplast and pass via plasmodesmata directly to the xylem or again permeate the apoplast of the stelar parenchyma before entering the xylem. According to the observations of Higinbotham et al. (86), passage across the wall of mature xylem vessels may also be restricted by suberization and by the loss of any remaining protoplasm and its associated plasmodesmata.

ROOT GROWTH AND WATER ABSORPTION

Water Absorption by Roots

The root systems of crops are usually more concentrated in the surface layers of soil which contain a high proportion of the available nutrients and are rewetted more frequently by rainfall or irrigation. Nevertheless, roots can be induced to proliferate deeper in the soil due to either prolonged drying of the surface soil or the presence of a water table within the rooting zone (87, 88). Two important features of crop root systems affect water uptake. The first is that high root densities can result in the rapid depletion of available water in surface soil layers. The second is that deep root penetration can make available a much larger volume of soil water for plant extraction than provided by the surface layers of soil. Water uptake by deep roots is in part influenced by the density of roots in the soil and by the resistances incurred by the long lengths of xylem needed to transport water up to the soil surface (89, 90).

After the xylem resistances to water transport are allowed for, water absorption by roots within any soil layer will then be largely influenced by the radial resistances to water movement from the soil to the root surface and then through the root tissues to the xylem (Fig. 3.3). For brevity we shall henceforth refer to water absorption as the processes involved in water movement radially from the soil into and through the root until it enters the xylem; this will include the effect of resistances both radially within the soil and within the root. Water transport will likewise refer to the processes involved in water movement axially through the xylem vessels of the plant, which will include Poiseuille-type resistances and perhaps gravity. The overall radial resistance to water absorption from soil is affected by root density in the following way:

$$\frac{1}{r_T} = \frac{a_1}{r_1} + \frac{a_2}{r_2} + \cdots + \frac{a_k}{r_k} \tag{3.6}$$

where r_T is the total resistance to water absorption from a given soil volume containing lengths of root a_1, a_2, a_3, etc., to which the resistances r_1, r_2, r_3, etc., per unit length of root can be attributed. If all the roots in a soil layer i can be said to have the same radial resistance r_i to water absorption then:

$$r_{T_i} = \frac{r_i}{L_i} \tag{3.7}$$

where L_i is now the total length of root in a given volume of soil in layer i, i.e.,

$$L_i = \sum_{j=1}^{k} a_{ij} \tag{3.8}$$

Hence, the resistance to water absorption is inversely proportional to root density.

This contrasts with the resistance to xylem water transport through a root system as given by:

$$f_T = \sum_{i=1}^{n} f_1 + f_2 + f_3 + \cdots + f_n \tag{3.9}$$

where f_T is the sum resistance to xylem water transport over n layers of soil. A value for the xylem resistance f across each soil layer of thickness h can be determined using Eq. (3.5), i.e., $f = 8h\mu/\pi x_e^4$. Since this summation of resistances is in series, the deeper the root system, the higher the resistance to water uptake. Furthermore, water may not be absorbed equally by all portions of the root system as indicated in Eq. (3.6). Consequently, the absorption and transport of water by root systems will be a complex interaction between resistances acting in parallel and in series.

The effectiveness of a root system is therefore greatly influenced by the relative absorption of water by different portions of the root, particularly where roots are sparse (24). Most evidence (76, 91, 92) indicates a preferential absorption of water by the section of root immediately behind the zone of elongation (Fig. 3.6). In certain species such as corn and barley, water and ion absorption can occur unhindered by endodermal suberization up to 20–30 cm from the root apex (62, 86): water absorption is generally found to reach a maximum at about 10 cm from the apex (93). Suberization of roots as they age will increase the resist-

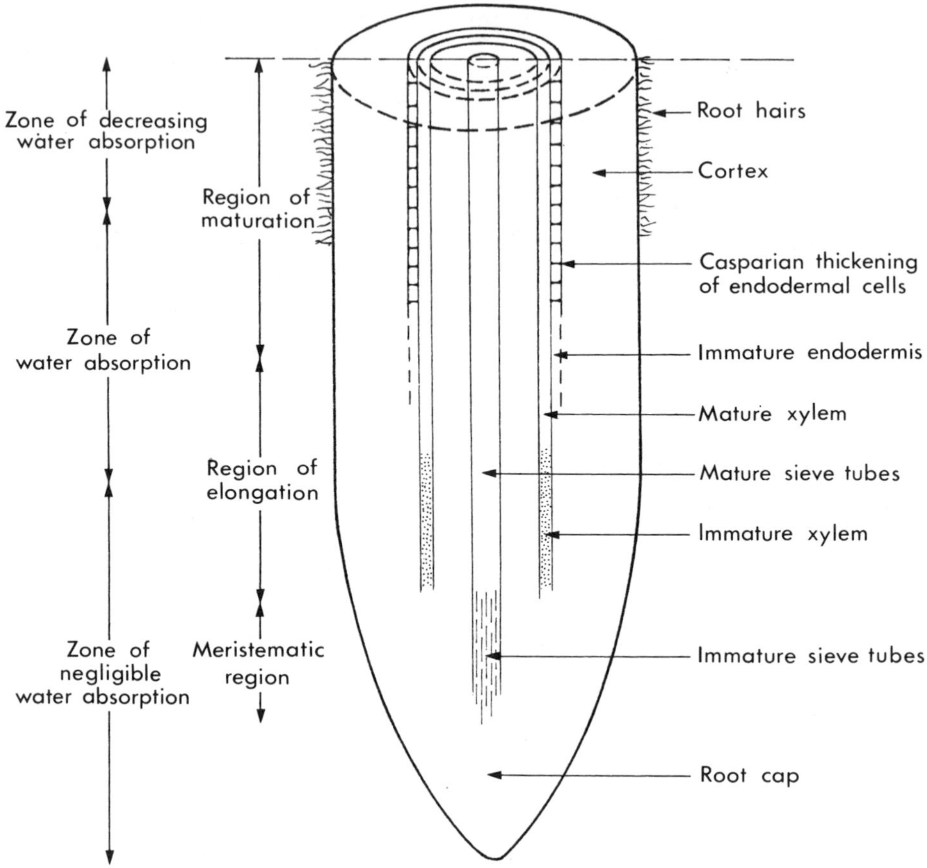

Figure 3.6. Diagram of a root tip showing the order of maturation of various tissues and the regions of water absorption.

ance to absorption of water and ions, particularly with suberization of the endodermis.

Table 3.1 provides estimates of the resistances incurred by alternate pathways for water absorption by a corn root. This table gives pathway resistances based on known conductivities of cell structures and compares the resistances incurred by pathways through the root with and without the effect of suberization, which essentially confines water movement to the symplasm pathway across the endodermis (Fig. 3.5). Pathways I, II, and III in Table 3.1 consider the resistances to free space movement and pathway IV gives the effect due to suberization of the endodermis on pathway I. Restricting water movement at the endodermis causes a 25-fold increase in the resistance to apoplastic water movement, i.e., from 0.2×10^5 to 5.5×10^5 bar sec cm^{-2}. If a similar calculation is made for the effect of suberization on pathways II and III the resistances would increase by more than 100% for each pathway. Incomplete suberization does, however, allow some apoplastic water movement to continue (60), particularly in lateral and nodal roots (62). Pathways V, VI, and VII have resistances that exceed most measured values (see Table 3.2) and are therefore assumed unimportant for water movement, provided the conductivity values used are applicable to these cells. This detailed knowledge on water movement within the root helps clarify the effect of preferential absorption by young roots on the overall pattern of water uptake by intact root systems from soil. This particularly applies where crop plants continue to initiate new lateral roots and maintain active root growth for the major part of their lives (94–96).

Published values for the radial resistance to water absorption per unit length of root indicate that most diversity exists between values obtained from intact root systems and those obtained from isolated root segments, rather than between values obtained for different species within either of the above categories (Table 3.2). This suggests that water is being absorbed at a slower rate by portions of root that are mature and suberized or, as is often the case, resistances elsewhere in the plant have been wholly or partly attributed to the radial pathway. In very wet soil or for roots in nutrient solutions, the majority of the water would probably be absorbed by the nonsuberized portions of the root. However, as the soil

Table 3.1. Calculated pathway resistances to radial water transport into a maize root 10 cm from the tip[a]

	Pathway for Water Movement	Radial Resistance for Unit Length of Root (bar sec cm^{-2})
I	Free space direct to the metaxylem passing between the cell wall and plasmalemma, but permeating the tangential cell walls	0.2×10^5
II	Free space but permeating intercellular spaces and radial cell walls between intercellular spaces	3.2×10^5
III	Free space but passing entirely in the radial cell wall	6.4×10^5
IV	Free space (as in I) but restricted to the symplasm pathway for passage through plasmodesmata at the endodermis	5.5×10^5
V	Vacuolar pathway traversing two pairs of membranes per cell	44×10^7
VI	Symplasm pathway passing across the tangential cell walls and associated plasmalemma	22×10^7
VII	Symplasm pathway via plasmodesmata	4×10^7

[a] Calculations of pathway resistances are based on the root dimensions of maize (Fig. 3.3) and on conductivities of cell walls, plasmalemma, and plasmodesmata provided by Newman (54) and Robards and Clarkson (60).

Table 3.2. Estimated values of radial root resistance to water absorption per unit length of root in intact root systems and in isolated root segments

Species	Radial Resistance for Unit Length of Root (bar sec cm^{-2})	Reference
	A. *Intact Root Systems*	
Wheat	$1.5–27.8 \times 10^7$	274, 275
Tomato	2.3×10^7	276
Sunflower	4.2×10^7	276
Maize	4.2×10^7	276
Clover	6.5×10^7	30
Fescue	6.5×10^7	30
Broad bean	8.0×10^7	276
Cotton	8.6×10^7	29
Dwarf bean	19.2×10^7	276
Italian ryegrass	31.5×10^7	277
	B. *Isolated Root Segments*	
Wheat	$0.2–6.6 \times 10^6$	76, 78, 278
Maize	$0.2–3.8 \times 10^6$	279–282
Pea	0.7×10^6	78
Broad bean	$1.1–10.0 \times 10^6$	283

surrounding the root becomes unsaturated, a substantial rhizosphere resistance can increasingly contribute to the overall resistance to water absorption, particularly around the more actively absorbing portions of the root (57). This resistance will tend to balance the rates of water absorption by younger and older portions of root for plants growing in soil. Consequently, it is often observed that water absorption is effectively uniform per unit length of root over the entire root system regardless of differences in root distribution between species or of differences in the distribution of younger roots within a soil profile (29, 30, 58). Assuming uniform absorption of water per unit length of root is usually an acceptable simplification of the complex interaction of resistances described above, except perhaps in species such as soybean and wheat, where absorption is complicated by the differential morphology of their root systems (85).

Root Development

An important difference in root architecture can be recognized between monocotyledons and dicotyledons. In dicotyledons, the primary root and its laterals constitute the entire root system, whereas in monocotyledons, the primary root and any seminal roots originating in the seed embryo are usually supplemented by adventitious roots (nodal roots) that develop from the basal nodes of the stem. Many crop species that are dicotyledonous develop a large tap root that can result in root penetration to considerable depth. For example, roots have been observed below 2 m in alfalfa (97) and in soybeans (89), a characteristic that allows such species to avert the effects of a sudden onset of drought. However, the lateral extension and relative density of this type of primary root system may somewhat restrict their competitiveness for water and nutrients in surface soil layers (98).

Although there are exceptions (99), the initiation of large numbers of nodal roots in monocotyledonous crops can give very high root densities in the surface soil, but result in fewer roots reaching any significant depth (24, 100–102). In cereal crops it has been suggested that this root architecture causes overexploitation of the surface soil and inadequate opportunity to capitalize on subsoil water reserves (103). In the event of sudden drought such crops rapidly dry the surface soil and immediately suffer the consequences of water stress, often without sufficient opportunity for stress adaptation to occur through changes in solute osmotic contribution or for maturity to be accelerated (104). Both of these responses could moderate the severity of any yield reduction due to water stress. Changing the cereal root system by selecting for either deep root penetration (103, 105), or for fewer seminal roots so that drying of the soil is prolonged (106), have been suggested as alternate strategies for improving crop yields in arid environments.

Root Extension

Root growth can be considered as a special case of the general theory of extension growth in plant cells described by Green (107) and described in greater detail later. Cell enlargement or elongation depends on increases in turgor potential exerted on the cell wall in the presence of auxin (108). Auxin-induced softening of the walls of meristematic cells will result in lower turgor potentials, leading to water absorption and cell enlargement.

Cell expansion growth in the root has to overcome the additional mechanical resistance imposed by the soil. Two aspects of soil resistance need to be considered: first the geometry of root expansion within the soil, and second, the pressure exerted by expanding cells during root expansion. In uniformly fine structured soils, the root mostly exerts a radial pressure so as to create a cylindrical root cavity. This radial pressure, P_r, can be estimated from the mechanical resistance to penetration of a finely tapered probe (109) using the relationships:

$$P_r \sim F_n = \frac{P_c}{1 + \tan\theta \cot\alpha} \qquad (3.10)$$

where F_n is the force normal to the surface of the probe tip and P_c is the pressure of the cross-sectional area of the probe, θ is the angle of soil probe friction, and α is the semiangle of the conical tip of the probe. Measured resistances to probe penetration give a complex function between soil strength and compressibility which depend on the soil bulk density and water content (110).

The effect of the soil resistance is to restrict the expansion growth of root cells by reducing their effective turgor. Greacen and Oh (111) found that the growth response of pea radicles to an increase in soil resistance or a decrease in soil water content was highly dependent on the extent of osmotic adjustment in expanding root cells. Osmotic adjustment allowed the root cells to maintain a gradient in water potential sufficient to overcome the effect of a high radial soil pressure or low soil water potential, thereby maintaining water uptake and cell expansion. Greacen and Oh (111) showed that osmotic adjustment fully maintained turgor as the soil water potential decreased from -2 to -8 bars, but failed to fully compensate for the mechanical resistance of the soil as it increased from -1 to -6 bars, thereby explaining the sensitivity of root growth to any mechanical impedence by soil.

Low soil temperatures, insufficient soil nutrients, salinity and waterlogging are additional factors that reduce root growth (112–117). Plant photosynthetic activity, partitioning of photosynthate and any loss of biomass, such as by grazing, all affect root growth. The general response to increasing soil water deficits, low soil temperatures, and a reduction in photosynthetic activity is usually an increase in the root-to-shoot ratio of plant biomass (114, 115, 118). Genotype also has a powerful influence on the relative change in this ratio (119). Favored root growth at the expense of leaf growth is a characteristic of plants that can tolerate drought as will be discussed in more detail in later sections on drought resistance.

ROLE OF SOLUTES IN PLANT WATER RELATIONS

The presence of solutes in cells provides the osmotic component of the total water potential of plant tissue. Even with an adequate supply of water, the concentration of solutes in plant cells varies widely: in general, xerophytic species have lower osmotic potentials than mesophytic species (120). The role of solutes in maintaining cell turgor has long been recognized (121), but the accumulation of solutes with water deficits and their role in maintaining cell expansion, particularly in crops, has now become better appreciated.

Osmotic Adjustment

As water is removed from the cell during the development of water deficits, the solutes in the cell are concentrated, leading to a lowering of the osmotic potential. If the cells behave as perfect osmometers then the osmotic potential (ψ_π) will change as a result of the passive concentration of solutes according to:

$$\psi_\pi = \frac{\psi_\pi^0 V^0}{V} \tag{3.11}$$

where V is the osmotic volume of the cell, and ψ_π^0 and V^0 are the osmotic potential and osmotic volume at a reference value such as full turgor or zero turgor. Equation 3.11 can be approximated by:

$$\psi_\pi = \frac{\psi_\pi^0 RWC^0}{RWC} \tag{3.12}$$

Where RWC and RWC^0 are the relative water contents of the cell at which ψ_π is measured and at the reference turgor, respectively. The change in osmotic potential arising from the concentration of solutes is represented by line ψ_π^1 in the Höfler diagram in Fig. 3.7.

It is apparent from Fig. 3.7 that cells with a lower osmotic potential, represented by line ψ_π^2, will have higher turgors (ψ_p^2 vs. ψ_p^1) at high water potentials and reach zero turgor at a lower relative water content, provided other cell characteristics do not change. Low osmotic potentials, therefore, provide a means of maintaining positive turgor at lower water contents and is the basis of drought tolerance in some species.

An alternative to having inherently low osmotic potentials is to regulate the osmotic potential of the cell as the water content decreases. This phenomenon, known as osmotic adjustment, osmoregulation, or turgor regulation has been widely recognized as a means whereby halophytes, algae, and microorganisms adjust their cell water relations to maintain cellular function as the osmotic concentration of the external medium varies (122).

Although the ability of plants to lower their osmotic potential in order to withstand salt stress has been known for some years (123–125), the ability of plants to lower their osmotic potential in response to water deficits, called osmotic adjustment (126), is a more recent observation (111, 126–128). The considerable evidence for osmotic adjustment in crop plants was summarized recently by Turner and Jones (126). Not all species or cultivars adjust osmotically (44, 128) and the degree of adjustment appears to be limited and readily reversible (126), but its importance in drought tolerance is clear. Just as an inherently low osmotic potential enables a plant to maintain positive turgor at low leaf water contents and leaf water potentials, so solute accumulation allows a plant to maintain positive turgor to lower water potentials (129, 130). Osmotic adjustment allows partial or full turgor maintenance depending on the species, rate of stress, and degree of stress (126).

A number of compounds appear to be involved in osmotic adjustment, depending on the species. In crop plants, the major contributors to osmotic adjustment appear to be soluble sugars, carboxylic acids, potas-

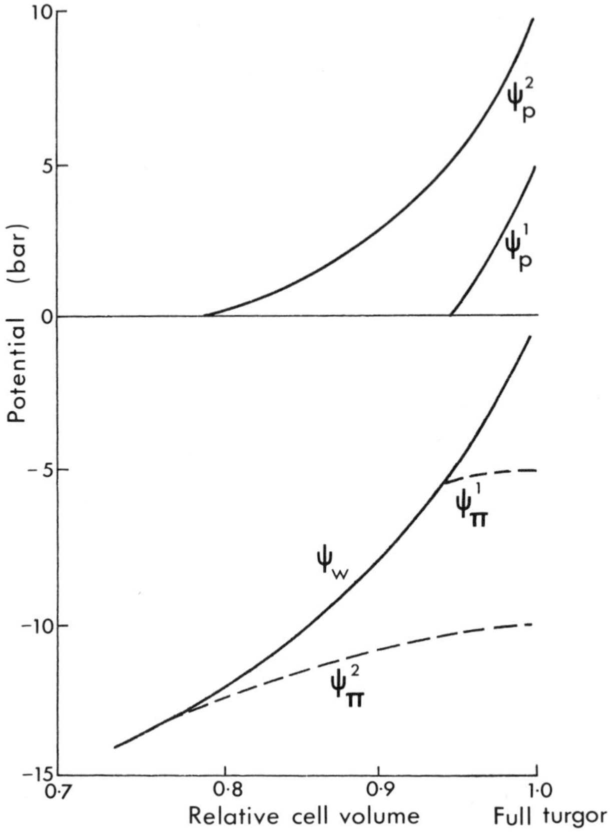

Figure 3.7. Modified Höfler diagram showing the variation in cell turgor potential (ψ_p), osmotic potential (ψ_π) and water potential (ψ_w) with decreasing relative cell volume for tissue at a high osmotic potential (ψ_π^1) and low osmotic potential (ψ_π^2).

sium, chloride, and amino acids, particularly proline (131–135). Sugars, potassium, and amino acids accounted for 60–100% of the osmotic adjustment in the apex and expanding leaves of wheat; and sugars, potassium, calcium, magnesium, amino acids, chloride, nitrate, and carboxylic acid accounted for 50–100% of the osmotic adjustment in sorghum and sunflower leaves (134, 135).

Role in Expansion Growth

The maintenance of turgor is particularly important in cell expansion because of its turgor dependence. The sensitivity of expansion growth to cell dehydration has been illustrated by the work of Green (107). In a subsequent study Green et al. (136) described the following equation for the relationships between turgor potential, ψ_p, and the cell elongation rate, E_r:

$$E_r = m(\psi_p - \psi_p^0) \quad (3.13)$$

where m is a variable to account for cell extensibility or "yielding tendency" and ψ_p^0 is the threshold turgor for growth to occur, i.e., the apparent "yield point." Therefore $(\psi_p - \psi_p^0)$ is the effective driving force for growth and if $\psi_p \leq \psi_p^0$, $E_r = 0$. An important contribution from this analysis was that growth can be controlled by a dynamic equilibrium between the three variables m, ψ_p and ψ_p^0, and not by a fixed value for ψ_p, as was previously suggested (137, 138).

If a water deficit is imposed on expanding cells then the model given in Eq. (3.13) can be used to account for the responses illustrated in Fig. 3.8. During the phase of declining turgor, growth almost ceases, but once turgor has stabilized at a lower level, ψ_p^0 also adjusts to correspond with the reduction in ψ_p and growth resumes accordingly. However, the rate of growth will be slower due to a reduced gradient in water potential between the expanding cell and the xylem as a consequence of the water deficit. Figure 3.8 shows that, over the time interval Δt, i.e., before growth resumes, the reduction in cell extension due to the decrease in cell turgor, relative to potential cell extension indicated by line a, is given by $\Delta \ln L_a$. After cell turgor is restored, cell extension accelerates for a short time before stabilizing at a new rate indicated by line c. The increment in cell extension due to the accelerated growth phase, $\Delta \ln L_b$ (shown as the difference between lines b and c), can be seen to approximate the magnitude of the initial growth reduction $\Delta \ln L_a$. This implies that cell rehydration following short duration water deficits can cause a dramatic and concurrent increase in cell extension sufficient to compensate for any reduction in growth due to the initial loss of turgor by cells. However, growth resumes at a slower rate at reduced levels of turgor and the differential between this growth and that possible at full turgor is never recovered on rehydration of the cell. In addition, the rate of growth after rehydration may not be equivalent to the original rate before cell turgor was lost, i.e., line b compared with line a, but this is influenced by other effects such as maturation of the cell and the duration of water deficits. This behavior was originally observed in *Nitella*, but other studies have confirmed that similar responses occur in crop plants (139, 140).

For turgor to be maintained in expanding cells there must be a movement of solutes or ions to these cells. McNeil (141) found sucrose to be the principal solute translocated to expanding cells for the synthesis of osmotic compounds and for dry matter accumulation. In an experiment using soybean hypocotyls, Meyer and Boyer (127) clearly demonstrated that gradually induced water deficits caused an increased transfer of solutes from the cotyledons to the hypocotyl. Thus, any solute accumulating at a meristem leading to osmotic adjustment will help in main-

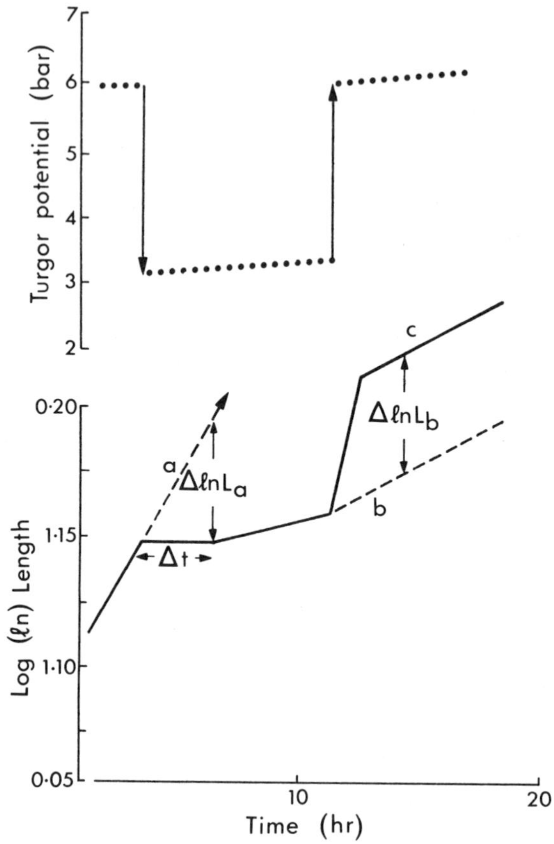

Figure 3.8. Influence of changes in turgot potential (. . .) on the elongation of a single cell of *Nitella* (———). A drop in the turgor potential at 3.5 hr caused a cessation of growth for a time Δt before growth was resumed. Restoration of turgor led to an acceleration of growth to a new rate of growth *c*. *a* is the initial rate of growth and *b* is the rate of growth equal to *c*. Vertical arrows indicate the change in turgor potential. [Adapted from Green et al. (136).]

taining a sufficient gradient in potential for cell absorption of water and, hence, in maintaining expansion growth. Once expansion growth ceases, there is apparently some mechanism which inhibits any further accumulation of sugar within the cell structure (141).

Auxins must be present at the cell wall before any pressure exerted on the cell wall by turgor induces cell elongation (108, 142, 143). Auxin is involved in the softening of the cellulose microfibril wall structure that is required before cell expansion arising from turgor can occur. Thus, there is an interaction in expansion growth between the chemical processes of auxin accumulation and solute accumulation and the physical effects of turgor arising from the accumulation of solutes.

WATER STRESS AND PLANT GROWTH

Development of Internal Water Deficits

In discussing plant water balance we showed that gradients of water potential provide the driving force for movement of water from the soil to the atmosphere through the plant; the steepness of the gradient in the various segments of the flow path depends on the rate of flow and the resistance in the various segments. A consequence of the gradient in water potential through the plant is that water not only flows through the xylem from the root, but is also drawn from other tissues in the leaves, stems, fruits, and roots that are at a higher water potential. Thus, diurnal monitoring of leaf thickness reveals a shrinkage during the morning as water is withdrawn from these tissues and a swelling in the evening as the plant rehydrates (144, 145).

As water is removed from the soil and not replenished, the water potential in the root zone decreases and, provided the resistances within the plant remain constant, the water potential throughout the plant must be similarly reduced in order to maintain the same rate of transpiration. Ultimately, the decrease in leaf water potential falls to a level at which processes such as leaf expansion and photosynthesis are decreased. The changes in leaf water potential, leaf osmotic potential, and the rate of net photosynthesis as the soil water potential at 25 cm decreased in a field-grown soybean crop are shown in Fig. 3.9. For the first 6 days following an irrigation or rain, the soil water potential in the root zone (i.e., to 1 m) remained high, the leaf water potential decreased diurnally from about -5 bars at sunrise to -15 bars in the early afternoon, leaf turgor remained positive as indicated by the difference between ψ_w^l and ψ_π^l, and photosynthesis followed a smooth diurnal pattern. However, by day 13 after irrigation or rain, the leaf water potential fell to about -18 bars and leaf turgor became zero in the early afternoon when ψ_w^l and ψ_π^l became equal (cf. Fig. 3.7) with a concomitant decrease in photosynthesis. By 27 days after irrigation or rain, the soil water potential at 25 cm was -5 bars, ψ_w^l fell to -20 bars in the middle of the day, and the integrated daily photosynthesis was only about 50% of that for plants in moist soil. By 33 days after irrigation or rain, the soil water potential at 25 cm was -10 bars, the leaves were at zero turgor for most of the day, and photosynthesis was severely depressed (24, 25, 44).

Methods of Measuring Water Deficits

The comprehensive review of methods of measuring water deficits by Slavík (146) provides the newcomer to the field with a baffling array of direct and indirect techniques for measuring a plant's water status. Since

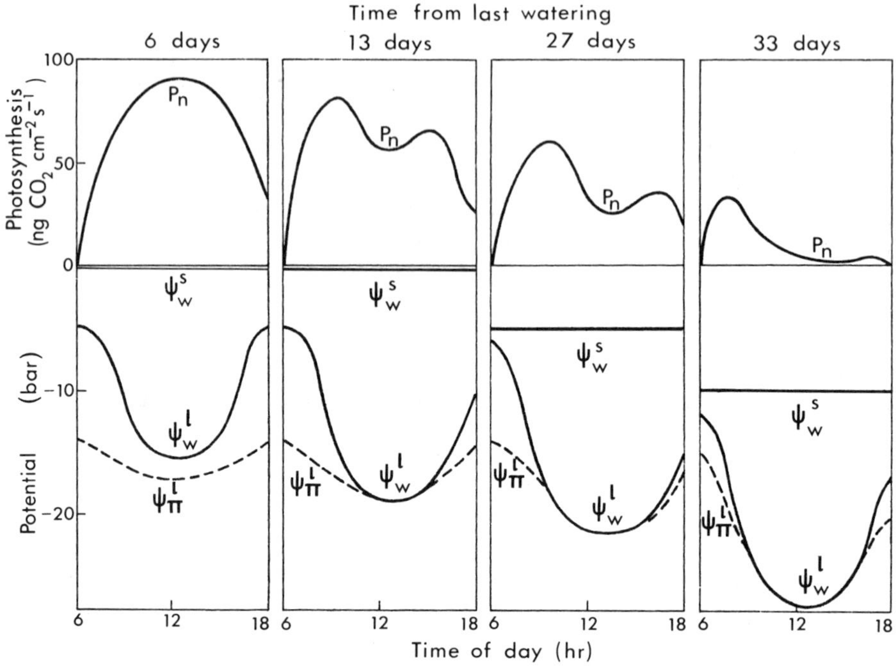

Figure 3.9. Diurnal variation in leaf water potential (ψ_w^l), osmotic potential (ψ_π^l), and photosynthesis (P_n) for soybean at various days after irrigation or rain as the soil water potential (ψ_w^s), measured at 25 cm, decreased. [Data from Burch et al. (24), Rawson et al. (25), and Turner et al. (44).]

a further comprehensive review in this chapter is inappropriate, only methods useful for the measurement of crop water deficits, particularly in the field, will be highlighted. For more detailed treatments of the subject, the reader is directed to Boyer (22), Barrs (147), Brown and Van Haveren (148), and Turner (149), in addition to the book by Slavík (146) referenced above.

One of the simplest methods for measuring crop water status is to measure the relative water content (*RWC*) or water saturation deficit (*WSD*):

$$RWC = \frac{\text{fresh weight} - \text{oven dry weight}}{\text{turgid weight} - \text{oven dry weight}} \times 100 \qquad (3.14)$$

$$WSD = 100 - RWC \qquad (3.15)$$

Samples from one or several plants are collected and weighed in a vial that can be tightly sealed. The samples are then rehydrated, usually by floating on water for several hours until fully turgid, surface dried, and reweighed. The *RWC* or *WSD* can then be determined after oven drying and reweighing. Care needs to be taken to ensure that prolonged rehydration of samples does not introduce errors due to weight changes from growth or respiration (150), but also that samples are fully rehydrated: the standard time of 4 hr (150) may be too short for very dehydrated samples (149).

The total water potential is most accurately measured by thermocouple psychrometry except where large concentrations of solutes occur on the leaf surface or are released into the psychrometer chamber (151). The thermocouple psychrometer is based on the principle that the relative vapor pressure above a solution or piece of plant tissue is related to its water potential (ψ_w) according to the following equation:

$$\psi_w = \frac{RT}{\bar{V}} \ln \left(\frac{e}{e_0}\right) \tag{3.16}$$

where R is the gas constant, T is temperature (°K), $\bar{V}$ the partial molal volume of water, e is the vapor pressure of water in the solution or tissue, and e_0 is the vapor pressure of pure water at atmospheric pressure. In the thermocouple psychrometer, a sample is sealed into the chamber and allowed to equilibrate and the wet bulb temperature of the air in the chamber is measured relative to the dry bulb temperature. Two basic types of thermocouple psychrometer are in current use. In the Peltier psychrometer (152), Peltier cooling is utilized to form a drop on the wet junction, whereas in the wet loop psychrometer (153), a silver ring or wire loop provides a holder for a permanent water droplet at the wet junction. Since a wet bulb depression of 1°C gives a water potential of −80 bars, facilities for strict temperature control and measurement of small temperature differences are required. Slavík (146), Barrs (147), and Brown and Van Haveren (148) provide details of improvements and refinements in the techniques of thermocouple psychrometry since the original methods were first described.

The requirement for strict temperature control limits the use of thermocouple psychrometers to situations where controlled temperature facilities are close at hand. Furthermore, long equilibration times (154) prevent frequent sampling unless a large automated system with many chambers is available. The development of temperature-compensated thermocouple psychrometers and dew point hygrometers (148) has reduced, but not eliminated, the need for temperature control and has led to the development of *in situ* psychrometers or hygrometers for the measurement of the water potential in roots (155), stems (156), and leaves (157). Although *in situ* psychrometers have been used in a number of field studies (156, 158–160), as pointed out by Begg and Turner (161), no rigorous comparisons with other techniques of measuring water potential have been carried out. Until such comparisons have been made, data from field studies using these techniques should be treated with caution, as equilibration times may be too long to sense the constantly changing values of leaf water potential (Fig. 3.8), and this together with changing leaf temperature could lead to spurious observations (149).

The pressure chamber technique provides the most suitable method for the routine measurement of leaf water potential in field studies: Ritchie and Hinckley (162) document its widespread use in ecological and agronomic research. With this technique a leaf or branch is cut and placed in a chamber that can be pressurized with the cut surface just protruding into the air through a rubber gland in the chamber. Pressure is slowly applied to the leaf or branch until the meniscus just returns to the cut surface: this balancing pressure approximates the leaf water potential in a wide range of species (162). The technique is rapid and largely insensitive to temperature (163), but since it is a destructive technique it cannot be used where only a limited amount of plant material is available. Furthermore, precautions need to be taken to minimize errors arising from rapid pressure increases (164, 165), exclusion errors (164, 166), and dehydration of leaf tissue between sampling and placing the sample in the chamber and while in the chamber itself (165, 167).

Vapor pressure and pressure chamber techniques are also used to measure the osmotic potential (146, 147, 149, 168). The vapor pressure osmometer, dew point hygrometer, and thermocouple psychrometer can all be used to measure the change in vapor pressure or dew point of a solution arising from the presence of osmotically active solutes. In order to measure the osmotic potential of plant tissue the turgor potential must be reduced to zero: this is usually achieved by killing the tissue by heat or cold. The osmotic potential is then measured on the cell sap expressed under pressure from the tissue. If samples are frozen, thawing must be sufficiently long for all the ice to melt otherwise a spuriously low osmotic potential will be recorded. Similarly, the killed sample cannot be left too long at room temperature otherwise insoluble sugars and starch may be converted to soluble sugars and a spuriously low osmotic potential again recorded (133). The cell sap from killed tissue is a mixture of both symplastic and apoplastic water: the latter generally has a very high osmotic potential and dilutes the solutes in the symplastic water. The volume of apoplastic to symplastic water at full turgor appears to vary from 10 to 40% (46, 169–171); as the volume of apoplastic water is unlikely to vary greatly as the tissue dehydrates, the dilution of symplastic water by apoplastic water in dehydrated tissue is much greater (170).

Some investigators have measured the osmotic potential on killed tissue rather than on sap expressed from killed tissue (128, 168, 172). This technique is designed to reduce sampling errors, but introduces an error associated with the flooding of cell sap into the cell wall matrix (168). It should be pointed out that this matric potential, which is usually small, but can be as large as -10 bars in some circumstances (172), is not the matric potential within the cell (21).

Scholander et al. (173) described a technique for measuring the osmotic potential of tissue in the pressure chamber. The turgor potential is reduced to zero by applying pressure directly to the leaves and the relationship between the osmotic potential and water content is established for successively lower water contents by forcing increments of water from the leaf cells. By extrapolation, the osmotic potential at any water content can be established. As cell membranes are not ruptured and as it can be reasonably assumed that little apoplastic water is removed over the range of pressures generally used (170), the method is not subject to the dilution errors associated with killed tissue. The technique has been widely used with trees (e.g., 46, 174), but has received only limited use with crops (169, 171, 175, 176). With crop species care should be taken to ensure that the xylem does not become constricted under pressure (176).

The matric potential of leaf cells has been measured using the pressure chamber (177) and thermocouple psychrometer (137, 172). However, the matric potentials so measured could well be artifacts resulting from the release of cell sap into the cell wall matrix (21, 121).

The water status of a crop can also be measured by several indirect methods. Visible wilting or color change may be sufficient for some purposes. Leaf temperature (178), leaf thickness (147, 179, 180), stem diameter (181, 182), and stomatal conductance or leaf permeability (35, 183) can all be correlated with the leaf water potential. However, all suffer from various disadvantages compared with direct measurement of leaf water potential: leaf temperature and leaf thickness are only poorly and nonlinearly related to leaf water potential; changes in stem diameter lag behind changes in leaf water potential (182, 184); stomatal conductance is not always linearly related to leaf water potential and the relationship between the two depends on light and stress history (35, 133, 185, 186).

Effects of Water Deficits on Physiological and Morphological Processes

In crop production one is concerned with the dry matter yield of the whole or part of the plant, i.e., the net accumulation of carbon dioxide and its partitioning. In simple terms the accumulation of carbon will depend on the leaf area, the photosynthetic rate per unit leaf area, and the rate of respiration. This section is confined to a consideration of the effects of water deficits on those physiological and morphological processes that affect dry matter accumulation directly, namely, leaf elongation and senescence, photosynthesis, respiration, and partitioning of the products of photosynthesis. For reference to the effects of water deficits on other physiological processes, see Gates (187) and Hsiao (188).

Leaf Expansion

Initial reports indicated that leaf expansion was very sensitive to water deficits. For example, data from maize grown in controlled environments indicated that leaf enlargement declined rapidly at leaf water potentials below −2 bars and ceased at potentials of −7 to −9 bars (138, 139, 189, 190). Similar sensitivity has been observed in a range of grasses (191) and in soybean and sunflower grown under controlled environment conditions (137, 138). This led Boyer (137) and Reicosky et al. (192) to conclude that growth must be confined mainly to the night.

Recent studies, however, suggest that leaf expansion under field conditions is less sensitive to water deficits than these earlier studies indicated. In Fig. 3.10, data for sorghum and soybeans are presented showing that the rates of leaf expansion were maximal in the afternoon when leaf water deficits reached a maximum. The strong correlation between extension growth and ambient temperature suggests that the rate of leaf expansion may have been determined largely by temperature and unaffected by leaf water deficits as low as −12 and −14 bars in the sorghum

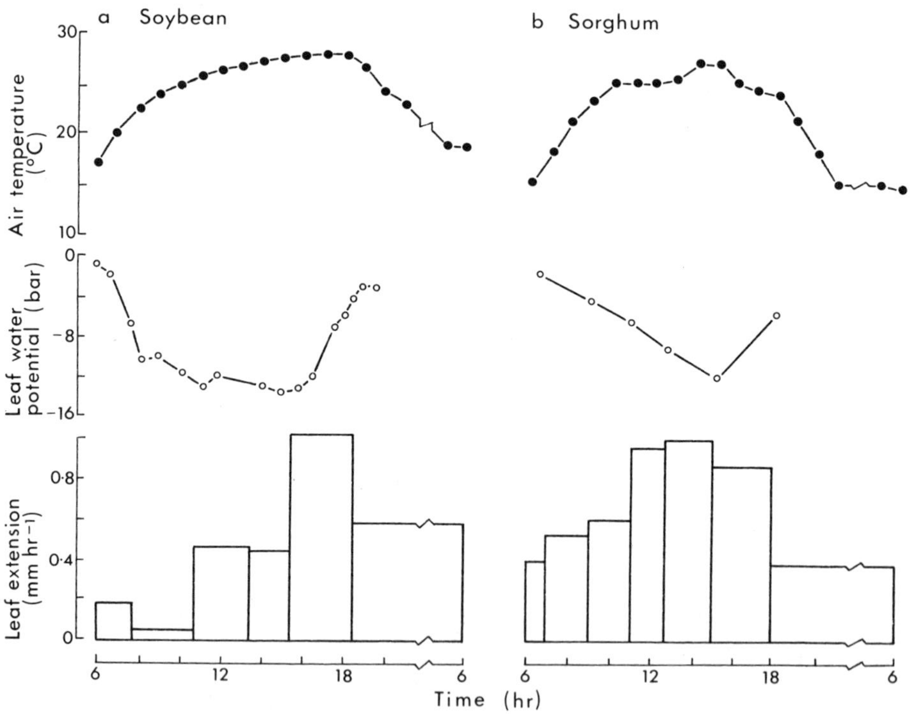

Figure 3.10. Variation throughout the day in the rate of leaf extension (histograms), leaf water potential (○), and ambient air temperature (●) for soybean and sorghum leaves under well-watered conditions in the field. [Adapted from Begg and Turner (unpublished) and Chu and Kerr (284).]

and soybean, respectively. Watts (193) and Hsiao et al. (190) provide similar evidence for maize.

In addition to affecting leaf expansion, water deficits can affect leaf area by the senescence and death of leaves during all phases of growth. Fischer and Kohn (194) showed a decrease in the leaf area duration of wheat with water deficits and Ludlow (195) showed that the percentage of dead leaf of *Panicum maximum* increased at all values of water potential below −15 bars, with all leaves being dead at −100 bars. Generally, leaf death occurs first in the lower, older leaves in the canopy and then progressively affects leaves higher in the canopy. Water deficits also reduce tillering and increase the death rate of tillers in multitillered species, thereby influencing the leaf area of a plant canopy (161).

Net Photosynthesis

The initial decrease in the rate of photosynthesis per unit leaf area as a result of water deficits is generally concluded to arise from stomatal closure. Considerable evidence has accumulated indicating that there is little change in stomatal resistance as leaf water potential decreases until a critical or threshold potential is reached below which the stomatal resistance increases markedly (161, 196). Turner (196) showed that the critical potential at which stomata closed varied with species and small differences between cultivars have also been reported (41, 42, 197). There is, however, no unique critical value for stomatal closure for a particular species or cultivar. The leaf water potential at which stomata close varies with position of the leaf in the canopy, leaf age, growth conditions, and rate of stress (35, 133, 185, 191, 198–200).

Evidence of a critical leaf water potential at which stomata close is not, however, universal (38). Some studies suggest a more gradual increase in leaf resistance as leaf water potentials decrease (186, 199, 201). The differences may in part be the result of differences in rate of stress and method of measurement. The rapid imposition of stress by excision of tissue and use of diurnal changes in stomatal resistance and concurrent changes in leaf water potential will lead to more rapid increases in resistance as water potentials decrease than in situations where stress is imposed more slowly and the plant can adjust osmotically to the water deficit (186).

The closure of stomata at low leaf water potentials is reflected in the rate of photosynthesis. This is clearly seen in Fig. 3.11 in which the leaf conductance, i.e., the inverse of leaf resistance, and the rate of photosynthesis of the same leaves are plotted as a function of the leaf water potential in two cultivars of soybean. In both cultivars, the leaf conductance and photosynthesis decreased approximately linearly between −12 and −24 bars.

Studies of the effects of water deficits on the photosynthetic pathway

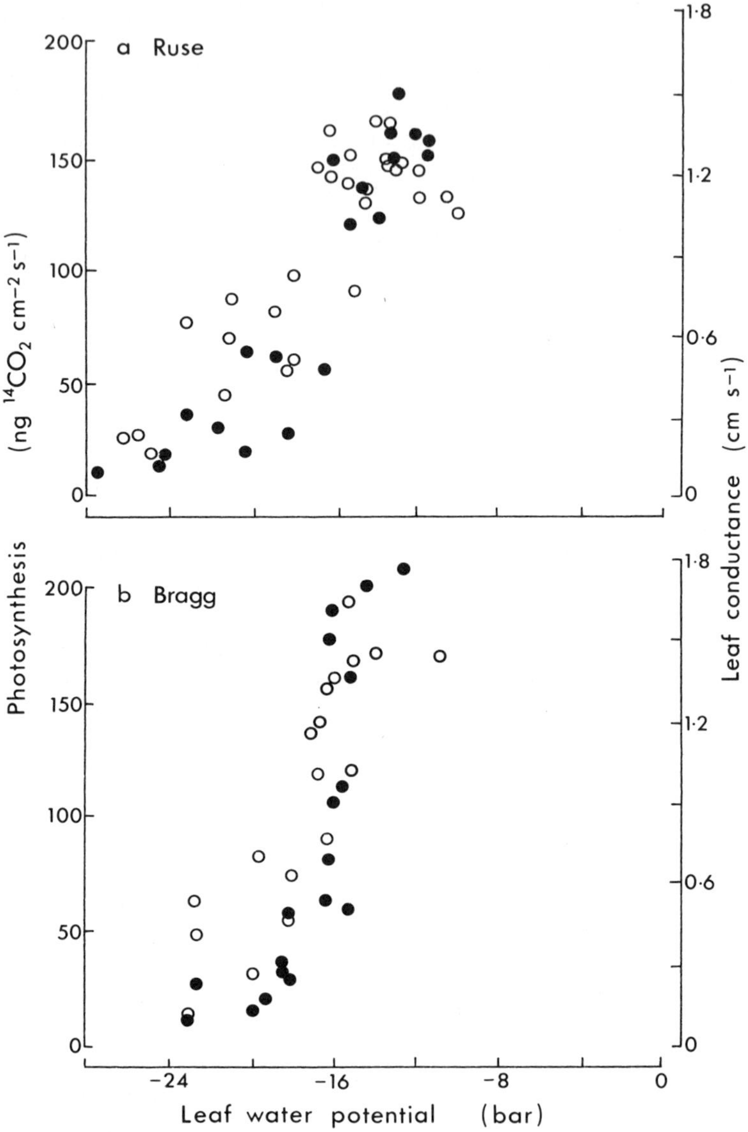

Figure 3.11. Variation in leaf conductance, i.e., 1/leaf resistance (●), and photosynthesis (○) with decreasing leaf water potential in (a) 'Ruse' and (b) 'Bragg' soybeans. [Adapted from Turner et al. (44).]

internal to the stomata have suggested that in many species the internal transport of carbon dioxide was unaffected until water potentials much lower than those required to close stomata (202) were reached. However, in some species, of which sunflower is a good example (203), stomatal and nonstomatal effects of water deficits occur at the same leaf water status. A recent analysis by Osmond et al. (204) suggests that nonstomatal and stomatal components of photosynthesis are adjusted in concert as water deficits develop, provided the rate of imposition of stress is not unrealistically rapid.

Water deficits sufficient to close stomata and reduce photosynthesis also decrease dark respiration, but the decrease in the rate of dark respiration is less than the decrease in photosynthesis. For example, as the leaf water potential of sunflower decreased from −4 to −18 bars, leaf photosynthesis decreased by 70%, whereas dark respiration decreased by 33% (203). Dark respiration decreased to the same degree as photosynthesis in shoots of sunflower, soybean, and maize (138), suggesting that dark respiration of stem and meristematic tissue may be reduced more than leaf tissue by water deficits.

From a review of the literature, Begg and Turner (161) concluded that photorespiration was unaffected by short-term stress in crop species, but that ultimately photorespiration decreased as the substrates for photorespiration were depleted. Recently, Powles and Osmond (205) have shown that when C_3 plants were deprived of carbon dioxide in the light their photosynthetic capacity in the absence of water stress was reduced due to photoinhibition, whereas the recycling of carbon dioxide by photorespiration prevented this photoinhibition. Osmond et al. (204) suggest that carbon dioxide recycling by photorespiration may be a mechanism for reducing the possible effects of photoinhibition when stomata close at low leaf water potentials.

Distribution of Assimilates

For plants in which the economic yield is only a proportion of the total yield, the effect of water deficits on the distribution of the assimilates, i.e., the products of photosynthesis, to the yield-producing organ is important. Current evidence indicates that the assimilate distribution pathway, the phloem, is highly resistant to water deficits (206). McWilliam (207) showed that movement of assimilate occurred from stem to roots and buds in dehydrated, dormant plants of *Phalaris tuberosa*. Likewise, Wardlaw (208–210) showed that the velocity of assimilate movement was unaffected by water deficits as low as −30 bars in wheat or *Lolium temulentum*.

Nevertheless, the overall translocation of assimilates is affected by water deficits because of the influence of these deficits on the rate of photosynthesis, the rate of utilization of the products of photosynthesis, or the loading and unloading of the phloem. For example, translocation in potatoes and wheat during grain filling was reduced due to a reduction in photosynthesis by the source leaves (208, 210–212). Wardlaw (209), on the other hand, concluded that it was the reduced utilization of assimilates by the apex, and not reduced photosynthesis, that reduced translocation in *Lolium temulentum*. Finally, Brevedan and Hodges (213) concluded that at grain filling in maize a reduction in the leaf water potential to −26 bars reduced vein loading.

The effects of water deficits on the actual distribution of assimilates to the various plant organs will depend on the stage of development of the plant, the degree of stress, the prehistory of stress, and the degree of sensitivity to stress of the various plant organs. Thus, for *Lolium temulentum* in the vegetative phase, Wardlaw (209) showed that leaf expansion was more sensitive to water deficits than photosynthesis and that the products of photosynthesis moved preferentially to young leaves, sheaths, and roots. However, in wheat during grain filling, photosynthesis by the leaves was more sensitive to water deficits than grain growth and assimilates moved preferentially from the lower leaves, stems, roots, and crown to the ear (208). Moreover, although the degree of stress may be sufficient to reduce the rate of net photosynthesis, it may not affect the pattern of distribution of assimilates (212). Where it does affect the distribution, the pattern may differ depending on the level of stress encountered: as the level of water stress increased, vegetative *Lolium temulentum* translocated more assimilates to the lower leaves and less to the crown.

Accumulation of assimilates in roots, leaves, and stems in the vegetative stage can provide a significant source of assimilate for later development. With adequate water, current photosynthetic products appear to provide most substrates for reproductive growth (214, 215), but a reduction in current photosynthesis due to water stress increases the proportion of stored assimilates transferred to the grain. For example, Gallagher et al. (216) have shown that approximately two-thirds of the carbohydrates in the grain of barley and wheat under extreme water stress were derived from assimilates accumulated prior to grain filling. Although the proportion of stored assimilates transferred to the grain under water stress may be greater, the absolute amount may be the same (215, 217): indeed there is some doubt as to the validity of the estimates by Gallagher et al. (216) in light of current evidence (215). Constable and Hearn (218) suggested that cultivars of soybean differed in the stem reserves that they transferred to the grain, and that water stress altered the proportion of stem reserves mobilized by the two cultivars.

Drought Resistance

Drought is defined in Webster's New World Dictionary as "prolonged dry weather." Drought resistance is the ability of a crop species or cultivar to grow and yield satisfactorily under drought. Whether a period of prolonged dry weather affects the physiological and morphological processes contributing to yield, and ultimately affects yield itself, depends on a number of factors. Not only will it depend on the timing of the dry weather in relation to the life cycle of the crop and on the water holding capacity of the soil in the root zone, but also on characteristics of the plants themselves. The plant characteristics that enable them to endure drought will be considered in this section.

There have been many attempts to define and categorize drought resistance (219–224). The one utilized here is that of May and Milthorpe (220). They identified three types of drought resistance:

(i) *Drought escape,* i.e., the ability of a crop species or cultivar to complete its lifecycle before serious plant water deficits develop.
(ii) *Drought tolerance at high plant water status,* i.e., the ability of a crop species or cultivar to tolerate prolonged dry periods while maintaining a high plant water status.
(iii) *Drought tolerance at low plant water status,* i.e., the ability of a crop species or cultivar to tolerate prolonged dry periods and tolerate low tissue water contents.

The characteristics of plants that enable them to escape or tolerate drought have been summarized by Turner (130) and are presented in Table 3.3.

Table 3.3. Characteristics of plants that enable them to escape or tolerate drought[a]

Drought Escape		
	(i)	Rapid phenological development
	(ii)	Developmental plasticity
Drought Tolerance at High Plant Water Status		
(a)		Reduction in water loss
	(i)	Increased stomatal and cuticular resistance
	(ii)	Reduced absorption of radiation
	(iii)	Reduced leaf area
(b)		Maintenance of water uptake
	(i)	Increased rooting depth and density
	(ii)	Increased hydraulic conductance
Drought Tolerance at Low Plant Water Status		
(a)		Maintenance of turgor
	(i)	Osmotic adjustment
	(ii)	Increased elasticity
	(iii)	Decreased cell size
(b)		Tolerance of dehydration or desiccation
	(i)	Protoplasmic tolerance
	(ii)	Cell wall properties

[a] Modified from Turner (130).

Drought Escape

The two characteristics that enable crop plants to escape drought and yield satisfactorily are rapid phenological development and developmental plasticity (Table 3.3). In situations in which the probability of drought increases with the life of the crop, the shorter the crop duration the greater the yield will be in most years. Reitz and Salmon (225), working over a 20 year period with from 7 to 14 cultivars of wheat grown at nearly 30 sites in the Great Plains of the United States, showed that each day of earliness imparted a yield advantage of 40–100 kg ha^{-1} over that of a standard late cultivar of winter wheat (226). Similarly, Fischer and Maurer (227) working with 53 cultivars of wheat, barley, and triticale and different dates of withholding irrigation in a rain-free environment in Mexico, showed that each day of earliness conferred a yield advantage of almost 30 kg ha^{-1}. Too great a degree of earliness may of course cause other constraints on yield to become limiting: Reitz and Salmon (225) showed that those cultivars that were more than 12 days earlier at heading than the standard Crimean type of wheat, "Kharkof", yielded poorly in some locations because of frost damage. Furthermore, in determinate crops there is a general relationship between yield and maturity date, with shorter maturity crops having generally lower yields (220). Thus, crops with too rapid a degree of phenological development may suffer in years of plentiful rainfall.

Developmental plasticity helps to overcome this disadvantage. Most crops have some measure of plasticity given favorable conditions: in wheat, for example, the time to reach anthesis and physiological maturity is extended, numbers of tillers producing an ear are greater, and the number of spikelets and florets per ear, together with the size of the grain, are all greater in the absence of water deficits compared with situations in which drought occurs (104, 228, 229). Furthermore, the transfer of stored assimilates to compensate for the downturn in current photosynthesis that has been reported for such species as wheat and barley subjected to water deficits is another aspect of developmental plasticity. Although developmental plasticity can exist in determinate crops, an indeterminate habit should provide the ultimate degree of developmental plasticity, i.e., the ability to flower and set seed as long as water is available and in the absence of other constraints. As pointed out by Turner (130) there has been a move toward selection for greater determinancy in indeterminate crops to aid in mechanical harvesting, but increasing determinancy may not be a desirable characteristic in water-limited environments and it may be preferable to select for other traits such as shattering resistance in order to aid mechanical harvesting.

The introduction of drought escape characteristics, particularly earliness, has undoubtedly brought the greatest advances in breeding for drought resistance to date. Nevertheless, crops in semiarid or even temperate and tropical areas (230, 231), are subjected to random droughts and therefore it would be advantageous to have some drought tolerance characteristics.

Drought Tolerance at High Plant Water Status

Crop plants can tolerate periodic drought and maintain a high plant water status by either preventing water loss or by maintaining the supply of water to the plant (Table 3.3).

To prevent water loss a crop plant must reduce transpiration by stomatal closure, reduce the absorption of radiation, or reduce the evaporating surface of the plant, or any combination of the three. As pointed out earlier, crop species close their stomata as water deficits increase and this reduces transpiration (232) and helps maintain a high plant water status, particularly if stomatal closure is complete and cuticular resistance is high. Succulents show the greatest degree of stomatal control of water loss and can maintain a high plant water status for months (233, 234). Pineapple is one crop that exhibits this form of stomatal control over water loss: the cuticular resistance of pineapple was observed to be in excess of 800 sec cm^{-1} and the midday stomatal resistance of a water stressed plant was frequently in excess of 600 sec cm^{-1} (235). Mesophytic crop plants do not have such a high cuticular resistance (34) or have such a degree of stomatal control, but nevertheless the stomata are important in reducing water loss.

A reduction in the radiation absorbed by leaves can be achieved by active or passive leaf movements, increased pubescence, or increased waxiness. Active movement of leaves parallel to the incoming irradiance has been observed in several crop species in response to water deficits (236). Passive wilting or rolling of leaves also reduces the radiation received by the crop, thereby reducing water loss and increasing water use efficiency to a greater extent than if the leaves remained horizontal or unrolled (237). Water stress increases leaf hairiness and/or waxiness in some crop species (238–240) and both of these adaptations can be expected to reduce water loss by increasing leaf reflectance (241, 242).

A notable effect of water deficits is the reduction in leaf area, either by a reduction in leaf expansion or an increase in leaf senescence or death (40), as described previously. Ritchie (243) and Kowal and Kassam (244) found an approximately linear decrease in crop evaporation as the leaf area index decreased below a value of 3. Therefore, a reduction in leaf area of the crop below this value can reduce crop evaporation, reduce soil water depletion, and slow the rate of development of low leaf water potentials.

An alternative means of tolerating drought at a high plant water status is to maintain the rate of water uptake: the two plant characteristics required for this are a prolific and deep root system and a low resistance to water flow through the plant. The importance of the density and depth of roots on the quantity of water available to the crop has already been emphasized: certainly water deficits increase the root-to-shoot ratio in crops and this can often be ascribed to an absolute increase in root weight (245) and root depth (87, 246).

Drought Tolerance at Low Plant Water Status

Many biochemical, physiological and morphological processes in the plant are sensitive to leaf turgor (188, 190, 247). Thus, maintenance of turgor as leaf water deficits increase should enable the maintenance of physiological function by the plant. Turgor can be partly or fully maintained by osmotic adjustment as leaf water potential decreases, by an increase in elasticity, or by a decrease in cell size (130). There is now considerable evidence that osmotic adjustment, leading to partial or full turgor maintenance, occurs in a range of crop species and cultivars. However, rapid development of stress and small degree of stress reduce the extent of osmotic adjustment (126, 129, 186).

Other things remaining equal, an increase in elasticity will increase the turgor potential at full turgor and at all water contents between full turgor and zero turgor (248). An increase in elasticity has been observed in expanding leaves of bean in response to water deficits (249), but it is unlikely that such increases occur in fully expanded leaves. Indeed elasticity apparently decreased or was constant in fully expanded leaves of sorghum and sunflower in response to water deficits (129, 250). Cell size also influences tissue elasticity; small cells are more elastic (251). Water deficits during leaf expansion generally reduce cell size (239) and may thereby increase elasticity and help maintain turgor.

The characteristics of drought tolerance discussed above have all been concerned with maintaining a high plant water status and maintaining plant function with a decreasing plant water status. These have been called dehydration postponing characteristics by Kramer (224) or drought avoidance characteristics by Levitt (222). We now turn to those characteristics that enable a plant to tolerate drought by tolerating dehydration. There is considerable evidence that crop species and cultivars differ in their ability to survive severe dehydration and recover (252, 253): although not explicitly checked, the test procedures that were used probably assessed dehydration tolerance rather than the postponement of dehydration.

The factors influencing the dehydration tolerance of plant tissues have not been elucidated although there have been many suggestions (130, 135). Most are at the protoplasmic level and suggest that differences in cellular structure, protoplasmic composition and viscosity, or enzyme activity are the basis of dehydration tolerance in plants (254–258). Additionally, the water retention properties of the cell wall have been suggested as important (13, 259).

Tolerance of dehydration is clearly important in survival of plants. However, as pointed out by Turner (130) and Fischer and Turner (260), survival of severe stress is less important in crop species than in natural communities, because crops that are severely stressed usually yield poorly. However, dehydration tolerance and its contribution to crop survival can be important in situations where random droughts occur or where the probability of rainfall increases with the life of the crop. Seedling survival and survival in the vegetative phase are important in obtaining yields under drought (230, 261), and differences in protoplasmic tolerance at both the seedling and later stages are known to exist among cultivars (252, 253).

WATER USE EFFICIENCY

The terms drought resistance and water use efficiency are sometimes used synonymously, although as pointed out by Hsiao and Acevedo (245) they are frequently unrelated. Drought resistance is the ability of a plant to survive and yield satisfactorily under conditions of drought, whereas water use efficiency (W) is simply the efficiency with which water is used to produce dry matter (D), i.e.:

$$W = \frac{D}{T_p} \tag{3.17}$$

where T_p is the transpiration. A number of recent publications have reviewed the concepts of water use efficiency, namely, Begg and Turner (161), Turner and Begg (191), Fischer and Turner (260), Pierre et al. (262), and Stone (263). Readers are referred to these publications for more detailed discussion of the subject.

For an individual leaf in a plant canopy, the uptake of carbon dioxide from the atmosphere in photosynthesis leads to an inevitable loss of water by transpiration because the pathway that permits the entrance of carbon dioxide through the stomata also permits the loss of water vapor. The actual rates of carbon dioxide uptake and water loss will depend on the concentration gradients for carbon dioxide and water vapor between the leaf interior and the atmosphere and the resistances to flow in the two pathways. Thus, from Eq. (3.3), the rate of transpiration (T_p) is given by:

$$T_p = \frac{\Delta e}{r_a + r_l} \tag{3.18}$$

the rate of photosynthesis (P_n) is given by:

$$P_n = \frac{\Delta c}{r_a' + r_l' + r_m'} \tag{3.19}$$

where Δe and Δc are the concentration gradients for water vapor and carbon dioxide, respectively, between the atmosphere and leaf interior, r_a, r_l, are the boundary layer resistances and leaf resistances for the diffusion of water vapor, and r'_a and r'_l are the corresponding resistances for carbon dioxide. r'_m is the mesophyll resistance which includes all the resistances to carbon dioxide diffusion between the intercellular air space and the chloroplast. The resistance $r'_a = 1.6\ r_a$ and $r'_l = 1.6\ r_l$ as a result of the differences in the diffusivity of carbon dioxide and water vapor. Thus, for a single leaf:

$$W = \frac{\Delta c(r_a + r_l)}{\Delta e(r'_a + r'_l + r'_m)} \tag{3.20}$$

This analysis at the single leaf level is useful because it demonstrates the importance of both environmental and plant factors on water use efficiency.

Environmental Factors Affecting Water Use Efficiency

The major environmental factor influencing the water use efficiency of a crop is atmospheric humidity. A lowering of the vapor pressure deficit of the atmosphere around a leaf will proportionately increase the transpiration rate of the leaf. Photosynthesis will not be affected, however, unless the stomata close as a result of a direct effect of humidity on the stomata or an indirect effect through a lowering of the leaf water potential. Thus, unless some compensatory closure of stomata occurs, a decrease in humidity will decrease the water use efficiency of the plant.

The level of carbon dioxide in the atmosphere will also influence the efficiency of water use by plants. By reference back to Eq. (3.20) it will be evident that an increase in the level of carbon dioxide concentration of the air will increase water use efficiency. Although the global concentration of carbon dioxide in the atmosphere is gradually increasing, there is little other possibility of increasing the level of carbon dioxide concentration in the atmosphere in field crops due to the rapid gaseous exchange between the crop and atmosphere (264).

Other environmental factors that influence water use efficiency are air temperature, irradiance, and soil water availability. Air temperature usually operates through its effect on atmospheric humidity (260). For maximum water use efficiency there is an optimum irradiance (265) which is usually less than the irradiance incident on a leaf oriented normal to the sun. Except when the mesophyll resistance is low and the boundary layer resistance is high, stomatal closure will lead to an increase in water use efficiency (265). Soil moisture stress, by closing stomata, might therefore be expected to increase the efficiency of water use. However, the effect of water deficits on the mesophyll resistance must also be considered and the current evidence suggests that, in general, water deficits increase or do not alter water use efficiency (266–268). However, evidence of a decrease in the efficiency of water use with water stress is available particularly with C_4 species (186, 269).

Plant Factors Affecting Water Use Efficiency

Differences in the ratio of stomatal resistance to mesophyll resistance [Eq. (3.20)] between plants is the major plant factor contributing to differences in the efficiency of water use. Plants have been divided into three categories depending on their method of utilizing carbon dioxide, viz., C_3 species, C_4 species, and those that demonstrate crassulacean acid metabolism (CAM). It is now clear that the three types of plants have very different water use efficiencies. C_3 plants have the lowest water use efficiencies because they have high mesophyll resistances and low stomatal resistances compared with C_4 plants which have low mesophyll resistances and tend to have somewhat higher stomatal resistances than C_3 plants (35, 270). The metabolism of CAM plants enables them to open their stomata at night when the vapor pressure gradient between the leaf and atmosphere is smaller than by day, thereby increasing the efficiency of water use (271). Indeed, the water use efficiency of CAM species can be several times greater than crop plants with C_3 metabolism.

Even within these broad categories there are differences in the efficiency of water use. For example, Rawson et al. (39) showed that jojoba (*Simmonsia chinensis* L.) had twice the water use efficiency of wheat even though both are C_3 species, but even so the water use efficiency of jojoba was less than that in several C_4 species studied under the same conditions.

Thus far, we have discussed environmental and plant factors affecting the efficiency of water use in the leaf. However, in field crops, water use efficiency must be integrated over the entire life of the crop. Thus, in agronomic terms water use efficiency (W) is defined as:

$$W = \frac{\text{total dry matter production}}{\text{water use}} \tag{3.21}$$

or:

$$W = \frac{\text{economic yield}}{\text{water use}} \tag{3.22}$$

These equations can preclude portions of the crop not recovered in the measurement of either dry matter production (e.g., roots) or economic yields (e.g., incomplete harvesting). Furthermore, in arriving at water use it is not usual to distinguish between water loss through the plant (i.e., transpiration) and water loss directly from the soil (i.e., evaporation). Thus, water use efficiencies derived from Eq. (3.21) and (3.22) will be lower than those found in leaf studies. Nevertheless, water-use efficiencies measured in the field often show similarities to those measured in the laboratory. Thus, the marked effects of humidity and differences among C_3, C_4, and CAM species in water use efficiency can be detected in the field. For example, comparisons have shown that sorghum a C_4 species, has a greater water use efficiency than soybean, a C_3 species, at the leaf level (25) and also in terms of dry matter accumulation and grain yield per unit of water lost by evapotranspiration (24, 272).

THE ROLE OF WATER IN PLANTS—CONCLUDING REMARKS

In this chapter we have shown first that the role of water is to maintain hydration of plant tissues. Adequate plant hydration is essential for physiological integrity and functioning of crop plants and for crop yield. Water deficits do affect the plant water status, but crops have various characteristics that enable them to escape and tolerate these deficits.

Plants also have an important role in the transfer of water from the soil to the atmosphere by transpiration. It has been pointed out that the water transpired by a crop on a daily basis is usually much more than that required to maintain the plant water status at levels that do not hinder physiological activity. The flow pathway through the plant is discussed in light of our present knowledge of resistances to flow in the root, xylem, and leaf. The plant, by having a variable resistance in the form of stomata in the vapor pathway of water movement, is able to maintain a high plant water status in the face of a decreasing water resource and at the same time transpire rapidly when water in the root zone is freely available. Resistances elsewhere in the pathway, notably in the root, also play a part in maintaining hydration.

From Eq. (3.17) it is clear that dry matter production of a crop is a simple function of the amount of water transpired and the water use efficiency of the crop (260, 273). Thus, where water is limited, the greater the proportion of water that passes through the plant the greater will be the dry matter production; and the greater the water use efficiency, the greater the dry matter production. Factors influencing the water use efficiency have been discussed.

In crop plants it is usually not the total dry matter production that is important, but that proportion of the dry matter that is harvested as economic yield. Thus, any increase in the proportion of harvestable yield to total dry matter, the greater the benefit to the farmer. It has been shown that cereal crops can compensate for lower total dry matter yields under water deficits by increasing the proportion of assimilates in the grain. The utilization of any genetic variability in this character is clearly of importance in increasing the yield of crops subjected to periodic water deficits. Thus, in conclusion, it can be seen that water is important not only in maintaining plant hydration, but also in transpiration, and because of the inevitable link between transpiration and dry matter production, transpiration is important for yield.

LIST OF SYMBOLS

ψ_w	total water potential or water potential
ψ_τ	matric potential
ψ_π	osmotic potential
ψ_p	turgor potential
ψ_w^l	leaf water potential
ψ_w^s	soil water potential
P_n	rate of photosynthesis
P	pressure
P_r	radial pressure
P_c	cross-sectional pressure
F_n	force normal to a probe tip
r_a, r_a'	diffusion resistance to gaseous exchange across the leaf boundary layer
r_l, r_l'	diffusion resistance to gaseous exchange across the leaf epidermis
r_m'	residual or mesophyll resistance to carbon dioxide exchange between the internal cell surfaces and the chloroplasts
r_c	diffusion resistance to gaseous exchange across the leaf cuticle.
r_s	diffusion resistance to gaseous exchange across the stomata
r_S	soil resistance to water movement
r_P	plant resistance to water movement
r_T	total resistance to water absorption by roots
f_T	sum resistance to water transport through the root xylem
T_p	transpiration rate
Q	flow rate
e, e_0	vapor pressure
R	gas constant
T	Kelvin temperature
C_i	water vapor concentration at internal evaporating surfaces of the leaf
C_a	water vapor concentration of bulk air
x	radius

γ	surface tension
h	height
y	contact angle at a liquid–solid interface
g	gravity
z	distance or depth
μ	viscosity coefficient
V	volume
$\bar{V}$	partial molal volume
E_r	cell elongation rate
m	cell extensibility
L	length
θ	angle of soil probe friction
α	semiangle of the conical tip of the probe
RWC	relative water content
WSD	water saturation deficit
D	dry matter production
W	water use efficiency

REFERENCES

1. A. B. Costin, in A. B. Costin and H. J. Frith, Eds., *Conservation* 2nd ed., Penguin, Harmondsworth, 1974, p. 77
2. A. B Hearn, *Outlook Agric.* **10**, 159 (1979).
3. P. G. Jarvis, in D. A. de Vries and N. H. Afgan, Eds., *Heat and Mass Transfer in the Biosphere,* I. *Transfer Processes in Plant Environment,* Scripta, Washington, 1975, p. 369.
4. R. O. Slatyer, *Plant–Water Relationships,* Academic Press, London, 1967.
5. P. J. Kramer, *Plant and Soil Water Relationships: A Modern Synthesis,* McGraw-Hill, New York, 1969.
6. O. L. Lange, L. Kappen, and E.-D. Schulze, Eds., *Water and Plant Life: Problems and Modern Approaches,* Springer-Verlag, Berlin, 1976.
7. T. T. Kozlowski, Ed., *Water Deficits and Plant Growth,* Vols. 1–5, Academic Press, New York, 1968–1978.
8. H. Mussell and R. C. Staples, Eds., *Stress Physiology in Crop Plants,* Wiley-Interscience, New York, 1979.
9. N. C. Turner and P. J. Kramer, Eds., *Adaptation of Plants to Water and High Temperature Stress,* Wiley-Interscience, New York, 1980.
10. P. J. Kramer, in H. J. Bogen and H. Ullrich, Eds., *General Physiology of the Plant Cell* (Encyclopedia of Plant Physiology, Vol. 1), Springer-Verlag, Berlin, 1955, p. 223.
11. J. D. Bernal, in G. E. Fogg, Ed., *The State and Movement of Water in Living Organisms, Symp. Soc. Exp. Biol.,* **19**, Cambridge University Press, Cambridge, 1965, p. 17.

12. P. M. Wiggins, *J. Theor. Biol.* **32,** 131 (1971).
13. D. F. Gaff and D. J. Carr, *Aust. J. Biol. Sci.* **14,** 299 (1961).
14. N. V. Churaev, V. D. Sobolev, and Z. M. Zorin, *Faraday Spec. Discuss. Chem. Soc.* **1,** 213 (1971).
15. D. F. Gaff, T. C. Chambers, and K. Markus, *Aust. J. Biol. Sci.* **17,** 581 (1964).
16. P. S. Nobel, *Introduction to Biophysical Plant Physiology,* Freeman, San Francisco, 1974.
17. R. O. Slayter and S. A. Taylor, *Nature (London)* **187,** 922 (1960).
18. D. C. Spanner, *J. Exp. Bot.* **24,** 816 (1973).
19. H. Meidner and D. W. Sheriff, *Water and Plants,* Blackie, Glasgow, 1976.
20. D. J. Connor, N. J. Legge, and N. C. Turner, *Aust. J. Plant Physiol.* **4,** 753 (1977).
21. J. B. Passioura, *J. Exp. Bot.* **31,** 1161 (1980).
22. J. S. Boyer, *Annu. Rev. Plant Physiol.* **20,** 351 (1969).
23. T. C. Hsiao, E. Fereres, E. Acevedo, and D. W. Henderson, in O. L. Lange, L. Kappen, and E.-D. Schulze, Eds., *Water and Plant Life: Problems and Modern Approaches,* Springer-Verlag, Berlin, 1976, p. 281.
24. G. J. Burch, R. C. G. Smith, and W. K. Mason, *Aust. J. Plant Physiol.* **5,** 169 (1978).
25. H. M. Rawson, N. C. Turner, and J. E. Begg, *Aust. J. Plant Physiol.* **5,** 195 (1978).
26. T. H. van den Honert, *Discuss. Faraday Soc.* **3,** 146 (1948).
27. I. R. Cowan, *J. Appl. Ecol.* **2,** 221 (1965).
28. A. E. Dimond, *Plant Physiol.* **41,** 119 (1966).
29. H. M. Taylor and B. Klepper, *Soil Sci.* **120,** 57 (1975).
30. G. J. Burch, *Aust. J. Agric. Res.* **30,** 279 (1979).
31. J. L. Hailey, E. A. Hiler, W. R. Jordan, and C. H. M. Van Bavel, *Crop Sci.* **13,** 264 (1973).
32. J. S. Boyer, *Planta* **117,** 187 (1974).
33. M. R. Kaufmann, in I. F. Wardlaw and J. B. Passioura, Eds., *Transport and Transfer Processes in Plants,* Academic Press, New York, 1976, p. 313.
34. P. Holmgren, P. G. Jarvis, and M. S. Jarvis, *Physiol. Plant.* **18,** 557 (1965).
35. N. C. Turner, in R. L. Bieleski, A. R. Ferguson, and M. M. Cresswell, Eds., *Mechanisms of Regulation of Plant Growth,* Bull. 12, Royal Society of New Zealand, Wellington, 1974, p. 423.
36. H. Meidner and T. A. Mansfield, *Physiology of Stomata,* McGraw-Hill, New York, 1968.
37. K. Raschke, *Annu. Rev. Plant Physiol.* **26,** 309 (1975).
38. A. E. Hall, E.-D. Schulze, and O. L. Lange, in O. L. Lange, L. Kappen, and E.-D. Schulze, Eds., *Water and Plant Life: Problems and Modern Approaches,* Springer-Verlag, Berlin, 1976, p. 169.
39. H. M. Rawson, J. E. Begg, and R. G. Woodward, *Planta* **134,** 5 (1977).
40. G. J. Burch and G. G. Johns, *Aust. J. Plant Physiol.* **5,** 859 (1978).
41. R. G. Henzell, K. J. McCree, C. H. M. Van Bavel, and K. F. Schertz, *Crop Sci.* **15,** 516 (1975).
42. A. Blum, *Crop Sci.* **14,** 361 (1974).
43. N. C. Turner and J. E. Begg, *Plant Physiol.* **51,** 31 (1973).
44. N. C. Turner, J. E. Begg, H. M. Rawson, S. D. English, and A. B. Hearn, *Aust. J. Plant Physiol.* **5,** 179 (1978).
45. J. E. Begg and N. C. Turner, *Plant Physiol.* **46,** 343 (1970).

46. J. Hellkvist, G. P. Richards, and P. G. Jarvis, *J. Appl. Ecol.* **11,** 637 (1974).
47. O. T. Denmead and B. D. Millar, *Agron. J.* **68,** 297 (1976).
48. J. A. Milburn, *Planta* **69,** 34 (1966).
49. J. A. Milburn, *Planta* **110,** 253 (1973).
50. J. S. Boyer, *Crop Sci.* **11,** 403 (1971).
51. D. T. Clarkson, A. W. Robards, and J. Sanderson, *Planta* **96,** 292 (1971).
52. D. T. Clarkson and J. Sanderson, *Agric. Res. Counc. Letcombe Lab., Annu. Rep.* **20,** 16 (1970).
53. E. I. Newman, in E. W. Carson, Ed., *The Plant Root and Its Environment,* University Press Virginia, Charlottesville, 1974, p. 363.
54. E. I. Newman, *Philos. Trans. R. Soc. London, Ser. B* **273,** 463 (1976).
55. W. R. Gardner, *Agron. J.* **56,** 41 (1964).
56. F. D. Whisler, A. Klute, and R. J. Millington, *Soil Sci. Soc. Am. Proc.* **34,** 382 (1970).
57. H. B. So, L. A. G. Aylmore, and J. P. Quirk, *Plant Soil* **49,** 461 (1978).
58. W. N. Herkelrath, E. E. Miller, and W. R. Gardner, *Soil Sci. Soc. Am. J.* **41,** 1039 (1977).
59. P. B. Tinker, *Philos. Trans. R. Soc. London, Ser. B* **273,** 445 (1976).
60. A. W. Robards and D. T. Clarkson, in B. E. S. Gunning and A. W. Robards, Eds., *Intercellular Communication in Plants: Studies on Plasmodesmata.* Springer-Verlag, Berlin, 1976, p. 181.
61. M. M. Caldwell, in O. L. Lange, L. Kappen, and E.-D. Schulze, Eds., *Water and Plant Life: Problems and Modern Approaches,* Springer-Verlag, Berlin, 1976, p. 63.
62. I. Karas and M. E. McCully, *Protoplasma* **77,** 243 (1973).
63. I. R. Cowan and F. L. Milthorpe, *Nature (London)* **213,** 740 (1967).
64. C. F. Jordan and J. R. Kline, *J. Appl. Ecol.* **14,** 853 (1977).
65. H. Meidner, *J. Exp. Bot.* **26,** 666 (1975).
66. D. W. Sheriff and H. Meidner, *J. Exp. Bot.* **25,** 1147 (1975).
67. M. J. Aston and M. M. Jones, *Planta* **130,** 121 (1976).
68. T. D. Pizzolato, J. L. Burbano, J. D. Berlin, P. R. Morey, and R. W. Pease, *J. Exp. Bot.* **27,** 145 (1976).
69. T. W. Tanton and S. H. Crowdy, *J. Exp. Bot.* **23,** 619 (1972).
70. J. S. Boyer, in D. H. Jennings., *Integration of Activity in the Higher Plant, Symp. Soc. Exp. Biol.* **31,** Cambridge University Press, Cambridge, 1977, p. 455.
71. T. W. Tanton and S. H. Crowdy, *J. Exp. Bot.* **23,** 600 (1972).
72. H. D. Barrs, in R. O. Slatyer, Ed., *Plant Response to Climatic Factors, Proc. Uppsala Symp. 1970,* UNESCO, Paris, 1973, p. 249.
73. R. Brouwer, *Acta Bot. Neerl.* **3,** 264 (1954).
74. R. Tinklin and P. E. Weatherley, *New Phytol.* **65,** 509 (1966).
75. R. Stoker and P. E. Weatherley, *New Phytol.* **70,** 547 (1971).
76. G. K. Hansen, *Acta Agric. Scand.* **24,** 37 (1974).
77. W. W. Emerson, *J. Agric. Sci.* **45,** 241 (1955).
78. P. Ponsana, *Drainage and Water Uptake Terms in the Water Balance,* Ph.D. Thesis, Univ. Adelaide, Australia, 1975.
79. G. F. Byrne, J. E. Begg, and G. K. Hansen, *Agric. Meteorol.* **18,** 21 (1977).
80. E. L. Greacen, in J. S. Russell and E. L. Greacen, Eds., *Soil Factors in Crop Production in a Semi-arid Environment,* University of Queensland Press, St. Lucia, Australia 1977, p. 163.

81. R. Giordano, A. Salleo, S. Salleo, and F. Wanderlingh, *Can. J. Bot.* **56,** 333 (1978).
82. R. J. Millington and D. B. Peters, *Scientia (Milan)* **105,** 26 (1970).
83. R. Van Dooren, *Phys. Fluids* **19,** 481 (1976).
84. J. J. Landsberg and N. D. Fowkes, *Ann. Bot. (London)* **42,** 493 (1978).
85. H. M. Taylor and B. Klepper, *Adv. Agron.* **30,** 99 (1978).
86. N. Higinbotham, R. F. Davis, S. M. Mertz, and L. K. Shumway, in W. P. Anderson, Ed., *Ion Transport in Plants,* Academic Press, London, 1973, p. 493.
87. B. Klepper, H. M. Taylor, M. G. Huck, and E. L. Fiscus, *Agron. J.* **65,** 307 (1973).
88. D. C. Reicosky, R. J. Millington, A. Klute, and D. B. Peters, *Agron. J.* **64,** 292 (1972).
89. S. T. Willatt and H. M. Taylor, *J. Agric. Sci.* **90,** 205 (1978).
90. W. S. Meyer, E. L. Greacen, and A. M. Alston, *J. Exp. Bot.* **29,** 1451 (1978).
91. R. Brouwer, *Proc. K. Ned. Akad. Wet., Ser. C* **56,** 106 (1953).
92. A. W. Robards, S. M. Jackson, D. T. Clarkson, and J. Sanderson, *Protoplasma* **77,** 291 (1973).
93. H. E. Hayward, W. M. Blair, and P. E. Skaling, *Bot. Gaz. (Chicago)* **104,** 152 (1942).
94. C. Hackett, *New Phytol.* **67,** 287 (1968).
95. D. R. Lungley, *Plant Soil* **38,** 145 (1973).
96. D. Tennant, *Aust. J. Agric. Res.* **27,** 183 (1976).
97. R. A. Kohl and J. J. Kolar, *Agron. J.* **68,** 536 (1976).
98. D. R. Cable, *Ecology* **50,** 27 (1969).
99. J. E. Begg, *J. Agric. Sci.* **65,** 341 (1965).
100. C. J. Walter and K. P. Barley, *Proc. Int. Congr. Soil Sci., 10th* p. 150 (1974).
101. E. L. Greacen, and C. T. Hignett, *Aust. C.S.I.R.O. Div. Soils, Tech. Pap.* **27,** 1 (1976).
102. W. C. Mayaki, L. R. Stone, and I. D. Teare, *Agron. J.* **68,** 532 (1976).
103. E. A. Hurd, *Agric. Meteorol.* **14,** 39 (1974).
104. J. F. Angus and M. W. Moncur, *Aust. J. Agric. Res.* **28,** 177 (1977).
105. W. R. Jordan and F. R. Miller, in N. C. Turner and P. J. Kramer, Eds., *Adaptation of Plants to Water and High Temperature Stress,* Wiley-Interscience, New York, 1980, p. 383.
106. J. B. Passioura, *Aust. J. Agric. Res.* **23,** 745 (1972).
107. P. B. Green, *Plant Physiol.* **43,** 1169 (1968).
108. R. Cleland, *Plant Physiol.* **47,** 805 (1971).
109. E. L. Greacen, K. P. Barley, and D. A. Farrell, in W. J. Whittington, Ed., *Root Growth,* Butterworths, London, 1968, p. 256.
110. K. P. Barley and E. L. Greacen, *Adv. Agron.* **19,** 1 (1967).
111. E. L. Greacen and J. S. Oh, *Nature (London) New Biol.* **235,** 24 (1972).
112. B. W. Eavis and D. Payne, in W. J. Whittington, Ed., *Root Growth,* Butterworths, London, 1968, p. 315.
113. D. J. Greenwood, W. J. Whittington, Ed., *Root Growth,* Butterworths, London, 1968, p. 202.
114. R. L. Davidson, *Ann. Bot. (London)* **33,** 561 (1969).
115. R. L. Davidson, *Ann. Bot. (London)* **33,** 571 (1969).
116. B. W. Eavis, *Plant Soil* **37,** 151 (1972).
117. B. C. Warnaars and B. W. Eavis, *Plant Soil* **36,** 623 (1972).
118. R. L. Davidson, *Ann. Bot. (London)* **33,** 579 (1969).
119. M. H. Salim, G. W. Todd, and A. M. Schlehuber, *Agron. J.* **57,** 603 (1965).

120. H. Walter and E. Stadelmann, in G. W. Brown, Jr., Ed., *Desert Biology,* Vol. 2, Academic Press, New York, 1974, p. 213.
121. P. E. Weatherley, *Adv. Bot. Res.* **3,** 171 (1970).
122. W. J. Cram, in U. Lüttge and M. G. Pitman, Eds., *Transport in Plants, II* (Encyclopedia of Plant Physiol., New Series, Vol. 2A), Springer-Verlag, Berlin, 1976, p. 284.
123. R. O. Slatyer, *Aust. J. Biol. Sci.* **14,** 519 (1961).
124. L. Bernstein, *Am. J. Bot.* **48,** 909 (1961).
125. L. Bernstein, *Am. J. Bot.* **50,** 360 (1963).
126. N. C. Turner and M. M. Jones, in N. C. Turner and P. J. Kramer, Eds., *Adaptation of Plants to Water and High Temperature Stress,* Wiley-Interscience, New York, 1980, p. 87.
127. R. F. Meyer and J. S. Boyer, *Planta* **108,** 77 (1972).
128. J. M. Morgan, *Nature (London)* **270,** 234 (1977).
129. M. M. Jones and N. C. Turner, *Plant Physiol.* **61,** 122 (1978).
130. N. C. Turner, in H. Mussell and R. C. Staples, Eds., *Stress Physiology in Crop Plants,* Wiley-Interscience, New York, 1979, p. 343.
131. J. A. Hellebust, *Annu. Rev. Plant Physiol.* **27,** 485 (1976).
132. J. M. Cutler, D. W. Rains, and R. S. Loomis, *Agron. J.* **69,** 773 (1977).
133. N. C. Turner, J. E. Begg, and M. L. Tonnet, *Aust. J. Plant Physiol.* **5,** 597 (1978).
134. R. Munns, C. J. Brady, and E. W. R. Barlow, *Aust. J. Plant Physiol.* **6,** 379 (1979).
135. M. M. Jones, C. B. Osmond, and N. C. Turner, *Aust. J. Plant Physiol.* **7,** 193 (1980).
136. P. B. Green, R. O. Erickson, and J. Buggy, *Plant Physiol.* **47,** 423 (1971).
137. J. S. Boyer, *Plant Physiol.* **43,** 1056 (1968).
138. J. S. Boyer, *Plant Physiol.* **46,** 233 (1970).
139. E. Acevedo, T. C. Hsiao, and D. W. Henderson, *Plant Physiol.* **48,** 631 (1971).
140. W. Wenkert, E. R. Lemon, and T. R. Sinclair, *Agron. J.* **70,** 761 (1978).
141. D. L. McNeil, *Aust. J. Plant Physiol.* **3,** 311 (1976).
142. M. C. Probine and R. D. Preston, *J. Exp. Bot.* **13,** 111 (1962).
143. P. B. Green and W. R. Cummins, *Plant Physiol.* **54,** 863 (1974).
144. T. T. Kozlowski, in T. T. Kozlowski, Ed., *Water Deficits and Plant Growth,* Vol. 3, Academic Press, New York, 1972, p. 1.
145. M. G. Huck, B. Klepper, and H. M. Taylor, *Plant Physiol.* **45,** 529 (1970).
146. B. Slavík, *Methods of Studying Plant Water Relations,* Springer-Verlag, Berlin, and Academia, Prague, 1974.
147. H. D. Barrs, in T. T. Kozlowski, Ed., *Water Deficits and Plant Growth,* Vol. 1, Academic Press, New York, 1968, p. 235.
148. R. W. Brown and B. P. Van Haveren, Eds., *Psychrometry in Water Relations Research,* Utah Agric. Exp. Stn., Logan, 1972.
149. N. C. Turner, *Plant Soil* **58,** 339 (1981).
150. H. D. Barrs and P. E. Weatherley, *Aust. J. Biol. Sci.* **15,** 413 (1962).
151. B. Klepper and H. D. Barrs, *Plant Physiol.* **43,** 1138 (1968).
152. D. C. Spanner, *J. Exp. Bot.* **2,** 145 (1951).
153. L. A. Richards and G. Ogata, *Science* **128,** 1089 (1958).
154. B. D. Millar, *J. Exp. Bot.* **25,** 1070 (1974).
155. E. L. Fiscus, *Plant Physiol.* **50,** 191 (1972).

156. H. H. Wiebe, R. W. Brown, T. W. Daniel, and E. Campbell, *BioScience* **20,** 225 (1970).

157. H. H. Neumann and G. W. Thurtell, in R. W. Brown and B. P. Van Haveren, Eds., *Psychrometry in Water Relations Research,* Utah Agric. Exp. Stn., Logan, 1972, p. 103.

158. G. J. Hoffman and S. L. Rawlins, *Science* **177,** 802 (1972).

159. G. S. Campbell and M. D. Campbell, *Agron. J.* **66,** 24 (1974).

160. O. T. Denmead and B. D. Millar, in D. A. de Vries and N. H. Afgan, Eds., *Heat and Mass Transfer in the Biosphere, I. Transfer Processes in Plant Environment,* Scripta, Washington, 1975, p. 395.

161. J. E. Begg and N. C. Turner, *Adv. Agron.* **28,** 161 (1976).

162. G. A. Ritchie and T. M. Hinckley, *Adv. Ecol. Res.* **9,** 165 (1975).

163. M. T. Tyree, J. Dainty, and D. M. Hunter, *Can. J. Bot.* **52,** 973 (1974).

164. R. H. Waring and B. D. Cleary, *Science* **155,** 1248 (1967).

165. P. W. Gandar and C. B. Tanner, *Am. Potato J.* **53,** 1 (1976).

166. B. D. Millar and G. K. Hansen, *Ann. Bot. (London)* **39,** 915 (1975).

167. N. C. Turner and M. J. Long, *Aust. J. Plant Physiol.* **7,** 527 (1980).

168. R. W. Brown, in R. W. Brown and B. P. Van Haveren, Eds. *Psychrometry in Water Relations Research,* Utah Agric. Exp. Stn., Logan, 1972, p. 198.

169. J. S. Boyer and J. R. Potter, *Plant Physiol.* **51,** 989 (1973).

170. M. T. Tyree, *Can. J. Bot.* **54,** 2738 (1976).

171. J. Grace and G. Russell, *J. Exp. Bot.* **28,** 268 (1977).

172. W. Shepherd, *J. Exp. Bot.* **26,** 465 (1975).

173. P. F. Scholander, H. T. Hammel, E. A. Hemmingsen, and E. D. Bradstreet, *Proc. Natl. Acad. Sci. U.S.A.* **52,** 119 (1964).

174. Y. N. S. Cheung, M. T. Tyree, and J. Dainty, *Can. J. Bot.* **53,** 1342 (1975).

175. N. C. Turner, *J. Exp. Bot.* **27,** 1085 (1976).

176. W. Wenkert, E. R. Lemon, and T. R. Sinclair, *Ann. Bot. (London)* **42,** 295 (1978).

177. J. S. Boyer, *Plant Physiol.* **42,** 213 (1967).

178. W. L. Ehrler, S. B. Idso, R. D. Jackson, and R. J. Reginato, *Agron. J.* **70,** 251 (1978).

179. F. S. Nakayama and W. L. Ehrler, *Plant Physiol.* **39,** 95 (1964).

180. P. G. Jarvis and R. O. Slatyer, *Science* **153,** 78 (1966).

181. L. N. Namken, J. F. Bartholic, and J. R. Runkles, *Agron. J.* **61,** 891 (1969).

182. F. J. Molz and B. Klepper, *Agron. J.* **64,** 469 (1972).

183. R. A. Fischer, M. Sanchez, and J. R. Syme, *Exp. Agric.* **13,** 341 (1977).

184. B. Klepper, V. D. Browning, and H. M. Taylor, *Plant Physiol.* **48,** 683 (1971).

185. C. L. Beadle, N. C. Turner, and P. G. Jarvis, *Physiol. Plant.* **43,** 160 (1978).

186. M. M. Jones and H. M. Rawson, *Physiol. Plant.* **45,** 103 (1979).

187. C. T. Gates, in T. T. Kozlowski, Ed., *Water Deficits and Plant Growth,* Vol. 2, Academic Press, New York, 1968, p. 135.

188. T. C. Hsiao, *Annu. Rev. Plant Physiol.* **24,** 519 (1973).

189. T. C. Hsiao, E. Acevedo, and D. W. Henderson, *Science* **168,** 590 (1970).

190. T. C. Hsiao, E. Acevedo, E. Fereres, and D. W. Henderson, *Philos. Trans. R. Soc. London, Ser. B* **273,** 479 (1976).

191. N. C. Turner and J. E. Begg, in J. R. Wilson, Ed., *Plant Relations in Pastures,* CSIRO, Melbourne, 1978, p. 50.

192. D. C. Reicosky, R. B. Campbell, and C. W. Doty, *Agron. J.* **67,** 380 (1975).
193. W. R. Watts, *J. Exp. Bot.* **25,** 1085 (1974).
194. R. A. Fischer and G. D. Kohn, *Aust. J. Agric. Res.* **17,** 281 (1966).
195. M. M. Ludlow, in R. Marcelle, Ed., *Environmental and Biological Control of Photosynthesis,* Junk, Hague, 1975, p. 123.
196. N. C. Turner, *Plant Physiol.* **53,** 360 (1974).
197. B. Roark and J. E. Quisenberry, *Plant Physiol.* **59,** 354 (1977).
198. A. B. Frank, J. F. Power, and W. O. Willis, *Agron, J.* **65,** 777 (1973).
199. K. J. McCree, *Crop Sci.* **14,** 273 (1974).
200. B. D. Millar and O. T. Denmead, *Agron. J.* **68,** 303 (1976).
201. J. A. Adams, F. T. Bingham, M. R. Kaufmann, G. J. Hoffman, and D. M. Yermanos, *Agron. J.* **70,** 381 (1978).
202. R. O. Slatyer, in R. O. Slatyer, Ed., *Plant Response to Climatic Factors. Proc. Uppsala Symp., 1970,* UNESCO, Paris, 1973, p. 271.
203. J. S. Boyer, *Plant Physiol.* **48,** 532 (1971).
204. C. B. Osmond, K. Winter, and S. B. Powell, in N. C. Turner and P. J. Kramer, *Adaptation of Plants to Water and High Temperature Stress,* Wiley-Interscience, New York, 1980, p. 139.
205. S. B. Powles, and C. B. Osmond, *Aust. J. Plant Physiol.* **5,** 619 (1978).
206. R. J. Fellows and J. S. Boyer, *Protoplasma* **93,** 381 (1978).
207. J. R. McWilliam, *Aust. J. Agric. Res.* **19,** 397 (1968).
208. I. F. Wardlaw, *Aust. J. Biol. Sci.* **20,** 25 (1967).
209. I. F. Wardlaw, *Aust. J. Biol. Sci.* **22,** 1 (1969).
210. I. F. Wardlaw, *Aust. J. Biol. Sci.* **24,** 1047 (1971).
211. R. Munns and C. J. Pearson, *Aust. J. Plant Physiol.* **1,** 529 (1974).
212. J. Moorby, R. Munns, and J. Walcott, *Aust. J. Plant Physiol.* **2,** 323 (1975).
213. E. R. Brevedan and H. F. Hodges, *Plant Physiol.* **52,** 436 (1973).
214. G. N. Thorne, in F. L. Milthorpe and J. D. Ivins, Eds., *The Growth of Cereals and Grasses,* Butterworths, London, 1966, p. 88.
215. F. Bidinger, R. B. Musgrave, and R. A. Fischer, *Nature (London)* **270,** 431 (1977).
216. J. N. Gallagher, P. V. Biscoe, and R. K. Scott, *J. Appl. Ecol.* **12,** 319 (1975).
217. H. M. Rawson, A. K. Bagga, and P. M. Bremner, *Aust. J. Plant Physiol.* **4,** 389 (1977).
218. G. A. Constable and A. B. Hearn, *Aust. J. Plant Physiol.* **5,** 159 (1978).
219. J. Levitt, C. Y. Sullivan, and E. Krull, *Bull. Res. Counc. Isr., Sect. D* **8,** 173 (1960).
220. L. H. May and F. L. Milthorpe, *Field Crop Abstr.* **15,** 171 (1962).
221. J. Parker, in T. T. Kozlowski, Ed., *Water Deficits and Plant Growth,* Vol. 1, Academic Press, New York, 1968, p. 195.
222. J. Levitt, *Responses of Plants to Environmental Stresses,* Academic Press, New York, 1972.
223. I. Arnon, in U. S. Gupta, Ed., *Physiological Aspects of Dryland Farming,* Oxford and IBH, New Delhi, 1975, p. 3.
224. P. J. Kramer, in N. C. Turner and P. J. Kramer, Eds., *Adaptation of Plants to Water and High Temperature Stress,* Wiley-Interscience, New York, 1980, p. 7.
225. L. P. Reitz and S. C. Salmon, *U.S. Dep. Agric. Tech. Bull.* **1192,** 1959, p. 1.
226. L. P. Reitz, *Agric. Meteorol.* **14,** 3 (1974).
227. R. A. Fischer and R. Maurer, *Aust. J. Agric. Res.* **29,** 897 (1978).

228. R. A. Fischer, in R. O. Slatyer, Ed., *Plant Response to Climatic Factors. Proc. Uppsala Symp., 1970,* UNESCO, Paris, 1973, p. 233.
229. R. O. Slatyer, in R. O. Slatyer, Ed., *Plant Response to Climatic Factors. Proc. Uppsala Symp., 1970,* UNESCO, Paris, 1973, p. 177.
230. J. C. O'Toole and T. T. Chang, in H. Mussell and R. C. Staples, Eds., *Stress Physiology in Crop Plants,* Wiley-Interscience, New York, 1979, p. 373.
231. T. Tazaki, K. Ishihara, and T. Ushijima, in N. C. Turner and P. J. Kramer, Eds., *Adaptation of Plants to Water and High Temperature Stress,* Wiley-Interscience, New York, 1980, p. 309.
232. N. C. Turner, *Plant Physiol.* **55,** 932 (1975).
233. S. R. Szarek and I. P. Ting, *Plant Physiol.* **54,** 76 (1974).
234. P. S. Nobel, *Oecologia* **27,** 117 (1977).
235. D. P. Bartholomew and S. B. Kadzimin, in P. de T. Alvim and T. T. Kozlowski, Eds., *Ecophysiology of Tropical Crops,* Academic Press, New York, 1977, p. 113.
236. J. E. Begg, in N. C. Turner and P. J. Kramer, Eds., *Adaptation of Plants to Water and High Temperature Stress,* Wiley-Interscience, New York, 1980, p. 33.
237. H. M. Rawson, *Aust. J. Plant Physiol.* **6,** 109 (1979).
238. J. D. Skoss, *Bot. Gaz. (Chicago)* **117,** 55 (1955).
239. S. A. Quarrie and H. G. Jones, *J. Exp. Bot.* **28,** 192 (1977).
240. J. D. Weete, G. L. Leek, C. M. Peterson, H. E. Currie, and W. D. Branch, *Plant Physiol.* **62,** 675 (1978).
241. N. J. Chatterton, W. W. Hanna, J. B. Powell, and D. R. Lee, *Can. J. Plant Sci.* **55,** 641 (1975).
242. J. Ehleringer, in N. C. Turner and P. J. Kramer, Eds., *Adaptation of Plants to Water and High Temperature Stress,* Wiley-Interscience, New York, 1980, p. 295.
243. J. T. Ritchie, *Agric. Meteorol.* **14,** 183 (1974).
244. J. M. Kowal and A. H. Kassam, *Agricultural Ecology of Savanna: A Study of West Africa,* Clarendon Press, Oxford, 1978.
245. T. C. Hsiao and E. Acevedo, *Agric. Meteorol.* **14,** 59 (1974).
246. R. S. Malik, J. S. Dhankar, and N. C. Turner, *Plant Soil* **53,** 109 (1979).
247. U. Zimmermann, *Annu. Rev. Plant Physiol.* **29,** 121 (1978).
248. M. M. Jones, *Physiological Responses of Sorghum and Sunflower to Leaf Water Deficits,* Ph.D. Thesis, Australian National University, Canberra, Australia, 1979.
249. J. Elston, A. J. Karamanos, A. H. Kassam, and R. M. Wadsworth, *Philos. Trans. R. Soc. London, Ser. B* **273,** 581 (1976).
250. M. M. Jones and N. C. Turner, *Aust. J. Plant Physiol.* **7,** 181 (1980).
251. E. Steudle, U. Zimmermann, and U. Lüttge, *Plant Physiol.* **59,** 285 (1977).
252. C. Y. Sullivan and J. D. Eastin, *Agric. Meteorol.* **14,** 113 (1974).
253. J. C. O'Toole, R. S. Aquina, and K. Alluri, *Agron. J.* **70,** 1101 (1978).
254. W. S. Iljin, *Annu. Rev. Plant Physiol.* **8,** 257 (1957).
255. O. Stocker, in *Plant Water Relationships in Arid and Semi-Arid Conditions: Review of Research,* UNESCO, Paris, p. 63 (1960).
256. O. Y. Lee-Stadelmann and E. J. Stadelmann, in O. L. Lange, L. Kappen, and E.-D. Shulze, Eds., *Water and Plant Life: Problems and Modern Approaches,* Springer-Verlag, Berlin, 1976, p. 268.
257. K. L. Giles, D. Cohen, and M. F. Beardsell, *Plant Physiol.* **57,** 11 (1976).

258. J. Vieira da Silva, in O. L. Lange, L. Kappen, and E.-D. Schulze, Eds., *Water and Plant Life: Problems and Modern Approaches*, Springer-Verlag, Berlin, 1976, p. 207.
259. T. S. Teoh, L. A. G. Aylmore, and J. P. Quirk, *Aust. J. Biol. Sci.* **20,** 41 (1967).
260. R. A. Fischer and N. C. Turner, *Annu. Rev. Plant Physiol.* **29,** 277 (1978).
261. D. A. Johnson, in N. C. Turner and P. J. Kramer, Eds., *Adaptation of Plants to Water and High Temperature Stress*, Wiley-Interscience, New York, 1980, p. 419.
262. W. H. Pierre, D. Kirkham, J. Pesek, and R. Shaw, Eds., *Plant Environment and Efficient Water Use*, Am. Soc. Agron. Soil Sci. Soc. Am., Madison, 1966.
263. J. F. Stone, Ed., *Plant Modification for More Efficient Water Use*, Elsevier, Amsterdam, 1975.
264. L. H. Allen, Jr., R. L. Desjardins, and E. R. Lemon, *Agron. J.* **66,** 609 (1974).
265. H. G. Jones, *J. Appl. Ecol.* **13,** 605 (1976).
266. R. O. Slatyer, in R. Jones, Ed., *The Biology of Atriplex*, CSIRO, Canberra, 1970, p. 23.
267. R. R. Johnson, N. M. Frey, and D. N. Moss, *Crop. Sci.* **14,** 728 (1974).
268. D. A. Johnson and M. M. Caldwell, *Oceologia* **21,** 93 (1975).
269. T. R. Sinclair, G. E. Bingham, E. R. Lemon, L. H. Allen, Jr., *Plant Physiol.* **56,** 245 (1975).
270. R. W. Downes, *Aust. J. Biol. Sci.* **23,** 775 (1970).
271. T. F. Neales, A. A. Paterson, and V. J. Hartney, *Nature (London)* **219,** 469 (1968).
272. I. D. Teare, E. T. Kanemasu, W. L. Powers, and H. S. Jacobs, *Agron. J.* **65,** 207 (1973).
273. J. B. Passioura, *J. Aust. Inst. Agric. Sci.* **43,** 117 (1977).
274. R. E. Andrews and E. I. Newman, *New Phytol.* **68,** 1051 (1969).
275. D. W. Lawlor, *New Phytol.* **72,** 297 (1973).
276. E. I. Newman, *New Phytol.* **72,** 547 (1973).
277. G. K. Hansen, *Acta Agric. Scand.* **24,** 83 (1974).
278. E. F. Cox, *Resistance to Flow of Water Through the Plant*, Ph.D. Thesis, Univ. Nottingham, England, 1966.
279. H. E. Hayward and W. B. Spurr, *Bot. Gaz. (Chicago)* **106,** 131 (1944).
280. C. R. House and N. Findlay, *J. Exp. Bot.* **17,** 627 (1966).
281. H. Ginsburg and B. Z. Ginsburg, *J. Exp. Bot.* **21,** 593 (1970).
282. H. B. So, L. A. G. Aylmore, and J. P. Quirk, *Soil Sci. Soc. Am. J.* **40,** 222 (1976).
283. R. Brouwer, in G. E. Fegg, Ed., *The State and Movement of Water in Living Organisms, Symp. Soc. Exp. Biol.* **19,** Cambridge University Press, Cambridge, 1965, p. 131.
284. A. C. P. Chu and J. P. Kerr, *N. Z. J. Agric. Res.* **20,** 467 (1977).

4

ALFALFA

Gary H. Heichel

INTRODUCTION

Alfalfa (*Medicago sativa* L.) is a long-lived perennial forage legume that evolved in the Near East and Central Asia. It was the first forage domesticated and has been grown for more than 3300 years. Alfalfa originated in a climate characterized by hot, dry summers and cold winters, and these climatic features are often typical of areas where alfalfa presently grows.

Crop Use and Yield Potential

The principal uses of alfalfa are as hay, silage, and pasture for ruminant animals, and as a source of protein and vitamin A for nonruminants like swine and poultry. After harvest, the crop may be preserved in the form of dry hay in bales of different sizes, chopped hay, dehydrated pellets, or compressed cubes. The crop is preserved as haylage or silage at higher moisture content. Currently, alfalfa sprouts sometimes garnish salads and sandwiches. However, alfalfa has potential as a source of purified plant protein for human diets.

Alfalfa is grown in monoculture, in rotations with grain crops, and in mixtures with various species of forage grasses. Approximately 10.9 million ha of alfalfa for hay and 19 million ha of alfalfa in mixture with other forage species were grown in the United States in 1976 (1). The value of alfalfa is difficult to estimate, because only 21% of the 1976 crop entered commerce; the remainder was used directly for animal feed. Using a recent market price of $66 per metric ton, the 64 million metric tons of alfalfa and alfalfa-based hay produced in 1976 would be valued at $4.2 billion before being marketed through animals.

The leading north central and great plains states in 1976 alfalfa hectarage were Wisconsin (1.22 million), South Dakota (0.93 million), Minnesota (0.89 million), Iowa (0.71 million), Nebraska (0.69 million), and North Dakota (0.66 million). The average yield of dryland alfalfa in these six states was 4.5 metric tons per hectare (1). In the eastern states where rainfall is more plentiful, yields in New York (0.40 million ha), Pennsylvania (0.33 million ha), and Ohio (0.22 million ha) averaged 6.3 metric tons per hectare. The irrigated nonhardy alfalfas in California (0.45 million ha) and Arizona (0.09 million ha) averaged 14.5 metric tons per hectare in 1976 (1). Nationwide, 2.0 million ha, or 22% of the alfalfa grown in 1969, were irrigated, the largest hectarage of any crop (2).

Adaptation

In North America, alfalfa is most productive between 35° and 45° north latitude, but it also prospers under irrigation in limited arid and semiarid regions. Elsewhere in the northern hemisphere, alfalfa is concentrated between 40° and 50° north latitude in Europe and in the Near East. In the southern hemisphere, alfalfa is adapted to central Argentina between 25° and 40° south latitude, to the southern tip of Africa between 25° and 35° south latitude, to the southeastern quadrant of Australia, and New Zealand. Although these are the major geographic regions of alfalfa production, the forage can be found wherever man has migrated.

Alfalfa is a genetically heterogeneous, outcrossing tetraploid with an enormous diversity of agronomic characteristics. The genetic diversity of alfalfa can be verified by scanning the range of inflorescence colors in a flowering field. Although alfalfa is often thought to have predominantly purple flowers, current cultivars adapted to temperate regions display many flower colors. The range of flower colors is evidence that cultivars contain germplasm from both purple-flowered *M. sativa* and yellow-flowered *M. falcata* sources. *M. falcata* has contributed winterhardiness, drought resistance, disease resistance, and creeping roots or rhizomes to present-day cultivars (3).

The New England colonists introduced alfalfa into the colonies about 1736, but the crop only prospered in limited areas of New York with calcareous soils. Spanish sources of alfalfa entered the southwestern United States about 1836 and were important in California in the 1850's. These strains of alfalfa lacked winterhardiness and their northward migration was limited to Iowa, Missouri, and Ohio until hardy selections descended from the population introduced into Minnesota from Germany in 1857 by Wendelin Grimm and introductions from Russia in 1894 to 1909 by N. E. Hanson permitted the expansion of the forage throughout the northern United States and southern Canada.

Availability of water affects the productivity of alfalfas that are adapted to the summer and winter temperature regimes of various production areas in the United States. Three rainfall regions important to alfalfa production are recognized. The humid climatic region, where rainfall is sufficient for normal growth, includes all the United States east of the 41 to 61 cm rainfall line which runs from northwestern Minnesota to south-central Texas. The semiarid climatic region, where rainfall restricts production to the early part of the growing season, ranges westward from the humid boundary to central Washington and Oregon, and California. The southern boundary of the semiarid region is set by the deserts of Nevada, Utah, Arizona, New Mexico, and southwestern Texas. These desert areas comprise the arid climatic region, where alfalfa succeeds only under irrigation.

Growth Habit

Adaptation to temperature constrains survival and productivity within each of the major rainfall areas. Non-winter-hardy cultivars have generally been derived from alfalfas originating in Peru, North Africa, and India. This group shares characteristics that include lack of cold tolerance with little or no winter dormancy, quick recovery after cutting, erect stems, and purple flowers.

The intermediate winter-hardy alfalfas can endure lower temperatures than those of the nonhardy group, but they lack the cold tolerance of winter-hardy cultivars. Cultivars that are intermediate in winterhardiness are characterized by fall and winter dormancy, tolerance of low temperatures, and purple flowers.

Winter-hardy alfalfas are characterized by an early fall and relatively long winter dormancy, and tolerance to very low temperatures. These alfalfas were developed from crosses between yellow- and purple-flowered species, as noted previously, and the resultant intercrossing has yielded purple, blue, green, yellow, bronze, and white flowers. In contrast to the erect-stemmed nonhardy cultivars, intermediate and winter-hardy cultivars exhibit progressively slower rates of regrowth after harvest and a more decumbent growth habit with the onset of short days and cool temperatures. Although there are gradations in temperature adaptation between regions, non-winter-hardy alfalfas are generally

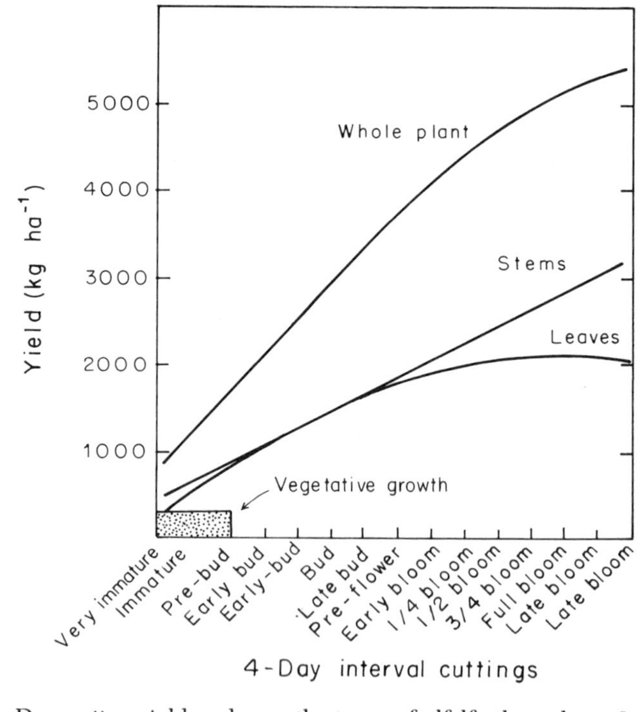

Figure 4.1. Dry matter yield and growth stages of alfalfa through an 8 week growing period. Whole plant = sum of leaves and stems. [Adapted from Kilchner and Hendrichs (4).]

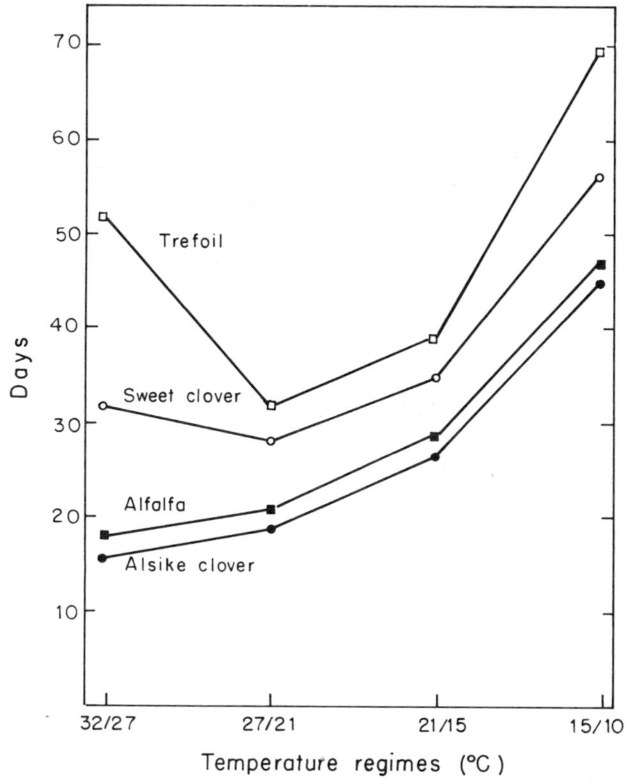

Figure 4.2. Days of growth of four forage legumes to early bloom stage in response to day and night temperatures during an 18 hr photoperiod. [Adapted from Smith (7).]

grown south of 34° north latitude, intermediate winter-hardy alfalfas between 34° and 40° north latitude, and winter-hardy alfalfas between 40° and 53° north latitude.

Established stands of alfalfa harvested for forage proceed through one to eight cycles of vegetative and early reproductive growth annually. In some years only one growth cycle may occur in the semiarid Great Plains where rainfall limits productivity. Three to four growth cycles occur throughout the Corn Belt states, with the first terminating in late May to early June and the last in mid-September to mid-October. In arid regions where the crop is grown under irrigation, as many as eight harvest–regrowth cycles can occur annually at 22–35 day intervals between April and November.

After growth commences in the spring from overwintering crown buds formed the preceding season, dry matter accumulation occurs linearly through the growth cycle until flowering, at which time vegetative growth starts to diminish (Fig. 4.1). Kilchner and Hendrichs (4) identified 13 stages of growth at 4–8 day intervals from very immature vegetative growth to late bloom. In current practice, only vegetative, early bud, bud, and early bloom are widely used since harvests are scheduled at 1–25% bloom, depending on the environment, for maximum production and greatest plant survival.

The timing of successive harvest cycles is keyed to the onset of flowering. Because alfalfa is a long-day plant, the timing of developmental events within a growth cycle and the duration of successive growth cycles are partly mediated by photoperiod. Lengthening photoperiods between April and late June lead to increasingly rapid development within a growth cycle, and progressively briefer duration of growth. Shortening photoperiods from July through October progressively increase the interval between early bloom and harvest (5, 6). Warm temperatures during long photoperiods and cool temperatures during short photoperiods complicate the apparent photoperiodic timing of developmental stages. In 18 hr days, alfalfa development and flowering after cutting are fastest at 32/27 and 27/21°C day/night temperatures, and slow dramatically in 15/10°C temperature regimes (Fig. 4.2) (7).

ROOT GROWTH AND FUNCTION

Germination and Emergence

Alfalfa grown for forage is seeded about 1.3 cm deep at 12–15 kg ha^{-1}. The seed imbibes all of the water needed for germination within 4–8 hr. Germination is practically inhibited at −12 to −15 bars osmotic potential (8), but above −9 bars extent of germination is little different from controls. In young alfalfa seedlings, shoot growth commences more rapidly than root growth, which is confined chiefly to the taproot.

Invasion of root hairs by the symbiotic bacterium, *Rhizobium meliloti*, results in the formation of oblong, finger-shaped structures called nodules 2 mm in diameter and 3–10 mm long on tap and lateral roots of seedlings. The structures contain specialized forms of the bacterium capable of converting gaseous nitrogen from the atmosphere into organic forms of nitrogen for plant metabolism. The onset of nodulation in alfalfa seedlings is strongly influenced by availability of nitrogen in the rhizosphere (9). Alfalfa sown in soils initially containing 80 kg ha^{-1} of nitrogen from mineralization of organic matter may not nodulate until late summer. Nodulation usually occurs within 3 weeks in soils containing less than 30 kg ha^{-1} available nitrogen.

Alfalfa is frequently established by sowing with a small grain companion crop, and initial seedling root growth is strongly dependent on the degree of shading by the grain (10, 11). Fifteen days after emergence, seedling shoot–root ratios range from 2.0 to 4.0 (12). When soil water is plentiful, the ratio declines to 1.7 within 8 weeks of emergence and

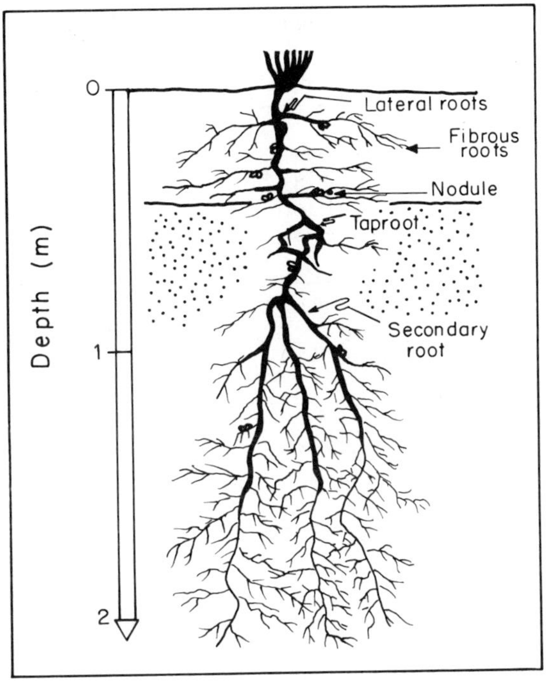

Figure 4.3. Root system of an excavated alfalfa plant. Root growth was restricted and distorted in the moderately cemented subsoil horizon, but was more extensive and deeper in the soil profile. [Adapted from Kohl and Kolar (13).]

varies between 0.8 and 1.0 for the remainder of the seeding year (11). Seedling root growth is diminished more than shoot growth by soil water stress, with stressed seedlings exhibiting a shoot–root ratio of 2.3 after 45 days of treatment compared with 1.8 for the controls (10).

Depth, Penetration, and Distribution

Rooting depth of alfalfa increases with age of stand. The root system is strongly dominated by a taproot that may branch into several secondary roots deep in the profile (13) (Fig. 4.3). Large differences among cultivars in the number of secondary roots and in the frequency of taproot branching have been observed (14). Fibrous roots proliferate in the upper 20 cm of the profile and bear the largest number of nodules. The quantity of fibrous roots and the effectiveness of the nitrogen-fixing root nodules are under genetic control (15, 16).

Taproots follow tortuous cracks and faults in heavily compacted subsoils (17), but resume normal or more finely branched growth at greater depths (13). When alfalfa roots encountered an impenetrable layer, lateral extension to 5 m or more has been observed (18). Thus, alfalfa roots

Table 4.1. Distribution of root mass by depth for two alfalfa cultivars differing in winterhardiness growing in two soil water regimes for 4 years[a]

Water regime[b] (cm deep)	Root mass (kg ha^{-1})			
	Moderately hardy 'Atlantic'		Nonhardy 'African'	
	Dry	Wet	Dry	Wet
0–8	2852	2395	1746	2818
8–15	1936	1139	752	1986
15–23	1145	1435	576	533
23–30	162	178	251	382
30–46	244	228	139	402
46–61	109	56	45	423
61–91	209	106	93	143
91–122	122	45	105	137
Total	6779	5582	3707	6824

[a] Adapted from Bennett and Doss (32).
[b] Dry—irrigation when 80% of the available soil water had been exhausted at 91 cm depth. Wet—irrigation when 30% of the available soil water had been exhausted at 91 cm depth.

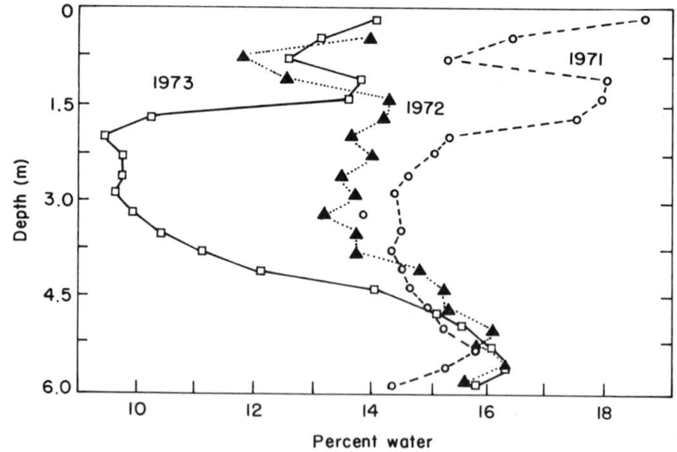

Figure 4.4. Water content initially and after 1 and 2 years of alfalfa growth after 45 metric tons ha^{-1} of manure was applied in 1968, 1969, and 1970. [Adapted from Mathers et al. (22).]

grow wherever water is available and penetrate deeply when the soil profile is free of obstructions and adequately supplied with nutrients (19).

Root penetration to 1.5 m by the second year of growth is common (14, 20) and rooting depths beyond 8 m in 4-year-old stands have been observed in Kansas (21). Root mass decreases logarithmically with depth, and 60–70% of the total root mass may be in the upper 15 cm of the soil profile (Table 4.1). The zone of soil water extraction is approximately that of root distribution. Soils on which alfalfa is grown become successively depleted of water to greater depths with age of stand (Fig. 4.4). When alfalfa in Kansas was grown for 4 years following 12 years of cereal crops, subsoil water reserves were exhausted to a depth of 5.5 m (21). Replenishment of subsoil water after alfalfa occurs slowly and is an important management consideration when long-term stands of alfalfa in semiarid regions are contemplated. The pattern of soil water extraction by alfalfa is dependent on length of growing season, rooting depth, and crop yield.

Seasonal Patterns of Root Growth

The developmental patterns of alfalfa roots in the field are poorly understood because of difficulties in sampling the entire root system. Rates of root proliferation over long periods have been inferred from the pat-

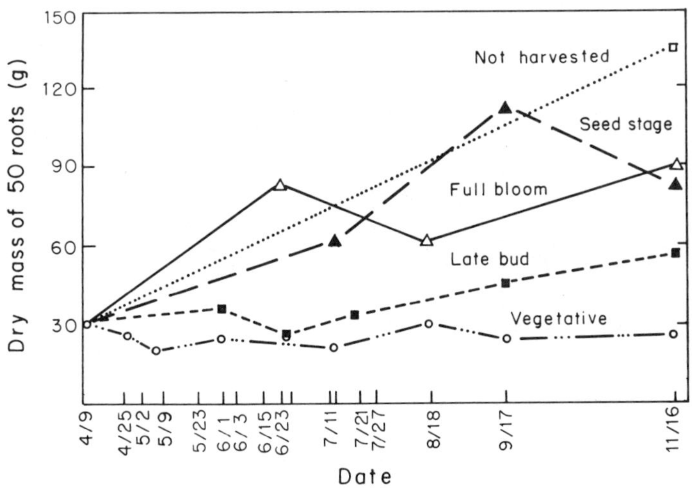

Figure 4.5. Effect of herbage harvest on alfalfa root growth in 1925 as measured by dry weights of 50 representative roots cut to a length of 7 in. [Adapted from Graber et al. (23).]

terns of soil water use described above, but root exploration of soil during the growing season remains uninvestigated.

Seasonal patterns of root dry matter accumulation are strongly modified by management practices such as frequency of grazing or harvesting. Alfalfa that remains uncut accumulates root dry mass steadily throughout the growing season (Fig. 4.5). Harvesting at the full-bloom stage in June and August, or in the seed stage in July and September, decreased root mass in November below that observed in unharvested plants. The usual management of alfalfa is periodic harvesting at late bud or early bloom stage (Fig. 4.1), a management practice that results in little net root mass increase throughout the growing season (23, 24).

The suppression of root growth by harvesting is also mirrored by nodular metabolism. Removal of 80% of the herbage on alfalfa seedlings reduced nitrogen fixation capacity by 90% within 24 hr (25). The reduction of nitrogen fixation by harvest varies with plant age and extent of defoliation. Established plants undergoing an initial harvest showed little reduction in nitrogen fixation when 75% of the leaf area was removed (Fig. 4.6). The second harvest, in which 85% of the leaf area was removed, reduced nitrogen fixation 46% within 2 days. Recovery of nodular function occurred concurrently with leaf area renewal, an observation that suggests the importance of photosynthetic capacity to symbiotic nitrogen fixation (26). Although nodules undergo temporary deterioration following herbage removal, they have the adaptive capacity

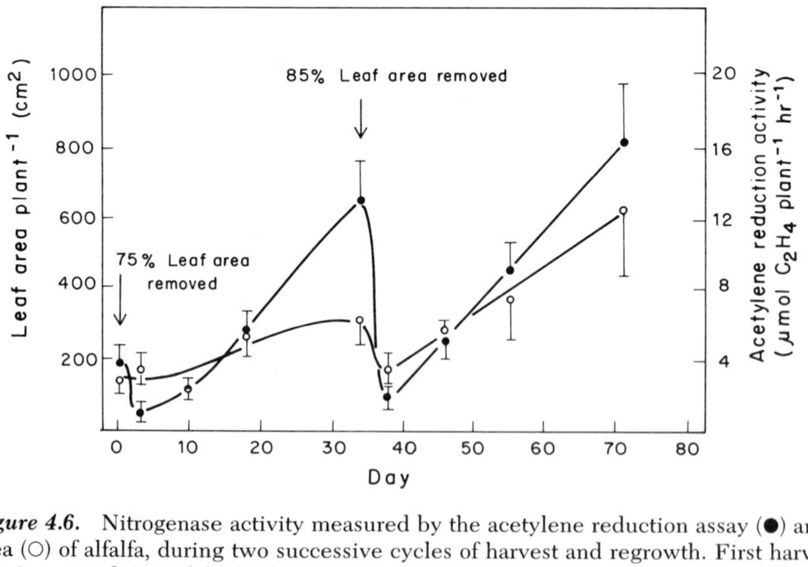

Figure 4.6. Nitrogenase activity measured by the acetylene reduction assay (●) and leaf area (○) of alfalfa, during two successive cycles of harvest and regrowth. First harvest on day 0 removed 75% of the leaf area, and the second harvest on day 34 removed 85% of the leaf area. [Adapted from Cralle (26).]

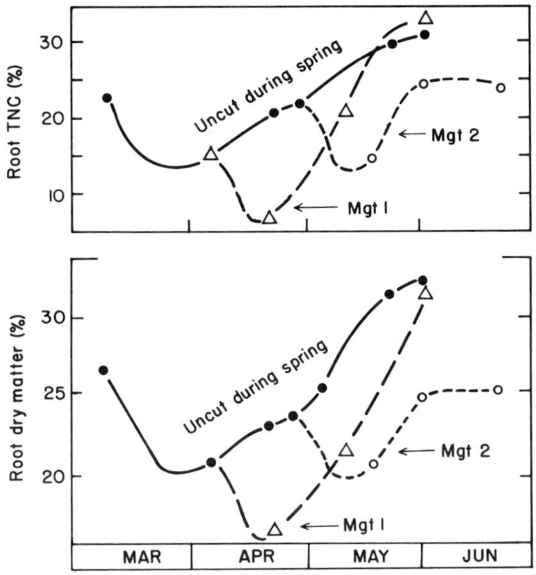

Figure 4.7. Response of total nonstructural carbohydrates (TNC) and dry matter in tap roots of alfalfa during initial spring growth to defoliation in early vegetative development (Mgt 1) and early bud development (Mgt 2). [Adapted from Wolf (27).]

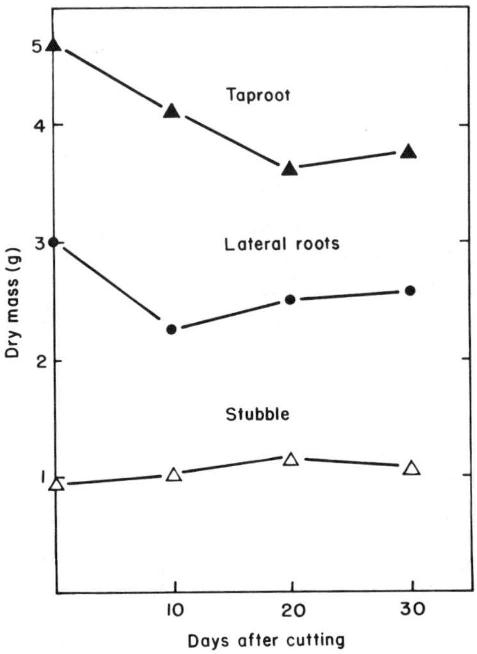

Figure 4.8. Changes in the dry weight of the taproot, lateral roots, and stubble (crown) during regrowth of alfalfa after complete herbage removal. [Adapted from Hodgkinson (28).]

to regain function, and there is little evidence of net loss of nodular mass after harvest (25).

It is well established that the regrowth of alfalfa herbage after harvest depends on remobilization of nonstructural carbohydrates deposited in roots before harvest. Carbohydrate and dry matter accumulation in roots directly reflects the severity and timing of defoliation, with early defoliation being more detrimental to root growth and carbohydrate storage than later harvest (27) (Fig. 4.7). Following alfalfa harvest, new growth of stems and leaves commences within 10 days and continues until the next harvest. Remobilization of carbohydrates leads to a net decline in mass of tap and lateral roots during this time, with the taproots being more strongly affected (28) (Fig. 4.8). The crown tissues of alfalfa show no great changes in mass during the regrowth cycle.

Water Relationships and Root Growth

Alfalfa is notorious for its susceptibility to excess soil water and to anoxia and impaired mineral absorption accompanying excessive soil water. Many root pathogens, especially *Phytophthora megasperma*, proliferate in wet soils. Although *Phytophthora* infestion may cease when the soil dries, the shallow lateral roots that develop from rotted taproots are often inadequate to restore plant vigor (29). Plants weakened by harvesting or from crown damage from vehicular traffic are especially susceptible to a complex of crown rots. Such rots can also influence water uptake and transport.

The response of alfalfa root growth to inadequate water supply is virtually unknown. Existing knowledge of root growth and water relationships comes from older irrigation studies where neither soil nor plant water potentials were measured. Increasing soil water stress decreased the growth of 'Ranger' alfalfa seedling roots more than that of seedling tops in comparison with adequately watered controls (10). Water stress also reduced seedling nodulation (30). Nodule nitrogenase activity was reduced 85% compared with controls at a leaf water potential of -15 to -20 bars, and cultivar differences in nodule recovery from water deficits were apparent (31).

Root growth of alfalfa apparently differs with cultivar winterhardiness characteristics and soil water regime (32) (Table 4.1). For the moderately hardy cultivar 'Atlantic,' the percentage of roots in the respective soil depth increments was not significantly altered by frequency of irrigation, but a greater total root mass was produced on dry plots receiving less-frequent irrigations. In the nonhardy 'African' cultivar, a greater proportion of the total root mass was in the upper 15 cm of the profile in wet plots than in dry plots, but the greatest root mass was produced in the wet plots. Thus, the winterhardiness characteristics of alfalfa clearly modify patterns of root growth, especially when soil water supply is a variable. Further investigations in this long-neglected area are needed to learn how soil and root water relationships limit root growth, distribution, penetration, and nutrient uptake. This is especially important in efforts to understand how symbiotic nitrogen fixation responds to water deficits, and how root infection by *Rhizobium meliloti*, nodule formation, and nodule persistence is modulated by water stress.

CANOPY GROWTH

Ontogeny and Dry Matter Accumulation

Herbage and Crown

Dry matter accumulation in the alfalfa canopy occurs in characteristic and predictable patterns during successive harvest intervals (Fig. 4.9). Dry matter in crowns, the tissue between the cotyledonary node and the normal cutting height of harvest, remains relatively invariant throughout successive harvests. Within each harvest interval, top dry matter increases to a maximum and then declines slightly as the crop begins to flower because of leaf fall in the lower canopy (Fig. 4.9). Herbage growth is less in each successive harvest interval throughout the northern United States, a response that is attributable to availability of soil water. In the arid Southwest, nonhardy alfalfas may yield less during the first harvest (mid-April to mid-May) than the second (mid-May to mid-June), and

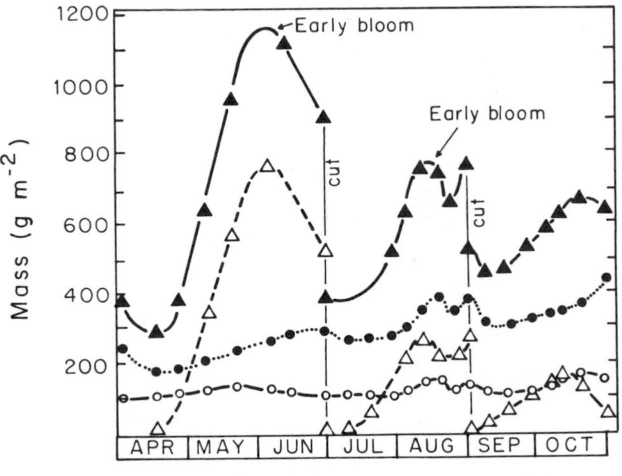

Figure 4.9. Dry matter distribution of alfalfa roots (upper 15 cm) (●) crowns (○), herbage (△), and whole plants (▲) for the 1964 growing season. [Adapted from Nelson and Smith (24).]

then successive harvests are progressively less (5). If harvest is delayed sufficiently, as in the second interval (Fig. 4.9), an upswing in herbage growth occurs because of release of crown and axillary bud dormancy after flowering.

Leaves and Stems

Within each harvest interval, dry matter distribution in leaves and stems occurs in a characteristic pattern (33) (Fig. 4.10). Stem dry matter is distributed in a pyramidal fashion from the soil surface upward. The thicker, coarser stems are lowest on the plant, so that 40–60% of the stem dry matter occurs within 20 cm of the soil surface. Stems become thicker and coarser as the season progresses, and a greater proportion of the stem mass occurs low in the canopy during later compared with earlier harvest periods (Fig. 4.10). Elsewhere in the Midwest later harvests often result in leafier and smaller-stemmed forage than early harvests.

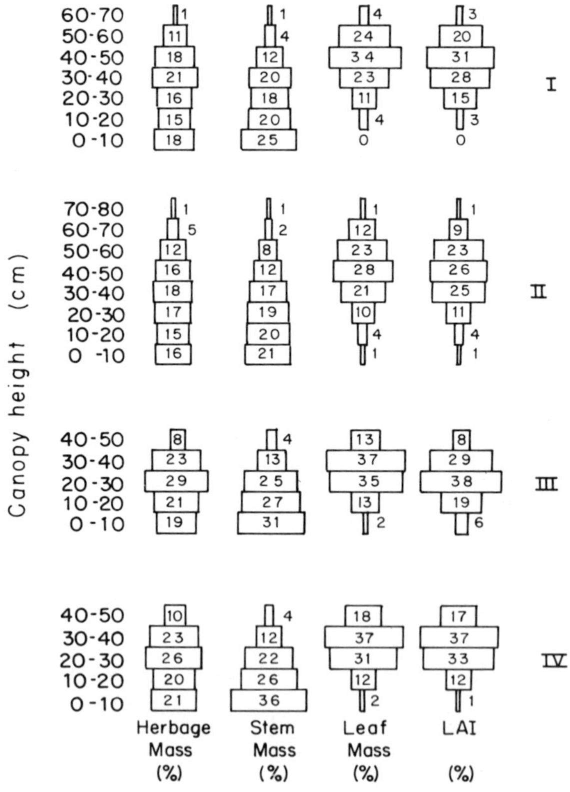

Figure 4.10. Distribution of herbage, stem, and leaf dry mass, and leaf area index (*LAI*) in 10 cm strata of the alfalfa crop canopy on the last sampling date of the first (I), second (II), third (III), and fourth (IV) growth periods in 1968. [Adapted from Bula et al. (33).]

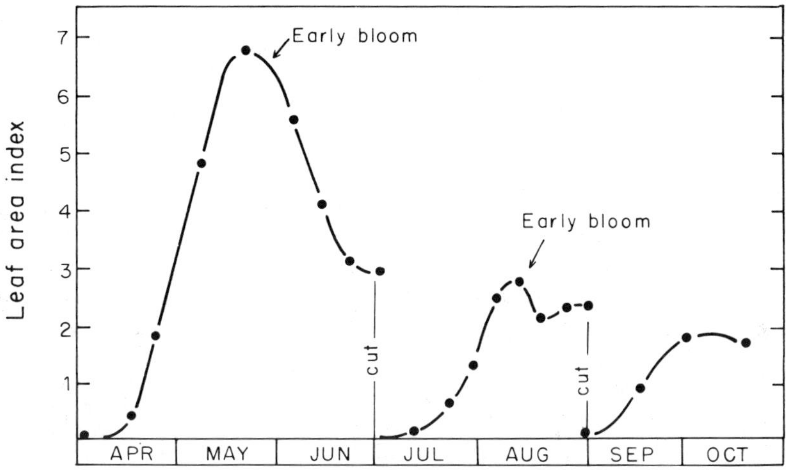

Figure 4.11. Leaf area indices of alfalfa during successive harvest–regrowth cycles after cutting at early bloom. [Adapted from Nelson and Smith (35).]

The greatest proportion of leaf mass is situated in the upper-middle portion of the canopy (Fig. 4.10). Less than 15% of the leaf mass occurs in the lower 20 cm of the canopy in humid areas because of leaf loss to mutual shading and foliar diseases. In arid areas nearly all leaves are retained on the stem. Leaf mass dwindles in the upper 20 cm of canopy because buds have changed from vegetative to reproductive.

Pods and Seeds

Seed development is normally precluded on alfalfa grown for forage because the crop is harvested at the late bud or early bloom stage. When alfalfa is sown for forage and then is allowed to fully flower with plentiful pollinating insects, 7–12% of the aerial mass of the crop develops into seed (34). Seed production is maximized at reduced seeding rates of 1.7–2.3 kg ha^{-1}, 12–15% of the rates used for establishing forage stands. The lower seeding rate promotes plant growth and development by reducing interplant competition for light and facilitates access to racemes by pollinators. Alfalfa grown for seed typically has more racemes per stem and pods per raceme than that grown for forage. Seed yields decline in older stands because of reduced plant density, reduced seed weight per plant, and fewer seeds per pod (34).

Leaf Area Index

Leaf area index (*LAI*), the ratio of leaf surface to land surface, is greatest during the first harvest interval and decreases during successive harvests (35) (Fig. 4.11). Within each harvest interval, *LAI* is greatest with onset of flowering. As flowering progresses, lower leaves succumb to mutual shading and foliar diseases, especially when lodging occurs, and *LAI* diminishes until harvest. The distribution of *LAI* within canopy strata mirrors the distribution of leaf mass (Fig. 4.10). Thus, the surface-to-mass ratio of an alfalfa canopy changes little throughout the season. Individual alfalfa leaflets are nearly horizontal in the upper 8 cm of canopy, and assume more vertical orientations with depth in the canopy (36). Reduction in radiation flux density in the leaf canopy generally follows Beer's law, but the *LAI* necessary for virtually complete light interception varies greatly with stage of herbage growth (37).

Stanhill (38) found that 33% of the incident photosynthetically active radiation was utilized by alfalfa harvested at 31 day intervals during 18 months. Lengthening the harvest interval to 48 days increased light utilization to 46%. Although total productivity increased with a longer harvest interval, it is likely that forage quality declined because of age of canopy and leaf loss.

Leaf and Canopy CO_2 Assimilation

The CO_2 exchange rate of brightly illuminated developing alfalfa leaves increases with ontogeny until the leaflets reach full expansion (39). Thereafter, CO_2 exchange rates decline with age to rates of about 20% or less of the maximum values 40 days after leaf initiation (Fig. 4.12).

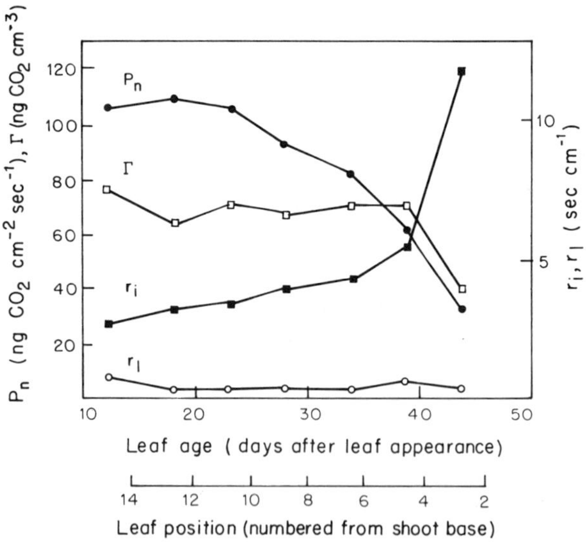

Figure 4.12. Net photosynthetic rate (P_n), CO_2 compensation concentration (Γ), residual (r_i), and leaf (r_l) resistances to CO_2 transport of leaves of different ages on an alfalfa stem. Photosynthesis was measured at an irradiance of 260 μE m^{-2} sec^{-1}, an ambient CO_2 concentration of 505 ± 4 ng CO_2 cm^{-3}, and 26°C leaf temperature. [Adapted from Hodgkinson (40).]

The decline in CO_2 assimilation with age may be attributable to increased residual resistance (r_i) to CO_2 diffusion in the liquid phase and to CO_2 fixation during carboxylation in mesophyll cells, rather than to increased stomatal diffusive resistance (r_l) to CO_2 (40). This explanation is tentative because the boundary layer resistance of the leaves in Fig. 4.12 was two to three times the stomatal resistance, which would contribute considerable imprecision to measurements of r_l. This may explain why the measured insensitivity of r_l with age contrasts with that of many crop species for which stomatal functioning also becomes increasingly impaired in older leaves. Older leaves within the canopy lose their CO_2 assimilation capacity owing to mutual shading (39). After herbage removal, older leaves remaining near the soil surface may regain some of their former CO_2 assimilation capacity. This is largely due to reduced resistance to CO_2 movement in the liquid phase and to increased carboxylation activity rather than to changes in stomatal diffusive resistance (40).

Water Relationships and Canopy Development

The physiological response mechanisms of alfalfa to water deficit have been conspicuously neglected, and there are few principles available for use in improving the production of this crop with limited water supply. Water relationships of alfalfa plants have been largely inferred from measurements of soil water stress, a parameter of limited usefulness in understanding how plant metabolism responds and adapts to water deficits. However, numerous irrigation trials (e.g., 41–45) have established that water supply is a major limitation to alfalfa productivity when rainfall is deficient during important stages of canopy development.

Plentiful moisture in the seedbed is essential for germination and uniform seedling establishment. Germination is largely unaffected by osmotic potentials down to -9 bars, but it is inhibited at -12 to -15 bars (8). Once seedlings become established, water deficits strongly modify the components of canopy growth. Reducing soil water potential from -0.5 to -10 bars in 4 weeks caused a 28% reduction in stem numbers (tillering) of alfalfa seedlings (46). The number of vegetative buds was reduced 31%, and plant height was decreased by 58% over the range of stresses imposed.

Response to water deficits varies with the cold-hardiness characteristics of the cultivar. Cold-hardy cultivars have more internodes per stem than nonhardy cultivars under nonstressed conditions. Reducing soil water from 100 to 49% of field capacity decreased the number of stems per plant and the number of internodes per stem on hardy and nonhardy cultivars (47). Internode length at several heights in alfalfa plants was greatly reduced by water deficit (Table 4.2), with the upper internodes being more strongly affected than the lower ones. A general pattern

Table 4.2. Internode lengths on the primary stem of two cold-hardy and two nonhardy alfalfa cultivars grown in the greenhouse under two soil water regimes and sampled at 5 to 10% bloom[a]

	Internode length (cm)							
	100% field capacity				49% field capacity			
Internode no.[b]	'Travois'[c]	'Vernal'	'Lahontan'	'Sonora'	'Travois'	'Vernal'	'Lahontan'	'Sonora'
2	2.50	2.60	3.69	4.25	1.50	2.30	3.17	3.42
3	4.20	5.20	5.45	4.97	2.55	3.22	3.35	2.85
4	4.45	4.87	4.77	4.30	2.41	2.40	2.30	1.72
5	4.27	4.95	4.57	4.07	2.22	1.94	1.75	1.22
6	3.92	4.62	4.10	3.87	1.62	1.42	0.87	0.65
7	3.50	4.10	3.35	3.37	1.20	1.02	0.65	0.35
8	3.27	3.50	3.00	3.00	0.90	0.77	0.50	0.25
% Internode length reduction by water deficit					53	56	56	62

[a] Adapted from Perry and Larson (47).
[b] Internode numbers counted from soil surface upward, with No. 2 being near soil surface and No. 8 near the top of the plant.
[c] 'Travois' and 'Vernal' are cold-hardy cultivars, and 'Lahontan' and 'Sonora' are nonhardy. The level of cold hardiness decreases from left to right.

Table 4.3. Canopy growth rates of alfalfa (cv 'Hairy Peruvian') after cutting with different timings of irrigation[a]

Days to Irrigation After Cutting	Canopy Growth Rate ($g\ m^{-2}\ day^{-1}$)	
	Days 1–10	Days 11–20
1	8.5	13.9
10	4.4	13.5
23	4.6	4.7

[a] Adapted from Cohen et al., (49).

emerged that the reduction of internode length by water deficit for nonhardy cultivars was greater than that for hardy cultivars. Thus, the classification of alfalfa cultivars according to hardiness may have implications in predicting cultivar response to water deficits.

A consequence of shortened internodes under water deficits is that partitioning of photosynthate in the aerial portion of the plant is changed. Alfalfa grown at −0.2 bars soil water potential partitioned 48% of its aerial dry matter into leaves, but at −2.0 bars soil water potential, 62% of the aerial dry matter was leaves (48). The canopy growth rate of irrigated alfalfa within a single cutting cycle declined 60–75% when soil water stress at 25–50 cm soil depth exceeded −2.0 to −2.5 bars. Irrigation resulted in partial recovery of canopy growth rate to levels about half of those occurring before the stress period (42). Total recovery of canopy growth rate would not be expected because of the decline in crop growth with canopy age.

Rate of canopy regrowth after forage harvest is strongly dependent on the availability of water. Regrowth of alfalfa in soil nearly at the permanent wilting percentage and irrigated to field capacity 1 day after cutting occurred at twice the rate of plants similarly irrigated 10–23 days after clipping (Table 4.3). Canopy regrowth of plants irrigated 23 days after clipping failed to attain the regrowth rates of plants irrigated sooner after cutting (49). Vegetative production of alfalfa is greatest when the soil water potential throughout the season averages −0.2 to −2.0 bars at 0.15–1.5 m depth (50). However, excessive water predisposes the crop to pathogens, and soil water levels greater than −2.0 bars may result in reduced yields.

In contrast to vegetative growth, seed production is optimal at soil water potentials between −2 and −8 bars (51). The reasons for this response are unclear, but deferring irrigation to reduce canopy growth during the vegetative stage increases the number of racemes (flower clusters) per stem (49). Avoidance of irrigation during flowering improves the effectiveness of pollinating insects (34). Finally, pod set in alfalfa is reduced as relative humidity increases from 10 to 90% at 21–32°C (52). Thus, a rather poorly defined combination of soil and atmospheric water stresses is conducive to maximum alfalfa seed production.

Evapotranspiration and Canopy Growth

Alfalfa has high evapotranspiration in comparison with other crops (e.g., 53), a characteristic that has been attributed to crop growth duration (54), depth of rooting (55), and extent of ground cover (56). When canopy closure is complete, energy dissipated by alfalfa evapotranspiration may exceed that gained from net radiation by 1.8 (57, 58) and be greater than that, e.g., 1.3, of grasses like barley and sorghum grown under similar conditions. Besides the crop characteristics listed previously, Fritschen (57) identified aerodynamic roughness of the canopy as a factor allowing greater sensible heat extraction from the atmosphere by alfalfa (rough canopy) than by grasses (smooth canopy).

In many alfalfa-growing areas in the Southwest and Great Plains states, local and regional advection of sensible heat from adjacent warmer land by horizontal air currents is an important energy source for evapotranspiration. In Nebraska, advection during May increased alfalfa evapotranspiration to a maximum of 12 mm day^{-1} compared with a maximum of 6.5 mm day^{-1} in April (58). Nocturnal evapotranspiration of 1 mm per night during temperature inversions also contributed to soil water depletion by the crop (58).

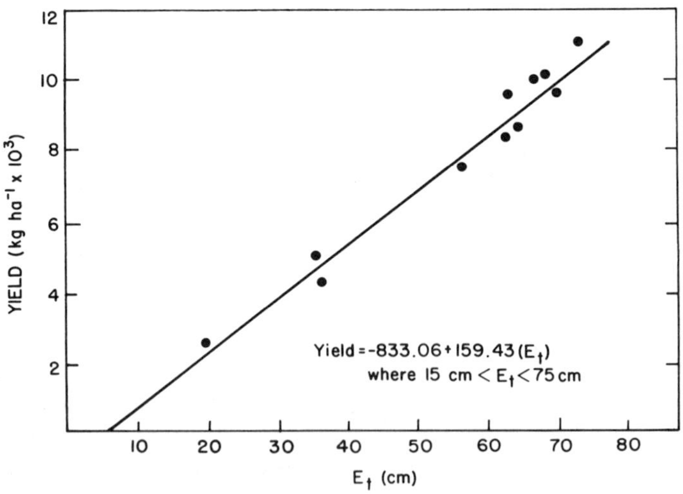

Figure 4.13. Alfalfa dry matter yields as a function of evapotranspiration (E_t) during the growing season. Each point is the mean of 24 replicates per irrigation treatment per year, 1973–1976. [Adapted from Bauder et al. (45).]

After irrigation of alfalfa with a completely closed canopy, the crop transpires freely, and there is little evidence that the plant exerts stomatal control over water loss until soil water is depleted to about −4 bars (59). Crop growth measured as the height of the canopy was reduced by 50% before evapotranspiration was reduced by stomatal closure that was induced by soil water potential (59). Presumably the *LAI* was sufficient for complete radiation interception. Complete canopy removal reduces evapotranspiration by 50% (56). The decline in evapotranspiration observed after cutting depends upon soil type and wetness.

Alfalfa yields are usually limited by a deficiency of water, and in this case cumulative forage productivity is linearly related to evapotranspiration when fertility is adequate (Fig. 4.13). Deviations from linearity early in the season have been explained by mobilization of stored carbohydrates from roots to foliage, whereas deviations late in the season have been attributed to storage of current photosynthate in roots rather than incorporation into foliage (60).

TRANSPIRATION AND THE REGULATION OF WATER LOSS

Depression of evapotranspiration below potential evapotranspiration or pan evaporation from a closed alfalfa canopy has been interpreted as stomatal control of water loss (Fig. 4.14). The evidence that canopy evapotranspiration is reduced sooner during the day and for a longer duration than stomatal resistance is increased leads to the speculation that resistance to water transport elsewhere in the plant may regulate water loss. However, recent field investigations of alfalfa water relationships using the pressure chamber technique have shown diurnal trends in leaf water potential (61), and a gradient in water potential between the roots and leaves (Table 4.4). The similarity in stem and root water potentials and their relatively small values compared with those of the leaf suggest that the major resistance to water flow in the alfalfa plant occurs in the leaf. Alfalfa stomata commence closure at soil water potentials near −4 bars (59), and at a leaf water potential near −10 bars (31, 62). However, there is a little information on stomatal response to water stress, seriously limiting our understanding of how alfalfa regulates its internal water status.

Other factors affecting stomatal action are more clearly understood. Alfalfa stomata open in response to light in much the same pattern as other C_3 dicots, although the minimum resistances obtained and the changes of diffusive resistance with age may differ strongly from values for other species. Stomata of alfalfa are more open in dim light than are those of C_4 species like corn or sorghum (63), and there is some evidence of incomplete stomatal closure in darkness (40). Stomatal diffusive resistances (to CO_2) in light as low as 0.2 sec cm^{-1} have been reported

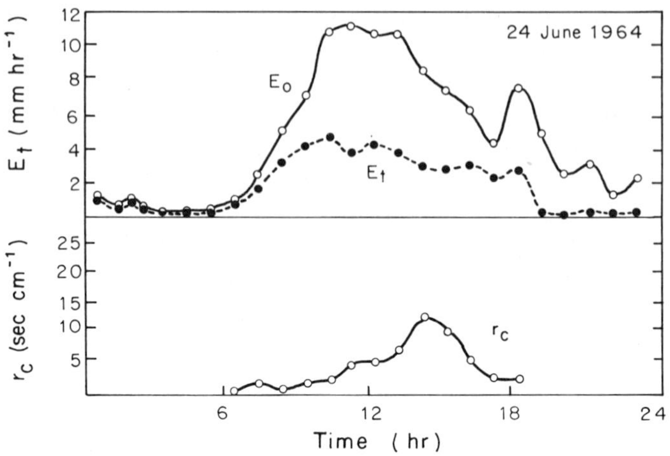

Figure 4.14. Evapotranspiration (E_t) and potential evapotranspiration (E_0) from alfalfa measured hourly in mountain standard time near Phoenix, Arizona, 27 days after irrigation. Hourly values of canopy resistance (r_c) are shown for daytime. [Adapted from van Bavel (59).]

(40), but much additional work on the response of alfalfa stomata to light is needed. There is little evidence of stomatal closure as leaves age and CO_2 exchange rate declines in alfalfa (Fig. 4.12). Indeed, this apparent lack of stomatal response may account in large part for differences in water use of well-watered alfalfa in comparison with other crops. For example, stomatal resistance in bright light of field-grown alfalfa was only 0.5 sec cm^{-1} compared to 2.4 sec cm^{-1} for the C_4 species *Paspalum* (64). At comparable vapor pressure deficits, transpiration rate per unit leaf area of well-watered alfalfa exceeded that of corn by threefold (65).

Stomatal density of alfalfa is greatest on the upper leaf surface (66, 67), but diffusive resistance of the upper epidermis is greater up to 0.3 full sunlight than that of the lower surface (63). This suggests that stomata

Table 4.4. Diurnal variation in water potentials of field-grown alfalfa[a]

	Water Potential (Bars)		
	Time of day		
Plant Part	1000	1400	1800
Leaf	−10.3	−12.4	−10.9
Stem	−1.3	−2.6	−1.8
Root	−1.7	−2.4	−2.4

[a] Adapted from data in Wali (61).

on the upper epidermis are not as widely open or have shorter aperture length than the stomata on the lower epidermis. Gindel's (66) observations of stomatal length of alfalfa leaves grown at three irrigation levels confirm that upper stomata are shorter than lower stomata. Stomatal density of alfalfa also varies substantially among cultivars, with a cold-hardy cultivar having 26 to 82% more stomata per mm^2 than several nonhardy cultivars (67). For example, the nonhardy cultivar 'Moapa' had 238 stomata mm^{-2} on the upper and 173 stomata mm^{-2} on the lower leaf surface, but the hardy cultivar 'Ladak' had 301 and 222 stomata mm^{-2} on the corresponding leaf surfaces (67). Comparably sized leaflets of alfalfa grown under increasing soil water deficit have shorter stomata and greater stomatal densities than well-watered controls. Leaflet area is reduced slightly by protracted water deficit (66), and the decrease in epidermal cell size suggests that water deficit decreases chlorenchyma cell size as well.

Although stomatal closure is the principal adaptation of alfalfa to brief water deficits, protracted drought and high temperatures induce dormancy in the crop with a significant reduction in evapotranspiration and soil water extraction (55, 68). Renewal of herbage growth occurs with very light rainfalls that are insufficient to thoroughly wet even the upper rooting zone but yet apparently stimulate release of bud dormancy and remobilization of root carbohydrates to new shoot tissue (69). The potential roles of leaf display and leaf abscission in reducing transpirational water loss are unknown.

IRRIGATION AND CROP WATER USE

Alfalfa is grown under irrigation in widely contrasting climates and for this reason generalizations about crop water use and irrigation have restricted value. The summaries by Stanberry (70) and Peterson (71) have guided the development of irrigation practices in many localities. In this section I will attempt to illustrate principles that govern the water needs of alfalfa grown under irrigation, but local authorities should be consulted for information most pertinent to alfalfa irrigation in specific areas.

Critical Water Demand Periods

Seedling establishment, forage regrowth, and seed production are the stages of alfalfa development requiring careful water management. Irrigation of the rooting zone to field capacity during seedbed preparation is preferred to sowing of seed in a dry seedbed followed by irrigation. This is especially important in saline soils where excess salts must be leached from the root zone before sowing. A moist seedbed before sowing facilitates firm contact between the soil and seed. Irrigation after seeding increases the risk of soil surface crusting and seedling attrition during establishment. Soil crusting necessitates frequent water applications to facilitate emergence. Water applications while the shallow-rooted seedlings are less than 10 cm tall should be avoided to prevent damage to stands by diseases like *Phytophthora* root rot (72). Following seedling establishment, the rooting zone of young stands should be maintained at about field capacity, as excessive water will increase diseases and retard crown and root development. Irrigation of dryland when the water table is 1.7–3 m from the soil surface is often of little value (73).

The amount of canopy regrowth after harvest is strongly influenced by the availability and timing of water supply. Light and infrequent applications of water insufficient to wet the upper roots may cause release of bud dormancy and regrowth of foliage (69). Irrigation soon after harvesting stimulates more rapid regrowth than deferring water for 10 to 23 days (Table 4.3). Scheduling of irrigation during vegetative growth varies with the season, rainfall, soil characteristics, and the necessity for machine harvests. A mean soil water potential of −0.2 to −2.0 bars at 0.15–1.5 m depth was optimal for seasonal vegetative production in Utah (50). The lack of more quantitative information on the response of alfalfa to water deficit has led to adoption of Stanberry's (70) empirical criterion of water stress as alfalfa plants showing a "bluish-green color of foliage" in projecting irrigation needs for the last 25 years (71, 74, 75).

Maximal alfalfa seed production is usually obtained by a slight water stress that is sufficient to suppress vegetative growth but insufficient to reduce pollination and seed development. The earliest results (51) suggested that a mean soil water potential over several depths in the profile of −2 to −8 bars was conducive to maximal seed yields under surface and sprinkler irrigation at several Utah locations. Alfalfa grown with subsurface drip irrigation in California (76) produced maximal seed yields when soil water potential was −0.5 bar at 23 cm depth in the row in a soil where roots were confined to 46 cm of profile. Subsequently, seed yields were greater with furrow irrigation on 7 day intervals after soil water potential exceeded −1 bar for 1–2 days at 30 and 46 cm depth (77). The results emphasize that method and frequency of irrigation, method of soil water stress characterization, and the depth of the root zone are important variables in managing water application for seed production.

Water Requirement and Consumptive Water Use

Projections of irrigation needs might be made from knowledge of the water requirement of growth, i.e., the mass of water lost by transpiration per mass of dry weight accumulated. The reciprocal of water requirement, water use efficiency, has been similarly used. When dry weight accumulation can be anticipated for a growing season, estimates of seasonal water needs of the crop can be calculated from knowledge of the water requirements of growth or of water use efficiency.

Numerous measurements have shown that the water requirement of alfalfa is greater than that of crops like corn or sorghum that photosynthesize by the C_4 pathway. Estimates of water requirement vary widely depending on variety, conditions of growth, and losses of water by soil evaporation or deep percolation. Briggs and Shantz (78) reported a water requirement range of 631 to 834 kg H_2O per kg dry matter for alfalfa grown in galvanized cans in lath houses. Shantz and Piemeisel (79) found that the water requirement of alfalfa varied from 890 to 957 kg H_2O per kg dry matter, depending on the cultivar. Cole et al. (80) measured variation in water requirement among cultivars and among genotypes within cultivars. Gifford and Jensen (81) observed a water requirement of 800 kg H_2O per kg dry matter for alfalfa growing in a wet soil water regime and 1360 for alfalfa growing in a dry soil water regime. When alfalfa is grown under field irrigation and all components of the water budget including soil storage, evaporation, and percolation are measured, the water requirement ranges from 512 to 663 kg H_2O per kg dry matter across years and among greatly differing climates (6, 45). These values suggest that alfalfa requires 5.6–7.3 cm ha^{-1} of water per metric ton dry forage to fulfill transpirational needs during the growing season.

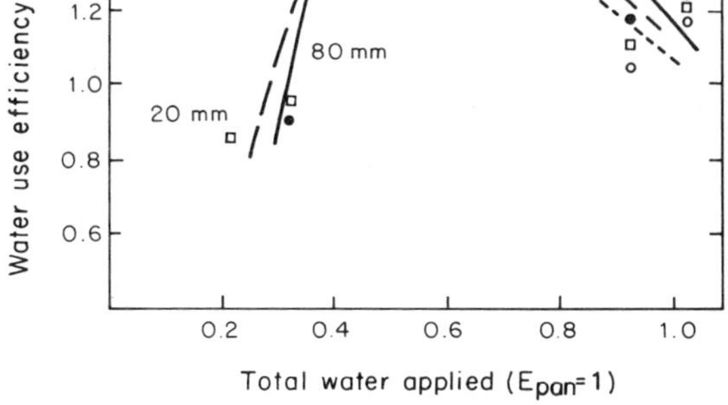

Figure 4.15. Water use efficiency (ratio of dry matter harvested to water applied) of alfalfa in response to total water application (expressed as a proportion of Class A pan evaporation, E_{pan}), and by frequency of irrigation with 5, 20, and 80 mm of water per application. [Adapted from Snaydon (69).]

The usefulness of water requirement values is lessened by the great variation in crop water use with season and with method of application. Water use efficiency of alfalfa was greatest when applied water was about 0.5 pan evaporation and least with smaller or larger water applications (Fig. 4.15). The optimal water use efficiency was for infrequent large irrigations rather than for more frequent small irrigations (69). Within harvest intervals, alfalfa water use is least when *LAI* is small and approaches maximum value when canopy closure is complete (54). Irrigation water is used most efficiently when crop growth is rapid (6), and in the cooler portions of the growing season (82). The water requirement of alfalfa is greatly increased when excessive or poorly scheduled irrigations enhance evaporation or deep percolation (5, 54). Although fertilization increases growth and thereby total water use, the least water is used per unit of crop growth when plant nutrition is adequate rather than limiting (70, 83).

Assessments of irrigation needs for an alfalfa crop are frequently made from measurements of soil water storage, percolation losses, rainfall, and seasonal patterns of consumptive water use (transpiration plus soil evaporation) for broad geographic areas (71, 75). This budgeting technique requires previous measurements of consumptive water use, which vary substantially among seasons for a particular locality owing to differences in growth temperatures and radiation. The pattern of consumptive water use at Tempe and Mesa, Arizona (Fig. 4.16), shows that crop water loss

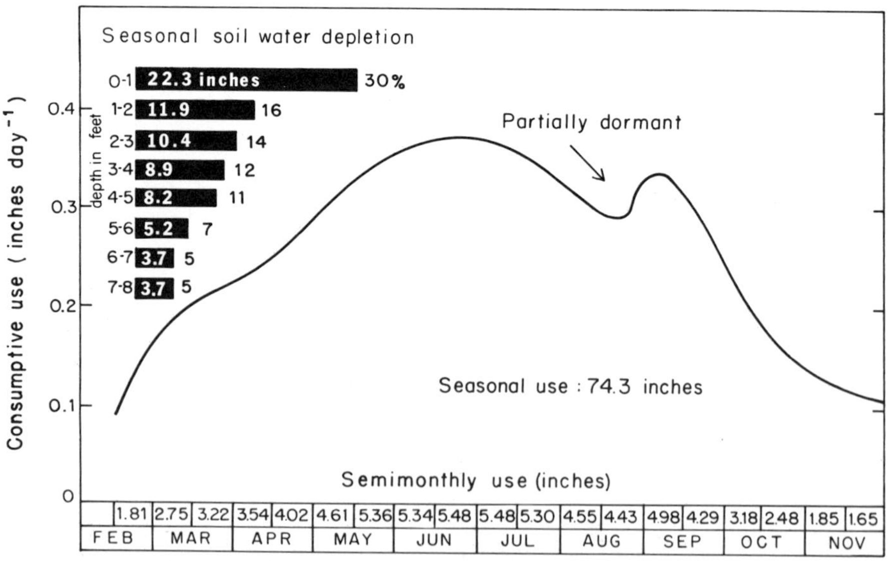

Figure 4.16. Mean consumptive use of water by alfalfa at Mesa and Tempe, Arizona (1946, 1950, 1962, 1963). Inset shows seasonal soil water depletion in inches and as a percentage of the total depletion with depth in the profile. [Adapted from Erie et al. (68).]

is greatest during the warmest periods of the year and least during crop dormancy. Because consumptive water use measurements are infrequently made in a given locality, baseline data are adapted to different seasons by transformations of Blaney and Criddle (84) based on mean monthly temperatures or daily maximum temperatures and hours of daylight.

Maximum evapotranspiration rates of alfalfa generally approach 7.6–9.0 mm day^{-1}, although extremes of 4.1 (53) to 12 mm day^{-1} (58) have been reported. Seasonal consumptive use ranges from 350 mm in short season mountainous regions of the western United States and 610 to 635 mm in the upper Great Plains to 1900 mm or more in the long-season arid Southwest. Length of growing season and maximum daily temperature are the primary determinants of seasonal consumptive use when water is limiting.

REFERENCES

1. United States Dept. of Agriculture, *Agricultural Statistics*, U.S. Govt. Printing Office, Washington, D.C., 1977.
2. N. L. Smith, *Opportunities for energy savings in crop production*, Report to the Subcommittee on Advanced Energy Technologies and Energy Conservation, Research and Development, U.S. House of Representatives. U.S. Govt. Printing Office, Washington, D.C. 1978.
3. D. K. Barnes, E. T. Bingham, R. P. Murphy, O. J. Hunt, D. F. Beard, W. H. Skrdla, and L. R. Teuber, *USDA Tech. Bull. No. 1571*, 1977.
4. M. R. Kilchner and D. H. Heinrichs, *Can. J. Plant Sci.* **54,** 739 (1974).
5. R. J. Joy and A. K. Dobrenz, *Prog. Agric. Arizona* **23,** 14 (1971).
6. R. J. Joy, H. T. Poole, and A. K. Dobrenz, *Prog. Agric. Arizona* **24,** 9 (1972).
7. D. Smith, *Agron. J.* **62,** 520 (1970).
8. R. Uhvits, *Am. J. Bot.* **33,** 278 (1946).
9. G. H. Heichel and C. P. Vance, *Crop Sci.* **19,** 512 (1979).
10. G. R. Gist and G. O. Mott, *Agron. J.* **49,** 33 (1957).
11. A. G. Matches, G. O. Mott, and R. J. Bula, *Agron. J.* **54,** 541 (1962).
12. G. R. Gist and G. O. Mott, *Agron. J.* **50,** 583 (1958).
13. R. A. Kohl and J. J. Kolar, *Agron. J.* **68,** 536 (1976).
14. D. Smith, *Agron. J.* **43,** 573 (1951).
15. D. R. Viands, D. K. Barnes, and G. H. Heichel, *USDA Tech. Bull., No. 1643* 1981.
16. D. R. Viands, C. P. Vance, G. H. Heichel, and D. K. Barnes, *Crop Sci.* **19,** 905 (1979).
17. J. B. Fehrenbacher, B. W. Ray, and W. M. Edwards, *Soil Sci. Soc. Am. Proc.* **29,** 591 (1965).
18. T. B. Paltridge, *CSIRO Bull. 274,* Canberra, Aust. 1955.
19. J. R. Simpson and J. Lippsett, *Aust. J. Agric. Res.* **24,** 199 (1973).
20. G. E. Van Riper, *Agron. J.* **56,** 45 (1964).
21. J. A. Hobbs, *Agron. J.* **45,** 490 (1953).
22. A. C. Mathers, B. A. Stewart, and B. Blair, *J. Environ. Qual.* **4,** 403 (1975).

23. L. F. Graber, N. T. Nelson, W. A. Luekel and W. B. Albert, *Univ. Wis., Agric. Exp. Stn. Res. Bull. 80,* 1927.
24. C. J. Nelson and D. Smith, *Crop Sci.* **8,** 21 (1968).
25. C. P. Vance, G. H. Heichel, D. K. Barnes, J. W. Bryan, and L. E. Johnson, *Plant Physiol.* **64,** 1 (1979).
26. H. C. Cralle, M. S. Thesis, Dept. of Agronomy and Plant Genetics, University of Minnesota, St. Paul, 1979.
27. D. D. Wolf, *Crop Sci.* **18,** 690 (1978).
28. K. C. Hodgkinson, *Aust. J. Biol. Sci.* **22,** 1113 (1969).
29. W. E. Lueschen, D. K. Barnes, D. L. Rabas, F. I. Frosheiser, and D. M. Smith, *Agron. J.* **68,** 281 (1976).
30. G. H. Heichel, *Plant Physiol.* **S59,** 128 (1977).
31. P. M. Aparico-Tego, M. F. Sanchez-Diaz, and J. I. Pena, *Physiol. Plant.* **48,** 1 (1980).
32. O. D. Bennett and B. D. Doss, *Agron. J.* **52,** 204 (1960).
33. R. J. Bula, D. A. Holt, R. G. May, and M. M. Schreiber, *Purdue Univ. Agric. Exp. Stn. Bull. 76,* 1975.
34. M. W. Pederson and W. P. Nye, *Utah Agric. Exp. Stn. Bull. 436,* 1962.
35. C. J. Nelson and D. Smith, *Crop Sci.* **8,** 25 (1968).
36. I. D. Teare, *Phyton* **29,** 37 (1972).
37. R. T. Wilfong, R. H. Brown, and R. E. Blaser, *Crop Sci.* **7,** 27 (1967).
38. G. Stanhill, *Neth. J. Agric. Sci.* **10,** 247 (1962).
39. D. D. Wolf and R. E. Blaser, *Crop Sci.* **12,** 23 (1972).
40. K. C. Hodgkinson, *Aust. J. Plant Physiol.* **1,** 561 (1974).
41. C. S. Scofield, *USDA Circular No. 735,* 1945.
42. W. D. Kemper and M. Amemiya, *Soil Sci. Soc. Am. Proc.* **21,** 657 (1957).
43. R. F. Lucey and M. B. Tesar, *Agron. J.* **57,** 519 (1965).
44. C. G. Janson, *N.Z. J. Exp. Agric.* **3,** 223 (1975).
45. J. W. Bauder, A. Bauer, J. M. Ramirez, and D. K. Cassel, *Agron. J.* **70,** 95 (1978).
46. E. R. Cowett and M. A. Sprague, *Agron. J.* **54,** 294 (1962).
47. L. J. Perry, Jr., and K. L. Larson, *Crop Sci.* **14,** 693 (1974).
48. L. R. Vough and G. C. Marten, *Agron. J.* **63,** 40 (1971).
49. Y. Cohen, H. Bielorai, and A. Dovrat, Crop Sci. **12,** 634 (1972).
50. S. A. Taylor, *Soil Sci.* **74,** 217 (1952).
51. S. A. Taylor, J. L. Haddock, and M. W. Pederson, *Agron. J.* **51,** 357 (1959).
52. C. O. Grandfield, *J. Agr. Res.* **70,** 123 (1945).
53. L. G. Sonmor, *Can. J. Soil Sci.* **43,** 287 (1963).
54. K. K. Krogman and E. H. Hobbs, *Can. J. Soil Sci.* **45,** 309 (1965).
55. N. H. Tadmor, O. P. Cohen, L. Shanan, and M. Evenari, Moisture Use of Pasture Plants in a Desert Environment, in *Proc. 10th. Int. Grassl. Congress,* Helsinki, Finland, 1966, pp. 897–906.
56. J. P. Hudson, *Exp. Agric.* **1,** 23 (1965).
57. L. J. Fritschen, *Agron. J.* **58,** 339 (1966).
58. N. J. Rosenberg, *Agron. J.* **61,** 879 (1969).
59. C. H. M. Van Bavel, *Agric. Meteorol.* **4,** 165 (1967).
60. J. I. Stewart and R. M. Hagan, Development of Evapotranspiration—Crop Yield Functions for Managing Limited Water Supplies, *in Proceedings of the 7th Congress,*

International Commission on Irrigation and Drainage, Mexico City, Mexico, 1969, pp. 23.505–23.530.
61. S-E. B. Wali, Ph.D. Thesis, Department of Agronomy, Purdue University, W. Lafayette, Ind. 1977.
62. M. Sanchez-Diaz and M. Sanchez-Marin. *An. Edafal. Agrobiol.* **33,** 743 (1974).
63. W. L. Ehrler and C. H. M. Van Bavel, *Plant Physiol.* **43,** 208 (1968).
64. J. P. Kerr and H. G. McPherson, *Proc. N.Z. Grassland Assoc.* **39,** 70 (1978).
65. B. J. Forde, K. J. Mitchell, and E. A. Edge, *Aust. J. Plant Physiol.* **4,** 889, (1977).
66. I. Gindel, *Physiol. Plant.* **21,** 1287 (1968).
67. D. F. Cole and A. K. Dobrenz, *Crop Sci.* **10,** 61 (1970).
68. L. J. Erie, O. F. French, and K. Harris, *Tech. Bull. 169,* University of Arizona, Tucson, 1968.
69. R. W. Snaydon, *Aust. J. Agric. Res.* **23,** 239 (1972).
70. C. O. Stanberry, Irrigation Practices for the Production of Alfalfa, in A. Stefferud, Ed., *Water.* The Yearbook of Agriculture, U.S. Govt. Printing Office, Washington, D.C., 1955.
71. H. B. Peterson, Water Relationships and Irrigation, in C. H. Hanson, Ed., *Alfalfa Science and Technology, Agronomy* **15,** 469 (1972).
72. F. I. Frosheiser and D. K. Barnes, *Crop Sci.* **13,** 735 (1973).
73. R. E. Campbell, W. E. Larson, T. S. Asheim, and P. L. Brown, *Agron. J.* **52,** 437 (1960).
74. R. E. Dennis and 22 coauthors, *Bulletin A16,* Coop. Ext. Serv., University of Arizona, Tucson, 1977.
75. H. D. Werner, *Miscellaneous Report 160,* Agric. Ext. Serv., Univ. of Minnesota, St. Paul, 1978.
76. R. W. Hageman, L. S. Willardson, A. W. Marsh, and C. F. Ehlig, *Calif. Agric.* **29,** 14 (1975).
77. R. W. Hageman, C. F. Ehlig, M. J. Huber, R. Y. Reynoso, and L. S. Willardson, *Calif. Agric.* **32,** 17 (1978).
78. L. J. Briggs and H. L. Shantz, *J. Agric. Res.* **3,** 1 (1914).
79. H. L. Shantz and L. N. Piemeisel, *J. Agric. Res.* **34,** 1093 (1927).
80. D. F. Cole, A. K. Dobrenz, M. A. Massengale, and L. N. Wright, *Crop Sci.* **10,** 237 (1970).
81. R. O. Gifford and E. H. Jensen, *Agron. J.* **59,** 75 (1967).
82. L. A. Daigger, L. S. Axthelm, and C. L. Ashburn, *Agron. J.* **62,** 507 (1970).
83. S. J. Bourget and R. B. Carson, *Can. J. Soil Sci.* **43,** 7 (1962).
84. H. F. Blaney and W. D. Criddle, *USDA Tech. Bull. No. 1275,* 1962.

5

BEAN

Anthony J. Halterlein

INTRODUCTION

The common bean (*Phaseolus vulgaris* L.) is an annual legume cultured for its green pod and dry seed, both of which are eaten. In common with all other legumes it is a good source of protein, but unlike most legumes it is a poor nitrogen fixer.

Although it originated in the semitropical regions of Central America, frosts do occur in the higher elevations of its habitat (2500–6000 feet above sea level) and so it is adapted to a wide climatic zone. Annual rainfall of its native environment ranges from 50 to 180 cm with about 80% of this falling from May to October. Flowering and pod maturation are controlled by the rainfall regime rather than by day length. In the wetter localities both flowering and maturation are later than in drier localities at the same latitude and elevation. Sunlight is necessary for good development since shading has been observed to stunt growth and reduce yields (1).

Most bean cultivars capable of attaining maturity in temperate climates have a dwarf, determinate habit. Best overall growth occurs between 20 and 25°C (2–4). Beans suffer from reduced yields when temperatures exceed 24.5°C (2). Growth is severely reduced at temperatures below 15°C (5). The plant requires a rich, well-drained soil with 50–100 cm of rainfall during the growing season for optimum germination and growth (1). Beans are described as being a salt-sensitive species (6).

In its native habitat the plant is sometimes a biennial vine, but under cultivation it grows as an annual. Habit, however, may be indeterminate as in pole types, determinate, or bush type, or semideterminate as represented by some types grown for the dry seed. Significant physiological growth stages approximate the following time schedule: emergence, 6–10 days after planting; development of first trifoliate leaves, 15 days after emergence; beginning of bloom, 34 days after emergence (at this time cover is about 50%); row closure 42 days after emergence; full cover 46 days after emergence (7).

WATER REQUIREMENTS

Bean plants are very sensitive to soil water conditions and quality and yield can suffer greatly from even brief periods of water shortage. Maurer, et al. (8) studied the effect of five soil water regimes on plants grown in weighing lysimeters. Plants irrigated when soil water fell to 88% of available water were taller, heavier, and produced more pods than plants irrigated when soil water was allowed to fall to 60% of available water. When soil water was allowed to drop to 32% of available water before irrigation, plants were small and produced low yields. Total water used by the plants growing on the high water regime was 116 liters and they produced a total of dry matter accumulation of above ground parts of 280 g. In the low water regimes 69 liters of water used produced a total dry matter accumulation of 172 g. In a medium soil water regime 87 liters of water produced dry matter of 206 g.

Although high yields depend heavily upon adequate soil moisture during the postbloom stage the need for high amounts of water during the prebloom stage is more controversial. MacKay and Eaves (9) found beans to be more sensitive than sweet corn to moisture stress at the preflowering stage. There are several other reports on the beneficial effects of irrigation during the vegetative stages of growth. Robins and Domingo (10) have shown that preflowering stress reduces the number of pods formed. Dreibrodt (11) also found that water stress during the early growth stage reduced yield. Bierhuizen and de Vos (12) found that if water stress occurred during the vegetative period plants were stunted, but this did not decrease seed yield if adequate soil water was maintained during flowering. Kriegbaum (13) found that irrigation applied during the period from germination until prior to flowering had no effect on yield.

Although there is no agreement on the effects of water stress during the prebloom growth stage of beans, there is a general agreement that when water deficits occur during flowering and pod development yields are reduced (8, 9, 14–19).

Nelson (20) showed that water stress during flowering and pod development reduced yields by an average of 22%. Vitkov (21) also found beans to be most sensitive to water stress during flowering and seed formation. Robins and Domingo (10) found water stress during flowering to reduce both pod number and the number of seeds per pod and water stress prior to harvest was observed to reduce seed weight. Parjol (22) reported that water stress occurring at anthesis resulted in reduced flower formation. Dubetz and Mahalle (24) also report that high stress at flowering resulted in inhibition of subsequent flower formation. Water stress imposed at the pod formation stage resulted in the highest percentage of empty seeds and the lowest seed protein content. Similarly water stress at pod development decreased yields and pod weight for three bean cultivars subjected to water stress during pod development (23), but time to harvest and length of pods were not affected.

Kriegbaum (13) reported that water applied 10 days before and again 10 days after the end of flowering resulted in the highest yield increase. This was shown in an increase in the number of seeds per pod and in the number of pods set. Mean seed weight was not significantly affected by irrigation during flowering, but when applied during the period of pod growth irrigation increased the mean seed weight by 6%. Wittmeyer (25) maintained that the critical water period for beans is just prior to blossoming and through pod development. Dubetz and Mahalle (24) subjected beans to a −8 bars soil water stress at the postflowering,

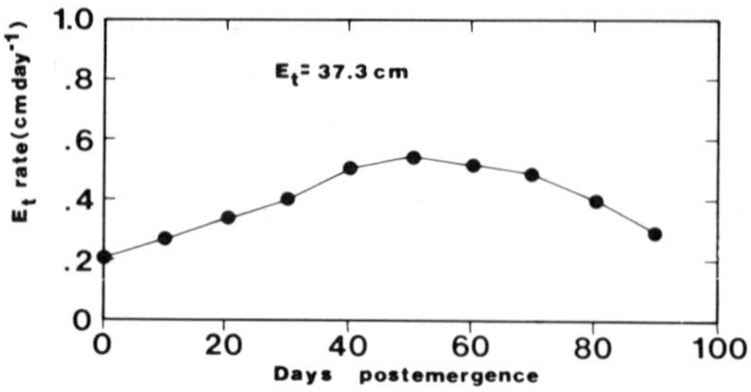

Figure 5.1. Average daily evapotranspiration (E_t) rate for pinto beans. [After Stegman and Olson (28).]

preflowering, and flowering stages and observed yield reductions of 35, 53, and 71%, respectively. High stress during flowering had the most detrimental effect on yield. But plants subjected to water stress at preflowering as well as those stressed at flowering were delayed 1 week in harvest when compared with plants under the postflowering water stress.

Although most work shows that water stress from the bloom stage through pod development reduces yield, Stoker (26) subjected beans to water stress at five different growth stages and obtained no yield reduction from water stress applied after flowering. Stress during flowering

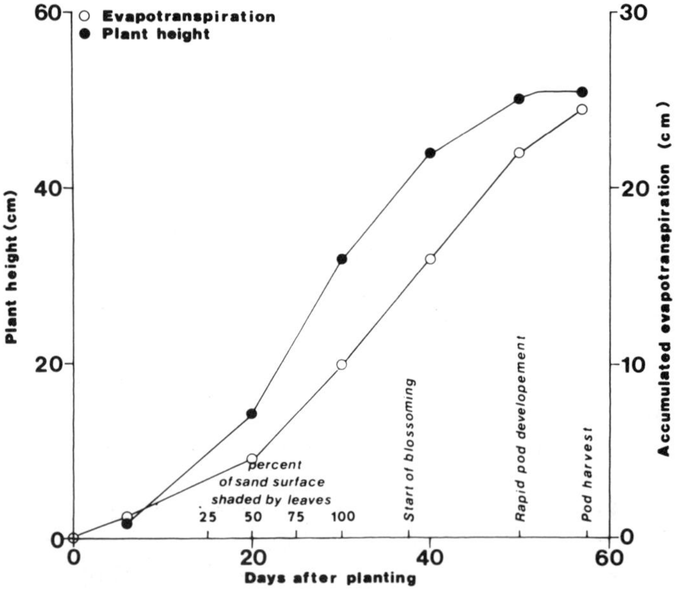

Figure 5.2. Snap bean plant height and accumulated evapotranspiration in relation to days after planting. Average of plantings made on June 6, June 20, July 1, July 14, and Aug. 8, 1966. [After Mielke and Peck (33).]

reduced yields most. Yield reduction was mainly in the form of fewer pods per plant, but number of seeds per pod was slightly less. Water stress at any of the five stages had no effect on the individual seed weight.

In summary, beans seem to suffer from water stress at all growth stages. But stress from just before flowering through pod filling apparently causes the most severe damage. The plants can, at least partly, overcome the effects of water deficits during the vegetative stages of growth.

An available soil water content of at least 70% has been shown by numerous investigators to be necessary for maximum growth and yield of beans (8, 27–29). Some evidence suggests that in the postflowering period available soil water should exceed 80% (29, 30). Deliblatov and Sarkizov (27) obtained highest yields when water in the root zone did not drop below 70% available soil water. A drop in available soil water to 65% resulted in severe yield reductions. According to Vittum and Gibbs (31) available soil water in the root zone should not drop below 50% of optimum growth is to be obtained.

Seasonal water requirements have been reported to be in the range of 3000–3780 $m^3\ ha^{-1}$ (27, 32) with about 2200 $m^3\ ha^{-1}$ being consumed during flowering. Deliblatov and Sarkizov (27) report maximum daily water consumption to reach 54 $m^3\ ha^{-1}$. Stegman and Olsen (28) estimated the seasonal water use of two pinto bean cultivars to average 37 cm from crop emergence to vine cutting date. Daily water use ranged from an average 0.18 cm day^{-1} at emergence to 0.53 cm day^{-1} at 50 to 60 days after emergence (Fig. 5.1). Vittum and Gibbs (31) calculated that on a hot, dry, windy day during the period of maximum water use (blossoming to harvest) up to 0.6 cm of water $plant^{-1}\ day^{-1}$ will be consumed by beans. Vitkov (21) calculated a mean daily water consumption from sowing to bud formation of 1219 $m^3\ ha^{-1}$. After this, consumption doubled. Total water consumption averaged over 3 years was 2460 $m^3\ ha^{-1}$ for a growing season of 95 days. Mielke and Peck (33) noted that evaporation rates increased with increased plant growth until leaf canopy was complete and then was nearly linear through harvest time (Fig. 5.2). Peck et al. (34), using a 41 day base season, observed that the seasonal evapotranspiration average for snap beans over 7 years was 16 cm and the maximum per day reached 4.3 mm (Fig. 5.3).

Excess soil water at different developmental stages was studied by Peev and Kolev (35). Beans were found to be most sensitive to excess soil water during flower bud formation and at flowering. However, MacKay and Eaves (9) found no detrimental effect of excess irrigation on yields.

The relative humidity of the atmosphere within the range of 35–100% was found to have no significant effect on the growth and yield of red kidney beans (36).

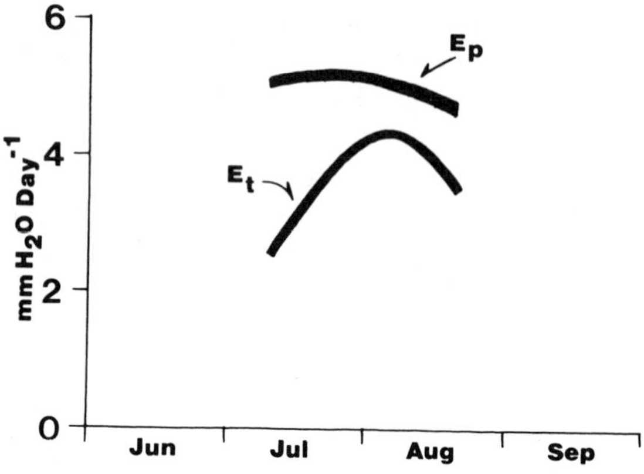

Figure 5.3. Overall standard quadratic regression curves for daily evapotranspiration (E_t) by beans and for evaporation of water from a standard weather bureau Class A pan (E_p) as dependent on day during the growing season. [After Peck et al. (34).]

Although the results of research on the effects of water stress on beans may be somewhat inconsistent, some useful conclusions can be made. (1) There is no agreement regarding the usefulness of applying irrigation water during the vegetative phase. (2)Irrigation added during flowering and pod development increases yield, while water deficits reduce yield. (3) It is not known if there is a variation in sensitivity to soil water with the three growth stages.

It must always be kept in mind that factors not directly related to the soil water level may also have an effect on the plant response to the water regime. Such factors include soil fertility, evaporative demand, plant diseases, and pests.

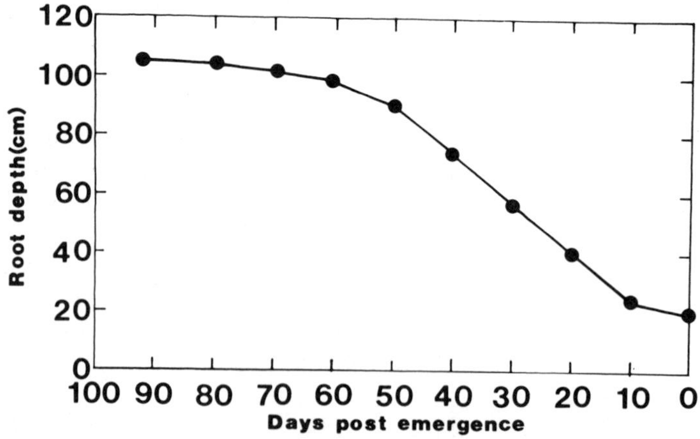

Figure 5.4. Average relationship of root depth with days after emergence for pinto beans. Zero represents depth at soil surface. [After Stegman and Olson (28).]

ROOT SYSTEM

Stegman and Olsen (28) have observed that the root zone depth of beans advances to near 91.4–106.7 cm by the time of full cover (Fig. 5.4). Beans have a shallow root system and so they are particularly responsive to frequent irrigation (8, 14, 18, 19, 24). Halterlein, Teare and Clayberg (unpublished) used a rhizotron to study the rooting habit of three bean cultivars. Maximum rooting depth for adequately watered plants was 81–102 cm. Maximum root extension occurred by full bloom for two cultivars, but for one only two-thirds of maximum extension had occurred by full bloom. At all growth stages at least 50% of roots were found to be in a zone bounded by a depth of 30 cm and lateral extension of 20 cm. Less than 10% of roots ever extended laterally beyond 20 cm. Likewise, less than 10% of roots ever extended to a depth exceeding 46 cm.

IRRIGATION

Low soil water tension should be maintained throughout the entire growth period of a bean crop. Beans are a rapidly growing crop. For some snap bean cultivars the time from planting to harvest may be as little as 45 days. Because of this, adequate soil water must be available constantly to ensure optimum growth and yield. Even water deficits of short duration can severely reduce yield and quality. For example, water stress at the preflowering, flowering, or pod development stages reduced yields almost 30% (37). Highest yields have been obtained when available soil water was maintained above 50% (14, 18, 31, 38). For optimum growth, available soil water should not drop below 50% in the root zone, even during the prebloom period. Kattan and Fleming (18) found that maintaining available soil water about 50% required the weekly application of 2.5 cm of irrigation water between planting and flowering and 3.8 cm from flowering to the end of harvest. Drought injury and water consumption were found to increase with the age of the plants. High levels of soil water stress during the periods of pod development and harvest was found to be most detrimental to yield, pod size, and quality even when preceded by optimum growing conditions. Irrigation during the period prior to flowering was not necessary provided that soil water was high at planting time. Plants were found to recover rapidly from stress when irrigation water was applied at flowering.

Barrios Gonzalez (39) showed that average yields were 85% higher in plots irrigated weekly when compared with plots depending on natural rainfall. He also found other advantages to weekly irrigation to be better plant stands and reduced incidence of fungus diseases. It also prolonged the life cycle of the plants by an average of 10 days.

Goodoy et al. (40) irrigated beans at intervals of 3, 6, 9, or 12 days in comparison with an unwatered control. The available soil water corresponding to these frequencies fell to about 75, 50, 25, and 0% of the maximum soil water. The plots were then irrigated to field capacity with each application. Yield was significantly higher with the 3 day interval than with any of the others. In the 6, 9, and 12 day intervals the results did not differ significantly from one another, but were superior to those of the controls. In a greenhouse pot experiment, Burman and Bohmont (41) found that beans irrigated at 0.5 atm had a maximum growth rate three times higher than beans irrigated at 4 atm.

The importance of irrigation at blossom time has been demonstrated by many workers as mentioned previously. Vittum et al. (42) have shown that a single irrigation applied at flowering may result in substantially improved yields. Higher yields result largely from increased pods per plant, but pod size also increased.

Pan evaporation data can be used to effectively schedule irrigation of beans under varied climatic conditions. An irrigation program that provided 8 mm of water at 23 mm of pan evaporation until one-third foliage cover, at 15 mm of pan evaporation until two-thirds foliage cover, then at 8 mm of pan evaporation during flowering and pod development and until harvest maintained a low level of soil water tension and resulted in the greatest pod yields (43). Jensen and Middleton (44) suggested that applied irrigation at full foliage cover should be equal to the amount of pan evaporation.

Smittle (45) observed visible leaf wilting when soil water tension was 40–50 centibars. Application of irrigation water with 8 mm of water evaporated from the Standard U.S. Weather Bureau Pan was sufficient to maintain soil water tension below 40 centibars. This regime increased yields by 72% over irrigation applied at 15 or 22 mm of pan evaporation.

Vitkov and Petkov (32) reported that irrigation to a depth of 60 cm gives the best results. Root growth observations made by Halterlein, Teare, and Clayberg (unpublished, discussed in the section on Root System) suggest that irrigation even to this relatively shallow depth may be unnecessary. Kiesling et al. (37) argued against infrequent heavy irrigations. They found that allowing water tension at a depth of 10.1 cm to reach 0.75 bar before irrigation for the entire duration of the growth period reduced yields by 80%. They argue that since root development is greatest near the soil surface early in the growth period infrequent irrigation results in fewer roots in surface layers of the soil because roots die from the resulting water deficit. They maintain that the deeper roots which result from infrequent heavy irrigations are not as effective in using applied irrigation as are surface roots. Thus, when water is applied new root growth is stimulated in surface layers and photosynthate is used for making new, but short-lived, roots (since they will die from lack of water) instead of producing marketable yield. Bascur and Fritsch (46) applied irrigation when tension meter readings at 20 cm depth was either

0.4 or 0.75 atm. Seed weight was increased significantly with the 0.4 atm frequency.

Singh (47) reported that from 0 to −4 bars (about 75% available soil water depletion) soil water stress, the relative turgidity and cell sap concentration of plant leaves were affected significantly. Thus, measurements of these factors can be used to determine the frequency of irrigation. Factors other than soil moisture stress, such as climatic factors, also affect relative turgidity and cell sap concentration of the leaves. Before they are used to schedule irrigation, measurements must be properly calibrated against climatic factors.

Method of irrigation may also be important. Yields were higher from strip irrigated beans than from those watered by overhead irrigation and water use was from 12 to 50% less. A spacing of 6 feet between strip tubes with the length of the materials no longer than 200 feet from the submain was recommended by Bryan et al. (48). Sanders and Miller (49) report bean yields increased as much as 300% by using an evaporative cooling (EC) method of irrigation with 3.8 mm of water applied when air temperature exceeded 29°C when compared with unirrigated beans. Irrigation by evaporative cooling also caused plants to flower sooner and produced yields 3–14 days earlier than in the nonirrigated beans. When compared with conventional irrigation (25.4 mm per week) yields were increased by 115% for fall crops receiving EC irrigation, but yields for spring crops were not increased.

Interlocular cavitation (the formation of cavities in the soft parenchymatous endocarp cells between seed locules) of snap beans was consistently associated with heavy irrigation during pod development and it has been found that irrigation favors the development of this physiological disorder (50, 51). Little or no interlocular cavitation is found when not more than 1.27 cm of water per week is applied. Lee et al. (51) found that pod yield and plant weight increased when irrigation frequency was increased from 1 to 2 times per week, but when irrigation frequency was increased from 2 to 3 times per week a very marked increase in interlocular cavitation was observed as well as a further increase in yield and plant weight.

In the cultivar "Bush Blue Lake 274" percentage dry weight of pods decreased and percentage ash weight increased with irrigation frequency increase. Seed number per pod also increased with irrigation frequency. P and K increased while Ca and Mg decreased with an increase in irrigation frequency (51).

DROUGHT BEHAVIOR

Under conditions of extreme drought and intense sunlight, bean leaves were observed by Dubetz (52) to orient themselves parallel to the incident light (parahelionastism), and to follow the course of the sun in that position. This is a reversible process since the leaves assumed normal positions by early morning. Unstressed plants do not show these nastic movements, nor do plants watered after being under stress. This type of leaf orientation and movement presumably is an adaptation that minimizes transpiration and perhaps prevents thermal death.

Wilson (53) has found that the primary factor inducing hardening against chilling injury in beans is water stress and not low temperature. He concludes that leaf dehydration and injury cannot be attributed to phase change in the membrane lipids of the leaf cell plasmalemma and other organelles. The most important changes occurring during drought hardening are those that occur in stomatal behavior. Drought hardening causes complete stomatal closure and prevents any increase in stomatal aperture on transfer to damaging cold temperatures. This does not occur under chill hardening. Under chill hardening stomata do not close entirely so that when chill hardened plants are moved to colder temperatures stomata are slightly open. The cause of stomatal closure during drought hardening is not known but it may be due to an increase in abscisic acid.

Todd (54) suggested that translocation might be a factor in relative drought hardiness among species. For example, he found that translocation in wheat plants (a relatively drought resistant species) was predominantly to new leaves and movements were unaffected by drought. In beans, however, translocation was into stems and meristematic regions and rates of movement were slower in plants receiving no water.

Couch and Todd (55) observed that bean plants subjected to drought showed a marked decline in proteins and nucleic acid.

STOMATAL CONTROL OF TRANSPIRATION

Kanemasu et al. (56) have observed that for any given set of conditions with respect to light, temperature, nutrition, etc., a reduction in soil water potential results in a reduction in leaf water potential and a concomitant increase in stomatal diffusion resistance. Kanemasu and Tanner (57) measured abaxial and adaxial stomatal resistance and leaf water potential of beans in both the field and the growth chamber. Their measurements show that the two surfaces of the leaflet react differently to water stress (Fig. 5.5). The number of stomata on the abaxial surface of the leaf is 7 times greater than on the adaxial surface, 28,100 vs. 4000 (58, 59). The abaxial stomata were not significantly affected as leaf water potential changed from 0 to −11 bars, but as leaf water potential changed from −11 to −12 bars resistance increased rapidly. The resistance of the adaxial stomata did not increase from 0 to −8 bars, but increased

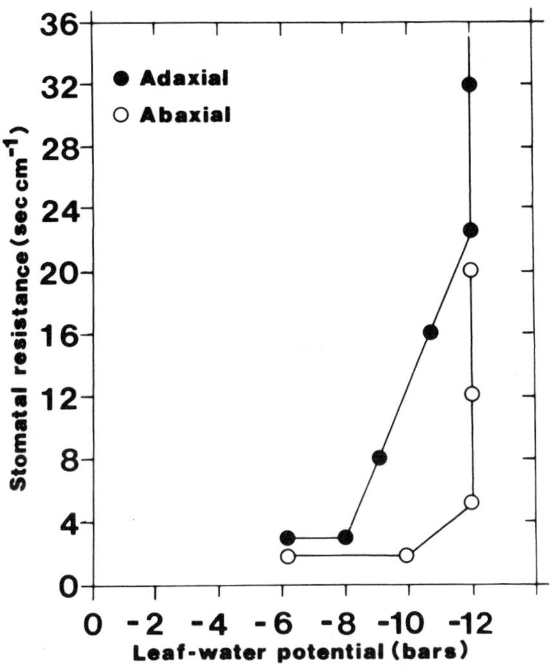

Figure 5.5. Relationship between adaxial and abaxial stomatal resistance and leaf water potential of upper canopy leaves of bean. [After Kanemasu and Tanner (57).]

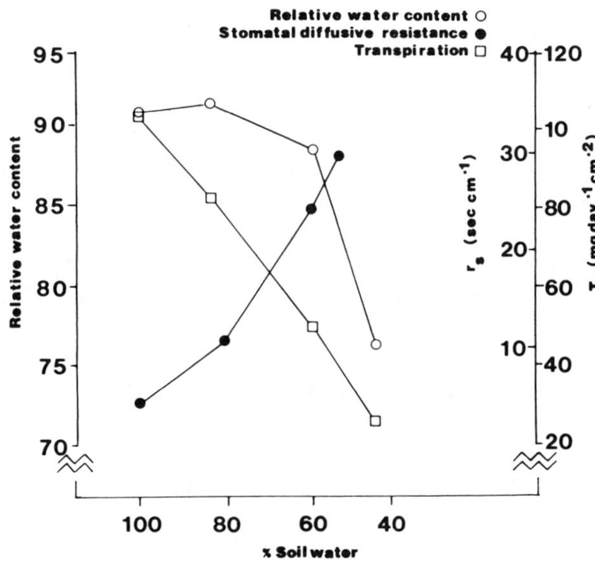

Figure 5.6. Effect of soil water content on the relative water content, stomatal diffusive resistance, and transpiration of beans. [After Al-Ani and Bierhuizen (60).]

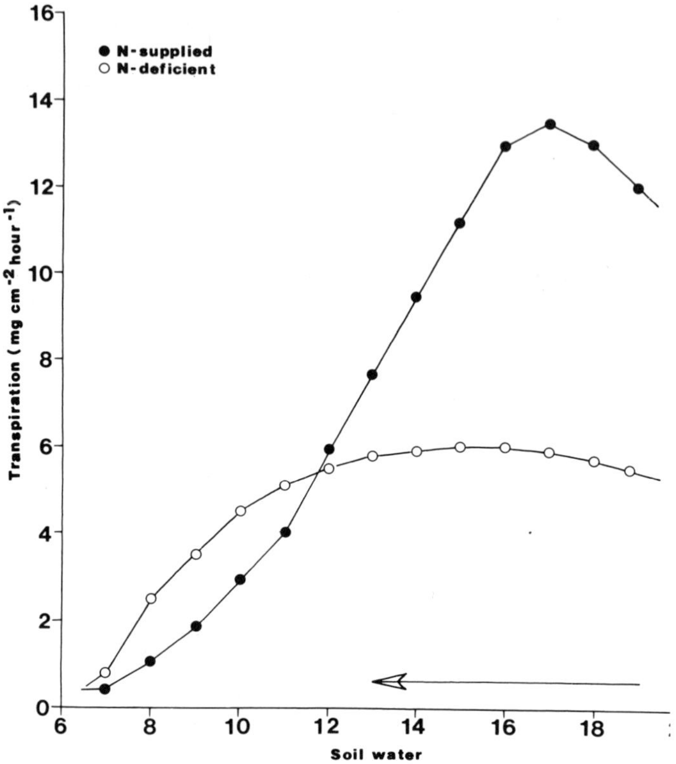

Figure 5.7. The effect of nitrogen supply and soil water on the transpiration rates of bean plants growing in pots. [After Shimshi (61).]

sharply from −8 to −12 bars. Thus, at leaf water potentials between −10 and −12 bars, the average stomatal resistance for both surfaces of the leaf acts as an on–off switch to prevent further decline in leaf water potential.

Millar and Gardner (19) also reported an increase in adaxial stomatal resistance at leaf water potentials lower than −8 bars. This was coincident with a rapid decrease in the dry matter production rate. Stomatal closure due to water stress resulted in a greater reduction of growth rate than in transpiration. Thus, at leaf water potentials of −8 bars and lower, stomatal resistance significantly affected CO_2 uptake and as a consequence severely reduced dry matter production. However, Solarova et al. (59) found that the response to progressive increases in water stress was parallel for both surfaces, and that the conductance of the abaxial epidermis was only 50% higher than the conductance of the adaxial epidermis. They also observed that the abaxial stomata are 25% shorter than the adaxial stomata.

An increase in stomatal resistance values with increasing soil water stress has been observed by Al-Ani and Bierhuizen (60). This effect is more pronounced when soil water content falls below 80% *ASW*. This phenomenon leads them to conclude that stomatal diffusive resistance is an excellent criterion for estimating the water deficit in the plant. Relative water content of the plant, however, was found to be a poor criterion since there is no measurable effect of water stress on relative water content up to severe water stress. Their results demonstrate that stomatal resistance can be used as a tool to predict the soil water–plant relationship. Transpiration rates were highest following irrigation when soil water content was at 100% of field capacity, then steadily declined with decreasing soil water, at the same time stomatal diffusive resistance increased. The relative water content of the leaves did not change significantly until a severe soil water stress occurred at 50% soil water depletion (Fig. 5.6).

Shimshi (61) found that nitrogen-deficient plants do not open their stomata as widely, nor close them as tightly, as plants supplied with adequate nitrogen. Thus, the transpiration rate of plants supplied with adequate nitrogen is higher than nitrogen-deficient plants when soil moisture is high. But as the soil dries the transpiration rate of nitrogen supplied plants drops below the rate of nitrogen deficient plants (Fig. 5. 7). Thus, the nitrogen status plays a significant role in stomatal control and transpiration.

CONTROL OF TRANSPIRATION BY ROOT RESISTANCE

Shirazi et al. (62) studied root resistance to water uptake by using a suction apparatus on the decapitated root system of kidney beans. This method simulates increasing or decreasing transpiration demand. They were able to show that water uptake increases exponentially with increasing suction. Decreases are also exponential. The magnitude of the water uptake in the descending path was shown to be higher than on the path of ascending suction. Thus, two values of water uptake were observed for each value of suction indicating a path-dependent nature of the water uptake rate (Fig. 5.8).

The effect of root temperature on the water uptake rates has also been studied (62). As temperature increased from 15 to 25°C, increased uptake was observed. Although uptake rate declined beyond 25°C a significant rise in uptake was observed from 30 to 35°C (Fig. 5.9). It was concluded that the rise in uptake may have been due to a breakdown of the protoplasmic membrane, thus eliminating the site of resistance in the roots. Such a sudden increase of water uptake has been noted in dead roots (63).

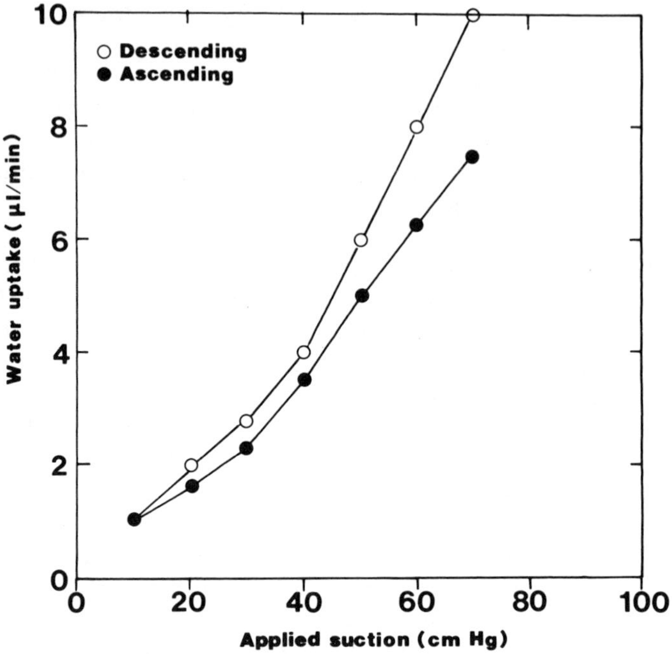

Figure 5.8. Water uptake rate of decapitated plants. Path of descending suction applications was always above ascending regardless of whether the starting point was high or low. [After Shirazi et al. (62).]

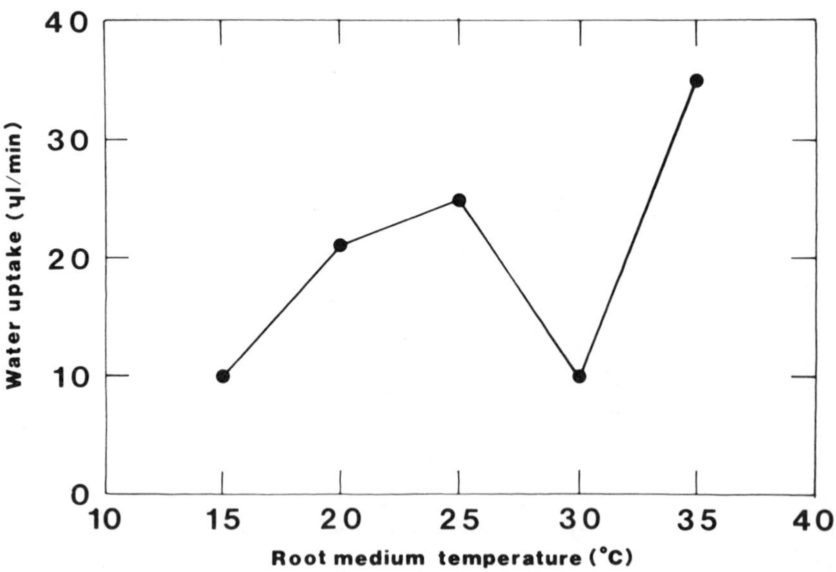

Figure 5.9. Water uptake rates of kidney beans as affected by temperature determinations made at a suction of 500 mm Hg. [After Shirazi et al. (62).]

ABSCISIC ACID

It has been proposed that stomatal closure is due to increased abscisic acid (ABA) levels since wilting under water stress has been observed to be accompanied by high ABA levels (64). Thus, there is strong evidence indicating that under water stress stomatal closure is controlled by endogenous ABA levels. The alleviation of stress may be one of the major functions of ABA.

Cis–trans ABA is a hormone synthesized in the leaf (65). The rate of synthesis of this hormone and consequently the concentration of the hormone in the leaf increases as the result of water stress (66). Cummins et al. (67) have found that ABA acts on stomata directly and is initiated within 5 min of application. Upon withdrawal of the hormone supply stomata begin reopening within 5 min.

The evidence that ABA regulates water loss under stress by modulating stomatal apertures includes the observations that in leaves and epidermal strips, stomata begin to close quickly when ABA is introduced into the transpiration stream and they reopen when ABA is removed. ABA levels in leaves rise rapidly and dramatically when either whole plants or single leaves are water stressed. Walton et al. (68) have shown that at a particular time after the initiation of a mild stress it is possible to obtain a good correlation between total leaf ABA content and stomatal resistance and that there is a threshold ABA concentration necessary to initiate stomatal closure. This was true for plants transferred to −3 bars solutions for 5 hr, but for plants transferred to −5 bars solutions there was no apparent relationship between total ABA content and stomatal resistance. Total ABA levels in leaves may not be helpful in understanding stomatal response to water stress and these workers suggest that a knowledge of ABA levels in guard cells might be more informative.

In wilted bean leaves ABA levels may be regulated in the following manner as proposed by Harrison and Walton (15): (1) wilting triggers an increased rate of ABA synthesis which is followed by an increased rate of metabolism; (2) ABA concentration increases and becomes relatively constant when the increased rates of synthesis and metabolism become approximately equal; and (3) ABA levels decrease as turgor increases due to decreased rates of ABA synthesis relative to metabolism. The increased rate of metabolism which occurs during wilting may result from increased levels or activities of metabolic enzymes due to water stress, or to the resultant elevated ABA concentrations.

Kriedemann et al. (69) have found that a twofold increase in ABA is sufficient to initiate stomatal closure. Stomata began closing by 8 min and were completely closed by 30 min. This suggests that ABA regulates a stomatal function. The work of Hiron and Wright (70) seems to firmly establish abscissic acid as the hormone responsible for alleviating the effects of water stress. Not only does it protect during periods of severe stress, such as drought, but it apparently enables plants to withstand smaller stresses for long periods by a process of adaptation. In this process when a continuous warm air stream was directed on the leaves, plants wilted and then recovered gradually. This was accomplished by an increased ABA level in leaves and an increase in leaf stomatal resistance, suggesting ABA-initiated stomatal closure enabled bean plants to regain full turgor.

Although there is considerable evidence indicating that ABA plays a causal role in regulating stomatal aperture in water stressed plants, Hsiao (71) cautions that there is no conclusive proof of this. This agrees with the conclusion of Walton et al. (68) who suggested that proof may have to include measurements of ABA fluxes to guard cells under conditions of both plant stress and nonstress.

DRY MATTER ACCUMULATION

Millar and Gardner (19) found that transpiration and dry matter production rates decreased curvilinearly with soil water potential. When soil water potential decreased from −0.28 to −0.40 bar there was a 47% reduction in dry matter production rates (Fig. 5.10). This behavior is related to the turgor pressure operated stomatal mechanism. Adaxial stomatal resistance increased at leaf water potential lower than −8 bars, which coincided with a rapid decrease in the dry matter production rate. Stomatal closure due to water stress resulted in a greater reduction of growth rate than in transpiration.

Soil water conditions were also shown by Maurer et al. (8) to affect plant height and dry matter production (Table 5.1). By using five different water regimes they found that the highest water regime produced the tallest plants with more branches and leaves than the lowest regime. The three other water treatments were intermediate in their effect on

Table 5.1. Effect of five soil water regimes on plant growth and water use[a]

Water Regime	Plant Height (cm)	Dry Matter of Plants (%)	Dry Matter Pods Plus Plants[b] (%)	Total Water Used (liters)
High	61	20	280	116
Low	46	19	172	69
Medium	51	19	206	87
High–low	55	19	200	80
Low–high	50	18	199	85

[a] Adapted from Maurer et al. (8), with permission of the Agricultural Institute of Canada, Ottawa, Canada.

[b] Calculated on the basis on % dry matter of plants and % dry matter of pods.

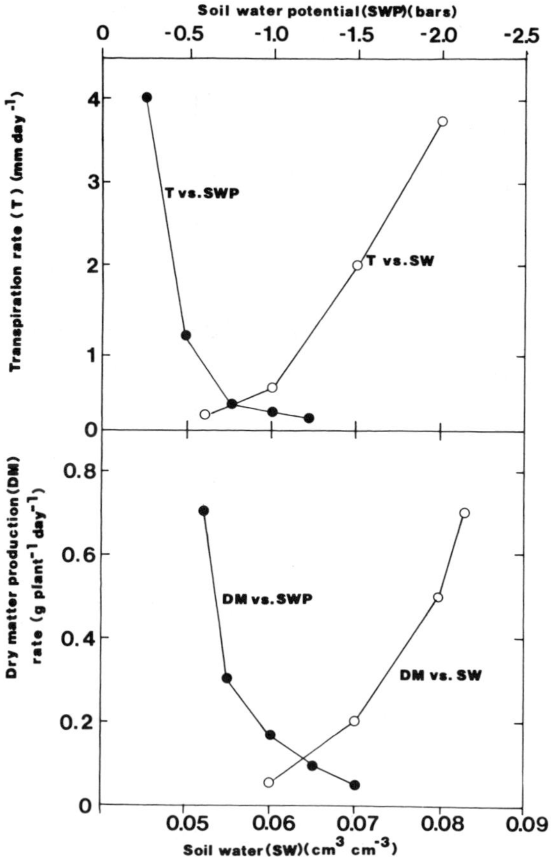

Figure 5.10. Transpiration rate and dry matter production rate of snap beans as a function of soil water and soil water potential. [After Millar and Gardner (19)].

plant growth. Similarly, dry matter production was significantly greater in the highest regime and least in the lowest regime. Although total water use was greatest in the high water regime, total dry matter produced per liter of water was not significantly different from the other treatments.

PHOTOSYNTHESIS

According to Moldau (72) there is a gradual decrease in photosynthesis with the onset of water stress which is brought about mainly by a decrease in the stomatal conductance. O'Toole et al. (73) have observed that net photosynthesis and transpiration rates begin to decrease while boundary layer and stomatal resistance increase at about -3 to -5 bars

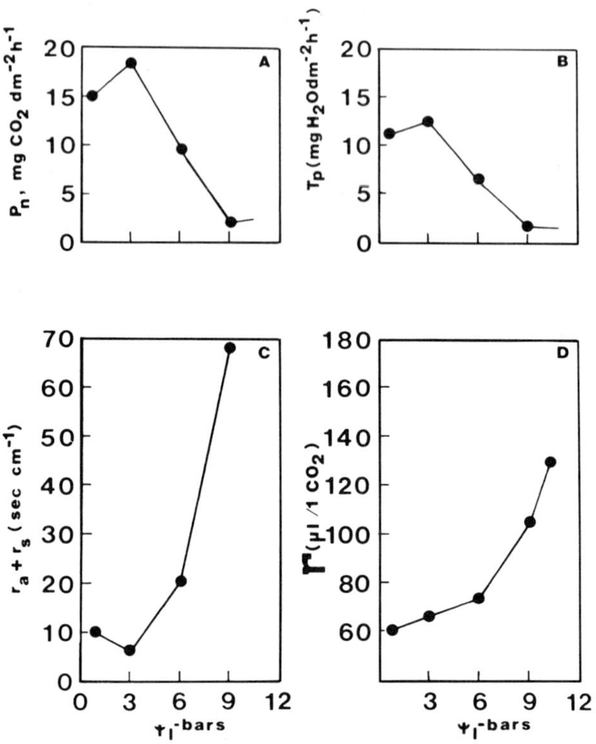

Figure 5.11. Response of (A) net photosynthesis (P_n), (B) transpiration (T_p), (C) boundary layer plus stomatal resistance to CO_2 diffusion ($r_a + r_s$), and (D) CO_2 compensation point (Γ) of bean leaflets at different leaf water potentials (ψ) . [After O'Toole et al. (73).]

leaf water potential. As the stress increased, net photosynthesis and transpiration rates were near 0 at -9 to -10 bars while CO_2 compensation point reached its highest level in this leaf water potential range (Fig. 5.11). This parallel decrease in net photosynthesis and transpiration rates strongly indicates stomatal closure as the principal causal factor in water stress mediated reduction of net photosynthesis. This basically agrees with the findings for other crop species in which the primary cause of decreased net photosynthesis due to water stress is closure of stomata in a critical leaf water potential range (74, 75). Although it has been demonstrated that photosynthetic rate declines quickly with the onset of severe water stress due to rapid stomatal closure, the translocation rate appears to be much less sensitive to water stress. After a brief transient increase in rate the translocation rate remains nearly constant for several hours following the initiation of water stress (76).

Although reports of in-phase changes in net photosynthesis and transpiration strongly implicate stomatal control as a primary causal factor in decreasing net photosynthesis under water stress conditions considerable evidence exists to support the supposition that nonstomatal factors are also involved. Increased mesophyll resistance has been associated generally only with severe stress levels. Some reports have shown that the enzyme ribulose bisphosphate (RuBP) carboxylase is sensitive to moderate levels of internal water stress. Activity of RuBP carboxylase was shown to be correlated positively with net photosynthesis by Neales et al. (77). The predominant part of the differences in net photosynthesis could not be attributed to changes in leaf stomatal resistance. They concluded that at saturating light intensities and CO_2 concentrations of 360 ppm the predominant rate limiting factor in net photosynthesis was the rate of carboxylation and not the rate of diffusive transfer of CO_2 to the photosynthetic sites.

Reports of reduced net photosynthesis prior to reduced transpiration also suggest that nonstomatal factors may play a significant role in the reduction (78–80). O'Toole et al. (73) have shown that decreased carboxylase activity also contributed to decreased net photosynthesis. The fact that mesophyll resistance increased and RuBP carboxylase activity decreased with lower leaf water potential (Table 5.2) indicates that both stomatal and nonstomatal factors were responsible for the reduction in net photosynthesis. Gale et al. (81) have shown that mesophyll resistance constitutes a significant percentage of the total resistance to photosynthesis under conditions of high vapor pressure differences and/or low leaf water potential. Mesophyll resistance is apparently independent of stomatal resistance.

O'Toole et al. (82) have shown that mesophyll resistance increases with decreasing leaf water potential. Concurrently, the CO_2 compensation point increases and RuBP carboxylase activity decreases as internal water stress increases. Thus, RuBP carboxylase may be a valuable biochemical indication of water stress. Pospisilova et al. (83) have provided indirect evidence that water stress not only inhibits CO_2 transport

Table 5.2. Response of mesophyll resistance (r_m) and ribulose bisphosphate carboxylase (RuBPc) activity to decreasing leaf water potential[a]

Range (bars)	Number of Observations	r_m ($sec\ cm^{-1}$)	RuBPc ($cpm \times 10^{-3}\ cm^{-2}$)
0 to − 3.0	15	7.0	389
−3.1 to − 6.0	14	9.4	364
−6.1 to −10.0	9	13.6	275

[a] Adapted from O'Toole et al. (73).

from atmosphere to carboxylation sites in chloroplasts, but also CO_2 conversion into photosynthate. They have observed that with decreasing water potential photorespiration rate decreases, but dark respiration rate is unchanged. Simultaneously, CO_2 compensation concentration increases.

Fraser and Bidwell (84) have demonstrated that there is an intrinsic ontogenetic pattern of gas exchange for each bean leaf which was modified by events taking place during development in other parts of the plant, such as subsequent leaf emergence or flowering. In the primary leaf, they found a brief transient increase in photosynthesis before or during the emergence of each subsequent leaf. Also, an increase in CO_2 uptake occurred in each leaf at the time of flowering. Photosynthesis was regulated by mesophyll resistance and weakly regulated by stomatal resistance. New leaves presumably did not constitute major sinks since their appearance causes only minor disturbances in rates of gas exchange while flowering may have been a more demanding sink, the effect of which was felt for a longer time.

Tanaka and Kikuchi (85) showed that the photosynthetic activity of the primary, first and third trifoliate leaves of bush type bean plants reached a brief peak then decreased rapidly with senescence. Leaves developing later and comprising the upper canopy had a slower developmental rate, however, and a more prolonged peak level of photosynthetic activity. Thus, rapid development and short duration of peak net photosynthetic rates in lower leaves may be part of the programmed developmental pattern of the bush type plant.

Completion of morphological development or attainment of full leaf area and length does not exactly coincide with maximum rates of net photosynthesis as the maximum net photosynthesis rate develops slightly before leaflet expansion. The increase in net photosynthesis and transpiration rates during leaf expansion occurs partially as a result of a decrease in stomatal resistance, but the increase in net photosynthesis may also be due partially to nonstomatal factors as indicated by decreased mesophyll resistance and increased RuBP carboxylase activity. O'Toole et al. (86) observed that net photosynthesis rate and RuBP carboxylase activity increased with leaflet expansion reaching peaks just prior to full expansion. Concurrently, boundary layer plus stomatal resistance, CO_2 compensation point and malate dehydrogenase activity decreased reaching lowest values at full leaflet expansion. Shortly after peak photosynthetic activity was reached all trends reversed as the physiological activity of the first trifoliate leaf of the determinate type plant changes to that of a senescent leaf. Treharne et al. (87) reported that synthesis of fraction 1 protein in 18-day-old bean leaves was complete much before full leaf expansion. Their results suggest that the increased activity of RuBP carboxylase during leaf expansion may be due to increased activation of the enzyme rather than to *de novo* synthesis.

PROLINE

Proline has been observed to accumulate in many species under stress conditions such as disease, wilting and nutrient deficiency. Stewart et al. (88) reported that wilted bean leaves accumulated 1.3 times as much proline as unwilted leaves after 24 hr of incubation. Four bean cultivars subjected to water stress behaved similarly and accumulated free proline, free amino acids, and soluble and insoluble proteins to a high level, but these decreased to the original level 24 hr after recovery from stress according to Rena and Masciotti (89). Changes in the content of free proline in dehydrated discs from the primary leaves of 20 bean cultivars showed that under conditions of both mild and severe osmotic dehydration all cultivars accumulated a high level of free proline compared with a turgid control (90). The cultivars could be divided into three groups: those with high, intermediate, and low capacity to accumulate proline. Under conditions of natural turgidity there were significant differences in free proline contents among cultivars.

Stewart (91) has shown that the decrease in proline utilization in nonstarved leaves due to wilting is clearly due to a decrease in the rate of protein synthesis. In starved leaves the rate of proline utilization was increased by wilting. Wilting causes an increase in the rate of proline formation in both starved and nonstarved leaves. It appears that this effect of wilting is not dependent on a large supply of carbohydrate in the leaf. On the basis of Stewart's work, the role of carbohydrates in proline accumulation appears to be the prevention of proline oxidation. Wilting caused an increase in nonprotein proline formation; it caused a decrease in proline utilization in nonstarved leaves by decreasing protein synthesis. In starved leaves wilting caused an increase in the rate of proline utilization but this is due to the higher content of proline in wilted leaves compared to the turgid leaves which causes more proline utilization by oxidation. The primary effects of wilting which lead to the accumulation of proline were to decrease protein synthesis and to increase proline formation. The source of the proline is not known, but the carbohydrate content of the leaf does not affect the increased formation of proline.

OTHER FACTORS IN WATER RELATIONS

Taerum (92) used a pressure chamber to measure the xylem water potential of leaves, shoots, and roots from beans grown with a 12 hr dark period and natural or artificial light conditions during the day. Water potentials were measured at the end of a dark period and during a light period. Measurements taken at the end of the dark period indicated normal potential gradients within the soil/plant system (leaf < shoot < root < soil) when the soil water potential was relatively high (above −0.02 bar) and the gradients remained normal during the day (under natural light conditions). When the soil water potential was −1.0 bar or lower in the morning the root xylem water potential was higher than the soil water potential; at very low soil water potential (<−4.0 bars) it remained higher during most of the day. From measurements of stomatal diffusion resistance, transpiration, relative water content of leaves, and of changes in the matric potential of soil water, it was concluded that when the matric potential of soil water was low water could be taken up by the plant against a water potential gradient. Because leaf xylem water potential was always lower than root xylem water potential the mechanism involved in the inversion of water potential gradient must be localized in the roots and probably related to ion uptake. Probably an active ion uptake with concomitant water uptake may be involved. If the ions were somehow deposited along the water pathway, the excess water would be carried upward by the transpiration stream along the potential gradient. In this way, water could be taken up by the root even if the water potential of the root xylem were higher than in the soil. This mechanism might, at least temporarily, enable the plant to survive drought conditions because the leaves remain under less water stress.

Shimshi (93) has shown that nitrogen deficiency impairs the ability of beans to adjust their water status to changes in soil water by regulation of stomatal transpiration and of sap solute concentration. When soil water was relatively high, the leaves of nitrogen deficient plants transpired less than those of plants supplied with sufficient nitrogen. The transpiration rate was closely associated with the chlorophyll content of the leaves of various ages. Stomatal width was much wider in nitrogen supplied plants than in nitrogen deficient plants. Throughout the available soil water range the water saturation deficit was higher than in nitrogen supplied leaves. In the dry range of soil water, chlorotic leaves transpired more than normal green leaves. Nitrogen deficient plants failed to show a sharp rise in the content of soluble metabolites in the sap when they approached the wilting range. The content of cell wall materials was higher in nitrogen deficient plants.

The relative turgidity and cell sap concentration measurements of bean leaves under field conditions reflect soil water conditions over the low range of soil water stress which also represents the range of more easily available water.

Moderate water stress (−6.0 to −8.0 bars leaf water potential) does inhibit several biochemical functions within the chloroplast and perhaps a key control point is decreased activity of RuBP carboxylase (82). Results with beans are in agreement with those obtained from cotton (94) and spinach (95) and leave little doubt that even mild to moderate stress can affect the biochemical processes of carbon fixation.

ROW SPACING

Several studies have investigated the effect of row spacing on yield. Results have shown that yields are significantly increased by closer row spacing and higher plant populations than are now used. However, the necessity of planting to accommodate mechanical harvesters is most likely to control row spacings. Mack and Hatch (96) have shown that snap bean yields were highest when plants were uniformly distributed in a square arrangement. At a 12.7 × 12.7 or 15.2 × 15.2 cm arrangement (430,000–620,000 plants per hectare) yields were 36% higher than for the same number of plants in 36 inch rows. Kueneman et al. (97) have shown that for a given density beans planted in narrow rows at more equidistant spacing will usually yield more than when planted in a more rectangular arrangement. Crandall et al. (98) reported yield increases of 64% with row spacings of 30.5 cm when compared with 91.4 cm rows at 8 plants per 30.5 cm of row in each case. They also observed that the increase was directly related to an increase in leaf area (64.4%). Chung and Goulden (99), Atkin (100), and Froussios (3) all reported similar results. Robins and Domingo (10) observed no interaction between four row spacings and three soil water regimes.

WATER RELATIONS OF DISEASE

Plants infested by *Uromyces phaseoli* were reported by Dunaway and Durbin (101) to be highly susceptible to drought due to excessive water loss through sporulating media and reduced root:shoot ratios. Plants wilted at soil water potentials greater than −1.0 bar, but healthy plants did not wilt until the soil water potential fell below −3.4 bars. Diseases that do not characteristically cause wilting are not generally thought to affect host water relations adversely, although there are indications that foliar infection, particularly by rust fungi, can cause unusual levels of plant water stress to occur. For example, rusted bean leaves had lower water content than healthy leaves under conditions conducive to water stress (102). This was also true for corn. Rust has also been shown to alter transpiration rates (103) and root growth (104). Rust infected bean plants wilted long before healthy plants when water was withheld (105).

Beans may fail to emerge in wet soil because of mechanical impedance, high soil temperatures, and microorganism attacks. Lack of oxygen in wet soil (soaking injury) is a less likely cause of poor emergence. Satisfactory emergence is obtained from seed sown at 2.5 cm depth and then irrigated provided that the soil is not allowed to dry or become excessively heated. Orphonos (106) found that seeds germinated but failed to emerge when planted deeper than 2.5 cm, probably due to the prevalence of mechanical injury, such as lost cotyledons and seedlings with thickened, flattened, and hollowed hypocotyls.

SOIL SALINITY

Over a 3 year period snap bean yields were reduced an average of 34% when irrigated with saline water. Of 12 crops studied, beans were the most severely affected by high water salinity (107).

Growth depression due to soil salinity was expressed in lower dry weight yields, smaller numbers of leaves per plant and smaller leaf area according to Meiri and Poljakoff-Mayber (108). Shoots and leaves were more severely affected by salinity than were roots. Daily increment in leaf area was found to be a sensitive measure of the effect of salinity with leaf expansion retarded immediately at the beginning of salinization proportionally to the rate of salinization. Gradual salinization did not inhibit leaf expansion completely, and when plants were transferred abruptly to a nonsaline medium an immediate and large increase in leaf growth rate resulted. When plants were transferred abruptly to a saline medium, leaf growth ceased almost completely, but recovered slowly with time. Water consumption of saline-grown plants was much lower than controls due to a lower transpiration rate and a smaller total transpiring area. This low transpiration rate has been reported to be due to partial closing of the stomata (81).

Jensen (109) found the maximum growth of top, root, and leaf areas occurred at an osmotic potential in the root medium of -0.9 atm. At -3.9 atm the growth of top dry matter was decreased about 42%. Net photosynthesis measured on single attached leaves in a leaf chamber was reduced 22% at a root medium osmotic potential of -3.9 atm as compared with the photosynthesis at 0.4 atm. Jensen found that about one-third of this reduction was due to an increase in leaf diffusive resistance in spite of osmotic adjustment of the leaves. The other two-thirds of the reduction was due to an increase in mesophyll resistance. Dark respiration decreased at the lower soil osmotic potential, as calculated per gram of leaf dry matter. But the light reactions of photosynthesis were not influenced by lowering soil osmotic potential.

The duration of exposure to salinity is the main factor influencing growth according to Meiri et al. (110). Salinity reduced tranpiration rate, but transpiration during exposure to salinity did not affect either salt accumulation in the leaves or any of the effects resulting from exposure to salinity. The plants were not able to fully adjust to salinity.

Increasing salinity led to greater leaf succulence and substantially higher ratios of mesophyll surface area to leaf area (111). But the increased internal surface for CO_2 absorption did not lead to higher rates of photosynthesis because the CO_2 resistance increased even more due to stomatal closure. Gale et al. (81) found that stomatal closure was responsible for most of the 30% reduction in photosynthesis of salinized beans. This was due to interference with CO_2 diffusion and could be overcome by raising the CO_2 concentration of the air. But at a later stage of growth salinity affected the light reaction of photosynthesis so that elevation of the air CO_2 concentration had little effect.

Nieman and Paulsen (112) found that high humidity consistently relieved the suppressive effect of salinity while a hot, dry environment has been shown to aggravate the effects of salinity (113); and hot dry periods during flowering, but not vegetative growth, decrease yield. Stomatal conductance was higher in a cool, humid rather than in a hot, dry environment for any value of the root medium osmotic potential. It seems likely that at least part of the growth reduction under saline conditions is due to the combined action of salinity and transpiration in reducing plant water potential. If transpiration were kept low by increasing atmospheric humidity, growth should improve. The observation that some crops are less sensitive to salinity in cool humid climates seems to support this hypothesis (30).

Meiri and Poljakoff-Mayber (114) found that increasing substrate salinity unfavorably affected the water balance of leaf tissue. Reduction in osmotic potential of the leaf tissue was slower than reduction in water potential and this caused a lag in adjustment. Consequently, a reduction in relative water content and in turgor pressure occurred. This lag was greater with rapid substrate salinization than it was with slow salinization. Cell sap concentration increased with increasing substrate salinity due to an increase in both the organic and inorganic constituents. Inorganic constituents, however, were predominant; the accumulation of chloride was especially marked. As cell sap concentration increased turgor pressure was gradually restored. Through exposure to constant salinity, turgor was completely restored and occasionally reached higher levels than in the mature leaves of control plants. Plants transferred from a saline to a nonsaline medium showed an increased gradient in water potential and increased water absorption by the tissues. As a result, cell sap was diluted, and the total water potential of the tissue, relative water content, and turgor pressure increased.

Frota and Tucker (115) have shown that protein synthesis is more severely reduced under salt stress than under water stress. The absorption rate of both NO_3^- and NH_4^+ has been shown to be impaired by osmotic stress due to saline conditions (116). Plants under water stress showed a higher absorption rate for NH_4^+ than for NO_3^-. Under saline conditions the same absorption rate prevailed for both nitrogen sources. The observed decrease in nitrogen uptake rate by beans under stress was related directly to the reduction in root permeability. Although nitrogen concentration was higher in stressed plants, unstressed plants contained more total nitrogen. Dry matter yield was more severely affected by stress than was nitrogen uptake. Under salt stress plants absorbed the same amount of nitrogen independent of source. Under water stress more NH_4^+ than NO_3^- was absorbed.

Beans growing under conditions of increasing salinity obviously suffer from a disturbed water balance. The effect is usually not permanent and is at least partially overcome by a process of adapting to the saline conditions. The results of recent research suggest that beans may not be as sensitive to saline conditions as previously thought.

SENESCENCE

Davis et al. (117) found that in a fully expanded leaf exposed to constant light the diffusive resistance of the whole leaf increased continuously with age until leaf abscission. This was also true for leaves that were continuously shaded at 15% of the full light level, but the resistances were higher. The effects of correlated factors such as shading, stage of plant development, and stomatal density were neglible. The leaf age effect occurred not only during the latter stages of senescence when the leaves were becoming chlorotic or necrotic, but as a continuing process that began at leaf maturity and continued until leaf abscission. These observations and those of Dumbroff et al. (118) suggest that changes in the level of abscisic acid in cotyledons is not directly related to advancing senescence, but may be a response to marked changes in water potential.

REFERENCES

1. H. S. Gentry, *Econ. Bot.* **23,** 55–69 (1969).
2. J. F. Davis, *J. Agric. Res.* **70,** 237–249 (1945).
3. G. Froussios, *Exp. Agric.* **6,** 124–141 (1970).
4. D. R. Viglierschio and F. W. Went, *Am. J. Bot.* **44,** 449–453 (1957).
5. R. B. Austin and M. S. M. MacLean, *Photosynthetica* **6,** 41–50 (1972).
6. L. A. Richard (Ed.), *USDA Handbook 60* (1954).
7. J. L. Wright and M. E. Jensen. *Trans. ASAE* **21,** 88–91, 96 (1978).
8. A. R. Maurer, D. P. Ormrod, and N. J. Scott, *Can. J. Plant Sci.* **49,** 271–278 (1969).
9. D. C. MacKay and C. A. Eaves, *Can. J. Plant Sci.* **42,** 219–228 (1962).
10. J. S. Robins and C. E. Domingo, *Agron. J.* **48,** 67–70 (1956).
11. L. Dreibrodt, *Acker Pflanzenbau* **95,** 353–359 (1952). (From P. J. Salter and J. E. Goode, Crop Responses to Water at Different Stages of Growth, *Res. Rev. No. 2 Comm. Bur. Hortic. Plant. Crops,* pp. 53–57 1967.)
12. J. F. Bierhuizen and N. M. DeVos. *Rep. Conf. Suppl. Irrig. Comm. VI Int. Soc. Soil Sci., Copenhagen,* pp. 83–92 (1959).
13. H. Kriegbaum, From P. L. Salter and J. E. Goode, *Res. Rev. No. 2,* pp. 53–57. Commonw. Bur. Hortic. Plant Crops, Maidstone, Kent, U.K., 1967.
14. W. N. Gableman and D. D. F. Williams, *Wis. Agric. Exp. Stn. Res. Bull.* **221,** 157 (1960).
15. M. A. Harrison and D. C. Walton, Plant Physiol. **56,** 250–254 (1975).
16. B. E. Jones, *Proc. Am. Soc. Hortic. Sci.* **51,** 457–462 (1948).
17. B. E. Jones and W. O. Drinkwater, *Conn. Agric. Exp. Stn. Bull.* **338** 82 pp. (1959).

18. A. A. Kattan and J. W. Fleming, *Proc. Am. Soc. Hortic. Sci.* **68,** 329–342 (1956).
19. A. A. Millar and W. R. Gardner, *Agron. J.* **64,** 559–562 (1972).
20. W. E. Nelson, *The Effect of Soil Moisture Stress at Critical Stages of Growth of Some Vegetable Crops*, Thesis, Rutgers State University, New Jersey, 1962.
21. M. Vitkov, *Rastenievud. Nauki.* **12,** 100–104 (1975).
22. L. Parjol, *Inst. Cercet. Cereale Plante Teh. Fundulea Sect Amelior Plant Probl. Genet. Teor Apl* **40,** 243–266 (1975).
23. A. Gonzalez and J. W. Williams, *Ark. Farm Res.* **27,** 3 (1978).
24. S. Dubetz and P. S. Mahalle, *J. Am. Soc. Hort. Sci.* **94,** 479–481 (1969).
25. E. C. Wittmeyer, *Am. Veg. Grower* **20,** 13–15 (1972).
26. R. Stoker, *N.Z. J. Exp. Bot.* **2,** 13–15 (1974).
27. S. Deliblatov and M. Sarkizov, *Rastenievud. Nauki* **11,** 123–132 (1974).
28. E. C. Stegman and H. M. Olsen, *N.D. Res. Rep. No. 61.* (1976).
29. M. Vitkov and T. Gruev, *Rastenievud. Nauki.* **10,** 99–104 (1973).
30. O. C. Magistad, A. D. Ayers, C. H. Wadleigh, and H. G. Gauch, *Plant Physiol.* **18,** 151–166 (1943).
31. M. T. Vittum and G. H. Gibbs, *Farm Res.* **27,** 15 (1961).
32. M. Vitkov and M. Petkov, *Rastenievud. Nauki.* **9,** 73–79 (1972).
33. L. N. Mielke and N. H. Peck, *Agron. J.* **59,** 602–604 (1967).
34. H. N. Peck, M. T. Vittum, and G. H. Gibbs, *Agron. J.* **60,** 23–26 (1968).
35. H. Peev and V. Kolev, *Rastenievud. Nauki.* **1,** 93–102 (1964).
36. J. W. O'Leary and G. N. Knecht, *J. Am. Soc. Hortic. Sci.* **96,** 263–265 (1971).
37. T. Keisling, D. Smittle, J. Stansell, and M. Walker, *Ga. Agric. Res.* **17,** 24–27 (1975).
38. C. W. Reynolds, *Rep. Md. Veg. Growers Assoc. 36th Annu. Meet.* pp. 5–8 (1955).
39. A. Barrios Gonzalez, *Mem. 6 as J. Agron. Maracaibo 1, Pap. 11,* p. 8 (1966).
40. O. P. Goodoy and others. *Rev. Agric. Pinocicaba.* **41,** 145–153 (1966).
41. R. D. Burman and D. W. Bohmont, *Agron. J.* **53,** 354–355 (1961)
42. M. T. Vittum, R. B. Alderfer, B. E. Janes, C. W. Reynolds, and R. A. Struchtmeyer, *N.Y. State Agric. Exp. Stn. Bull. 800,* 63 pp. (1963).
43. D. A. Smittle, *J. Am. Soc. Hortic. Sci.* **101,** 37–40 (1976).
44. M. C. Jensen and J. E. Middleton, *Wash. Agric. Exp. Stn. Circ.* **527,** 1–13 (1970).
45. D. A. Smittle, *Ga. Agric. Res.* **16,** 6–8 (1974).
46. B. G. Bascur and F. N. Fritsch, *Agric. Tec.* **35,** 147–152 (1975).
47. R. Singh, *J. Indian Soc. Soil Sci.* **18,** 215–220 (1970).
48. H. H. Bryan, W. M. Stall, J. D. Dalton, and H. W. Ford, *Proc. Fl. State Hortic. Soc.* **88,** 190–196 (1975).
49. D. C. Sanders and C. H. Miller, *Am. Veg. Grower* **25(9),** 26, 52 (1977).
50. J. M. Lee, P. E. Read, and D. W. Davis, *Annu. Rep. Bean Improv. Coop.* **15,** 65–68 (1972).
51. J. M. Lee, P. E. Read, and D. W. Davis, *J. Am. Soc. Hortic. Sci.* **102,** 276–278 (1978).
52. S. Dubetz, *Can. J. Bot.* **47,** 1640–1641 (1969).
53. J. M. Wilson, *New Phytol.* **76,** 257–270 (1976).
54. G. W. Todd, *Abstr. 11th Int. Bot. Congr. Seattle.* p. 220 (1969).
55. R. Couch and G. W. Todd, *Proc. Okla. Acad. Sci.* **52,** 58–61 (1972).
56. E. T. Kanemasu, G. W. Thurtell, and C. B. Tanner, *Plant Physiol.* **44,** 881–885 (1969).
57. E. T. Kanemasu and C. B. Tanner. *Plant Physiol.* **44,** 1547–1552 (1969).

58. B. S. Meyer, D. B. Anderson, R. H. Bohning, and D. G. Fratianne, *Introduction to Plant Physiology*, D. Van Nostrand, New York, 1973, pp. 72–76.

59. J. Solarova, J. Vaclavik, and J. Pospisilova, *Biol. Plant* **19,** 59–64 (1977).

60. T. A. Al-Ani and J. Bierhuizen, *Acta Bot. Neerl.* **20,** 318–326 (1971).

61. D. Shimshi, *New Phytol.* **69,** 405–412 (1970).

62. G. A. Shirazi, J. F. Stone, L. S. Croy, and G. W. Todd, *Physiol. Plant.* **33,** 214–218 (1975).

63. P. D. Kramer, *Am. J. Bot.* **20,** 481–491 (1933).

64. S. T. C. Wright and R. W. P. Hiron, in D. J. Carr, Ed., *Proceedings of the 7th International Conference on Plant Growth Substances*, Springer-Verlag, New York, 1970, pp. 291–298.

65. B. V. Milborrow and R. C. Noddle, *Biochem. J.* **119,** 727–734 (1970).

66. S. T. C. Wright. *Planta* **86,** 10–20 (1969).

67. W. R. Cummins, H. Kende, and K. Raschke, *Planta* **99,** 347–351 (1971).

68. D. C. Walton, E. Galson, and M. A. Harrison, *Planta* **133,** 145–148 (1977).

69. P. E. Kriedemann, B. R. Loveys, G. L. Fuller, and A. C. Leopold, *Plant Physiol.* **49,** 842–847 (1972).

70. W. P. Hiron and S. T. C. Wright, *J. Exp. Bot.* **24,** 769–781 (1973).

71. T. C. Hsiao, *Annu. Rev. Plant Physiol.* **24,** 519–570 (1973).

72. H. Moldau, *Photosynthetica* **7,** 1–7 (1973).

73. J. C. O'Toole, J. L. Ozbun, and D. H. Wallace, *Physiol. Plant.* **40,** 111–114 (1977).

74. P. A. Dube, K. R. Stevenson, and G. W. Thurtell, *Can. J. Plant Sci.* **57,** 765–770 (1974).

75. R. R. Johnson, N. M. Frey, and D. N. Moss, *Crop Sci.* **14,** 728–731 (1974).

76. J. Hoddinot, D. L. Ehret, and P. R. Gorham. *Can. J. Bot.* **56,** 768–776 (1979).

77. T. F. Neales, K. J. Treharne, and P. F. Wareing, *Photosynthesis and Photorespiration*. Wiley-Interscience, New York, 1971, pp. 89–96.

78. J. S. Boyer, *Plant Physiol.* **40,** 229–234 (1965).

79. A. J. Redshaw and H. Meidner, *J. Exp. Bot.* **23,** 229–240 (1972).

80. D. Shimshi, *Plant Physiol.* **38,** 713–721 (1963).

81. J. Gale, H. C. Kohl, and R. M. Hogan, *Isr. J. Bot.* **15,** 64–71 (1966).

82. J. C. O'Toole, R. K. Crookston, K. J. Treharne, and J. L. Ozbun, *Plant Physiol.* **57,** 465–468 (1976).

83. J. Pospisilova, S. Ticha, J. Catsky, and J. Solarova, *Biol. Plant.* **20,** 368–372 (1978).

84. D. E. Fraser and R. G. S. Bidwell. *Can. J. Bot.* **52,** 2561–2570 (1974).

85. A. Tanaka and K. Kikuchi, from Ref. 86.

86. J. C. O'Toole, P. M. Ludford, and J. L. Ozbun, *New Phytol.* **78,** 565–571 (1977).

87. K. J. Treharne, J. L. Stoddard, J. Pughe, K. Paranjothy, and P. F. Waring, *Nature (London)* **228,** 129 (1970).

88. C. R. Stewart, C. J. Morris, and J. F. Thompson, *Plant Physiol.* **41,** 1585–1590 (1966).

89. A. B. Rena and G. Z. Masciotti, *Rev. Ceres.* **23,** 288–301 (1976).

90. R. C. R. Machado, A. B. Rena, and C. Viera, *Rev. Ceres.* **23,** 302–309 (1976).

91. C. R. Stewart, *Plant Physiol.* **51,** 508–511 (1972).

92. R. Taerum, *Physiol. Plant.* **28,** 471–475 (1973).

93. D. Shimshi, *New Phytol.* **69,** 413–424 (1970).

94. H. G. Jones, *New Phytol.* **72,** 1095–1105 (1973).

95. Z. Plaut, *Plant Physiol.* **48,** 591–595 (1971).
96. H. J. Mack and D. L. Hatch, *Proc. Am. Soc. Hortic. Sci.* **62,** 418–425 (1969).
97. E. Kueneman, R. Sandsted, D. Wallace, A. Bravo, and H. Wien, *Agron. J.* **71,** 419–424 (1979).
98. P. C. Crandall, M. C. Jensen, J. D. Chamberlain, and L. G. James, *Hortic. Sci.* **6,** 345–347 (1971).
99. J. H. Chung and D. S. Goulden, *N.Z. J. Agric. Res.* **14,** 227–234 (1971).
100. J. D. Atkin, *N.Y. Agric. Exp. Stn. Farm Res.* **27,** 13 (1961).
101. J. M. Dunway and R. D. Durbin, *Phytopathology* **60,** 1290 Abstr. (1970).
102. C. E. Parwood, *Phytopathology* **55,** 1373–1374 (1965).
103. J. M. Dunway and R. D. Durbin, *Phytopathology* **61,** 114–119 (1971).
104. A. I. Zaki and R. D. Durbin, *Phytopathology* **55,** 528–529 (1965).
105. J. M. Dunway and R. D. Durbin, *Plant Physiol.* **48,** 69–72 (1971).
106. P. I. Orphanos, *J. Hortic. Sci.* **52,** 447–455 (1977).
107. F. E. Robinson, W. F. Lehman, and K. S. Mayberry, *Calif. Agric.* **31,** 14 (1977).
108. A. Meiri and A. Poljakoff-Mayber, *Soil Sci.* **109,** 26–34 (1970).
109. C. R. Jensen, *Acta Agric. Scand.* **25,** 3–10 (1975).
110. A. Meiri, E. Mor, and A. Poljakoff-Mayber. *Ann. Bot.* (*London*) **34,** 383–391 (1970).
111. D. Longstreth and P. Nobel, *Plant Physiol.* **63,** 700–703 (1979).
112. R. H. Nieman and L. L. Paulsen, *Bot. Gaz.* **128,** 69–73 (1967).
113. G. Hoffman, J. Jobes, Z. Hanscom, and E. Maas. *Trans. ASAE* **21,** 713–718 (1978).
114. A. Meiri and A. Poljakoff-Mayber, *Isr. J. Bot.* **18,** 99–112 (1969).
115. J. Frota and T. C. Tucker. *Soil Sci. Soc. Am. J.* **42,** 743–746 (1978).
116. J. Frota and T. Tucker, *Soil Sci. Soc. Am. J.* **42,** 753–756 (1978).
117. S. D. Davis, C. H. M. van Bavel, and K. J. McCree. *Crop Sci.* **17,** 640–645 (1977).
118. E. Dumbroff, D. Brown, and J. Thompson, *Bot. Gaz.* (*Chicago*) **13,** 261–265 (1977).

6

CORN

Richard P. Waldren

INTRODUCTION

Corn or maize (*Zea mays* L.) is grown throughout the temperate and tropical regions of the world. It ranks third, after wheat and rice, in the world production of cereal crops. It is probably the most completely domesticated of all field crops and cannot exist for very long as a wild plant.

There are many vegetative types of corn that are adapted to a wide range of environmental conditions under cultivation. In the Western Hemisphere, it is cultivated from latitude 58°N in Canada to 35–40°S in South America. It is grown from below sea level to altitudes of 4000 m. Plants can vary from less than 1 m tall with eight or nine leaves and maturity in 50 days, to those that grow over 6 m tall with 42–44 leaves and maturity in 330 days.

Climatic Requirement

Corn does best in areas that have a minimum of 120 frost-free days and an average of 2100–2200 or more growing degree days. Highest yields are obtained when temperatures are moderate and water is plentiful. The ideal temperature for corn production is generally from 24 to 32°C. At 10°C corn grows very little. As the temperature rises to 27 to 32°C, corn grows faster if moisture is adequate. Above 32°C the roots have increasing difficulty taking in water fast enough to keep the plant cells turgid. When soil water is short, the optimum temperature is less than 27°C (1).

Contrary to popular belief, corn does not grow best during hot, humid nights. On hot, humid nights, respiration is greatly increased and thus more energy is used. Under such conditions, corn plants may actually lose dry matter during the night. Cell division and enlargement may continue in darkness when the temperature exceeds 14°C (2).

Large quantities of water are essential for producing top corn yields. In the United States the best nonirrigated corn regions have an annual precipitation of 60–100 cm. Generally dryland corn production is limited to areas with an annual precipitation of 38 cm or more (2).

Crop Use and Yield Potential

Yield potential in corn is closely related to water availability, either from rainfall or irrigation. Although other factors such as plant population and fertilizer use can affect the yield potential of corn, these are usually adjusted according to the amount of water the crop will have during the growing season. In areas where annual precipitation is 75 cm or less, yield potential of dryland corn is closely tied to average annual precipitation, and can vary from 3700 kg ha^{-1} in the drier areas to 10,000 kg ha^{-1} in the wetter areas. In areas where annual precipitation is greater than 75 cm, or under irrigation, the yield potential is usually determined by factors other than water, such as soil fertility. In these areas, yield potentials can exceed 12,500 kg ha^{-1}.

In all areas where corn is grown, whether dry land or irrigated, the potential of a soil to produce corn under conditions of adequate fertility and a high level of management is largely determined by the soil's capacity to store and supply water to the crop (3). This is determined by the amount of available stored soil water per centimeter of soil and the depth that the roots can easily penetrate the soil. Generally, the yield potential of corn is highest on deep, well-drained silt loam soils. High yields are possible with other soil types but management is more critical. For instance, it is possible to obtain high yields of corn on sprinkler irrigated sandy soils but water and nitrogen fertilizer must be added frequently and in smaller amounts due to the soil's inability to store adequate amounts of each.

Growth Stages of Corn

Hanway (4) has developed a set of growth stages for corn that are widely used by people who work with corn. The stages are described in Table 6.1, and are designed for field corn grown in the Corn Belt of the United States. Knowledge of how a corn plant develops is important when studying water relations in corn, especially with regard to the time that water deficits occur during the growing season.

***Table 6.1. Growth stages of corn*[a]**

Stage	Description
0.0	Plant emerges from soil.
0.5	Two leaves fully emerged (collars visible).
1.0	Four leaves fully emerged.
1.5	Six leaves fully emerged.
2.0	Eighth leaf fully emerged.
2.5	Tenth leaf fully emerged.
3.0	Twelfth leaf fully emerged.
3.5	Fourteenth leaf fully emerged.
4.0	Sixteenth leaf fully emerged. Tip of tassel emerges from whorl.
5.0	Silks emerge. Tassel fully emerged and shedding pollen.
6.0	Grain at blister stage. Cob, husks, and shank fully developed.
7.0	Dough stage. Grain filling rapidly.
8.0	Beginning dent stage. Embryo rapidly developing.
9.0	All kernels fully dented.
10.0	Physiological maturity. Maximum dry matter accumulation.

[a] After Hanway (3).

ROOT GROWTH AND FUNCTION

Corn plants have an extensive fibrous root system typical of most grasses. Since corn is a fast growing plant, its root system is also quite vigorous. Although the first roots to develop are seminal roots, the bulk of the plant's root system consists of secondary or adventitious roots which include coronal or crown roots and brace or aerial roots.

Germination

Corn seed will proceed to germinate rapidly when the soil temperature at a 5 cm depth reaches 15°C. Germination occurs in 2 or 3 days. The radicle emerges first followed by seminal roots. These roots serve to initially anchor the seedling and absorb nutrients. They do not develop extensively and may or may not live to maturity of the plant.

The secondary or main root system develops from the crown or first node and later from subsequent nodes. These roots begin to elongate about 2 weeks after germination when the plant has four to seven leaves.

Depth and Distribution

The size and depth of the root system is determined genetically and increases with that of the top growth. Compared with a small type, a large type of corn may have a 50% greater spread, 10% deeper penetration, 65% more functional main roots, 92% greater combined length of main roots per plant, 311% greater root weight, and 29% larger diameter of main roots (2). The usual spread of corn roots is about 1.25 m and the depth of many of the roots can be 1.6 m with some approaching 2 m.

In irrigated corn, 64% of the root dry matter is in the upper 30 cm and 92% in the upper 90 cm. In dryland corn, however, 39% of the total root dry matter is in the upper 30 cm of the soil and 70% in the upper 90 cm (5). This indicates that nonirrigated corn will expand its root system deeper in the soil during periods when soil water is less than optimum in comparison with irrigated corn. This increases the supply of available soil water to the plant by tapping reserves located deeper in the soil. The fact that corn will expand its root system when faced with less than optimum soil moisture indicates that the corn root system possesses a potential drought avoiding mechanism.

In a deep, well-drained soil there is a linear relationship between soil water depletion and root density, where root density is defined as length of roots per unit volume of soil ($cm\ cm^{-3}$) (6). Studies in Iowa have shown that 70–75% of the water used by nonirrigated corn plants from a 1.3 m profile was extracted from the top half of the profile (7). This is illustrated in Table 6.2 (8) which shows water extraction by nonirrigated

Table 6.2. Water extraction from the soil profile at different depths during the growing season of nonirrigated corn[a]

	% E_0 or E_t				
	Depth (cm)				
Dates	0–30	31–61	62–91	92–122	123–152
To June 7	100 (0–15cm)				
June 8–14	100				
June 15–27	67	33			
June 28–July 4	60	20	20 (62–76cm)		
July 5–11	60	20	20		
July 12–18	60	15	15	10 (92–107)	
July 19–25	60	15	15	10	
July 26–Aug. 1	60	10	10	10	10[b]
	60	15	15	10	
After Aug. 1	60	10	10	10	10[b]
	60	15	15	10	

[a] Values for each date are given as the percentage of evaporation (E_0) or evapotranspiration (E_t) that occurs from each depth. After Shaw (8).
[b] Used only if 1–122 can have less than 50% available moisture.

corn from the soil profile throughout the growing season at several locations in Iowa. There are many factors that affect root growth and density but the most important are soil temperature, aeration, water supply, and physical resistance of the soil to root penetration.

In soils with a water table within reach of the corn roots, total water uptake by the plant may not be as closely correlated to root density as in a well-drained soil. Once the roots reach the capillary fringe above the water table, most of the water is absorbed by the roots in this zone even though they may make up a small fraction of the total roots (9). Until the roots reach the capillary fringe the plant must depend on stored soil moisture above it and water uptake is correlated with root density as mentioned before.

Physical resistance of the soil to root penetration can greatly affect corn root depth and distribution. Roots take up space in the pores of the soil and the size and number of these pores affect the ability of roots to penetrate the soil. Corn roots 5 weeks old are unable to penetrate a subsoil compacted to a bulk density of 1.5 g cm^{-3}. However, they can grow profusely in subsoil with a density of 1.2 g cm^{-3} (10), and can grow satisfactorily in subsoils of 1.3–1.4 g cm^{-3}.

Resistance to penetration can affect lateral root growth as with interrow compaction from tractor tires; or it can affect the depth of roots due to a compacted horizontal layer in the soil. Sometimes cultivation can help alleviate problems with interrow compaction (11) and sometimes deep tillage can aid roots in penetrating deeper in the soil (12). However, many times such deep tillage does not affect the yield of corn even though it may result in better distribution of roots in the soil.

Temperature has a great influence on root development. Dry weights of shoots and roots at different growth stages decrease significantly with increasing soil temperature, with 20°C being the optimum soil temperature for dry matter accumulation in corn seedlings (13). During emergence, there is a linear relationship between rate of elongation of roots and shoots, and temperature over the range of 10–30°C (14). When the temperature exceeds 32°C there is a sharp decline in the rate of elongation of both roots and shoots and a constant temperature of 40°C can be lethal to the seedling. At temperatures below 9°C there is little if any elongation of roots.

The amount of water in the soil can alter the effect of soil temperature on root growth. This is especially true with minimum tillage where soil temperatures will likely be lower when compared to conventional tillage systems. Under conditions of lower soil temperature and high moisture, raising interrow temperature by removing mulch between the rows can increase branching and elongation of roots. However, under conditions of higher soil temperature and moderate water, raising interrow temperature has little effect on root growth (15). The presence of mulch between the rows can increase root weight and lateral root proliferation when soil temperature is high and soil water is low (16). These effects are confined primarily to the upper 30 cm of the soil. Mulch can decrease soil temperature in the upper 10 cm by as much as 25°C depending on the amount of solar radiation reaching the soil surface. By reducing the extremely high temperatures that are possible on the soil surface, root growth in the surface 5 cm of the soil can be doubled (17).

Fertilizer can affect root growth and water use significantly. Well-fertilized corn uses water more efficiently as it gets more water through deeper roots, and the roots have a slightly greater capacity to extract water from smaller pores and the films surrounding soil particles. Corn with adequate nutrients, especially nitrogen, will produce deeper and more extensive root systems during early growth than unfertilized corn (18). Since water utilization is closely related to root growth, plant nutrition has a very positive effect on water use efficiency, especially under conditions of less than optimum soil moisture.

Root Growth in Relation to Physiological Stages

After the secondary roots emerge at the four to seven leaf stage, rapid growth and proliferation of roots occurs. By the time the plant has eight fully emerged leaves, the roots will have penetrated to a depth of 45 cm.

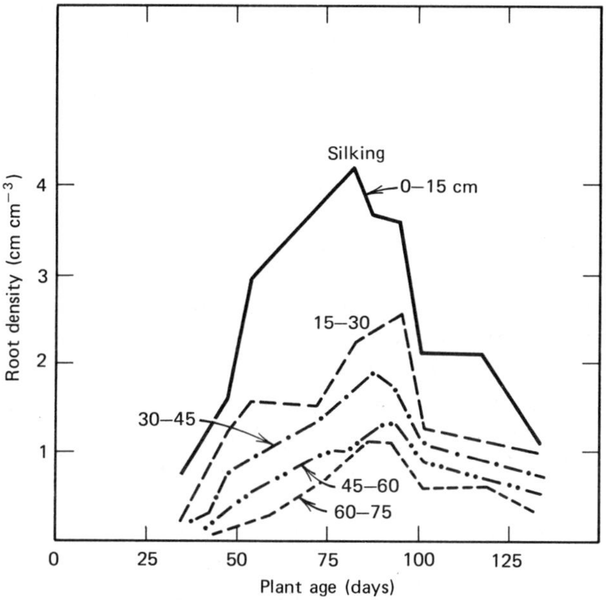

Figure 6.1. Relation of plant age to root density at five soil depths during corn growth. [After Mengel and Barber (20). Reprinted from *Agronomy Journal,* Volume **66,** page 343, 1974, by permission of the American Society of Agronomy, 677 South Segoe Road, Madison, WI 53711].

Brace roots develop from the lower nodes of the stalk about the time tasseling occurs. These brace roots provide support of the stalk and as well as functioning in water and nutrient uptake.

Root growth consists of a series of overlapping stages which are associated with stages of top growth. Early root growth occurs largely in a downward diagonal direction, followed by extensive lateral growth which is completed 1 or 2 weeks before tassel emergence. Brace roots appear near the completion of lateral root development (19). Most root development occurs prior to tasseling so the length of time to tasseling affects the size of the root system. Earlier types that tassel sooner will not have as large a root system as later maturing types which tassel later. This would lead one to think that later maturing types would be better under conditions of inadequate soil moisture. However, this is not the case as the later types are usually larger plants and use considerably more water. When exposed to only a short period of drought, later maturing corn will yield more than earlier types because it readily taps additional soil water with its larger root system. Under prolonged drought, however, earlier corn can yield more than later types because it does not exhaust the soil water supply as rapidly.

As shown in Fig. 6.1, the size of the root system of a corn plant increases rapidly until tasseling and then declines during grain filling

(20). During early growth stages, roots grow rapidly and few, if any, die, so the size of the root system increases exponentially. As the plant reaches flowering, roots begin to die as fast as new roots are produced so that the size of the root system remains constant. As the plant proceeds through later growth stages, the number of roots dying exceeds the number being produced and the overall size of the root system declines.

Root Water Relations and Dry Matter Accumulation

Although genetics affects the potential depth of the root system, there is a close relationship between corn root elongation and the water content of the soil as shown in Fig. 6.2. Not only does root elongation increase with decreasing soil water tension, but also increases with increasing soil water content at the same tension (21). This illustrates the effect of soil texture on root growth as it affects the water holding capacity of the soil. As was mentioned earlier, there is a significant correlation between corn yields and the amount of available stored soil water. Since most of the cell elongation in young roots is directly due to water uptake, it becomes evident that yield potential is higher on soils which can hold more water because of the potential for an increased root system.

Increasing stored soil water in sandy soils through the use of a subsurface barrier can also increase dry matter accumulation in corn (22). When barriers such as asphalt prevent deep percolation of soil water by

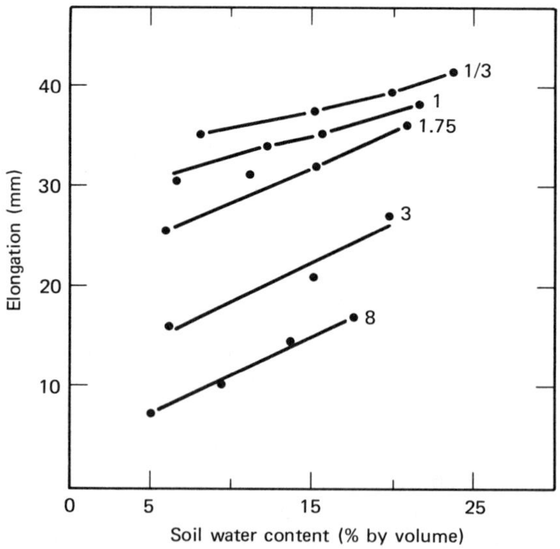

Figure 6.2. Elongation of corn seedling roots in a 24 hr period as a function of soil water content. Numbers next to lines refer to respective soil water tensions (−bars). [After Peters (21)].

holding it in the root zone, root distribution and water uptake are similar to that found with a high water table.

Roots deep within the soil are often considered to be less effective in absorbing water than more shallow roots because they occur at greater distances from the plant. However, many times these deep roots are more efficient in water uptake because they are younger, are usually in wetter soil, and are less crowded than roots found closer to the surface (23). This helps explain why total root weight declines after flowering while plant water use is still quite high.

TOP GROWTH

For a given hybrid, the yield of corn generally increases as population is raised until one or more factors such as water supply, available plant nutrients, or others, become limiting. Corn grain yield can be expressed as a product of grain per plant and plant population. Corn, however, is much more sensitive to plant population pressure than many other crops. There is wide variation in population response between diferent hybrids which indicates genetic control of this trait. Barrenness is the major factor limiting corn grain yields at high plant population under favorable environmental conditions. Barrenness may be a direct result of shading which causes poor corn silk development (24). There are several morphological and physiological traits that affect barrenness. Density tolerant corn genotypes are characterized by rapid completion of silk extrusion, coincidence of pollen shedding and silk extrusion, rapid growth of the first ear and its silks, prolificacy, reduced tassel size, and efficient production of grain per unit leaf area (25).

Dry Matter Accumulation and Physiological Stage

Dry matter accumulation is the balance between photosynthesis and respiration. Therefore, any practices or factors that increase photosynthesis and decrease respiration will usually increase yield. From a practical standpoint it is much easier to affect the rate of photosynthesis than respiration. Such things as soil fertility, leaf area, leaf angle, water supply, and improved breeding can affect photosynthesis and can be controlled by man. On the other hand, few of the factors that affect respiration are easily controlled.

Corn is an efficient plant on the basis of carbon dioxide fixed per unit of leaf area. The leaves of the plant are produced first followed by the leaf sheaths, stalks, husks, ears, and finally the grain. By stage three enough leaves are exposed to sunlight so that the rate of dry matter accumulation will continue at a nearly constant daily rate until near

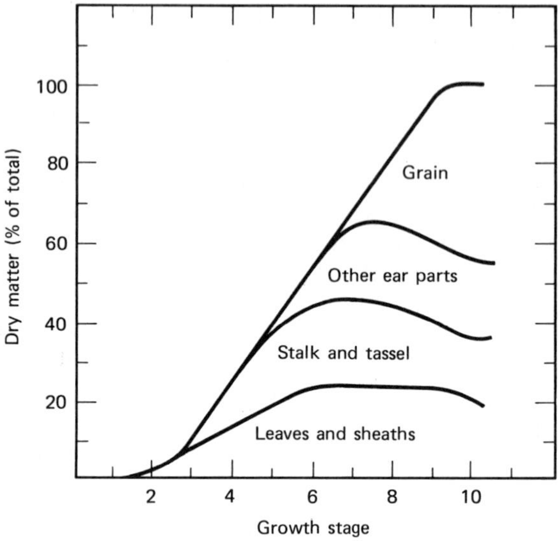

Figure 6.3. Dry matter accumulation in different parts of a corn plant in relation to growth stage. [After Hanway (3). Reprinted from *Soil Science Society of America Proceedings,* Volume **21,** page 482, 1957, by permission of the Soil Science Society of America, 677 South Segoe Road, Madison, WI 53711.]

maturity as shown in Fig. 6.3 (3). All the leaf dry matter is produced early and stalk dry matter peaks shortly after flowering. Dry matter accumulation in the grain is very rapid and accounts for practically all of the increase in dry matter after silking. There is even a loss in dry matter in other parts of the plant, some of which is translocated to the grain. The fact that about 45% of the plant's total aboveground dry matter is in the grain attests to the tremendous sink of the developing grain for photosynthate and other translocated materials.

The maximum potential rate of photosynthesis decreases as leaves get older in corn as well as in most other plants. Since the oldest leaves on a corn plant are at the bottom of the stalks, there are usually wide differences in the photosynthetic rate of individual leaves due to both age and shading by other leaves. These differences are less when the plants are young and the canopy has not yet closed, and are greatest as the plant nears maturity.

The rate of photosynthesis in a corn canopy under full sunlight is a function of leaf area and plant age. Canopy photosynthesis increases as leaf area increases until the canopy closes, usually just before tasseling. Rates of canopy photosynthesis will usually peak about silking or early grain filling and then begin to decrease as the leaves continue to age. There appear to be differences between genotypes of corn as to when and to what extent canopy photosynthesis declines as the plants mature

(26), and corn breeders and physiologists are studying this trait in an effort to increase yields.

Numerous researchers have shown a direct relationship between leaf area index (*LAI*) and plant population. *LAI* increases linearly as the plant population of corn increases from 34,000 to 69,000 plants ha^{-1}. The leaf area per plant decreases, however, as the plant population increases (27).

One study showed that the combined leaf area per plant for the primary ear leaf with the first leaf above and below the ear decreased from 2300 cm^2 at 39,500 plants ha^{-1} to 2150 cm^2 at 59,300 plants ha^{-1} (28). However, the leaf area produced on the main stalk of corn plants does not decrease in inverse proportion to an increase in plant density (29); thus area changes must be attributed to differences in silking with density.

An *LAI* of 2.6 is optimum for grain yield in hybrid corn while 2.0 is considered optimum for inbreds as shown in Fig. 6.4 (30). At optimum *LAI* about 90% of the incoming solar radiation is intercepted by the crop. Yields of corn grain tend to be linearly related to *LAI* at silking and to *LAI* days over the grain formation period when *LAI* is less than 3.4 and *LAI* days are less than 156 (31). LAI values greater than 4.0 substantially reduce the depth of light penetration into normal leaf canopies of corn and greatly reduce yield.

Dry matter accumulation is the balance between photosynthesis and

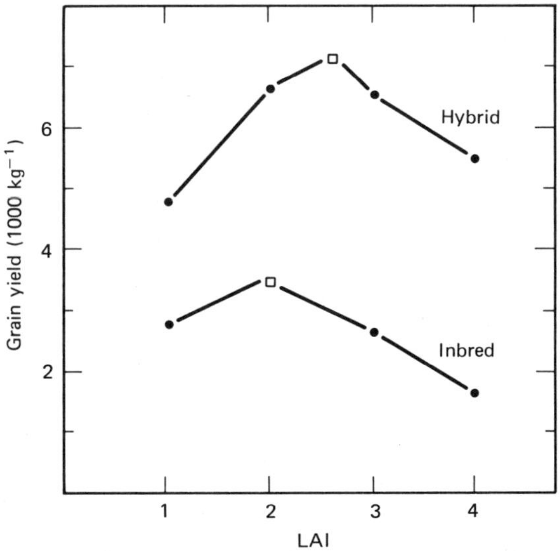

Figure 6.4. Grain yield of inbred and hybrid corn at different leaf area indices indicating optimum *LAI* (□). [After Major et al. (30).]

respiration. Dry matter produced in corn with a *LAI* of 3.3 by the top, middle, and bottom leaves is of the ratio of 4:21:1 (32). The low rate in the bottom leaves is probably due to shading by the above leaf canopy at a high *LAI*, and leaf age. The high rate in the middle is probably due to the close proximity of the developing grain which provides for a large sink for photosynthate. Two-thirds of the corn hybrids in one study reached maximum dry matter accumulation 54 to 57 days after pollination (33) with the average for the Corn Belt being about 55 days after 75% of the plants have visible silks (34). At maximum dry matter accumulation the moisture content of the grain will range from 28 to 38%.

Water Relations and Top Development

Early in the season much of the water loss from a corn field is by evaporation from the soil surface since only sparse cover exists and the *LAI* is low. A partial crop cover may reduce evapotranspiration (E_t) because the soil surface becomes dry, resulting in a lower rate of loss due to evaporation. The changing crop cover during the growing season for corn causes changes in total E_t in the field. Using open pan evaporation as the potential value (E_0), E_t averages only 35% of E_0 early in the season. It then increases in a sigmoidal manner to a maximum of 80% shortly after silking, and decreases as the crop approaches maturity, as shown

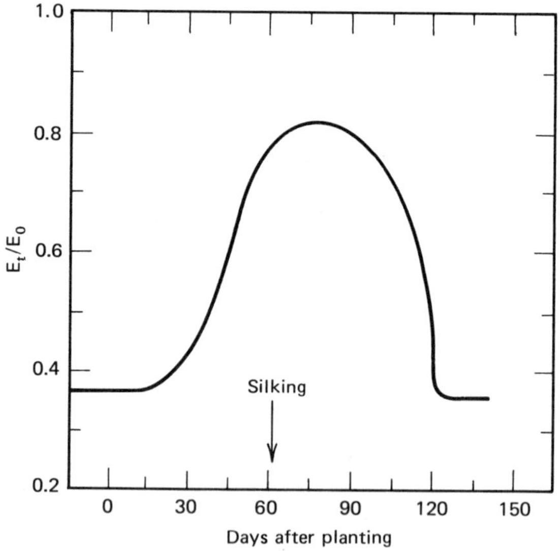

Figure 6.5. Ratio of evapotranspiration (E_t) from corn to open pan evaporation (E_0) during the growing season. [After Denmead and Shaw (35). Reprinted from *Agronomy Journal*, Vol. **51**, p. 726, 1959, by permission of the American Society of Agronomy, 677 South Segoe Rd, Madison, WI 53711.]

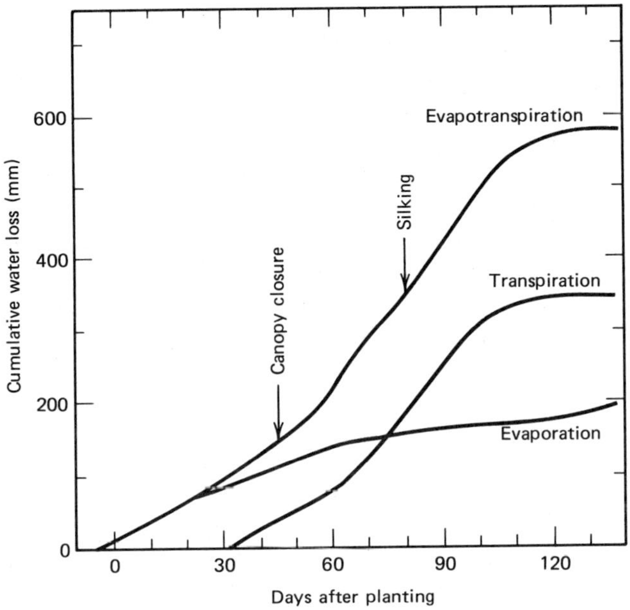

Figure 6.6. Cumulative evapotranspiration and its components in irrigated corn. [After Rosenthal et al. (36). Reprinted from *Agronomy Journal*, Vol. **69**, p. 463, 1977, by permission of the American Society of Agronomy, 677 South Segoe Rd, Madison, WI 53711.]

in Fig. 6.5 (35). Figure 6.6 shows the cumulative E_t and its components in irrigated corn for a growing season (36). Notice that evaporation begins to level off as leaf area begins to rapidly increase as evidenced by the rate of transpiration. This occurs shortly after canopy closure.

Water consumption by corn varies greatly throughout the growing season and is affected significantly by climate. Daily water use by corn is greater in the southern latitudes of the United States than in the northern latitudes. Daily changes in the weather also affect E_t. Corn may use up to twice the water on a hot, sunny day as it uses on a cool, cloudy day. Early in the season, per plant water consumption will average about 0.25 cm day^{-1} and will peak shortly after silking at about 1 cm day^{-1} on the average (8).

Top development can be affected by soil temperature which affects the ability of the roots to take up water. Active water uptake by roots is indirectly affected by temperature because of the effect of temperature on the metabolism of the root. When soil temperature drops below 28°C there is a decrease in the leaf elongation rate due to restricted water uptake by the roots. When plant water potential reaches -9 bars, leaf elongation ceases due to the decline of leaf turgor pressure, although transpiration and net photosynthesis may not be affected until plant water potential reaches -12 or -13 bars (37). Therefore, the primary effect of suboptimal soil temperature on canopy growth is the decrease in leaf elongation rate as affected by restricted water uptake.

This decrease in leaf elongation rate also shows up whenever the plant experiences water stress. When the soil water potential decreases from −0.35 to −2.5 bars, leaf elongation rate decreases 44% and soluble carbohydrate concentration increases 42%. At the same time, total dry matter accumulation decreases 26% and transpiration 24% (38). During periods of mild water stress the plant maintains a positive carbon balance in the leaves partly in the form of an abundant supply of soluble carbohydrates in order to maintain metabolic processes. The sensitivity of leaf elongation rate to mild water stress appears to be an important growth determining factor.

Row Spacing and Evapotranspiration

Many studies have shown that decreasing row spacing to less than 90 cm can increase grain yield of corn. Assuming that moisture, soil fertility, and population are optimal, the next growth-limiting factors will be light penetration, carbon dioxide exchange, and other factors that affect the rate of photosynthesis.

When corn is planted in narrower rows at the same population, the plants are more uniformly distributed over the soil surface. This makes the crop more effective in intercepting solar radiation and shading weeds. The canopy will usually close sooner and result in lower soil temperatures, thus reducing evaporation from the soil surface. Studies with sorghum have shown that shading the soil sooner with narrow rows can reduce the sensible heat load, and subsequently the E_t rate in the canopy (39). This is probably also true for corn canopies. More uniform distribution of plants will also help reduce the effect of raindrop impact on soil particles by intercepting more drops with the leaves. This results in higher infiltration rates and more effective rainfall utilization (40, 34).

Since narrower rows can increase grain yield and light interception, and decrease E_t, the net result is an indirect increase in water-use efficiency. Water use does not significantly change but the yield of grain per unit of water does significantly increase. However, the total forage yield usually does not increase so narrow rows are not necessarily an advantage when corn is raised for silage (41).

Under less than optimal moisture, narrow rows do not necessarily result in yield increases (42, 34). Even though soil evaporation may still be reduced with narrow rows and water use efficiency may increase, the yield potential is still limited by water which greatly reduces the effects of narrow rows.

Uniform distribution of plants within the row is also important in maximizing yields. Plants bunched together in the row react to the increased competition with each other. When viewed on an individual plant basis the effective population is much greater than that found on a field basis. This may also account for studies which show that corn randomly spaced such as when planted with an airplane does not yield as well as corn planted in rows (43).

TRANSPIRATION

Transpiration in corn accounts for a considerable portion of water loss during peak consumption periods. Studies have shown that it can account for 50–70% of the total water lost during peak use (44, 45). The water pathway in a plant acts as a continuous hydraulic system. Any change in the water potential of the roots is reflected by a change in leaf water potential of a similar magnitude (46). This, in turn, will affect stomatal diffusive resistance and ultimately the rate of transpiration.

Regulating Water Loss

Soil water is essential to maintain turgidity, meet transpiration requirements, and serve as a constituent of plant cells. It is hard to separate the reactions of the plant to high temperatures which usually accompany water stress from the reduced cooling effect due to reduced evaporation of water caused by stomatal closure.

When young corn plants lack adequate water, stomata close, the leaves roll, and become dull in appearance. Leaf rolling is caused by a lack of turgidity in the bulliform cells that are found along the midrib of the leaf blade and occurs as the leaf water potential approaches -15 bars. At the onset of drought, corn plants of all ages will wilt in the afternoon and recover turgidity at night. In succeeding days as drought progresses, wilt occurs at earlier periods during the day until the leaves remain wilted all day. As the soil approaches the permanent wilting point the plants will fail to recover turgidity at night.

The stomata of corn plants exposed to severe drought for 3 or 4 days resume much of their apparent normal behavior after a 1 or 2 day recovery period. But, severe drought over longer periods of 1 week or more produces marked changes in stomatal behavior and they never seem to be able to fully open (47). This indicates that stomata are severely damaged by long periods of drought which will cause subsequent reductions in yield even if soil water levels remain adequate following the drought.

Corn with its amphistomatous leaves has a slightly greater stomatal density on the lower leaf surface (8929 and 10,800) than on the upper leaf surface (7037 and 9800) [stomatal densities are from Miller (48) and Meidner and Mansfield (49), respectively]. Stomatal density also changes in relation to position on the blade. Density is lowest near the sheath and increases toward the tip of the blade (48).

With adequate water, the lower side of the leaf will have slightly lower stomatal diffusive resistance than the upper side. As leaf water potential decreases, the resistance of the lower stomata will increase faster than the upper. When the potential approaches −15 bars the resistance of both surfaces will be about the same. When rewatered, the resistance of both sides will decrease rapidly, but the lower stomata will again have lower resistance once the leaf regains full turgor. There appears to be no difference between the upper and lower surface regarding the leaf water potential at which the stomata begin to react to changes in potential (50).

At sunrise in well-watered plants, the stomata on the lower side of the leaf open much more quickly than the upper stomata. The lower stomata will be about two-thirds open by the time the upper ones begin to open (51).

The rate of photosynthesis is not affected by leaf water potential changes from 0 to −3.5 bars (52). At −3.5 bars stomatal diffusive resistance will begin to decrease. Between −3.5 and −10 bars, the rate of photosynthesis is directly related to stomatal resistance. At potentials below −10 bars, other factors begin to affect photosynthesis although their effect is small (53). Rapid changes in stomatal resistance occur when leaf water potential reaches −10 to −12 bars regardless of light intensity (54).

There are differences between crops in stomatal control. Although the rate of photosynthesis in corn is directly related to stomatal diffusive resistance at leaf water potentials between −3.5 and −10 bars, in soybeans this direct relationship occurs between −11 and −16 bars (52). This supports the findings of many researchers that the stomata of C_4 species such as corn are more sensitive to environmental changes than are the stomata of C_3 species such as soybeans and wheat. Under similar conditions of high light intensity, C_4 plants will usually have a higher rate of photosynthesis and similar transpiration rates resulting in more dry matter per unit of water transpired. At low light intensity, the stomata of C_4 plants are more likely to be closed resulting in less water loss. Thus, the greater responsiveness of stomata of C_4 plants to changes in environmental conditions, coupled with the greater potential productivity of those leaves, accounts at least in part for the greater water-use efficiency usually exhibited by C_4 plants (55).

The lower leaves of corn close their stomata at a higher water potential than the upper leaves (56). This can be partly attributed to the high osmotic potential of the lower leaves at full turgor and not because of differences in the degree of turgor potential. Light is no doubt also a factor. Stigter (60) found there is no difference in stomatal diffusive resistance in sun leaves at different canopy heights, but resistance in shaded leaves increases from the upper canopy to the lower and is directly related to light intensity. Also, stomata react to changes in carbon dioxide concentration in the air. Stomatal diffusive resistance decreases and rate of transpiration increases with increasing CO_2 concentration. The response is greater at low light intensity than at high light intensity (57). The stomata on the lower side of the leaf react less to changes in CO_2 concentration than those on the upper side (58), and stomatal response can occur in as little as 10 sec with the movement completed within 6–10 min (59). Stomatal resistance is not affected by leaf age until the leaf begins to yellow (61).

Stomatal resistance at a particular time of the day is a function of the radiation on the leaf, the degree of senescence of the leaf, and the water potential of the leaf. Small changes in the water content of turgid leaves are associated with relatively large changes in leaf water potential. Leaf water potential, osmotic potential, and turgor potential are the highest at sunrise and then decrease to reach minimum values near midday. They will then increase again in the afternoon. At the same time, stomatal resistance will increase from sunrise to midday and then decrease in the afternoon. This diurnal pattern of stomatal diffusive resistance, leaf water potential, and other parameters occurs in both stressed and nonstressed plants, the only difference being the size of values and the degree of change (62–64).

This shows the dependence of leaf water potential and, hence, stomatal resistance, on the energy flux in the field, which, in turn, affects the rate of transpiration. Because of resistances to water flow in the plant and the soil, the plant cannot absorb water fast enough to replace that lost by transpiration under conditions of high transpirational demand. In this situation, the water potential gradient will increase and reflect the balance between environmental demand and the rate of soil water extraction, which is controlled by the root system.

Since energy flux has a diurnal pattern, it follows that leaf water potential will also vary according to a diurnal pattern. This, in turn, will affect the rate of growth of the corn plant since leaf cell elongation is directly affected by leaf water potential. Leaf growth may stop during high energy flux and will occur mostly during the night (65).

Young, rapidly expanding leaves of corn plants are major photosynthetic sinks (66). Leaf elongation is severely restricted by leaf water potentials of about -7 to -9 bars. However, the rate of photosynthesis is not severely affected until the leaf water potential decreases to about -12 bars (67). In a rapidly growing corn plant this difference in inhibition by drought between the sink (young leaves) and the source (photosynthesis) could result in a feedback inhibition of photosynthesis. This feedback inhibition would only occur if root growth was also inhibited due to low soil temperature or another cause, thus reducing the other major sink for photosynthate in a young corn plant.

The relationship between leaf water potential and stomatal resistance for five corn genotypes is shown in Fig. 6.7 (68). A genotype in which

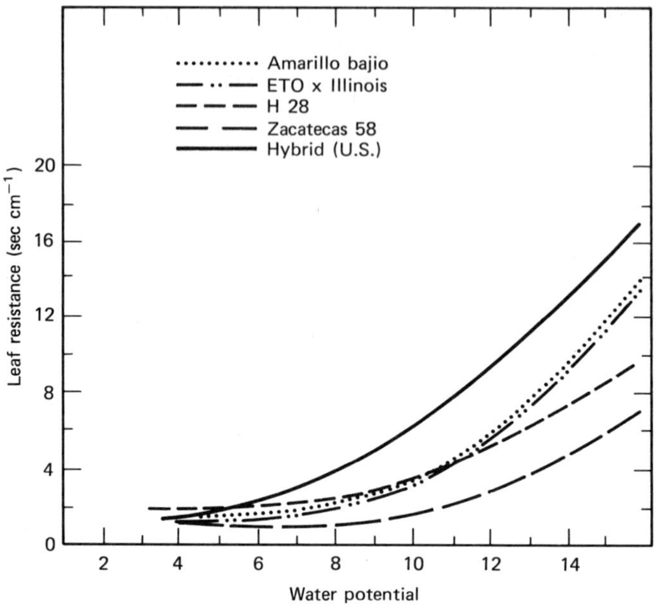

Figure 6.7. Relationship between leaf water potential and stomatal resistance for five corn varieties. [After Goldsworthy (68).]

the stomata remain open at low water potentials would have an advantage in a short period of drought stress since the diffusion of carbon dioxide for photosynthesis and leaf cooling can proceed with minimum interruption. However, this type of stomatal behavior would be detrimental under extended drought as the plant would exhaust the soil water supply rapidly. Under irrigation it could be advantageous to have the stomata remain open at lower water potentials since drought stress would not be expected to be of long duration. Dryland corn raised in areas that can expect protracted periods of drought would have a greater chance of failure if it exhibited this type of stomatal behavior, however.

Lower concentrations of adenosine triphosphate (ATP) are found in water stressed cells of leaves. This is probably due to a decrease in respiration and/or an uncoupling of oxidative phosphorylation. However, it could also be due to increased utilization of ATP. Water stress could influence ATP production by metabolic and/or biophysical mechanisms. The large decrease in cell turgor at the onset of water stress, which can be as much as 90% in the first 3 hr, is the most probable biophysical factor. Cell water activity, water content, and ionic strength all change less than 4% in the first 3 hr and are unlikely to affect ATP production (69).

The theory that ATP activity and not ATP synthesis is regulated is supported by the fact that corn leaves released from a short, mild water stress undergo a short period of very rapid compensatory growth followed by a long period of delayed growth. ATP may act as a modulator of metabolic activity within the cell during short periods of drought stress such as the diurnal water stress cycles frequently experienced by field plants.

Other physiological changes occur in plant cells during drought stress. Nitrate reductase activity drops in corn leaves under both drought and heat stress (70) as does polyribosomal content (71). Abscisic acid level increases as the cells become drought stressed (72). Other physiological indicators of drought stress, such as changes in proline concentration, have been found in other crops and there is every indication that corn will react in the same way, although little research has been done with corn.

There is a change in minimum leaf water potential values as the corn plant changes from vegetative to reproductive growth. Leaf water potential is lower after tasseling than before under the same environmental conditions and is apparently associated with physiological changes in the plant. Therefore, leaf water potentials during the reproductive stages may not indicate the same levels of stress as the same leaf water potentials during vegetative stages (65). It is thought that leaf water potential during vegetative stages are good indicators of drought stress, but interpretation of leaf water potential data during reproductive stages is uncertain. This is unfortunate since the corn plant is very sensitive to drought stress during reproductive stages.

Preferential Flow of Water

A study of drought resistance in two corn inbreds of differing drought resistance indicated a higher resistance to water flow in the xylem of the drought susceptible line (73). A subsequent study with the same two inbreds found that differences in both lines may be attributable to dark respiration and cuticular resistances to water vapor diffusion (74). Dark respiration occurred at a greater rate in the drought susceptible line. Cuticular resistance to water vapor diffusion was initially higher in the drought resistant line and increased at a faster rate as the drying cycle progressed.

A greater conduction inhibition exists in the leaf sheath and ligule area of the younger leaves of drought susceptible lines of corn (75). This suggests an inability of water conduction to compensate for high transpirational losses under high atmospheric demand. Rapid increases in both total resistance to water vapor and mesophyll resistance to carbon dioxide occur within a narrow range of water potential in corn with simultaneous decreases in transpiration and photosynthesis (75).

Corn is not considered to be a saline tolerant plant and a common problem associated with high transpirational demand in arid and semiarid regions is the probable presence of excess soluble salts in the root zone. Large quantities of these salts move along with the soil water which flows to the roots to meet the transpirational demand of the plant. This flow of salt along with water can raise the concentration of salts at or near the corn root surface to a much higher level than the rest of the soil (76). Maximum accumulation occurs in the apparent free space of the roots, followed by the closely adhering soil (77). During low transpirational demand these excess salts can diffuse back in the soil bulk due to the concentration gradient that is present between the root surface and the soil.

The high concentration of salts at the soil–root interface can affect the osmotic potential and decrease the ability of the roots to take up water. The salinity tolerance limit of corn plants is greatly influenced by the amount of transpirational demand occuring during the growing season.

APPLIED ASPECTS OF IRRIGATION

As was shown in the previous section on transpiration, water use by corn changes with the age of the crop. Under irrigated conditions, these changes in water use by the crop become important in maintaining soil water levels adequate for maximum production of corn grain or forage.

About 80% of the available water in a deep soil can be extracted by corn before stomatal regulation of transpiration begins (78). However, most irrigation scheduling programs use 50% soil water depletion as the point at which water is added to the field. Adding water at 50% depletion allows for maximum efficiency of irrigation plus a safe margin of water to cover periods of high water demand and mechanical failure in the irrigation system which could delay the application of water for a few days.

Critical Water Demand Periods

As stated previously, peak water use by corn occurs about the time of silking or shortly thereafter. Much research has shown that water deficits at the time of tasseling and silking also cause the greatest reduction in yield. Water stress reduces grain yield by 25% when prior to silking, by 50% when occurring at silking, and by 21% after silking (79). Length of the stress period is also important. Soil water depletion to the wilting percentage for 2 days during the tasseling or pollination period can result in as much as a 22% decrease in yield, while a 6–8 day period of depletion can cause a yield reduction of about 50% (80).

Studies with sweet corn have shown that 85% soil water depletion during silking results in 40% yield loss, reduced plant height, and increased incidence of stalk rot. It does not significantly reduce ears per hectare and seed quality, including seed size and germination (81). This illustrates the importance of adequate soil water, either from precipitation or irrigation, 2 weeks prior to and following tasseling.

Severe drought stress during grain filling in corn can completely stop photosynthesis just as it can at any other growth stage. However, dry matter accumulation in the developing grain may not completely stop. When corn was subjected to leaf water potentials of -18 to -20 bars throughout the grain filling period, photosynthesis was virtually zero but grain yield was reduced only 47–69% (82). This indicates that dry matter may be translocated from other parts of the plant. Other studies have indicated that kernel weight does not increase once stress becomes severe enough to prevent recovery during the night, but resumes again when stress is relieved (83). As long as drought does not disrupt the flowering process, however, total photosynthetic accumulation during the growing season can control the final yield of corn.

Corn grown under limited irrigation benefits most from water applied just prior to tasseling. Water applied before planting, either in the fall or spring, appears to have little effect on yields except when an occasional very dry spring occurs. Although yields often are higher with full irrigation, water-use efficiency is highest when water is applied about the time tasseling occurs (84).

Water stress at tasseling not only hinders the plant's ability to flower and shed pollen, but also can greatly affect the viability of corn pollen, especially when the drought is accompanied by high temperatures as is usually the case. Water stress at silking can impair extrusion of the silks from the husk and cause desiccation of the silks, reducing the number of seeds set on the ear. Water stress before floral initiation can reduce the number of kernels that can potentially be produced by the plant. Stress after floral initiation can reduce the potential size of the kernels.

Water Requirement of Corn

Although corn has a high water requirement, it is one of the most efficient crops in producing dry matter with the water it uses. Corn requires about 372 unit weights of water per unit weight of dry matter produced, compared with 271 for sorghum, 505 for wheat, 562 for cotton, and 858 for alfalfa (2). Water use efficiency in corn increases as yields increase, and higher populations give decreased efficiency during seasons of adverse soil water depletion (85, 86). Early maturing corn hybrids are more efficient in using water for grain production while the later maturing hybrids use water more efficiently for forage production. However, water use efficiency does not give the total picture. Later maturing hybrids will usually produce more grain than earlier hybrids even though they may be less efficient in water use.

Under less than optimal water conditions, corn will use all the water available and water use efficiency will be closely correlated with yield. But, if stress occurs at flowering, yields will be greatly reduced and may not be correlated to water use efficiency.

Stored soil water, growing season precipitation, and added irrigation water are all equally effective in enhancing corn production and can be substituted for each other (87). Therefore, the addition of irrigation water only when needed can greatly increase irrigation efficiency, especially in subhumid climates where precipitation can satisfy most of a corn plant's requirements.

Heavy irrigations that saturate the soil to a depth of at least 2 feet are more effective than several light irrigations, except on clay pan soils and sandy soils. On sandy soils, yields and water-use efficiency are highest when water is added in light sprinkler irrigations of about 1.3 cm per application when needed as determined by tensiometers placed 37–50 cm deep. Further increase in yield and water use efficiency on sandy soils is obtained when fertilizer is applied through the sprinkler system in small increments throughout the growing season (88). On clay pan soils, light sprinkler irrigations of about 1.3 cm per application as needed help maintain soil aeration and capacity for rainfall while supplying needed water. During tasseling and grain filling, 2.5 cm of water should be added as needed (89).

Irrigating for about 3 weeks, beginning just before tasseling and extending through silking, should have first priority. Although highest yields are obtained with full irrigation throughout the growing season, these yields are not substantially higher than those obtained with partial irrigation at critical periods of flowering.

REFERENCES

1. S. R. Aldrich, W. O. Scott, and E. R. Leng, *Modern Corn Production*, 2nd ed., A & L Publications, Champaign, Ill., 1975.
2. J. H. Martin, W. H. Leonard, and D. C. Stamp, *Principles of Field Crops Production*, 3rd ed., MacMillan, New York, 1976.
3. R. A. Leeper, E. C. A. Runge, and W. M. Walker, *Agron. J.* **66,** 723 (1974).
4. J. J. Hanway, *How a Corn Plant Develops*, Special Report No. 48, Iowa State Univ. Coop. Ext. Ser., Ames, 1971.
5. W. C. Mayaki, L. R. Stone, and I. D. Teare, *Agron. J.* **68,** 532 (1976).
6. D. W. Grimes, R. J. Miller, and P. L. Wiley, *Agron. J.* **67,** 519 (1975).
7. C. W. Carlson, J. Alessi, and R. H. Mickelson, *Soil Sci. Soc. Am. Proc.* **23,** 242 (1959).
8. R. H. Shaw, *Estimation of Soil Moisture Under Corn*, Res. Bull. 520, Iowa Agric. and Home Econ. Exp. Stn., Ames, 1963.
9. R. F. Follett, R. R. Allmaras, and G. A. Reichman, *Agron. J.* **66,** 288 (1974).

10. A. R. Bertrand and H. Kohnke, *Soil Sci. Soc. Am. Proc.* **21**, 135 (1957).
11. M. R. Chaudhary and S. S. Prihar, *Agron. J.* **66**, 350 (1974).
12. D. C. Reicosky, R. B. Campbell, and C. W. Doty, *Agron. J.* **68**, 499 (1976).
13. E. G. Beauchamp and P. J. Lathwell, *Agron. J.* **59**, 189 (1967).
14. W. M. Blacklow, *Crop Sci.* **12**, 647 (1972).
15. R. R. Allmaras and W. W. Nelson, *Agron. J.* **65**, 725 (1973).
16. R. R. Allmaras and W. W. Nelson, *Soil Sci. Soc. Am. Proc.* **35**, 974 (1971).
17. J. J. Onderlank and V. W. Ketcheson, *Soil Sci. Soc. Am. Proc* **37**, 904 (1973).
18. D. L. Linscott, R. L. Fox, and R. C. Lipps, *Agron. J.* **54**, 185 (1962).
19. H. D. Foth, *Agron. J.* **54**, 49 (1962).
20. D. B. Mengel and S. A. Barber, *Agron. J.* **66**, 341 (1974).
21. D. B. Peters, *Soil Sci. Soc. Am. Proc.* **21**, 481 (1957).
22. W. K. Robertson, L. C. Hammond, G. K. Saxena, and H. W. Lundy, *Agron. J.* **65**, 866 (1973).
23. H. M. Taylor and B. Klepper, *Agron. J.* **65**, 965 (1973).
24. D. N. Moss and H. T. Stinson, Jr., *Crop Sci.* **1**, 416 (1961).
25. L. L. Buren, J. J. Mock, and I. C. Anderson, *Crop Sci.* **10**, 9 (1970).
26. D. M. Victor, R. P. Ariyanayagam, and R. B. Musgrave, *Crop Sci.* **17**, 567 (1977).
27. A. F. Numez, *Crop Sci.* **8**, 303 (1968).
28. F. C. Stickler, *Agron. J.* **56**, 438 (1964).
29. L. T. Evans, *Crop Physiology—Some Case Histories*, Cambridge University Press, London, 1975.
30. D. J. Major, R. B. Hunter, L. W. Kannenberg, T. B. Daynard, and J. W. Tanner, *Can. J. Plant Sci.* **52**, 315 (1972).
31. E. Kalju and J. J. Hanway, *Agron. J.* **58**, 16 (1966).
32. P. Hoyt and R. Bradfield, *Agron. J.* **54**, 523 (1962).
33. M. T. Hillson and L. H. Penny, *Agron. J.* **57**, 150 (1965).
34. R. L. Mitchell, *Crop Growth and Culture*, 2nd ed., Iowa University State Press, Ames, 1970.
35. O. T. Denmead and R. H. Shaw, *Agron. J.* **51**, 725 (1959).
36. W. D. Rosenthal, E. T. Kanemasu, R. J. Raney, and L. R. Stone, *Agron. J.* **69**, 461 (1977).
37. E. W. R. Barlow, L. Boersma, and J. L. Young, *Agron. J.* **69**, 95 (1977).
38. E. W. R. Barlow, L. Boersma, and J. L. Young, *Crop Sci.* **16**, 59 (1976).
39. E. W. Chin Choy and E. T. Kanemasu, *Agron. J.* **66**, 98 (1974).
40. J. W. Pendleton, "Increasing Water Use Efficiency by Crop Management," in W. H. Pierre et. al, Eds., *Plant Environment and Efficient Water Use*, Am. Soc. Agron., Madison, Wis., 1966.
41. D. G. Cummins and J. W. Dobson, Jr., *Agron. J.* **65**, 240 (1973).
42. J. Alessi and J. F. Power, *Agron. J.* **66**, 316 (1974).
43. J. J. Mock and L. C. Heghin, *Agron. J.* **68**, 577 (1976).
44. D. B. Peters and M. B. Russell, *Soil Sci. Soc. Am. Proc.* **24**, 526 (1960).
45. C. B. Tanner, A. E. Peterson, and J. R. Love, *Agron. J.* **52**, 373 (1960).
46. R. A. Nulsen, G. W. Thurtell, and K. R. Stevenson, *Agron. J.* **69**, 951 (1977).
47. J. Glover, *J. Agric. Sci.* **53**, 412 (1959).
48. E. C. Miller, *Plant Physiology*, 2nd ed., McGraw-Hill, New York, 1938.

49. H. Meidner and T. A. Mansfield, *Physiology of Stomata*, McGraw-Hill, New York, 1968.
50. M. F. Sanchez-Diaz and P. J. Kramer, *Plant Physiol.* **48,** 613 (1971).
51. W. Domes and A. Bertsch, *Planta* **86,** 84 (1969).
52. J. S. Boyer, *Plant Physiol.* **46,** 223 (1970).
53. J. S. Boyer, *Plant Physiol.* **46,** 236 (1970).
54. C. L. Beadle, K. R. Stevenson, H. H. Neumann, G. W. Thurtell, and K. M. King, *Can. J. Plant Sci.* **53,** 537 (1973).
55. S. Akita and D. N. Moss, *Crop Sci.* **12,** 789 (1972).
56. N. C. Turner and J. E. Begg, *Plant Physiol.* **51,** 31 (1973).
57. S. Akita and D. N. Moss, *Crop Sci.* **13,** 234 (1973).
58. W. Domes, *Planta* **89,** 47 (1971).
59. K. Raschke, *Plant Physiol.* **49,** 229 (1972).
60. C. J. Stigter, *the Epidermal Resistance to Diffusion of Water Vapour: An Improved Measuring Method and Field Results in Indian Corn* (*Zea mays*), Agric. Res. Reports 831, Centre for Agric Publ. and Doc., Wageningen, Netherlands, 1974.
61. N. C. Turner, *Crop Sci.* **9,** 303 (1969).
62. N. C. Turner, *Plant Physiol.* **53,** 360 (1974).
63. N. C. Turner, *Plant Physiol.* **55,** 932 (1975).
64. N. C. Turner and J. E. Begg, *Plant Physiol.* **51,** 31 (1973).
65. D. C. Reicosky, R. B. Campbell, and C. W. Doty, *Agron J.* **67,** 380 (1975).
66. E. W. R. Barlow and L. Boersma, *Agron J.* **68,** 923 (1976).
67. J. S. Boyer, *Plant Physiol.* **46,** 233 (1970).
68. P. R. Goldworthy, *Proceedings of the Seminar on World Wide Maize Improvement in the 70's*, CIMMYT, Mexico, 1974.
69. E. W. R. Barlow, T. M. Ching, and L. Boersma, *Crop Sci.* **16,** 405 (1976).
70. J. A. Amos and R. L. Scholl, *Crop Sci.* **17,** 445 (1977).
71. C. A. Marilla, J. S. Boyer, and R. H. Hageman, *Plant Physiol.* **51,** 817 (1973).
72. M. F. Beardsell and D. Cohen, *Plant Physiol.* **56,** 207 (1975).
73. D. A. Dube, K. R. Stevenson, and G. W. Thurtell, *Can. J. Plant Sci.* **54,** 765 (1974).
74. D. A. Dube, K. R. Stevenson, G. W. Thurtell, and R. B. Hunter, *Can. J. Plant Sci.* **55,** 565 (1975).
75. T. C. Kilem and R. H. Andrew, *Agron. J.* **61,** 669 (1969).
76. B. K. Sinha and N. T. Singh, *Agron. J.* **68,** 346 (1976).
77. B. K. Sinha and N. T. Singh, *Agron. J.* **66,** 557 (1974).
78. J. T. Ritchie, *Agron. J.* **65,** 893 (1973).
79. O. T. Denmead and R. H. Shaw, *Agron. J.* **52,** 272 (1960).
80. J. S. Robins and C. E. Dominga, *Agron. J.* **45,** 618 (1953).
81. D. J. Makus, M. M. El-Forgany, and W. R. Simpson, *Water Stress and Sweet Corn Seed Production*, Current Information Series No. 421, Univ. of Idaho Coop. Ext. Serv., Moscow, Idaho, 1977.
82. H. G. McPherson and J. S. Boyer, *Agron. J.* **69,** 714 (1977).
83. C. M. Woodruff, *Stretching a Short Water Supply*, Proceeding of the Irrigation Conference, University of Missouri, Columbia, Mo, 1977.
84. R. Gwin, *Spring Field Day Report*, Tribune Branch Experiment Station, Tribune, Kan., 1978.

85. D. R. Timmons, R. F. Holt, and J. T. Maroghan, *Agron. J.* **58,** 429 (1966).
86. T. C. Olson, *Agron. J.* **63,** 104 (1971).
87. J. F. Power, J. J. Bond, W. A. Sellner, and H. M. Olson, *Agron. J.* **65,** 464 (1973).
88. F. M. Rhoads, R. S. Mansell, and L. C. Hammond, *Agron. J.* **70,** 305 (1978).
89. C. M. Woodruff, *Irrigating Corn on Claypan Soils in Missouri*, Science and Technology Guide, University of Misouri–Columbia Ext. Div., Columbia, Mo., 1975.

7

COTTON

Wayne R. Jordan

INTRODUCTION

General

The early work by Balls (1) served as the beginning for physiological studies of cotton in its roles as both a unique plant and a major agricultural crop. Driven by an increasing demand for cotton fiber, research efforts multiplied dramatically and delved into every phase of growth, development, and productivity. Since only those aspects dealing with water will be dealt with here, readers are referred to reviews on various other aspects of cotton physiology and production (2–9). Of particular interest are several recent efforts that combine basic research information regarding plant response to light, temperature, water, and nutrition into models that describe the dynamics of plant–environment interaction in terms of activity of some growth process or overall growth and yield of the crop. A few important examples of such models have been reported for seedling emergence (10–12), stomatal action (13, 14), evapotranspiration during crop development (15), respiration and carbon balance (16), and environmental control of morphogenesis (17). The most extensive effort in simulation of dynamics of growth and yield resulted in development of the SIMCOT model described by Duncan (18).

Origin, Evolution, and Distribution of *Gossypium*

The genus *Gossypium* is included in the family Malvaceae (Mallow) and is currently comprised of 30 or more tropical and subtropical species. Of these, four species have economic importance in agriculture. *G. herbaceum* L. and *G. arboreum* L. are diploid ($n = 13$) species grown commercially in the Old World, while *G. hirsutum* L. and *G. barbadense* L. are amphidiploid ($n = 26$) species grown in the United States and other portions of the New World.

The evolutionary account of Fryxell (19) serves to summarize several events through which the original species diversified from its tropical origins and spread to its present distribution. The first event was believed to occur in the Mesozoic era resulting in the formation of spatially isolated groups. This is borne out by cytological investigations which have distinguished five (20, 21) or six (22, 23) separate genome groups (A–F) at the diploid level, and one separate amphidiploid group which combines the A and D genomes (24). Several authors have suggested this ancient diversification may be attributed to continental drift. Later, during the Tertiary period, climatic changes led to expansion of arid environments forcing adaptation of the genus to desert habitats (19).

The New World amphidiploid species arose from hybridization between a diploid representative of Asian–African origin bearing the A genome and the D genome representative widely spread throughout middle America. This event is believed to have occurred between 100,000 and 1 million years ago.

Domestication of diploid and amphidiploid species occurred during the past 1000 years and is strongly associated with its cultivation as an annual crop. Wild species are sensitive to low temperature and flower only under short days, characteristics that effectively limited the genus to tropical and subtropical distribution. With cultivation, man provided selection pressures that led to development of the day-neutral flowering response permitting adaptation toward the annual habit necessary for production in more temperate climates (19).

General Anatomy, Morphology, and Phenology

The following brief description applies principally to current American upland cottons which are grown extensively in temperate regions around the world.

Leaves

Leaves are usually palmately lobed and are typically dorsiventral with often only a single layer of palisade cells which occupy one-third to one-half of the blade thickness (25). While leaf thickness may vary considerably with age and environment, a typical thickness for field-grown leaves is about 0.21 mm (26). Dark-pigmented, gossypol-laden, lysigenous glands are found in substantial numbers in the leaf mesophyll as well as most other organs (22).

Leaf size also varies considerably with plant age and environmental conditions. The area of mature leaves commonly varies between 1 and 3 dm^2. Stomata are found on both surfaces with densities ranging between 50 and 300 per mm^2 (8). While densities may vary with changes in leaf size, the ratio of abaxial to adaxial densities is near 2.0 over a wide range of leaf sizes and growth conditions.

Other distinctive features include the presence of trichomes and nectaries which occur on the abaxial surface of the main ribs of the leaf. A single abscission zone forms near the juncture of the petiole and stem.

Stems

Stem anatomy of cotton is similar to most dicotyledonous plants which exhibit secondary thickening. The stems are quite woody when mature and vary in diameter between 0.5 and 1.5 cm depending on stand density and growth conditions. The xylem is endarch and since the secondary xylem vessels are not heavily lignified the tissue rapidly disintegrates following harvest (27).

Typically, three vascular strands of the stem are involved in the transport of water to a single leaf (28). The three strands do not remain distinct as they pass through the petiole, often anastamosing to a considerable extent so the individual vascular traces are complex. This vascular arrangement may present significant resistance to flow of liquid water, and may be largely responsible for the gradient in leaf water potentials that exists between leaves at different insertion levels on the main stem when water flux is high (29). Phyllotaxy of American upland cottons is commonly $\frac{3}{8}$ while that of Old World cottons is $\frac{1}{3}$.

Roots

Cotton develops a single tap root with extensive lateral branching. The primary xylem is usually tetrarch in structure giving rise to four rows of lateral roots (8). Secondary thickening begins early in ontogeny and the cortex and epidermis slough off to be replaced with a multilayered periderm. The vascular anatomy of the transition zone between shoot and root is complex (27). Considering the root system as a whole, there appears to be little axial resistance to water flow (30).

Fruit

The developing fruit is a leathery capsule containing four or five compartments or locks in which develop 20–40 seeds. Enlargement of the fruit, termed a boll, occurs rapidly reaching a final size of about 25–35 mm. Boll growth during most of this period is steady and the swelling and shrinking of other tissues in response to temporal water deficits are not observed (31).

Growth Stages and Habit

Cotton is a long-season crop, requiring a growing season (planting to harvest) of 150–220 days depending on cultivar and climatic conditions. A generalized illustration of growth and development of an irrigated crop is shown in Fig. 7.1.

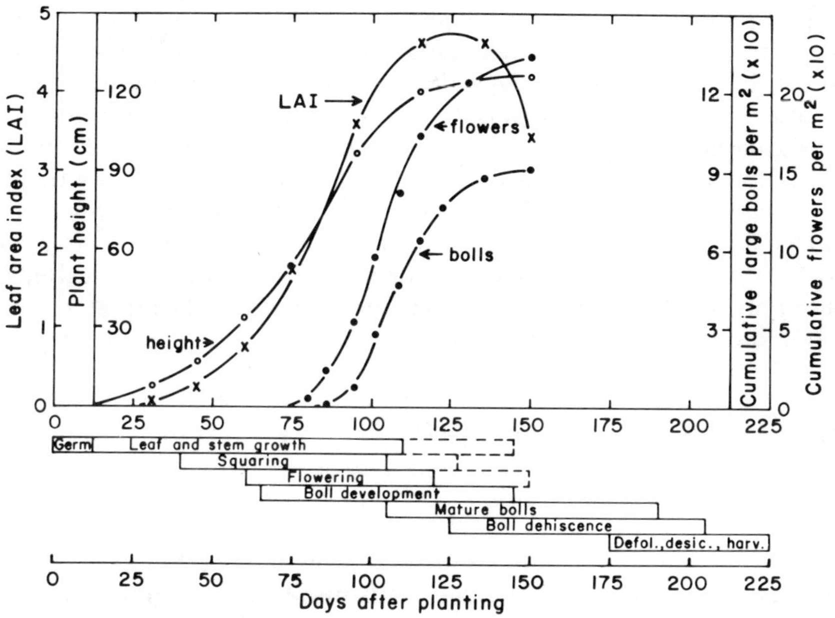

Figure 7.1. Vegetative and reproductive growth for an irrigated cotton crop (upper) and a generalized crop growth stage calendar (lower). Data are for Acala SJ-2 cotton grown under full irrigation in Israel (Danield Wallach, unpublished data). Crop growth stage calendar generalized from personal observations and published reports by Stockton et al. (114), Longenecker and Erie (143), and Halevy (107).

Cotton is cold sensitive with tissue damage occurring at temperatures of 5–10°C. Current recommendations are to delay planting until minimum soil temperatures exceed 16°C for 3 consecutive days. Cotton seeds planted 4–7 cm deep when soil temperatures are above 16°C emerge within 5–10 days. The cotyledons are carried above the soil surface by the elongating hypocotyl and expand rapidly to about 2 to 5 times their preemergent size within the next 7 days and become photosynthetic.

The main stem has an indeterminate growing point, but the degree of indeterminacy is variable among present-day cultivars. Leaves on the main stem are formed at the rate of one every 3–4 days depending on temperature (7). Two buds form at the base of each leaf, one of which may develop into either a vegetative or fruiting branch (32, 33). Additional branches may develop from axillary buds on either vegetative or fruting branches.

The first fruiting branch usually develops from the sixth to ninth node above the cotyledons. As the fruiting branch grows, the meristem produces a leaf and a floral bud which consumes the meristem of the axis. The axillary bud of the leaf begins to grow as the floral bud begins to develop. Each successive internode develops from a new branch axis producing the typical zigzag form of the fruiting branch (5, 32).

The first floral bud (square) becomes visible about 30 days after plant emergence, with anthesis occurring some 20–30 days later. Peak flowering rates occur about 30 days after first bloom, but flowering proceeds irregularly during a total period of 40–60 days. Bolls retained by the plant enlarge rapidly to reach a final size of 25–35 mm by 20–25 days after anthesis. Secondary thickening of lint fibers is complete by about 35–40 days after anthesis. Early formed, mature bolls at lower nodes dehisce and begin to dry while bolls at higher nodes remain immature. Frost terminates growth in some portions of the Cotton Belt, but for a major portion of the cotton producing areas growth is terminated and the crop is prepared for harvest with chemical defoliants and desiccants.

GROWTH AND DEVELOPMENT

Germination and Emergence

It is convenient to separate the initial growth of cotton seedlings into two phases. The first phase deals with imbibition, resumption of metabolic activity, and growth of the radicle to rupture the seed coat. The second phase deals with growth processes in both hypocotyl and radicle resulting in the emergence and establishment of a vigorous seedling. The two phases are visually separable and have been studied in some detail to allow dynamic simulation of the germination and emergence processes as influenced by the environment surrounding the seed.

Soil Water Potential and Germination

When placed into a favorable seedbed environment, water is rapidly imbibed by dry seeds. Two conditions are necessary for imbibition of seeds in the field: (1) a water potential gradient must exist between the seed and the soil, and (2) substantial contact must be established between soil particles and seed for the transfer of sufficient water to occur. The total water potential of air dried seeds is often on the order of −1000 bars. Thus, the condition of existence of a water potential gradient is met in all but the most rigorous situations. Under favorable soil water and temperature conditions, imbibition is normally complete within the first 36–48 hr. During this period, sufficient water is absorbed to increase seed weight by about 80% and seed volume by 50%. Metabolic activity is low during this phase of germination and water uptake occurs with a low temperature dependence typical of physical rather than biochemical processes.

Water potential differences between seed and soil, and the resistance to water movement to and into the seed, determine the rate of seed water uptake. Present information with seeds of a number of species indicates the rate of water movement through soils at potentials greater than −15 bars should not limit seed water uptake (34, 35). Observed water uptake curves for cottonseed from soils at −10 bars were not much different from those of −0.3 bar (36), suggesting that the capacity of soil to transmit water does not seriously delay seed hydration within this range of soil water potentials.

Since water uptake by seeds is a dynamic process, a static definition of the soil water potential in the region immediately adjacent to seeds may be insufficient to describe the limiting factors controlling uptake (37). The water potential of bulk soil may be a substantial overestimate of the "effective water potential" at the seed surface because of rapidly reduced soil water content near imbibing seeds. The difference between the water potential of bulk soil and that at the seed surface increases as the bulk soil water potential decreases. The net effect of reduced water content or potential of the sphere of soil surrounding the seed is to greatly slow the flow of water since hydraulic conductivity of soil is related to water content in an exponential manner (38).

Water entry into seeds may be seriously hampered by impermeable seed coats ("hard seeds") prevalent in some wild species of *Gossypium* (39). The acid delinting process whereby remnant fuzz and lint fibers are removed with sulfuric acid apparently serves to greatly reduce this hard seed characteristic so that most commercial seed lots freely imbibe water. The diffusivity of cottonseed increases as seed moisture content increases, further suggesting that water movement through soil does not limit seed water uptake (40). The development of higher diffusivities as seeds become hydrated may reflect coverage of an increasing percentage of the seed surface with a liquid water film as imbibition proceeds. The slowing of seed water uptake as high seed water contents are approached is probably the result of a greatly reduced water potential gradient between seed and soil.

Hydration of the seed is followed quickly by resumption of metabolic activity signaling a more temperature sensitive phase of germination. Polysomes increase in the now turgid cells and nucleic acid synthesis allows successful transcription and translation of stored genetic information necessary for the resumption of cell division and organ growth.

Emergence

Emergence requires sufficient hypocotyl growth to raise the cotyledons above the soil surface, usually a distance of 5–8 cm. Successful establishment also requires concomitant growth of the radicle to anchor the plant and provide sufficient absorptive surface to supply the developing shoot with water and mineral nutrients. Growth of both shoot and root is sensitive to microclimatic and physical characteristics of the seedbed, especially temperature, water potential, and soil strength.

Soil Microclimate and Seedling Growth. The temperature dependence of hypocotyl and radicle growth has been illustrated in several reports (36, 41, 42). Based on the early work of Arndt (41), Wanjura et al. (11) derived a temperature-dependent expression for hypocotyl growth rate with an optimum at 34.4°C and predicted upper and lower growth limits of 41.9 and 14.4°C, respectively. Similar temperature optima are indicated for elongation of both hypocotyl and radicle.

When soil temperature is near optimal the effect of lower soil water potentials is to reduce overall growth rates, but hypocotyl elongation is reduced to a relatively greater extent than root elongation as shown in Fig. 7.2. Maximum growth rate for the radicle at $\psi_s = -10$ bars was reduced only 28% while the reduction in growth rate for the hypocotyl was 87%. Seedlings already germinated required 3 days to emerge from a soil depth of 5 cm with $\psi_s = -0.3$ bar, while those in soil with $\psi_s = -3.0$ bars required 7 days. Seedlings growing in soil with $\psi_s = -10$ bars did not emerge within 13 days (42). In a similar study with a silt loam soil, Jensen (43) found emergence could be related to soil matric potential according to the equation $y = 74.9 + 3.61x + 0.877x^2$, where y = percent emergence after 4 weeks, and x = absolute value of soil matric potential in bars. The strong dependence of emergence on matric potential may be a reflection of either a greatly reduced capacity of soil to transmit water at rates sufficient to support rapid cell extension or to a direct inhibitory effect of low tissue water potential on hypocotyl growth.

Soil Strength, Emergence, and Establishment. Emergence may be delayed or prevented by soil crusting even if water and temperature conditions at seeding depth are favorable. Actual strength of soil crusts are determined by a complex interaction of soil texture, soil structure, and water content. To emerge through a crust the seedling must be capable of producing enough growth pressure (force per area) at the hypocotyl hook to fracture the crust. The influence of soil impedance (strength) on growth of cotton seedlings after germination is illustrated in Fig. 7.3.

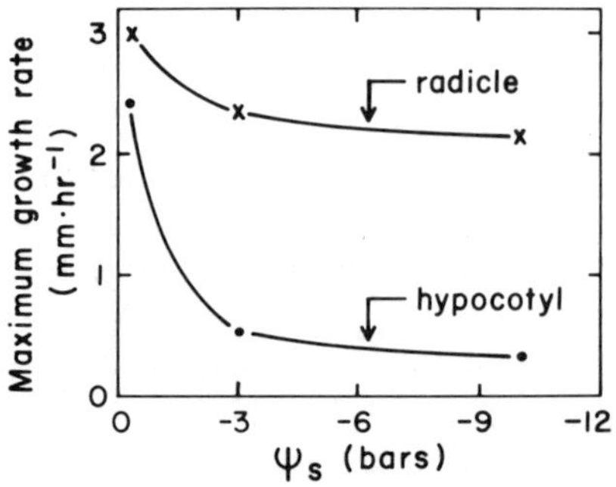

Figure 7.2. Maximum rates of hypocotyl and radicle elongation of unemerged cotton seedlings as influenced by soil water potential (Ψ_s) at constant temperature (32.2°C) and soil impedance (0.23 bar). From data of Wanjura and Buxton (42).

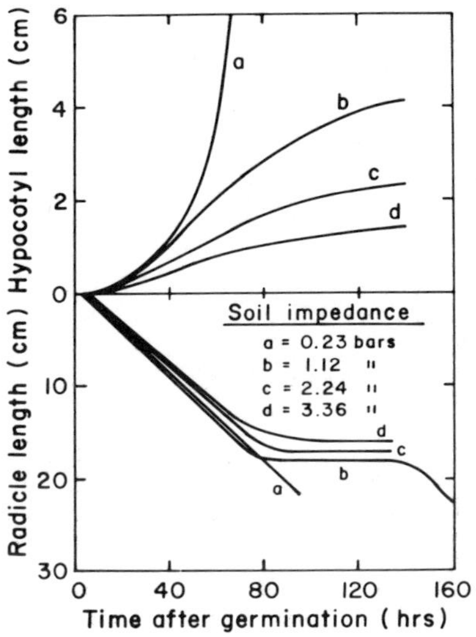

Figure 7.3. Hypocotyl and radicle lengths of unemerged cotton seedlings with time after germination as influenced by soil impedance at constant temperature (32.2°C) and soil water potential (−0.3 bar). Reproduced from *Agron. J.* **64**, 431–434 (1972) by permission of the American Society of Agronomy (42).

Hypocotyl growth was greatly reduced while radicle growth was largely unaffected by soil strengths (measured with a penetrometer) up to 3 bars (42). The maximum emergence force exerted by a single cotton seedling is about 600g (44).

Even in the absence of soil crusts, soil strength plays an important part in successful establishment of a vigorous plant in soils where compacted layers (plow pans) occur below planting depth. These compacted layers may reduce root growth by providing a mechanical impedance to the growing root tip. Soil strength changes not only with water content and bulk density, but also with types and amounts of cations, the number of particle-to-particle contacts, the type of clay material, and the amount and type of organic materials (45). An illustration of interactions among soil strength, bulk density and cotton root growth is given in Fig. 7.4 based on data of Taylor and Gardner (46) and Taylor and Ratliff (47). These results are similar to those of Wanjura and Buxton (42) (Fig. 7.3) in that root growth is little affected at soil strengths below 3 bars. Further increases in soil strength effectively reduced root elongation so that no growth occurs near 14 bars (by extrapolation to zero growth in Fig. 7.4). Taylor et al. (48) also reported that in four different soils root penetration was prevented by strengths near 20 bars.

Turgor pressure in the elongating cells of the hypocotyl and root provides the driving force to overcome constraints to growth provided by the cell walls and external materials. The pressure (force per area) available for a hypocotyl or root to accomplish work against an external constraint is termed growth pressure (49). Eavis et al. (50) and Taylor and Ratliff (47) provided evidence that the maximum axial pressure generated by cotton roots was near 15 bars, a value that corresponds well

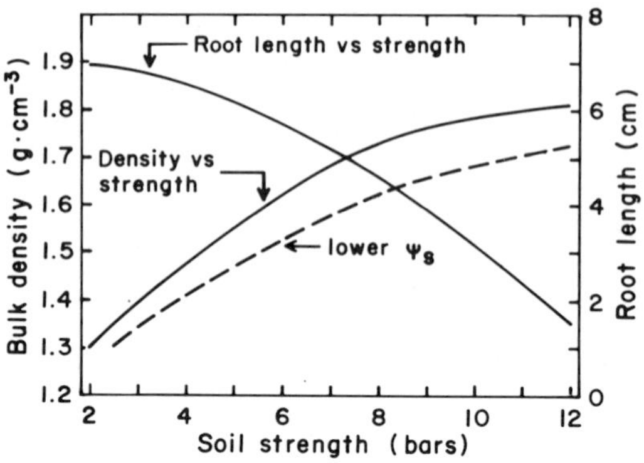

Figure 7.4. An illustration of typical interactions among soil strength, bulk density, and soil water potential (ψ_s) on growth of cotton roots. The dashed curve illustrates the effect of lower ψ_s on the density vs. strength relation. Reproduced by permission from the *Proc. Soil and Crop Sci. Soc. Florida* **36,** 1–4 (1977) (174).

with that obtained from growth in soil (Fig. 7.4.). Results presented in Fig. 7.3 suggest that the hypocotyl is not capable of generating growth pressures equivalent to those of roots, although this interpretation may be in question because of differences in the mechanism of growth between the two organs. Less adequate lateral support for the hypocotyl compared to the root, leading to "instability" failure, may also be one reason for this difference.

The mechanisms by which the hypocotyl or root senses an external mechanical impedance and transduces this stimulus into growth pressure are indeed interesting. Working with pea epicotyls, Goeschl et al. (51) found that a sharp increase in ethylene evolution and in radial expansion occurs when elongation of the epicotyl was mechanically impeded. Both the increase in endogenous ethylene and the associated increase in diameter appeared to be proportional to the degree of mechanical impedance thereby providing a ready mechanism to adjust the emergence force (growth pressure) of the seedling to prevailing soil conditions. Whether a similar mechanism is present in cotton hypocotyls is not known, but exogenous ethylene which induces radial swelling of pea epicotyls also causes swelling of cotton hypocotyls (52). Emergence force of cotton seedlings is strongly related to hypocotyl diameter (44). Root diameters do not appear to increase in response to ethylene, suggesting the root may utilize other stimuli to induce higher growth pressures.

Axial extension may not be the only form of root growth affected by compacted layers. Roots may be able to penetrate these layers when wet, but subsequent radial growth may be greatly restricted as the soil dries and strength increases (53–55). These conditions cause girdling of the taproot and may restrict the capacity for transport of water and mineral nutrients. Mathers and Welch (55) used acrylic plastic clamps to restrict radial growth of cotton taproots and found that a restriction of 0.79 cm or less for a period as brief as 2 weeks reduced lint yield. The magnitude of the yield reduction depended on severity and duration of the root restriction.

Much of the aforegoing discussion of root growth and soil strength was supported by evidence obtained with uniform substrate materials compacted under carefully controlled conditions. Nearly all field soils have some structural development which creates discontinuities in the soil matrix. Soils of moderately high densities may swell and shrink creating vertical planes of weakness along which roots may be concentrated (45). Strong ped development may further create a three-dimensional network of fracture planes. Large roots may be restricted to these planes of weakness, but smaller secondary or tertiary roots may penetrate the peds.

Plant Water Deficit and Vegetative Growth

Evaporative Demand and Crop Water Use

Evaporative Demand. In its common usage, evaporative demand implies a potential rate of water vapor loss from a surface with only atmospheric restrictions to vapor flow. Thus, evaporative demand may be considered as the upper limit or potential evapotranspiration (E_0) for a crop. The crop canopy influences evaporative demand through its effect on net radiation, temperature, and aerodynamic properties of the evaporating surface (56, 57).

Crop Water Use. Total water use, consumptive use, and actual evapotranspiration (E_t) are terms used to describe the total water lost from soil and crop surfaces from the time of planting until harvest. Prior to plant emergence all water loss occurs directly via soil evaporation, with actual rates of loss dependent on the frequency of wetting by rainfall or irrigation. Soil surfaces evaporate at or near the potential rates for only brief periods (less than 1 to a few days) after wetting. Thereafter, actual evaporation rates decrease steadily as the surface dries and soil resistance to water flow increases.

Following plant emergence, the crop canopy plays an increasing role in the source and rate of water lost from the crop surface. Once the crop canopy develops an *LAI* of 3, and if water remains freely available to the roots, E_t/E_0 attains values near 1.0 suggesting the canopy acts as a completely wet surface wherein essentially all net radiation is dissipated as latent heat flux (vaporization of water) (15, 58, 59). As available soil water is depleted, the crop canopy begins to restrict water vapor exchange so that the canopy no longer behaves as a wet surface and E_t

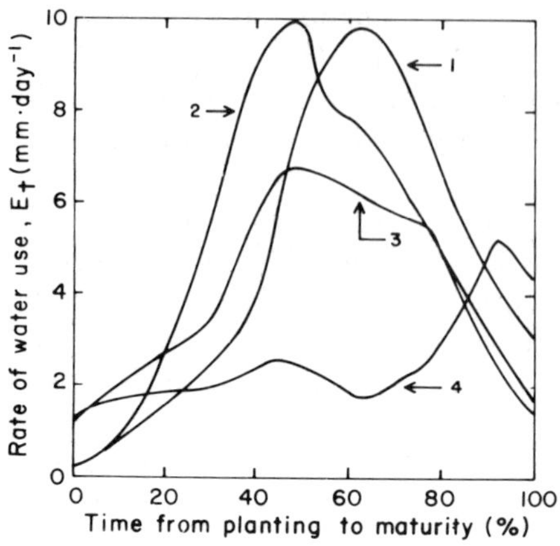

Figure 7.5. Rates of soil water use (E_t) during the growing season for cotton crops grown at various locations. Curve 1 from data of Thaxton and Swanson (175) for irrigated cotton on the Texas High Plains; curve 2 from data of Kittock (132) for irrigated Pima S-4 cotton at Phoenix, Arizona; and curves 3 and 4 from data of Ritchie and Burnett (60) for dryland cotton at Temple, Texas, in 1968 and 1969, respectively.

falls below E_0 (60). Seasonal patterns of water use for cotton are illustrated in Fig. 7.5 for four different locations and cultural conditions. These patterns serve to illustrate the no simple curve may represent E_t for a specific crop.

Soil Water Availability. Older treatments of this topic considered all soil water in the root zone at potentials above -15 bars as available, although not all this water had the same degree of availability. Recent studies have sought to quantify both the total available soil water and the degree of availability. Based on an analysis of many field studies, the lower limit of available soil water which would support crop water loss at potential rates ($E_t = E_0$) ranged between 65 and 75% (15, 61; C. B. Tanner and J. T. Ritchie, unpublished data, 1974). This has been termed "freely available" soil water as illustrated in Fig. 7.6.

The relation between "freely available" soil water and total soil water above -15 bars is illustrated in Fig. 7.7 for a Houston Black Clay soil

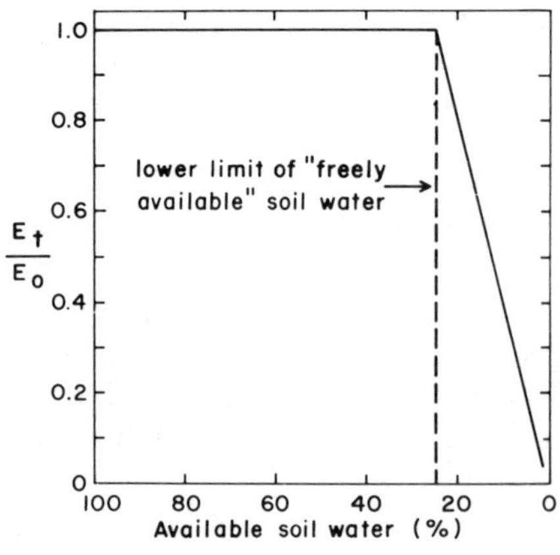

Figure 7.6. Ratio of actual (E_t) to potential (E_0) evapotranspiration for a cotton crop with a full canopy as influenced by available soil water remaining in the profile. From data of Ritchie et al. (59).

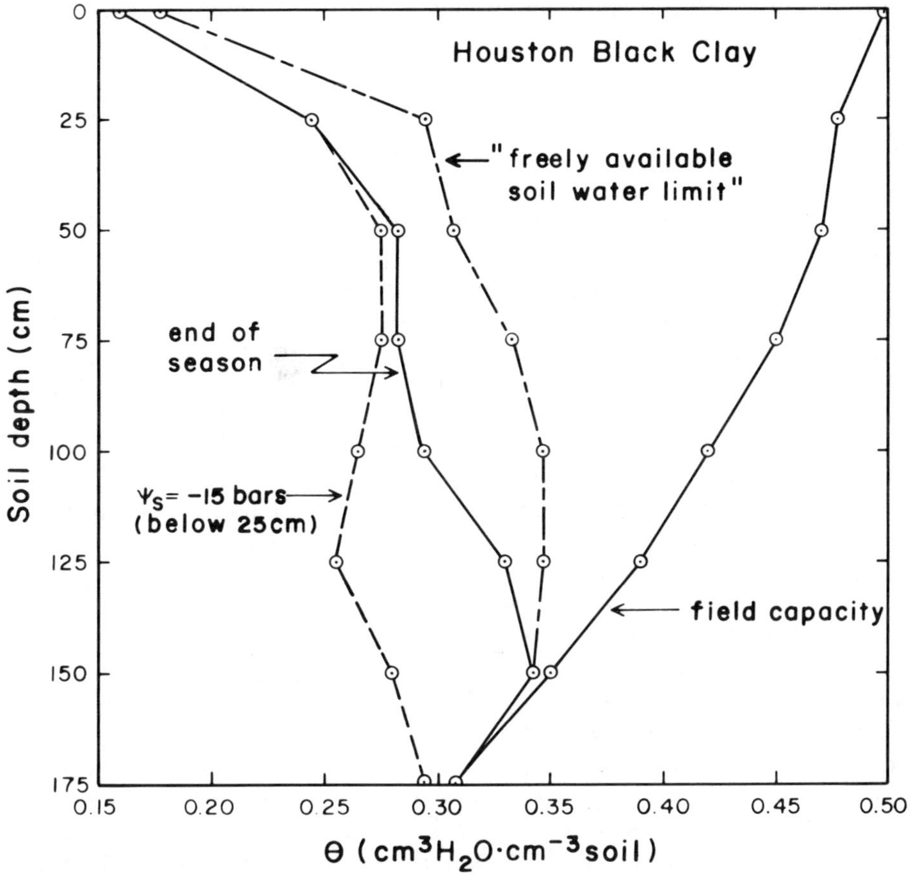

Figure 7.7. An illustration of the distribution of soil water content with soil depth for Houston black clay on which a cotton or grain sorghum crop was grown. The "freely available soil water limit" curve represents the water content distribution when E_t falls below E_0 for a crop with a full canopy (Fig. 7.6). Total water extracted between field capacity and the lower limit of freely available soil water, water remaining at seasons end, and −15 bars was 18 cm, 25 cm, and 31 cm, respectively. Reproduced from *Moving up the Yield Curve: Advances and Obstacles,* special publication No. 39, 1980, pp. 1–23 by permission of the American Society of Agronomy (176).

and is typical for either cotton or sorghum. As is the case with many soils, water stored deep in the profile may not be completely utilized by the growing crop.

As currently used, all available soil water concepts require a subjective evaluation of the soil depth that constitutes the root zone. For soils with well-defined physical or chemical barriers that limit root growth, the rooting volume is readily defined. This is not the case with many deeper soils where effective rooting volume may vary with crop and season. Readers must be aware of this important limitation of the available or extractable water concept, especially since many of the current crop improvement programs strive to increase effective rooting depth.

Origin of Plant Water Deficits and Sources of Resistance to Water Transport

Almost without exception, crop plants develop internal water deficits on a daily basis. The degree or magnitude of these diurnal deficits depends on (1) E_0, (2) the internal resistance to liquid water or water vapor transport within the plant, and (3) the resistance to liquid water flow from the soil to the root interior (xylem) (30, 62). Generalized interactions among evaporative demand (E_0), leaf (ψ_l) and root xylem ($\psi_{r,x}$) water potentials, and ψ_s are illustrated in Fig. 7.8. These curves represent typical data for cotton growing in a field environment (63–71).

Maximum leaf water potentials occur before dawn and range between -1 and -2 bars for cotton plants growing in soils with high water potentials ($\psi_s < -1$ bar) in the rooting zone to -10 bars or more for plants growing in drying soils (66, 72). Minimum leaf water potentials correspond with peak rates of evapotranspiration (usually between 1300 and 1500 hr) and may range between -8 and -15 bars and -15 and -30 bars for plants growing in moist and drying soils, respectively (Figs. 7.8.1 and 7.8.3). Root xylem water potentials are plotted relatively close to ψ_l suggesting that a major source of resistance is encountered in the flow of water radially from the root surface to the xylem (68, 70, 73). Therefore, for cotton, major sources of resistance to water transport are concentrated in the tissue region between the root–soil interface and the root xylem and between the leaf interior and bulk air surrounding individual leaves, although hydraulic resistances may occur in stems and leaves. A "canopy resistance" may also be important in a complete description of movement of water vapor from bulk air within the canopy to some point above the crop (8).

Resistances between Bulk Soil and Root Xylem. Most current analyses of water transport in this portion of the pathway assume that water moves radially from a cylinder of soil surrounding each root to the root xylem (74). In doing so water must pass through soil, and soil–root interface, and root tissues external to the xylem vessels. Total "radial" resistance in this pathway thus contains components of both soil and plant. Relative magnitudes of each component change with soil water content and metabolic activity (permeability) of living root tissues. Little is known about absolute magnitudes of resistance in the radial pathway for cotton, particularly since cotton roots may swell and shrink in concert with diurnal plant water deficits (75). Estimates of Taylor and Klepper (30) and Meyer and Ritchie (73) place the radial resistance per unit root length in the range of 1174–2802 cm^{-3} day bar cm for plants grown in soil at high ψ_s.

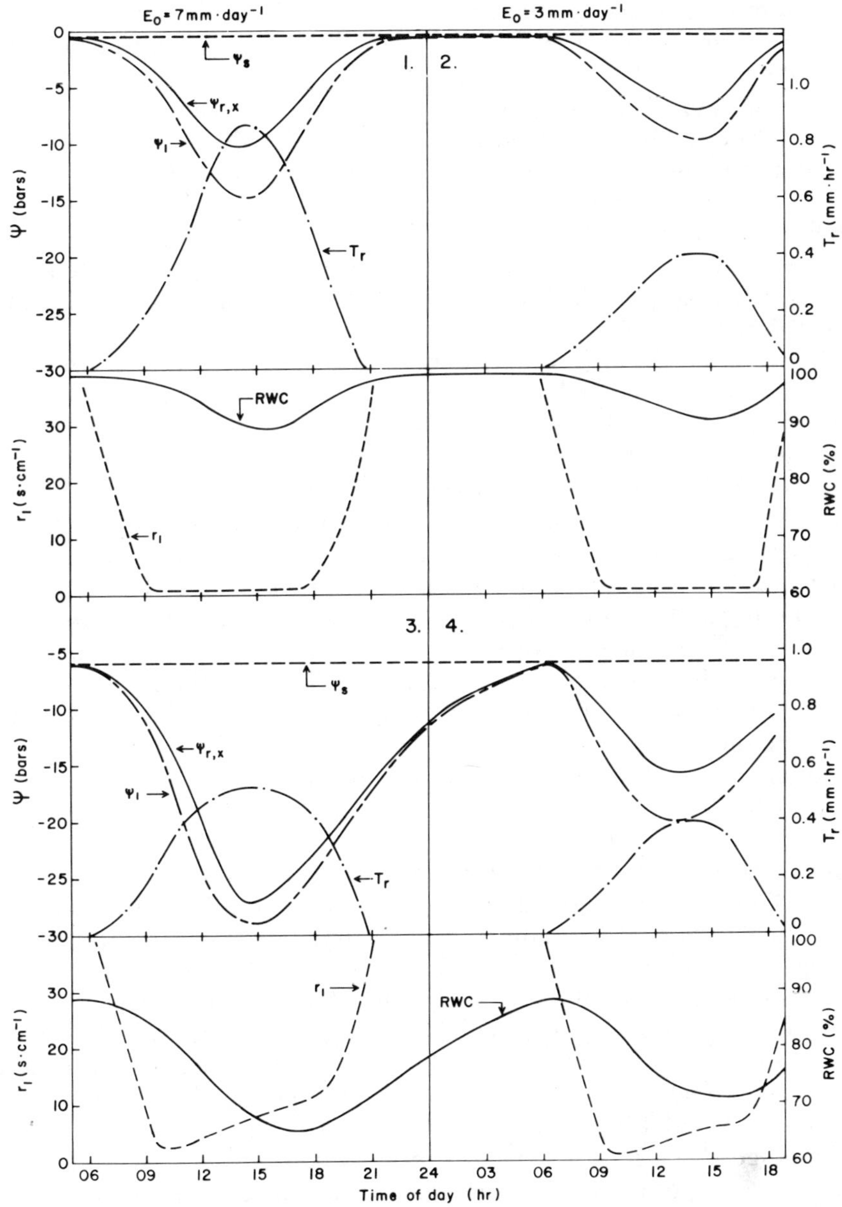

Figure 7.8. A generalized illustration of diurnal variations in water potentials of soil (ψ_s), the xylem in the root ($\psi_{r,x}$), and the leaf (ψ_l), in transpiration rate (T_r), in leaf resistance (r_l), and in relative water content (RWC) of leaves of a cotton crop for 4 days. Parts (1) and (3) are for a day with moderately high potential evapotranspiration (E_0 = 7 mm day^{-1}), and parts (2) and (4) are for a day with a low potential evapotranspiration (E_0 = 3 mm day^{-1}). Reprinted by permission from Agri. Meteorol. **14,** 183–198 (1974) (176).

Hydraulic Resistances within the Plant. Axial root resistance to water transport from points of absorption to the soil surface appears small or insignificant for cotton (68), but the absolute value may depend on rooting volume (73). However, significant frictional resistances may occur within tissues of the shoot leading to a measurable decrease in leaf water potential among leaves at successively higher insertion levels on the main stem (29, 76). Specific resistances associated with either stems or petioles are not available, but reports by Barrs (77) and Stoker and Weatherly (78) provide evidence that, for cotton, total plant resistance to liquid flow is constant at high flow rates and ranges between about 2 and 7 bars $g^{-1}dm^2$ hr (79).

The major point(s) of resistance in the shoot may lie between the stem xylem and leaf mesophyll. Depending on transpiration rate and irradiance of individual leaves, differences in ψ_l between lower and upper leaves are commonly on the order of 5–10 bars, with points of measurement about 12–15 nodes apart. Assuming constant resistance within the stem and within each leaf, these observations suggest a decrease in ψ_l of less than about 0.8 bar per node. Measurements of ψ_l on adjacent leaves, one of which was covered so as to prevent water loss (and presumably to come into water potential equilibrium with the stem xylem at that point), produce differences of several bars (73 and personal observations). These results suggest a relatively larger potential drop occurs between stem xylem and leaf mesophyll than within the stem xylem between successive nodes. The complex arrangement of vascular elements at the juncture of petiole and stem may be the major source of this resistance (28).

Vapor Phase Resistance and Stomatal Regulation of Water Loss. Transport of water vapor from the leaf interior to bulk air external to the leaf occurs principally through stomatal pores, but direct loss through the cuticle also occurs and may be important when stomata are closed. Examination of the magnitude of individual component resistances to vapor transfer provides an insight into their relative importance in changing environments. Boundary layer resistances vary with wind speed, leaf shape, and leaf surface morphology, but usually range between 0.1 and 1.5 sec cm^{-1} (80). Cuticular resistances are typically in the range of 30–200 sec cm^{-1} (81–83), while stomatal resistance varies within the range of 1.0–30.0 sec cm^{-1} depending on illumination and plant water status (66, 81, 83–86). Since stomatal and cuticular resistances act as resistors in parallel, leaf resistance is dominated by stomatal resistance so long as stomata are open. Also, since total resistance to water vapor flow is represented by the sum of the boundary layer and leaf resistances, a proportional reduction in water flux with increases in stomatal resistance may not occur in still air where the boundary layer resistance may equal or exceed leaf resistance.

CONSEQUENCES OF PLANT WATER DEFICITS

We must assume crop plants have adapted in various ways to the normal range of diurnal water deficits encountered when grown in moist soils. This does not assume that physiological process are unaffected by deficits of this normal magnitude, but a realistic "baseline deficit" must be established, at least conceptually, by which to evaluate consequences of greater than normal deficits. This baseline is herein defined as the minimum plant water potential, usually evaluated as the water potential of upper, fully exposed leaves, which develops diurnally when water is freely available to the roots. As defined, this baseline deficit is not constant, but fluctuates with the evaporative demand and may range between -8 and -15 bars.

Because of the central role of photosynthesis to growth and yield, the rate of photosynthate production and subsequent utilization will be discussed in detail below. Other metabolic consequences of water deficits have been recently reviewed (87–89).

Photosynthesis and Dry Matter Production

One of the most striking and general effects of water deficits is a reduction in the rate of cell expansion (88, 90, 91) because of its dependence on turgor. Reduced rates of cell expansion are immediately reflected in reduced rates of leaf area development and vegetative growth in general, and are eventually reflected as reductions in the rate of dry matter accumulation. It is important to stress the sequential nature of this chain of events to better understand how plants transduce water deficits into alterations in the rates of individual growth responses.

Net Photosynthesis and Plant Water Deficit

Estimates of net photosynthesis (P_n) vary widely depending on species, plant age, and environment during growth. In a survey of 26 *Gossypium* species, El-Sharkawy et al. (92) found 20 species to have maximum P_n of 40 mg CO_2 dm^{-2} hr^{-1} or greater, with a range of 24–45 mg CO_2 dm^{-2} hr^{-1}. An earlier report (93) found no measurable variation in P_n among cultivars of *G. hirsutum* and *G. barbadense* where P_n in optimum conditions was 45 mg CO_2 dm^{-2} hr^{-1}.

Even through reductions in E_t and P_n with increased r_s may be quantitatively different, the expected result is that both processes should be reduced progressively as developing water deficits cause r_s (or r_l) to increase as shown in Fig. 7.9 (94). Similar relations between water deficit and P_n have been shown by El-Sharkawy and Hesketh (95) and between water deficit and water flux by Ehlig and Gardner (96). These results, all with potted plants subjected to rapid water deficits, suggest that r_s

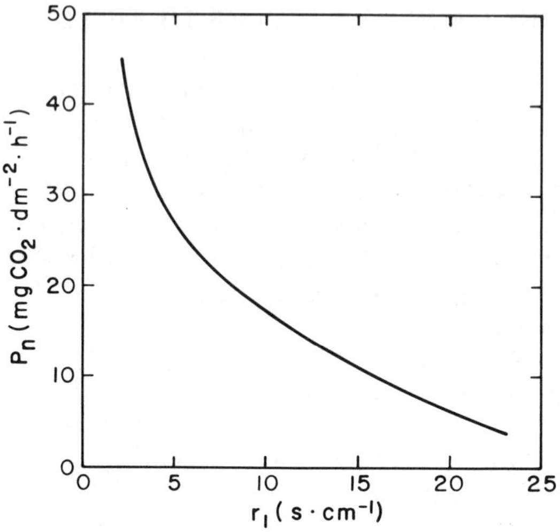

Figure 7.9. Net photosynthesis (P_n) of cotton leaves in relation to leaf diffusive resistance (r_l). Reprinted by permission from *Aus. J. Biol. Sci.* **22**, 289–302 (1969) (86).

is a dominant factor in the regulation of P_n. Under these conditions r_s increases rapidly between about -12 and -20 bars (66, 96, 97).

Stomata of plants growing in field environments do not exhibit the same sensitivity to water deficits as do potted plants, presumably because of larger diurnal deficits associated with higher irradiance and a slower development of greater deficits because of larger rooting volumes (66, 97) or because of exposure to repeated stresses which allows "hardening" (98–100). The effect of water deficit on P_n and leaf resistance for field grown plants is shown in Fig. 7.10 (64). In this instance, large reductions in P_n occurred in the absence of parallel increases in r_l suggesting non-stomatal control of P_n. Although the source of control is not known for the example shown in Fig. 7.10, several reports suggest mesophyll resistance may increase with decreasing ψ_l. Troughton (86) suggests mesophyll resistance may increase as the relative water content (*RWC*) decreases below about 75%. This *RWC* was reached at about -20 bars for the results in Fig. 7.10, but decreases in P_n were observed at much higher potentials. Fry (101, 102) presented evidence from potted plants that Hill reaction activity of chloroplasts isolated from stressed cotton leaves showed a steady decline of about 1.8% per bar as ψ_l decreased from -8 to -30 bars. Reduced Hill activity and measured P_n may both be related to the loss of chloroplast integrity and increased hydrolytic activity associated with dehydration discussed by Vieira da Silva (89) for cotton.

Photosynthate Utilization in Growth

Respiratory processes, wherein photosynthate is utilized in maintenance of existing physiologically active tissues and synthesis of new cells, have

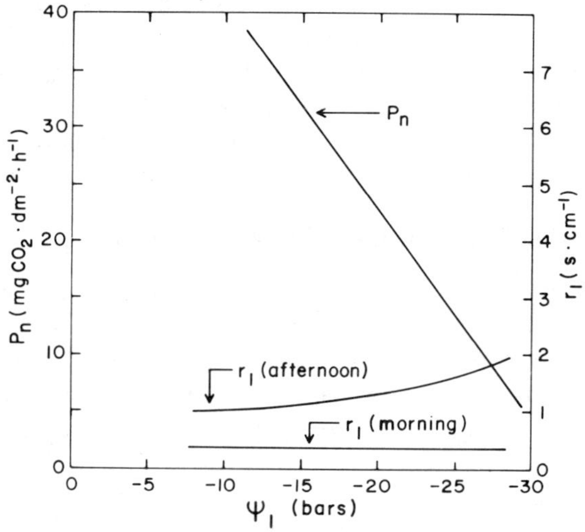

Figure 7.10. Rates of net photosynthesis (P_n) and leaf diffusive resistance (r_l) measured in late morning or mid-afternoon in relation to leaf water potential (ψ_l) for field-grown cotton. Reproduced from *Crop Sci.* **17**, 81–84 (1977) by permission of the Crop Science Society of America (63).

not been studied in great detail with cotton. Recent research efforts have undertaken a more complete, quantitative description of both photosynthesis and respiration (7, 16, 103). Maintenance and growth components of total respiration may be expressed as:

$$R_t = aP + bW = R_s + R_m \qquad (7.1)$$

where R_t is total daily plant respiration, a is the fraction of gross photosynthesis (P) respired during the synthesis of new cellular material, and b is the fraction of total plant dry weight (W) respired each day for maintenance of existing tissues. Thus, aP represents growth respiration (R_s) and bW is maintenance respiration (R_m). Photorespiration is not included in this analysis but may be estimated from CO_2 exchange rates in the absence of O_2. Baker et al. (103) reported that light respiration in field stands during the boll-filling stage varied between 8 and 15% of gross photosynthesis.

Essentially no information is available regarding effects of water deficits on respiration rates of cotton. Referring to Eq. (7.1) above, the effects of water deficits can be assessed in a qualitative manner. In the short term, growth respiration may be reduced if deficits become large enough to reduce gross photosynthesis by either stomatal or nonstomatal means. Prolonged deficits of this severity would also reduce maintenance respiration relative to the well-watered control situation because of reduced rates of increase in total dry matter. These comments represent an oversimplification of a complex interaction between physiological function and environment. Since plant temperature would also rise as stomata close, respiration rate would increase. Also, the above analysis only holds true so long as the chemical nature of material being respired does not change, an assumption that may not be valid during severe, prolonged water stress. Finally, plant dry weight (W) may decrease as a result of water deficit-induced shedding of fruit or leaves and root senescence. These latter aspects will be enlarged upon in a later section.

Root Growth and Distribution in the Soil Profile

Rooting patterns vary greatly among crop species, stage of growth, and environments, both aerial and subterranean. The characteristic rooting pattern for each species is genetically determined but environmentally modified. Following germination, the taproot of cotton elongates rapidly, often penetrating to depths of 15–25 cm before plant emergence. Downward root growth varies greatly with cultivar and environment, but rates up to 6 cm day^{-1} are reported (47, 104, 105). Final rooting depth also varies greatly, but may easily exceed 2 m provided chemical or physical barriers do not limit penetration. Total root length reaches a maximum about the same time plant height slows, presumably because of the increasing requirement by bolls for photosynthate (106). At maturity, as little as 5–10% of total plant dry matter is located in roots of an irrigated crop (107) as shown in Fig. 7.11.

The normal rooting pattern for cotton growing in an undisturbed soil can usually be described as exponentially decreasing downward from

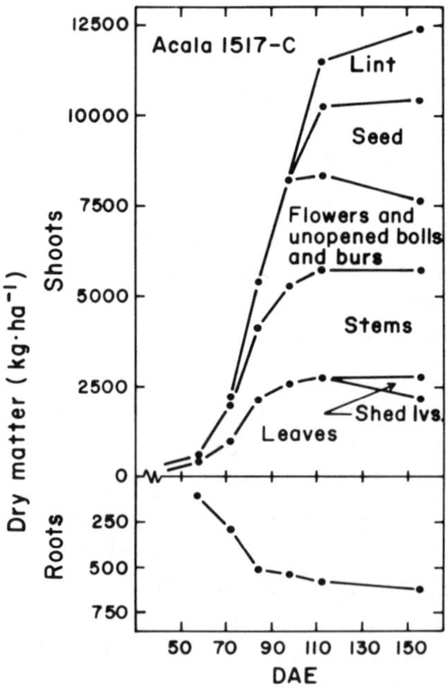

Figure 7.11. Dry matter distribution for an irrigated cotton crop during the growing season. Data for root dry matter are underestimates since all small roots were not recovered. Reproduced from *Agron. J.* **68**, 701–705 (1976) by permission of the American Society of Agronomy (107).

soil layers near the soil surface (102, 108). Often, more than 90% of total root dry matter is found within the uppermost 30 cm soil layer.

The root length density (L_v, cm root per cm^3 soil) pattern may vary from the dry weight pattern because of the high weight per unit length of large roots near the surface. Values of L_v range from approximately 4.0 near the surface to less than 1.0 deep in the profile, but this normal rooting pattern may be altered greatly if the profile dries gradually from the surface. In a drying soil profile, highest root length densities may be observed at greater depths than in the well-watered profile (Fig. 7.12). The dynamics of root growth and senescence in natural soil profiles are not well understood, but it is clear from data shown in Fig. 7.13 for two cotton varieties grown on Houston Black Clay soil during a long drying cycle and work at the Auburn rhizotron that (1) roots are formed and lost simultaneously even in moist soil, (2) roots may be rapidly lost in portions of the profile allowed to dry, and (3) root proliferation may occur rapidly within deeper soil layers as the profile dries (30, 67–70, 72). Thus, while total root dry matter may remain stable or even decrease during the latter stages of the growing season, actual effectiveness of the root system in terms of water uptake may be maintained by continued growth into unexplored regions of the profile. These deep roots appear to be as effective in water uptake as roots nearer the soil surface (69). Root length densities as low as 0.2 cm cm^{-3} appear to be as effective as greater densities when sufficient time is allowed for water extraction.

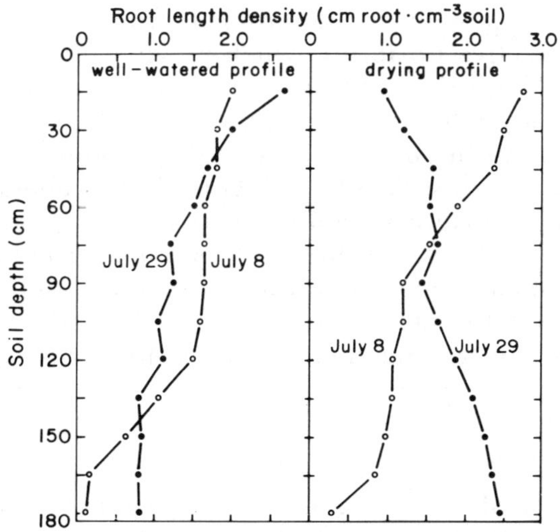

Figure 7.12. Root length density profiles with soil depth for cotton growing in rhizotron compartments that were either well-watered (left) or allowed to dry (right). Reproduced from *Agron. J.* **65**, 307–310 (1973) by permission of the American Society of Agronomy (67).

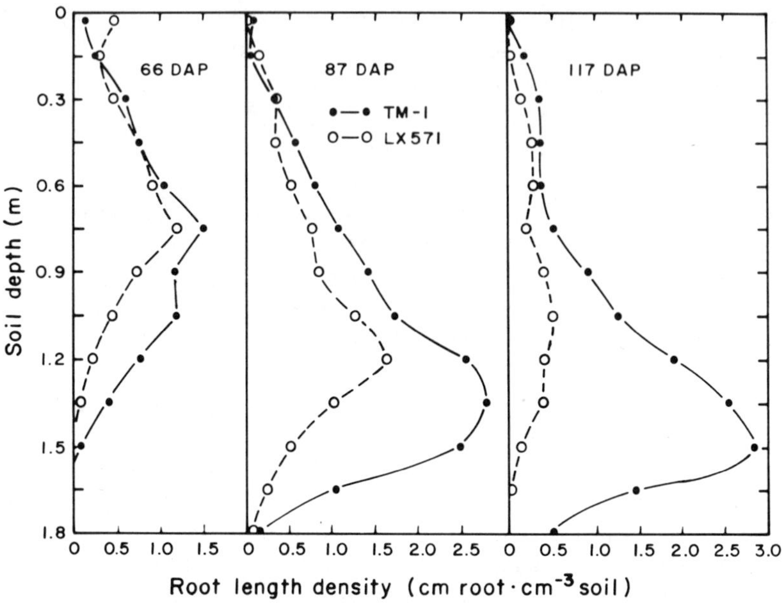

Figure 7.13. Root length density profiles with soil depth for cotton cultivars "TM-1" and "Lankart LX571" grown on Houston black clay soil at Temple, Texas in 1979 during a long drying cycle. DAP is days after planting.

Vegetative Growth and Yield

Crops, such as cotton, with an indeterminate fruiting pattern have a high degree of reproductive flexibility which provides the crop with a rapid means to adjust the fruit load in response to water supply. As pointed out earlier, cotton produces a large excess of fruiting forms even under optimum growth conditions. A portion of these forms (squares or young bolls) abscise and are lost so as to maintain what is generally considered to be a delicate balance between vegetative and reproductive growth. The specific means by which the fruit load is adjusted in response to soil water limitations depends on timing, severity, and duration of the resulting plant water deficit. In general, water deficits may alter (1) the potential number of fruiting sites, (2) the shedding of squares, and (3) the shedding of young bolls.

Production of Potential Fruiting Sites

As pointed out by Mauney in a recent symposium (33), the production of potential fruiting sites is dependent on vegetative growth. The rate of formation and total number of fruiting sites are determined by the rate of production of successive nodes and by the ratio and location of vegetative and reproductive fruiting branches. Since reductions in vegetative growth rates and total dry matter production are a common consequence of soil water deficits (29, 109), the potential number of fruiting sites must also be reduced. Provided the ratio of vegetative to fruiting branches does not change as a result of water deficit, the production of potential fruiting sites should exhibit a similar sensitivity to water supply as does total shoot growth. Data of Bruce and Romkens (110) and Hancock (111) support this conclusion.

Abscission of Squares and Bolls

High rates of square and boll shed are associated with periods of low soil water availability under both rain-fed and irrigated situations. However, significant differences in the timing, duration, magnitude, and rate of development of plant water deficits do occur between these two situations which may bear heavily on abscission pattern within the plant and actual numbers of squares or bolls lost.

Because of near continuous high soil water availability with an irrigated crop, the rate and final amount of growth are usually much greater than those for a rain-fed crop (60, 112). Maximum leaf area indices (*LAI* values) for rain-fed cotton are often near 3 and usually range between 1 and 4, while for a fully irrigated crop *LAI* values over 5 are common. Rapid canopy development results in a high rate of water use since the crop will use water at rates equal to E_0 after an effectively full canopy is developed (*LAI* near 3) and soil water remains freely available in the root zone. This high rate of water use, coupled with the low water holding capacity of many soils in areas under full-season irrigation, create conditions favorable for the rapid development of plant water deficits if the soil profile is not frequently wetted. Wilting may be frequently observed between irrigations when scheduled on 10–14 day cycles, whereas noticeable symptoms of stress may take several weeks to develop in rainfed areas.

Based on analyses of boll retention patterns reported from a number of studies, Hall (113) and more recently Mauney (33) concluded that fruiting forms produced at the first position (closest to main stem) on each fruiting branch had a much higher probability of retention than those at second, third, or more distant positions. This result probably reflects competition for photosynthate within individual fruiting branches with the older bolls having the advantage during their rapid growth phase. Once flowering has begun, the period between flowers on the same fruiting branch is about 6–8 days, creating a situation of unequal "sink strength" between adjacent bolls because of greatly different growth rates. Elongation rates of branches as well as the main axis are usually lower for the rain-fed crop; therefore, not only are there fewer total fruiting sites, but a higher proportion of existing sites are "first position" sites and are favored for retention. This leads to a higher percentage retention for a crop grown under limited water supply compared with a well-watered or irrigated crop (114).

To better understand the effect of a rapid, moderate, or severe water stress on square or boll shed it is necessary to review current concepts related to control of the abscission process. These concepts, outlined by Eaton (5), and more recently by Guinn (6) and McArthur et al. (7), form the bases for the nutritional and hormonal hypotheses of fruit shed.

In the absence of other external limitations, the balance of growth among vegetative and reproductive structures is closely regulated by the carbohydrate and nitrogen status of the plant. The dynamic nature of this balance is suggested by the rapid increase in fruit shed associated with cloudy weather or artificial shading (5, 6, 115, 116), or by brief periods of plant water deficit (114, 117, 118). While the effect of reduced solar radiation on carbohydrate supply is easily rationalized, any proposed effect of water deficit is more difficult to envision, particularly at deficits insufficient to cause stomatal closure. At least two explanations may be proposed to account for water deficit-induced abscission (119).

Fruit shed associated with periods of water deficit between irrigations may be indirectly associated with a nutritional stress rather than being a direct result of the water deficit. Vegetative growth is highly sensitive to internal water deficits because of the requirement for turgor to support cell expansion. Cell division may continue at plant water potentials that substantially inhibit cell expansion, creating a situation whereby leaf primordia form but do not expand. Following reapplication of water, turgor is reestablished and the total demand for carbohydrate and nitrogen to support renewed vegetative growth as well as growing bolls may be sufficient to trigger shedding of squares and small bolls.

Plant water deficits may also act directly to induce fruit and leaf shed. Most work on the sequence of events in water deficit-induced abscission has been done on leaves rather than bolls, but current evidence suggests the process and control of abscission is similar in leaves and bolls. Assuming equivalence in control of abscission of these two organs, plant water deficits may act to trigger several changes in the rate of supply of phytohormones to the abscission zone. Simply stated, hormonal control of abscission is believed to be associated with the rate of supply of juvenile (auxin) and aging or senescence hormones from the leaf blade or fruit to the abscission zone (5, 120, 121). The separation layer in the abscission zone is maintained in a normal state when juvenile hormone activity predominates, but should the hormone balance in this zone shift in favor of aging hormones, then irreversible changes are triggered in cells of the separation layer which result in eventual cellular breakdown and loss of the organ.

Changes in concentrations of several phytohormones in cotton tissues are associated with plant water deficits of a magnitude sufficient to induce leaf and boll abscission. Ethylene production increases (115, 122,

123) as does the concentration of abscisic acid (ABA) (76, 124, 125). Ethylene appears to play multiple roles in abscission; auxin transport to the abscission zone is reduced, hydrolytic enzyme synthesis is induced, and these enzymes are secreted into walls of cells in the separation layer (120, 126, 127). It is not now known if all actions of ethylene are involved in water deficit-induced abscission. A strong association exists between plant water deficit and reduced auxin transport (128, 129) but whether this effect is independent of ethylene action is not known. Transport of other growth regulators (ABA, kinetin, gibberellic acid) is not altered by plant water deficits as low as −22 bars (124, 130). Direct induction of abscission as described above may result from sudden, rather severe plant water deficits occurring near the ends of irrigation cycles. Rapid development of deficits may limit the capacity of plants to acclimate or harden; thus, rate of onset and magnitude of water deficit may interact in a complex manner to govern specific growth and abscission responses.

Problems Arising from an Oversupply of Soil Water

Discussions in previous sections stressed the importance of plant water deficits as a deleterious condition resulting in reductions in vegetative growth, production of fewer potential fruiting sites, increased square and boll shed, and lower yields. While not normally considered harmful to most crops past seedling stage, an oversupply of water provided by excessive rainfall or irrigation may result in serious yield reductions as illustrated by Grimes et al. (131). Plant height was positively associated with yield where water was a limiting factor in growth, but excessive water additions created conditions favoring excessive vegetative growth. Excessive vegetative growth resulted in reduced yields, presumably because leaf expansion has a higher priority for available carbohydrate than does boll growth under these conditions. Similar results have been reported from studies in Arizona (132) and Texas (133, 134).

Relation of Water Supply to Quality of Seed and Lint

The literature related to environmental modification of seed constituents and lint properties is quite extensive and no attempt will be made to review all pertinent reports. Many results are contradictory, but a few findings appear to be consistent with concepts developed earlier in this chapter regarding effects of water deficit on processes contributing to growth.

It is difficult to separate direct effects of water supply on seed and lint quality from those effects arising indirectly as a result of alterations in crop maturity. Depending on the specific pattern of soil water deficit the crop experiences, maturity may be hastened or delayed. For example, many rain-fed crops are subjected to a progressively more severe soil water deficit as the season progresses. If early season rains are sufficient to allow early bolls to be set, plant water deficits may eventually slow or stop additional growth and prevent retention of later-formed bolls, thereby promoting earliness. However, if an early water deficit causes excessive fruit shed, and rainfall occurs during midseason, the indeterminate growth habit may allow formation of a later crop. Similar circumstances may influence maturity in irrigated crops. Changes in maturity are important to seed and lint quality because temperature during boll development of early and late season bolls may be quite different. In general, greatest fiber length occurs in bolls set during the early to peak flowering period and decreases thereafter (111, 135, 136). Patterns of fiber coarseness (micronaire) and strength with maturity are less predictable, but both are reduced for lint from late season bolls of most cultivars (135, 136). When night temperatures were controlled in field chambers during the entire season, reduced temperatures were associated with lower micronaire, shorter fiber length, and generally higher fiber strength (137, 138). Low night temperatures also reduced oil and nitrogen contents of seeds.

The extent to which temperature interactions have confounded results from water deficit studies is not known. Lint from plants subjected to soil water deficit is usually stronger, but increased strength is associated with reduced length (5, 139) and higher micronaire (139). Oil content of seeds produced under water deficient conditions is usually reduced and nitrogen content may be slightly increased (5). Small changes in lint and seed properties occur from year to year under field conditions, but a direct association with plant or soil water deficit is difficult to establish and may vary with cultivar (140).

EFFICIENT WATER USE IN COTTON PRODUCTION

Earlier discussions in this chapter have sought to illustrate effects of sub- or supraoptimal water supply on specific physiological processes which place an upper limit on the biological efficiency of dry matter accumulation and yield of lint and seed. Many factors may act as practical constraints in achievement of the highest use efficiency of the total water resources available to the crop. Several of these constraints deal with management decisions associated with the complete farming system, not just to those decisions that directly affect the growing crop.

Yield of cotton lint is very tightly coupled to total water use as illustrated in Figs. 7.14.1 and 7.14.2 (131, 141). Data for full-season irrigation (Fig. 7.14.1) suggest maximum lint production was obtained when total water use was 68 cm. The same general result was obtained for both

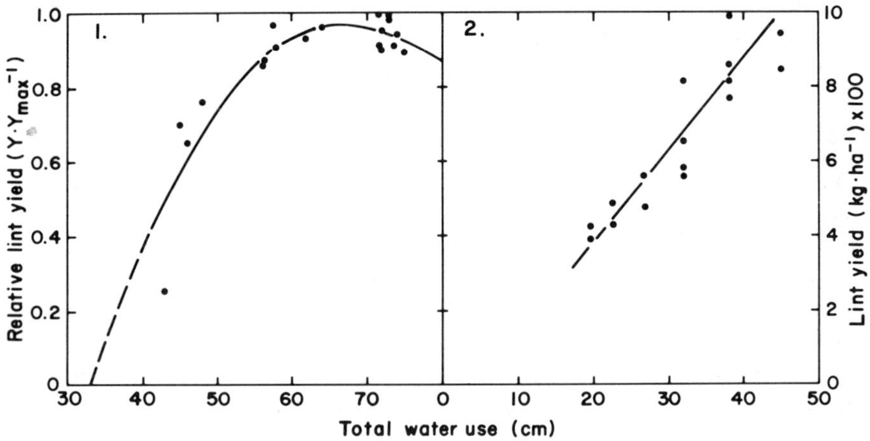

Figure 7.14. The relation of yield to total water use for cotton crops grown with greatly different water supplies. Part (1) represents the relation for a full-season irrigated crop, while part (2) is for a dryland crop in a stressful environment. Drawn from data of Gerard et al. (141) and Grimes et al. (139).

Hesperia sandy loam and Panoche clay loam soils, even though the yield potential for crops produced on these soils differed by about 400 kg lint ha^{-1}. Extrapolation of the curvilinear relationship suggests approximately 33 cm of water were required before a measurable yield was produced. Excess water (more than 68 cm) resulted in rank vegetative growth and reduced lint yield.

Results from rain-fed or supplementally irrigated crops differ slightly from those for the fully irrigated situation (Fig. 7.14.2). Yield of lint was linearly related to total water use, but the total range of these water use data are below those shown in Fig. 7.14.1. Also, lint yields of about 400 kg ha^{-1} were obtained with as little as 20 cm water use.

This linear nature of the relationship between yield and total water use suggests several possibilities for assuring most efficient use of water from all sources. The most important of these would appear to be (1) reduce soil evaporation, (2) increase soil water storage, (3) maximize the harvest index, and (4) improve the photosynthetic efficiency, thereby increasing the slope of the curve shown in Fig. 7.14.2. A recurrent theme throughout this section will be how specific management decisions and current research programs affect either the total amount of water available to the crop or the efficiency with which the water is used.

Crop Management and Production Practices

Management decisions are associated with the entire cropping system as well as the specific crop grown in any one season. Decisions related to cropping sequence, fertility rates for previous crops, and specific production practices associated with previous crops may have a large impact on water resources available to the current crop.

Preplanting Management Decisions

Cropping Sequence. Many factors enter into decisions related to actual cropping sequences used by individual producers. The most common problems are related to weed control, increased populations of insects and pathogens, and market status as well as water supply. These and other unnamed factors enter into the producer's profit–loss analysis associated with each management decision.

For most rain-fed, summer crop production systems, stored soil water and early growing season precipitation constitute the only major water resources. Previous cropping history may have a great influence on the soil water storage provided winter and preplant spring rainfall are insufficient to refill the profile to some minimum level. This is well illustrated by the work of Bloodworth (142, cited in 143) shown in Fig. 7.15. The most striking feature of these data is the promotion of deep rooting by cotton following either Harding grass or alfalfa. In terms of seed cotton, the additional water available to the crop was worth about 50 kg ha^{-1} cm^{-1}. These results with cotton are not unique and apply, in principle, to several other crops. Results from a 36 year dryland study on the Texas High Plains suggested that lint yields from cotton grown in continuous cotton or cotton–sorghum rotations were not different, but grain yields from cotton–sorghum rotations were slightly higher than

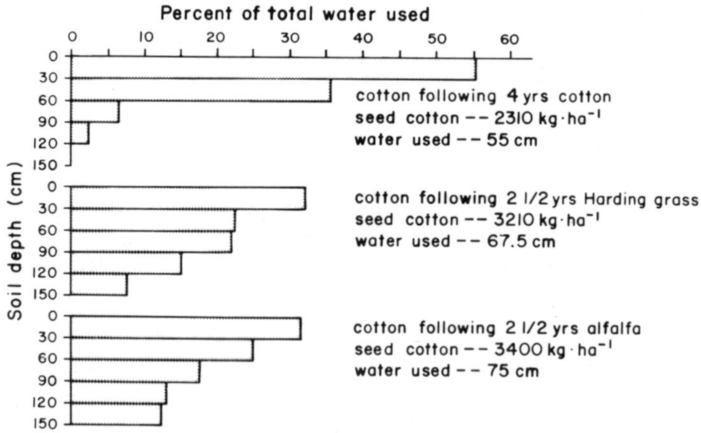

Figure 7.15. The distribution of soil water extraction with soil depth for a cotton crop grown in rotation with cotton (upper), Harding grass (middle), or alfalfa (lower). Based on the data of Bloodworth (142). Reprinted by permission from *Advances in Production and Utilization of Quality Cotton,* Fred C. Elliott, Marvin Hoover, and Walter K. Porter, Jr., Eds., Iowa State University Press, Ames, Iowa (143).

from the continuous sorghum cropping system (144). An interaction between cultivars and previous cropping history was found for cotton grown in the Rio Grande Valley of Texas (134). Yields of "Tamcot SP-37," an early cultivar, were lower following a winter crop compared with yields after fallow. No similar yield advantage was found for the later maturing cultivar, "Stoneville 213," possibly because this cultivar was better able to utilize late season rainfall.

Preplant Irrigation. The importance of adequate levels of soil water at planting time are also illustrated by studies with single preplant irrigations (145, 146). In the Texas High Plains, a small amount (75–150 mm) of additional water applied before planting resulted in high probabilities for doubling lint yields from about 250 to 500 kg ha^{-1} (147). This yield advantage is associated with high fruit set early in the reproductive cycle. Additional benefits from preplant irrigation are greater latitude in planting date and a higher probability of avoiding extremely low yields in years of drought.

Row Spacing and Plant Population. Traditionally, cotton has been planted in single rows spaced from about 80 to 120 cm apart. Desired plant populations are achieved by varying the distance between plants within rows and may range between 60×10^3 and 500×10^3 plants ha^{-1}. For the most part wide row widths were used for convenience of subsequent tillage and harvest operations. Because of a number of factors, current production trends are toward more narrow rows and higher plant populations. Perhaps the primary benefit is a higher yield potential, but earlier maturity, reduced cost of insect control (directly associated with maturity), reduced cost of weed control, and once-over harvest also enter into the more favorable economic picture for narrow-row, high-population plantings. Several cultivars appear capable of using soil water throughout the soil profile at row widths ranging between 25 and 100 cm provided populations are high enough. Apparently, high populations (approximately 500×10^3 plants ha^{-1}) are necessary before cultivars adapted to narrow-row culture efficiently utilize soil water throughout the entire soil profile (148).

Although benefits of these recent production trends are clearly indicated, water supply may still dictate planting pattern and population in many cotton producing regions. In arid regions dependent solely on rainfall, experience has shown that lower plant populations arranged in various skip-row patterns produced acceptable yields with lower risk of complete failure compared with solid plantings. Several variants are in use today ranging from 1 row in −1 row out, 2 in −1 out, 2 in −2 out and 1 in −2 out. The basic principle illustrated from these studies is that the cotton root system is capable of exploring large soil volumes both vertically and laterally. Row omission, coupled with adequate weed control, provides greater soil volumes and more available water to existing plants. Actual water use per plant in skip-row culture may be higher than in conventional plantings because of reduced interrow competition for light, greater advected energy load from bare soil between rows, and greater air turbulence within the canopy. Additional water stored in the larger soil volume provides reserves to be used during the critical flowering period when actual yields are determined.

Postplanting Management Decisions

Irrigation Timing. The importance of maintaining an adequate supply of "freely available" soil water to support vegetative growth and develop high yield potential has been discussed. Actual procedures to accomplish this result are varied and depend on several factors including total water supply, rate of supply, soil storage capacity, soil infiltration rate, and stage of crop growth. Possible interactions among these variables are too numerous to cover in detail here, but many studies have contributed to development of a few general principles regarding irrigation.

Where quantities of water are limited to the extent that only supplemental irrigation is possible, timing and quantity of water applied are extremely important. The goal should be to provide water during the most critical stages of growth while allowing greater deficits to occur at less critical stages. Current evidence suggests larger yield reductions occur when deficits occur during the peak flowering period compared with reductions from comparable deficits occurring either earlier or later in the flowering period (149).

In studies on an alluvial clay soil in the coastal plain of Israel, Marani and Fuchs (145) found that a single application of water sufficient to wet the soil profile to a depth of 90 cm (150 mm water) at the beginning of flowering resulted in optimum yield for "Pima 32" cotton. A larger amount of water (200 mm) which wet the soil to greater depths induced more vegetative growth and flowering, but lint yields were not increased. In other experiments conducted under these same conditions, equivalent yields were obtained when the depth of soil wetting by each irrigation varied from 45 to 120 cm so long as the same total water was supplied. Under a wider variety of conditions, the depth to which the soil profile should be wet should closely match the rooting depth which may vary greatly with soil type (134).

Where more frequent irrigations are possible, proper timing is also important for efficient water use. Using information derived from yield responses in irrigation timing studies, Hiler and co-workers have proposed a Stress Day Index concept wherein the crop susceptibility to water deficit is combined with an actual measure of plant or soil water status to better time irrigations (150, 151). Increased efficiency of water use (up to 38%) resulted from a lower total amount of water added rather than increased yield, illustrating that existing water supplies may be "stretched" in certain situations without loss of yield.

Much work has been done with timing of irrigations and total water supply for fully irrigated crops. One well-illustrated example is provided by Grimes et al. (131, 139) wherein relative lint yield on two soils with variable nitrogen fertility could be predicted from evapotranspiration. A single quadratic equation of the form

$$\frac{Y}{Y_0} = -266.4 + 10.72\,E_t - 0.079\,E_t^2 \tag{2}$$

where Y is actual yield, Y_0 is maximum yield, and E_t is actual water use provided an adequate description of the yield response of "Acala SJ-1" cotton grown on Panoche clay loam and Hesperia fine sandy loam soils. Nitrogen fertilization increased water use efficiency (kg lint per cm water used) in some but not all cases, supporting earlier results of Scarsbrook et al. (152).

Significant savings in water and energy may be achieved by properly timing the final irrigation as well. Knowledge of water holding capacity of the soil coupled with an assessment of crop maturity enables producers to anticipate cut-off dates and eliminate one or more irrigations (153).

Antitranspirants. A number of substances are now available, which, when sprayed on leaves, increase the leaf resistance to water vapor loss and thus reduce transpiration. Antitranspirants fall roughly into two categories based on how these materials induce increased resistances, i.e., film formers (plastics, wax emulsions) that coat the epidermis and plug stomatal pores, and metabolic inhibitors such as abscisic acid and phenylmercuric acid (PMA) that reduce stomatal apertures.

Even though antitranspirant activity of several materials has been studied for a number of years, none is in general commercial use on cotton today. Ideally, an antitranspirant should reduce the rate of water loss but allow relatively high CO_2 exchange rates. This may be possible in some cases since transpiration and photosynthesis are not equally sensitive to changes in stomatal resistance. With cotton, PMA at concentration of 10^{-4} to 10^{-5} M was the only substance of several tested which caused proportionately greater reductions in transpiration than photosynthesis (154). During a 25 day period, the transpiration ratio (g water lost per g dry matter produced) was 245 for controls and 180 for PMA-treated plants. The effectiveness of single applications of PMA are reduced with time. No work is available to adequately demonstrate that present materials could be economically used to improve yield of cotton under conditions where water is the primary factor limiting production. While PMA may seem promising, it is doubtful that any mercurial compounds would be brought into general usage because of inherent dangers to the environment. At present, antitranspirants do not appear to be a viable management tool for use on cotton.

Crop Improvement through Plant Breeding

Development of improved management practices often demands or is accompanied by improvement of cultivars better able to take advantage of specific benefits derived from intensive management. Several recent genetic advances may have significant effects on total water use by the crop when incorporated into complete, integrated production systems.

Short-Season Cultivars

Research during the past decade has shown that high population plantings in narrow rows has potential for enhanced cotton production efficiency (155–158). Higher populations reduce the number of bolls per plant required to achieve a target yield. Reduced row width allows greater distances between plants within rows, thereby providing a more favorable planting geometry at a given plant population. In combination, high populations in narrow rows provides more rapid ground cover, thereby capturing more sunlight for photosynthesis and reducing costs for weed control. A major benefit is also found in the potential for early maturity. Because fewer bolls per plant are needed, a shorter fruiting period is required to achieve a target yield. Since a larger leaf area is achieved earlier in the season, early season fruiting progresses at a faster rate than normal. Production efficiency is increased with early maturity (140–160 vs. 180–210 days) because of reduced production costs associated with insect and weed control, irrigation, and once-over harvesting. One recent study (159) conducted in the Lower Rio Grande Valley of South Texas projected increased net returns of $45.71 and $24.88 per hectare for short-season production with and without irrigation, respectively. Primary benefits were realized through fewer irrigations (1 vs. 3) and lower insecticide costs when compared with full-season production.

Not all standard genotypes are suitable for narrow-row, high-population culture. Baker (160) reported that "Coker 310" could not achieve higher yield potentials at high populations because of excessive shedding during the first 3 weeks of blooming. Burch (161) also found that high populations delayed maturity, reduced yields and fiber strength, increased lodging, and promoted boll rot for 10 upland varieties.

In recognition of these problems, breeders have defined plant types suitable for narrow-row, short-season production (157, 162). Adapted cultivars are generally shorter, have short fruiting branches and fewer vegetative branches. In relatively diminutive and determinate genotypes, fruiting potential may be reduced in comparison with standard upland types, but this reduction may be more than offset by higher populations. New cultivars are capable of higher blooming and maturation rates which permit earlier harvesting and greater overall uniformity of the crop with respect to quality characteristics. At present, the primary constraint to narrow row culture appears to be mechanical, i.e., the existing equipment for cultivation and harvest are not adaptable to narrow rows. The high cost of machinery replacement has undoubtedly slowed adoption of the culture system.

Canopy Architecture and Leaf Shape

Variations in leaf shape have been known to exist for more than 100 years, but specific beneficial effects for okra and superokra shapes have been studied in detail only recently. Lines with the okra leaf character, L^o, have deeply cleft and narrowly lobed leaves. Leaves of superokra (L^s) lines are even more deeply cleft and may even be reduced to a simple willowlike leaf. Areas of individual leaves may be only 20–60% of those of normal leaves, therefore L^o and L^s lines should have a more open canopy which allows better light distribution and air movement throughout the canopy (163). Since leaves subtending bolls are the primary sources for photosynthate required for boll growth, increased irradiance on these leaves may be expected to improve the carbohydrate status in regions of the canopy normally shaded thereby promoting boll retention and higher yields. In general, however, consistently higher yields have not been achieved with either L^o and L^s characters. Boll retention is lower than for normal types, suggesting a possible limitation in carbohydrate supply from the subtending leaf because of reduced area. At equal plant densities, L^o and L^s lines produced lower *LAI* values than the normal line, although on a ground area basis rates of water and carbon exchange did not differ among leaf types (164). Earliness appears to be promoted by both L^o and L^s, and boll rot is reduced in both (165, 166). These latter characters may be useful in short-season production systems utilizing higher than normal plant populations as a means to increase production efficiency, but a specific advantage in water use efficiency apart from that associated with earlier maturity is not apparent.

Improved Drought Resistance

Where water is the primary constraint to production, a very small additional increment of available soil water may increase cotton lint yield dramatically (131, 142, 147, 167, 168). Preceding sections have stressed the importance of management as a means to maximize stored soil water and to match crop water use to that available from all sources. Additional benefits may be realized by increasing the overall drought resistance of the crop, although this is no easy task.

In a recent review, Quisenberry (169) summarized much of the information related to current efforts in breeding for drought tolerance as means of achieving more efficient water use. Several elements are necessary for success of such an improvement program, i.e., the pattern of drought must be known in detail so as to indicate the type of tolerance mechanism(s) of most value, traits known to contribute to drought tolerance must exist within *Gossypium* spp., variability in these traits must exist and they must be heritable, and means must exist or be developed to allow rapid, quantitative determination of each trait. Such a program is clearly beyond the scope of any single research scientist; progress will require the efforts of one or more multidisciplinary research teams comprised of not only geneticists, but also engineers, soil scientists, and plant physiologists.

To date, crop maturity is the only trait known to be consistently associated with drought tolerance for a number of crop species. Early maturity is correlated with high yields when upland cotton is forced to grow on stored soil water only, but the correlation fails and may be reversed if timely rainfall occurs during the growing season (168). When yield is used as the sole selection criterion to identify superior cultivars for water stress environments care must be taken to insure these cultivars retain the capability to respond to favorable conditions should they occur. Current efforts with short-season cultivar development may overcome this limitation.

Once alterations of crop maturity have been fully exploited, additional improvement for drought tolerance must come from other traits. In general, these traits allow cultivars to either maintain a high internal plant water status (water potential) by increasing water absorption or reducing water loss, or they allow the plant to function even though the plant water potential is lowered.

Water absorption from the soil profile is closely associated with the nature and extent of root development. Though much effort has been devoted to root study in recent years, no means are currently available by which to define a root ideotype for cotton. Evidence with other crops suggests that the capacity for deep root growth may be beneficial in times of drought provided deep soil water exists and is replenished each year (170, 171). In cotton, *G. hirsutum* L. "Acala" extracted more water from soil depths of 180 to 210 cm than did *G. barbadense* L. "Pima," but no detailed studies have been reported to suggest that genetic variability really exists in this trait (172). Root densities required to extract water from deep soil layers are quite low at 0.2 cm cm^{-3} or less (69) and are likely to be achieved by all cultivars provided they can root to those depths. Results from a greenhouse study with exotic, nonflowering cotton cultivars suggests that genetic variability in tap root extension and lateral branching does exist (Table 7.1). The successes achieved with other crops strongly suggest improvement for deep rootedness may also

Table 7.1. Root growth characteristics of 16 cotton lines grown in a glasshouse in Peatlite mix contained in 7 × 240 cm acrylic plastic tubes[a]

Entry	Race	Taproot Length (cm)	Number of Laterals[b]
T 1[c]	*palmeri*	191 abcd[d]	4.0 bc
T 25	*punctatum*	196 ab	9.3 a
T 102	*latifolium*	169 abcde	4.6 b
T 111	*marie-galante*	166 bcde	0.8 c
T 133	*morrilli*	177 abcde	2.7 bc
T 141	*marie-galante*	173 abcde	1.3 bc
T 147	*morrilli*	176 abcde	4.6 b
T 169	*latifolium*	193 abc	3.9 bc
T 174	*latifolium*	147 e	1.0 c
T 254	*morrilli*	163 cde	1.1 c
T 265	*morrilli*	165 bcde	3.0 bc
T 284	*morrilli*	172 abcde	1.7 bc
T 292	*morrilli*	178 abcde	1.8 bc
T 295	*morrilli*	198 a	2.2 bc
T 461	*richmondi*	175 abcde	3.4 bc
Blightmaster	*latifolium*	178 abcde	2.3 bc
$\bar{x}$		176	3.0

[a] Data were collected after 35 days growth.
[b] The number of strong, downward-growing laterals observed against the face of the acrylic tube 1 m below the surface.
[c] Texas designation of the strain.
[d] Means within columns followed by different letters are significantly different at the 0.05 probability level based on Duncan's multiple range test.

be possible for cotton and such investigations should be given high priority.

Reductions in water loss may be achieved through partial or complete stomatal closure, changes in leaf orientation and leaf area, and changes in surface albedo and in characteristics of the cuticle. Little experimental information is available to suggest how each of these means of restricting water loss may be of value to the plant in overall drought tolerance. It is clear that stomatal closure restricts water loss, but photosynthesis is also reduced. Recent field studies suggest that stomatal closure in field plants may not occur until low leaf water potentials develop (63, 64, 66, 97); therefore, this may be a trait primarily associated with survival during extreme drought. Genetic variability in stomatal closure does exist among upland cottons (173) and, although heritability is rather low, selection appears possible (169).

When the function of most cotton crops is reduced to mere survival levels, little yield may result even if rainfall eventually comes. Under conditions of severe drought. vegetative growth slows or stops and much leaf area is eventually shed. However, if the indeterminate growth habit remains and rainfall eventually occurs, growth and fruiting may occur and some yield may be realized. The degree of indeterminacy differs among upland cottons, and those more indeterminate lines appear more drought tolerant in these situations (168).

For stomatal closure to be an effective means of reducing water loss, the cuticle must provide a large resistance to water vapor transport. This is apparently true in cotton since, as previously discussed, most estimates of cuticular resistance are in the range of 60–100 sec cm^{-1}. Even though resistances are high, substantial water may be lost through the cuticular pathway during periods of high evaporative demand when leaf temperatures are high. Genotypic variability in cuticular resistance may exist, but little is known of the importance of this variability or of the heritability of this trait. High cuticular resistances may be associated with high levels of epicuticular waxes, but high wax content in itself does not necessarily impart high resistance.

Another means of reducing both water loss and heat load is through leaf shed. Large genotypic variations in sensitivity to drought stress occur for this character in upland cottons (personal observation), but this character has not been quantified in a manner such that its importance to drought tolerance may be evaluated. While reduced leaf area reduces water loss, less photosynthetic area remains to support renewed growth once water becomes available again.

In summary, little actual time, effort, and money have been devoted to systematic attempts to improve overall drought resistance of upland cottons. The economic advantages associated with superior performance under drought conditions are tremendous. Such advantages will become more and more apparent as supplies of high quality water diminish and energy to supply water becomes more expensive. Considerable management technology now exists that would allow us to stretch available water supplies so additional acres could be irrigated with the same amount of water. More efficient groundwater use for irrigation may preserve a valuable resource for years to come. Development of cultivars capable of more efficient soil water use should become a high priority objective as a means to enhance and complement other conservation efforts.

REFERENCES

1. W. L. Balls, *The Cotton Plant in Egypt*, MacMillan, London, 1912.
2. H. B. Brown and J. O. Ware, *Cotton*, McGraw-Hill, New York, 1958.
3. H. R. Carns and J. R. Mauney, Physiology of the Cotton Plant, in F. C. Elliott, M. Hoover, and W. K. Porter, Jr., Eds., *Advances in Production and Utilization of*

Quality Cotton: Principles and Practices, Iowa State University Press, Ames, Iowa, 1968, pp. 41–73.
4. R. H. Dastur and R. D. Ansana, Physiology, *Cotton in India, a monograph*, The Indian Central Cotton Committee, Bombay, Vol. 2, 1960, pp. 1–105.
5. F. M. Eaton, *Annu. Rev. Plant Physiol.* **6,** 299 (1955).
6. G. Guinn, *Causes of Square and Boll Shedding in Cotton*, USDA-ARS Tech. Bull. TB1672, 1982 (in press).
7. J. A. McArthur, J. D. Hesketh, and D. N. Baker, Cotton, in L. T. Evans, Ed., *Crop Physiology: Some Case Histories*, Cambridge Univ. Press, Cambridge, 1975, pp. 295–325.
8. G. Stanhill, Cotton, in J. L. Monteith, Ed., *Vegetation and the Atmosphere*, Vol. 2, Academic Press, New York, 1976, pp. 121–150.
9. P. E. Tollervey, *Cotton Gr. Rev.* **47,** 245 (1970).
10. C. G. Coble and H. D. Bowen, *Trans. ASAE* **13,** 162 (1970).
11. D. F. Wanjura, D. R. Buxton, and H. N. Stapleton, *Agron. J.* **62,** 741 (1970).
12. D. F. Wanjura, D. R. Buxton, and H. N. Stapleton, *Trans. ASAE* **16,** 227 (1973).
13. I. R. Cowan, *Planta* **106,** 185 (1972).
14. D. W. DeMichele and P. J. H. Sharpe, A Morphological and Physiological Model of the Stomata, in C. E. Murphy, Ed., *Modeling the Growth of Trees*, International Biological Program EDFB-IBP #72-11, Oak Ridge Natl. Lab., Oak Ridge, Tenn., 1972.
15. J. T. Ritchie, *Water Resources Res.* **8,** 1204 (1972).
16. J. D. Hesketh, D. N. Baker, and W. G. Duncan, *Crop Sci.* **11,** 394 (1971).
17. J. D. Hesketh, D. N. Baker, and W. G. Duncan, *Crop Sci.* **12,** 436 (1972).
18. W. G. Duncan, SIMCOT: A Simulator of Cotton Growth and Yield, in C. E. Murphy, Ed., *Modeling the Growth of Trees*, International Biological Program EDFB-IDP #72-11. Oak Ridge Natl. Lab., Oak Ridge, Tenn., 1972.
19. P. A. Fryxell, *Adv. Frontiers Plant Sci.* **10,** 31 (1965).
20. J. O. Beasley, *J. Hered.* **31,** 39 (1940).
21. J. B. Hutchinson, R. A. Silow, and S. G. Stephens, *The Evolution of Gossypium and the Differentiation of the Cultivated Cottons*, Oxford University Press, New York, 1947.
22. A. A. Bell, R. D. Stipanovic, D. H. O'Brien, and P. A. Fryxell, *Phytochemistry* **17,** 1297 (1978).
23. L. L. Phillips, *Am. J. Bot.* **53,** 328 (1966).
24. J. O. Beasley, *Am. Nat.* **74,** 285 (1940).
25. P. R. Morey, J. E. Quisenberry, and B. Roark, *Crop. Sci.* **14,** 595 (1974).
26. H. W. Gausman, W. A. Allen, C. L. Wiegand, D. E. Escobar, R. R. Rodriques, and A. J. Richardson, *The Leaf Mesophylls of Twenty Crops, Their Light Spectra, and Optical and Geometrical Parameters*, ARS Technical Bulletin **1465,** 1973.
27. H. E. Hayward, *The Structure of Economic Plants,*, MacMillan, New York, 1938, pp. 411–450.
28. R. R. Hoverson and J. Hacskaylo, The Influence of Xylem Arrangement and Transpiration on the Distribution of Solutes in Cotton, *Tex. Agric. Exp. Stn.* MP-889 (1968).
29. W. R. Jordan, *Agron. J.* **62,** 699 (1970).
30. H. M. Taylor and B. Klepper, *Soil Sci.* **120,** 57 (1975).
31. D. B. Anderson and T. Kerr, *Plant Physiol.* **18,** 261 (1943).

32. J. R. Mauney, Morphology of the Cotton Plant, in F. C. Elliott, M. Hoover, and W. K. Porter, Jr., Eds., *Advances in Production and Utilization of Quality Cotton: Principles and Practices*, Iowa State University Press, Ames, Iowa, 1968, pp. 23–40.
33. J. R. Mauney, Flowering, Fruiting and Cutout: Production of Fruiting Points, *Proc. Beltwide Cotton Prodn. Res. Conf., 1979*, pp. 256–261.
34. A. Hadas and D. Russo, *Agron. J.* **65,** 643 (1974).
35. A. Hadas and D. Russo, *Agron. J.* **65,** 647 (1974).
36. D. F. Wanjura and D. R. Buxton, *Agron. J.* **64,** 427 (1972).
37. A. Hadas and E. Stibbe, An Analysis of Soil Water Movement Toward Seedlings Prior to Emergence, *Ecological Studies, Analysis and Synthesis*, Vol. 4, Springer-Verlag, New York, 1973, pp. 97–106.
38. R. O. Slatyer, *Plant-Water Relationships*, Academic Press, New York, 1967.
39. M. N. Christiansen and R. P. Moore, *Agron. J.* **51,** 582 (1959).
40. R. E. Phillips, *Agron. J.* **60,** 568 (1968).
41. C. H. Arndt, *Plant Physiol.* **20,** 200 (1945).
42. D. F. Wanjura and D. R. Buxton, *Agron. J.* **64,** 431 (1972).
43. R. D. Jensen, *Agron. J.* **63,** 766 (1971).
44. C. J. Gerard, *Agron. J.* **72,** 473 (1980).
45. H. M. Taylor, Root Behavior as Affected by Soil Structure and Strength, in E. W. Carson, Ed., *The Plant Root and Its Environment*, University of Virginia Press, Charlottesville, 1974, pp. 271–291.
46. H. M. Taylor and W. R. Gardner, *Soil Sci.* **96,** 153 (1963).
47. H. M. Taylor and L. F. Ratliff, *Soil Sci.* **108,** 113 (1969).
48. H. M. Taylor, G. M. Robertson, and J. Parker, Jr., *Soil Sci.* **102,** 18 (1966).
49. W. R. Gill and G. H. Bolt, *Agron. J.* **47,** 166 (1955).
50. B. W. Eavis, L. F. Ratliff, and H. M. Taylor, *Agron. J.* **61,** 640 (1969).
51. J. D. Goeschl, L. Rappaport, and H. K. Pratt, *Plant Physiol.* **41,** 877, (1966).
52. J. D. Goeschl and S. K. Kays, *Plant Physiol.* **55,** 670 (1975).
53. J. J. Taubenhaus, W. N. Ezekiel, H. E. Rea, *Plant Physiol.* **6,** 161 (1931).
54. H. M. Taylor, A. C. Mathers, and F. B. Lotspeich, *Agron. J.* **56,** 328 (1964).
55. A. C. Mathers and N. H. Welch, *Agron. J.* **56,** 313 (1964).
56. T. A. Howell, W. R. Jordan, and E. A. Hiler, Evaporative Demand as a Plant Stress, in B. J. Barfield and J. F. Gerber, Eds., *Modification of the Aerial Environment of Plants*, Monograph 2, Am. Soc. Agric. Eng., St. Joseph, Mo., 1979, pp. 97–115.
57. C. B. Tanner, Evaporation of Water from Plants and Soils, in T. T. Kozlowski, Ed., *Water Deficits and Plant Growth*, Vol. 1, Academic Press, New York, 1968, pp. 73–106.
58. J. T. Ritchie, *Agron. J.* **63,** 51 (1971).
59. J. T. Ritchie, E. Burnett, and R. C. Henderson, *Agron. J.* **64,** 168 (1972).
60. J. T. Ritchie and E. Burnett, *Agron. J.* **63,** 56 (1971).
61. S. Al-Khafaf, P. J. Wierenga, and B. C. Williams, *Agron. J.* **70,** 912 (1978).
62. E. I. Newman, *Phil. Trans. R. Soc. London, Ser. B* **273,** 463 (1976).
63. R. C. Ackerson, D. R. Krieg, C. L. Harring, and N. Chang, *Crop Sci.* **17,** 76 (1977).
64. R. C. Ackerson, D. R. Krieg, T. D. Miller, and R. E. Zartman, *Crop Sci.* **17,** 80 (1977).
65. H. Bielorai and P. A. M. Hopmans, *Agron. J.* **67,** 629 (1975).
66. W. R. Jordan and J. T. Ritchie, *Plant Physiol.* **48,** 783 (1971).
67. B. Klepper, H. M. Taylor, M. G. Huck, and E. L. Fiscus, *Agron. J.* **65,** 307 (1973).

68. H. M. Taylor and B. Klepper, *Aust. J. Biol. Sci.* **24,** 853 (1971).
69. H. M. Taylor and B. Klepper, *Agron. J.* **66,** 584 (1974).
70. H. M. Taylor and B. Klepper, *Adv. Agron.* **30,** 99 (1978).
71. R. E. Young and V. D. Browning, *Trans. ASAE* **20,** 673 (1977).
72. V. D. Browning, H. M. Taylor, M. G. Huck, and B. Klepper, Water Relations of Cotton: A Rhizotron Study, *Agric. Exp. Stn., Auburn Univ. Bull.* **467,** (1975).
73. W. S. Meyer and J. T. Ritchie, *Agron. J.* **72,** 577 (1980).
74. W. R. Gardner, Availability and Measurement of Soil Water, in T. T. Kozlowski, Ed., *Water Deficits and Plant Growth,* Vol. 1., Academic Press, New York, 1968, pp. 105–135.
75. M. G. Huck, B. Klepper, and H. M. Taylor, *Plant Physiol.* **45,** 529 (1970).
76. W. R. Jordan, K. W. Brown, and J. C. Thomas, *Plant Physiol.* **56,** 595 (1975).
77. H. D. Barrs, Controlled Environment Studies of the Effect of Variable Atmospheric Water Stress on Photosynthesis, Transpiration and Water Status of *Zea mays* L. and Other Species, *Proc. Symp. Plant Response to Climatic Factors,* UNESCO, Uppsala, Sweden, 1970.
78. R. Stoker and P. E. Weatherly, *New Phytol.* **70,** 547 (1971).
79. J. L. Hailey, E. A. Hiler, W. R. Jordan, and C. H. M. van Bavel, *Crop Sci.* **13,** 264 (1973).
80. R. O. Slatyer, *Agric. Meteorol.* **3,** 281 (1966).
81. P. G. Jarvis and R. O. Slatyer, *Planta* **90,** 303 (1970).
82. R. O. Slatyer and J. F. Bierhuizen, *Aust. J. Biol. Sci.* **17,** 115 (1964).
83. C. H. M. van Bavel, F. S. Nakayama, and W. L. Ehrler, *Plant Physiol.* **40,** 535 (1965).
84. W. L. Ehrler, *Agron. J.* **65,** 404 (1973).
85. P. J. H. Sharpe, *Agron. J.* **65,** 570 (1973).
86. J. H. Troughton, *Aust. J. Biol. Sci.* **22,** 289 (1969).
87. J. S. Boyer, Water Deficits and Photosynthesis, in T. T. Kozlowski, Ed., *Water Deficits and Plant Growth,* Vol. 4, Academic Press, New York, 1976, pp. 153–190.
88. T. C. Hsiao, *Plant Physiol.* **24,** 519 (1973).
89. J. Vieira da Silva, Water Stress, Ultrastructure and Enzymatic Activity, in O. L. Lange, L. Kappen, and E. D. Schulze, Eds., *Ecological Studies, Analyses and Synthesis,* Vol. 19, Springer-Verlag, 1976, pp. 207–224.
90. T. C. Hsiao, E. Acevedo, E. Fereres, and D. W. Henderson, *Phil. Trans. R. Soc. London, Ser. B* **273,** 479 (1976).
91. T. C. Hsiao, E. Fereres, E. Acevedo, and D. W. Henderson, Water Stress and Dynamics of Growth and Yield of Crop Plants, in O. L. Lange, L. Kappen, and E. D. Schulze, Eds., *Ecological Studies, Analyses and Synthesis,* Vol. 19, Springer-Verlag, New York, 1976, pp. 281–305.
92. M. El-Sharkawy, J. D. Hesketh, and H. Muramoto, *Crop Sci.* **5,** 173 (1965).
93. H. Muramoto, J. Hesketh, and M. El-Sharkawy, *Crop Sci.* **5,** 163 (1965).
94. J. H. Troughton and R. O. Slatyer, *Aust. J. Biol. Sci.* **22,** 815, (1969).
95. M. A. El-Sharkawy and J. D. Hesketh, *Crop Sci.* **4,** 514 (1964).
96. C. F. Ehlig and W. R. Gardner, *Agron. J.* **56,** 127 (1964).
97. W. J. Davies, *Crop Sci.* **17,** 735 (1977).
98. K. W. Brown, W. R. Jordan, and J. C. Thomas, *Physiol. Plant.* **37,** 1 (1976).
99. J. M. Cutler, D. W. Rains, and R. S. Loomis, *Agron. J.* **69,** 773 (1977).
100. J. M. Cutler and D. W. Rains, *Crop Sci.* **17,** 329 (1977).

101. K. E. Fry, *Plant Physiol.* **45,** 465 (1970).
102. K. E. Fry, *Crop Sci.* **12,** 698 (1972).
103. D. N. Baker, J. D. Hesketh, and W. G. Duncan, *Crop Sci.* **12,** 431 (1972).
104. D. W. Grimes, R. J. Miller, and P. L. Wiley, *Agron. J.* **67,** 519 (1975).
105. P. Marani, R. Kaiser, and P. Villemin, *Cot. Fib. Trop.* **33,** 291 (1978).
106. R. W. Pearson and Z. F. Lund, *Agron. J.* **60,** 442 (1968).
107. J. Halevy, *Agron. J.* **68,** 701 (1976).
108. M. E. Bloodworth, C. A. Burleson, and W. R. Cowley, *Agron. J.* **50,** 317 (1958).
109. C. H. Wadleigh and H. G. Gauch, *Plant Physiol.* **23,** 485 (1948).
110. R. R. Bruce and M. J. M. Romkens, *Agron. J.* **57,** 135 (1965).
111. N. I. Hancock, *J. Am. Soc. Agron.* **33,** 590 (1941).
112. D. A. Ashley, B. D. Doss, and O. L. Bennett, *Agron. J.* **57,** 61 (1965).
113. W. C. Hall, Physiology and Biochemistry of Abscission in the Cotton Plant, *Tex. Agric. Exp. Stn.* MP-285, 1958.
114. J. R. Stockton, L. D. Doneen, and V. T. Walhood, *Agron. J.* **53,** 272 (1961).
115. G. Guinn, *Crop Sci.* **14,** 291 (1974).
116. G. Guinn, *Crop Sci.* **16,** 89 (1976).
117. G. Guinn, *Plant Physiol* **57,** 403 (1976).
118. B. L. McMichael, W. R. Jordan, and R. D. Powell, *Agron. J.* **65,** 202 (1973).
119. W. R. Jordan, Role of Plant Water Deficit in Control of Fruiting, Flowering and Cutout of Cotton, *Proc. Beltwide Cotton Prodn. Res. Conf., 1979,* pp. 297–301.
120. E. M. Beyer, Jr., and P. W. Morgan, *Plant Physiol.* **48,** 208 (1971).
121. H. R. Carns, *Annu. Rev. Plant Physiol.* **17,** 295 (1966).
122. W. R. Jordan, P. W. Morgan, and T. L. Davenport, *Plant Physiol.* **50,** 756 (1972).
123. B. L. McMichael, W. R. Jordan, and R. D. Powell, *Plant Physiol.* **49,** 658 (1972).
124. T. L. Davenport, W. R. Jordan, and P. W. Morgan, *Plant Physiol.* **59,** 1165 (1977).
125. B. L. McMichael and B. W. Hanny, *Agron. J.* **69,** 979 (1977).
126. F. B. Abeles, *Plant Physiol.* **44,** 447 (1969).
127. F. B. Abeles, L. E. Craker, and G. R. Leather, *Plant Physiol.* **47,** 7 (1971).
128. T. L. Davenport, P. W. Morgan, and W. R. Jordan, *Plant Physiol.* **59,** 554 (1977).
129. P. W. Morgan, W. R. Jordan, T. L. Davenport, and J. I. Durham, *Plant Physiol.* **59,** 710 (1977).
130. T. L. Davenport, W. R. Jordan, and P. W. Morgan, *Plant Physiol.* **63,** 152 (1979).
131. D. W. Grimes, W. L. Dickens, and W. D. Anderson, *Agron. J.* **61,** 769 (1969).
132. D. W. Kittock, *Agron. J.* **65,** 498 (1973).
133. C. J. Gerard and L. N. Namken, *Agron. J.* **58,** 39 (1966).
134. C. J. Gerard, B. W. Hipp, and S. A. Reeves, Yields, Growth, and Management Requirements of Selected Crops as Influenced by Soil Properties, *Tex. Agric. Exp. Stn. Bull.* **1172,** (1977).
135. W. R. Meredith, Jr., and R. R. Bridge, *Crop Sci.* **13,** 307 (1973).
136. L. M. Verhalen, R. Mamaghani, W. C. Morrison, and R. W. McNew, *Crop Sci.* **15,** 47 (1975).
137. J. R. Gipson and H. E. Joham, *Agron. J.* **60,** 292 (1968).
138. J. R. Gipson and H. E. Joham, *Agron. J.* **60,** 296 (1968).
139. D. W. Grimes, W. L. Dickens, and W. D. Anderson, *Agron. J.* **61,** 773 (1969).

140. W. L. Parks, J. R. Overton, and C. R. Graves, Cotton Irrigation, *Univ. Tenn. Agric. Exp. Stn. Bull.* **581**, (1978).

141. C. J. Gerard, D. G. Bordovsky, and L. E. Clark, Water Management Studies in the Rolling Plains, *Tex. Agric. Exp. Stn. Bull.* **B-1321**, (1980).

142. M. E. Bloodworth, Cotton Production as Influenced by Soil and Water Management, Paper, American Cotton Congress, Harlingen, Texas, 1955.

143. D. E. Longnecker and L. J. Erie, Irrigation Water Mangement, in F. C. Elliott, M. Hoover, and W. K. Porter, Jr., Eds., *Advances in Production and Utilization of Quality Cotton: Principles and Practices*, The Iowa State Univ. Press, Ames, Iowa, 1968, pp. 321–345.

144. J. D. Bilbro, *Agron. J.* **64,** 140 (1972).

145. A. Marani and Y. Fuchs, *Agron. J.* **56,** 281 (1964).

146. A. Marani and D. Levi, *Agron. J.* **65,** 637 (1973).

147. J. D. Bilbro, *Agron. J.* **66,** 833 (1974).

148. C. W. Wendt and L. L. Ray, Influence of Row Spacing on the Yield and Soil Moisture Utilization of Several Upland Cotton Varieties, *Proc. Beltwide Cotton Prodn. Res. Conf., 1971,* pp. 25–28.

149. D. W. Grimes, R. J. Miller, and L. Dickens, *Calif. Agric.* **24,** 4 (1970).

150. D. G. Bordovsky, W. R. Jordan, E. A. Hiler, and T. A. Howell, *Agron. J.* **66,** 88 (1974).

151. E. A. Hiler and R. N. Clark, *Trans. ASAE* **14,** 757 (1971).

152. C. E. Scarsbrook, D. L. Bennett, and R. W. Pearson, *Agron. J.,* **51,** 718 (1959).

153. D. W. Grimes and W. L. Dickens, *Agron. J.* **66,** 403 (1974).

154. R. O. Slatyer and J. F. Bierhuizen, *Aust. J. Biol. Sci.* **17,** 131 (1964).

155. R. E. Johnson, R. G. Curley, A. George, O. D. McCutcheon, V. T. Walhood, C. R. Brooks, and P. Young, *Calif. Agric.* **28,** 6 (1974).

156. I. W. Kirk, A. D. Brashears, and E. B. Hudspeth, Jr., Influence of Row Width and Plant Spacing on Cotton Production Characteristics on the High Plains, *Tex. Agric. Exp. Stn.* MP-2937, 1969.

157. G. A. Niles, Development of Plant Types with Special Adaptation to Narrow-Row Culture, *Proc. Beltwide Cotton Prodn. Res. Conf., 1970,* pp. 63–64.

158. E. L. Whiteley, C. A. Rinn, C. E. Bolton, and O. Smith, Narrow-Row Cotton in the Central Blacklands, *Tex. Agric. Exp. Stn.* MP-1407, 1979.

159. J. L. Larson, R. D. Lacewell, J. E. Casey, L. N. Namken, M. D. Heilman, and R. D. Parker, Impact of Short-Season Cotton Production and Producer Returns, Insecticide Use and Energy Consumption, *Tex. Agric. Exp. Stn.* MP-1204, 1975.

160. S. H. Baker, *Agron. J.* **68,** 85 (1976).

161. T. A. Burch, Effects of Plant Spacing on Varieties, Yield, Fiber Properties and Growth Habits of Cotton, *Proc. Beltwide Cotton Prodn. Res. Conf., 1970,* p. 54.

162. L. L. Ray, Breeding Cotton Varieties for Narrow Row Production, *Proc. Beltwide Cotton Prodn. Res. Conf., 1970,* p. 57.

163. D. N. Baker and D. L. Myhre, *Physiol. Plant.* **22,** 1043 (1969).

164. E. J. Pegelow, Jr., D. R. Buxton, R. E. Briggs, H. Muramoto, and W. G. Gensler, *Crop Sci.* **17,** (1977).

165. J. A. Andries, J. E. Jones, L. W. Sloane, and J. G. Marshall, *Crop Sci.* **10,** 403 (1970).

166. T. A. Kerby and D. R. Buxton, *Agron. J.* **70,** 535 (1978).

167. C. E. Fisher and E. Burnett, Conservation and Utilization of Soil Moisture, *Tex. Agric. Exp. Stn. Bull.* 767, 1953.

168. J. E. Quisenberry and B. Roark, *Crop Sci.* **16,** 762 (1976).

169. J. E. Quisenberry, Breeding for Drought Resistance and Water Use Efficiency, in M. N. Christiansen and C. F. Lewis, Eds., *Breeding Plants for Less Favorable Environments*, Wiley-Interscience, New York, 1981, pp. 193–212.

170. E. A. Hurd, Plant Breeding for Drought Resistance, in T. T. Kozlowski, Ed., *Water Deficits and Plant Growth*, Vol. 4, *Breeding for Drought Resistance*, Academic Press, New York, 1976, pp. 317–353.

171. H. M. Taylor, E. Burnett, and G. D. Booth, *Z. Acker—und Pflanzenbau* **146,** 33 (1978).

172. K. E. Fry and W. L. Ehrler, Water Stress Compared in Upland and Pima Cotton Varieties, *Proc. Beltwide Cotton Prodn. Res. Conf., Las Vegas, 1976.*

173. B. Roark and J. E. Quisenberry, *Plant Physiol.* **59,** 354 (1977).

174. J. M. Davidson and L. C. Hammond, Soil Physical Aspects of Root Development, *Proc. Soil Crop Sci. Soc. Fla.* **36,** 1–4, 1977.

175. E. L. Thaxton, Jr., and N. P. Swanson, Guides in Cotton Irrigation on the High Plains, *Tex. Agric. Exp. Stn. Bull.* **833,** (1956).

176. J. T. Ritchie, *Agric. Meteorol.* **14,** 183 (1974).

8

PEANUT

K. J. Boote

INTRODUCTION

The diversity of climates at the center of origin and at current areas of world peanut production may indicate the relative adaptation of peanuts to water deficits. Peanut, *Arachis hypogaea* L., is an indeterminate warm season legume originating from South America. The geographic center of origin of the cultivated species is presumed to be the eastern slopes of the Andes in Bolivia; however, relatives of the genus, *Arachis*, are widely dispersed throughout Brazil, Argentina, Bolivia, Paraguay, and Peru (1). The climate in this area is tropical and subtropical with approximately 100–150 cm of rainfall per year primarily during the summer season (2).

CROP USE, PRODUCTION AREAS, AND YIELD POTENTIAL

Peanut-producing areas have changed considerably over the past 100–200 years. The European colonists distributed peanuts from South America to Africa, Asia, Europe, the Pacific Islands, and eventually to the present Southeastern United States (3). Table 8.1 show the current major peanut producing countries (4). India and China harvest a little more than 9 million hectares each year and produce nearly half of the world's peanuts. The United States ranks sixth in terms of hectareage, but high yields per hectare give it the third highest total production.

World peanut-producing climates vary from wet and humid to semiarid, while temperatures are commonly warm during the growing season. India and West-Central Africa are typical of much of the world's peanut production area and have short summer rainy seasons in a dryland savannah climate. Short-season cultivars are grown there, usually without irrigation. Table 8.1 shows relatively low yields in India and the West-Central African countries due in part to low rainfall in combination with low fertility and poor leafspot (*Cercospora*) control. In the mid-1970's a major drought cycle in West and Central Africa reduced yield considerably. Yields in Nigeria, for example, were reduced from the "normal" production level of 900 kg ha^{-1} in 1969–1971 to 433 kg ha^{-1} in 1974–1976. By 1978, yields in this region had mostly recovered to

Table 8.1. Average harvested hectares and yield of peanuts in world peanut-producing countries (1977–1979)[a]

Country	Harvested Area (× 1000 ha)	Yield (kg ha^{-1})	Production (× 1000 MT)
India	7259	839	6091
China	2390	1110	2654
Sudan	1013	966	977
Senegal	1000	903	889
Nigeria	673	717	457
USA	614	2878	1768
Burma	552	758	419
Indonesia	512	1401	718
Zaire	457	684	312
Argentina	396	1403	547
Brazil	256	1436	369
Other	3884	875	3398
World	19006	979	18599

[a] After FAO Production Yearbook (4).

"normal" levels. In spite of greater rainfall, yields are low in Zaire and Burma, probably because of subsistence agriculture. Yields in the United States (2878 kg ha^{-1}) reflect not only greater rainfall, but better technology and longer season cultivars. Using similar cultivars and technology, but with irrigation, Israel's average yield was 3740 kg ha^{-1}. This difference is an approximation of the "normal" effect of insufficient water on U.S. yields. However, a severe drought in southeastern United States in 1980 caused U. S. average yield to fall to 1830 kg ha^{-1}. Peanut yields in the dryland areas of the southwestern United States are normally much less than in the southeastern United States (5) (Table 8.2).

Country by country comparisons indicate that short erratic rainy seasons, for example, have reduced rain-fed peanut yields in West Africa and India. However, these countries also have infertile soils and poor leafspot control; thus, these yields are difficult to compare with rain-fed yields in other countries. A comparison of dryland yields to irrigated peanut yields within a region or state such as Texas is more desirable. In Texas in 1976–1977, dryland peanut production averaged 1001 kg ha^{-1} while irrigated production averaged 2494 kg ha^{-1} (A. M. Schubert, personal communication). Experimental data are available from individual research reports for specific irrigation management regimes. Chema et al. (6, 7) reported experimental rainfed yields in India of 1300 and 1840 kg ha^{-1} for 421 and 377 mm of water used, respectively. Yields of 485, 1275, and 1920 kg ha^{-1} were obtained for cultivar "C 55-437" in Senegal and Nigeria when rainfall totaled 156, 242, and 296 mm, re-

Table 8.2. Average peanut yield during 1977, 1978, and 1979 in states of the United States producing more than 20,000 hectares of peanuts.[a]

State	Pod Yield ($kg\ ha^{-1}$)
Texas	1678
Oklahoma	2328
North Carolina	2907
Alabama	3051
Virginia	3156
Florida	3643
Georgia	3507

[a] After Agricultural Statistics 1980 (5).

spectively, during the growing season (8). In Oklahoma, rain-fed Spanish peanuts yielded 239, 1368, 2978, and 1137 $kg\ ha^{-1}$ in four seasons during which crop water use equaled 109, 464, 492, and 987 mm, respectively (9).

Yield potential of full-season peanut cultivars is high with irrigation, leafspot control, and good cultural practices. Peanut yields as high as 6070 and 7200 $kg\ ha^{-1}$ were cited by McCloud (10) for experimental and producer fields, respectively, in the United States. The highest reported peanut yield was 9600 $kg\ ha^{-1}$ in 1974 by a producer in Rhodesia, B. Huntsman Williams (D. E. McCloud, personal communication). These high peanut yields are well in excess of the highest yields for soybean (*Glycine max* L.) and indicate that peanuts may have a greater yield potential than soybean.

GROWTH AND DEVELOPMENTAL STAGES

Cultural practices for fertilizer, fungicide, insecticide, and herbicide application, as well as water management, change with growth stage of the peanut plant; therefore, it is important to be able to identify the various developmental stages. In Figure 8.1, I have identified recognizable phenological events for *Arachis hypogaea* spp. *hypogaea* cv.[6] Florunner' in relation to its dry matter accumulation (11) and its growth phases (12). These visually observable phenological events were observed and subsequently proposed by Boote (12) as part of a system of uniform crop growth stage descriptions for peanut. Williams et al. (13) defined 10 somewhat arbitrary phases of peanut growth and development but placed less emphasis on visually recognizable phenological events. Phenological events appear at varying times depending on temperature, available soil water, and genetic makeup of the plant, but all events occur in order unless the peanut life cycle is terminated. Shorter season Spanish-Valencia cultivars (*Arachis hypogaea* spp. *fastigiata*) differ primarily by having shorter intervals between events. More complete information on growth, development, and botanical aspects are available in *Peanut Culture and Uses* (1).

Consider the hypothetical effect of water deficit during three different growth phases of Florunner peanut. Moderate water deficit during the first 50 days would reduce vegetative development; however, vegetative growth is frequently excessive (14, 15). Moderate early water deficits

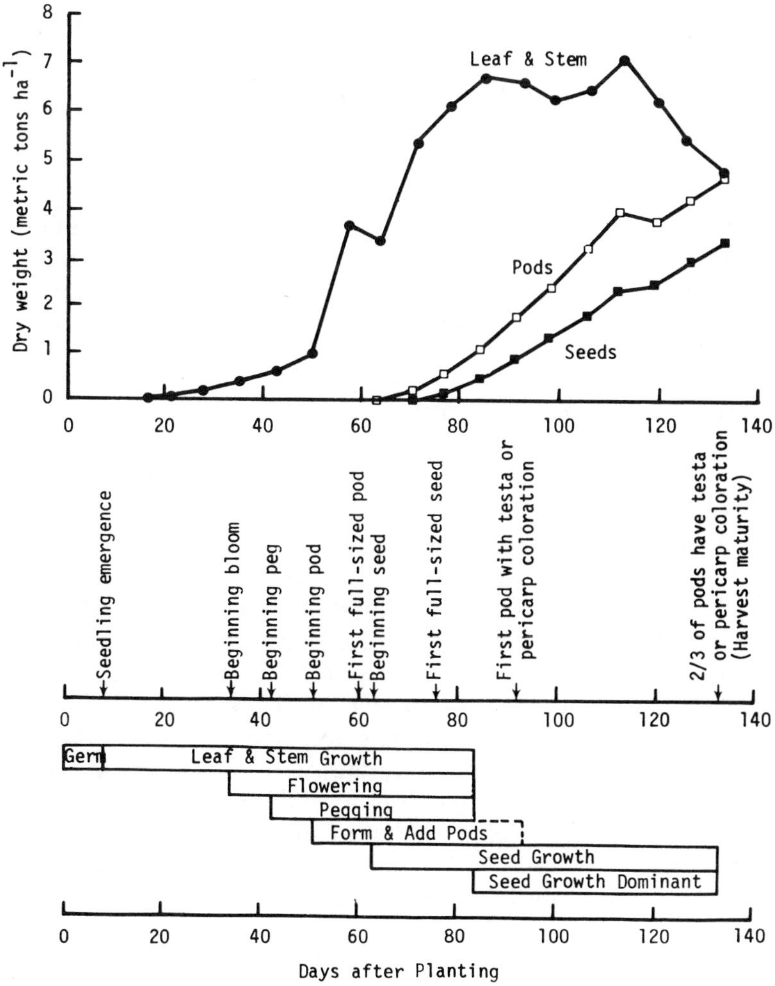

Figure 8.1. Phenological growth stages of Florunner peanut plant relative to growth and development processes. [Phenological stages were developed and observed by Boote (12), with dry matter accumulation curves after McGraw (11).]

relieved after 50 days will probably have a small effect on yield if adequate leaf area index (*LAI*) is achieved. Reduction in flowering or pegging from water deficit prior to 50 days is insignificant because early flowering is slow and few fruits are formed by 50 days (11).

By contrast, water stress during the pod formation period, between 50 and 80 days, may seriously interfere with the rate of fruit addition (16, 17) in several different ways: First, pegging and pod formation are adversely affected by dry soil in the fruiting zone (18, 19) which reduces Ca uptake by fruits (20, 21, 22). (This problem will be discussed later in this chapter.) Second, the process of pegging and fruit formation requires turgor pressure which will be much reduced if the soil and plant water potentials are low (16, 23). Third, and possibly most important, water deficit reduces photosynthesis (23), thereby limiting the number of fruits carried and decreasing final yield. Shading studies have shown that under conditions which limit photosynthate supply, especially between 65 to 86 days (24), peanut plants will cease adding fruits when the growth of existing fruits is sufficient to utilize the current photosynthesis supply.

Water deficit after a full fruit load (about 85 days) will reduce yield by causing the smaller and younger fruits to terminate growth and possibly by slowing growth of other fruits. Harvest quality and yield will be reduced by increased incidence of shriveled and aborted seeds.

ROOT GROWTH AND FUNCTION

Germination

Peanut seeds are fairly slow to germinate and usually require about 7 days for emergence in warm soils and up to 3 weeks in cool soils. Producers lacking irrigation normally plant their peanuts after rain to ensure adequate soil water for germination and emergence. In the semiarid tropics, the planting date varies considerably depending on the first significant rainfall of the monsoon season. The capacity of peanuts to emerge from deeper depths than many crop seeds may be an advantage to ensure germination when planted deeply to obtain soil water in dryland areas. The maximum soil water potential for germination of peanuts is not known. Mixon (25) observed that peanut seed germination (radicle emergence) began when average seed moisture achieved 35% and that all seeds had initiated radicles by a seed moisture of 55–60%.

Root Depth and Distribution

Peanuts are weakly taprooted plants which often have profuse small lateral roots on which nitrogen-fixing nodules develop. Root growth is predominant during seedling growth, reaching 10–16 cm in 5.5 days and a depth of 30 cm by 11.5 days (26). At this stage taproots and lateral roots were well developed, but only four leaves of the main axis had begun to develop, and only two cotyledonary branches had been initiated. Taylor and Ratliff (27) observed that some taproots of 6-day-old plants were 24 cm long in noncompacted soil at 32°C. At this stage, only a few leaves were present.

Data on the seasonal pattern of peanut root growth are sparse, probably because of the difficulty in digging roots. According to Lenka and Misra (28) the majority (77%) of the root weight is in the top 15 cm of soil, partly because of the thick taproot. The weight of fibrous roots below 15 cm is probably small but important because of rooting length density with respect to soil depth. Robertson et al. (29) reported that 39% of the total rooting length density was in the top 15 cm of soil and 55% in the top 30 cm. Average rooting length density of "Florunner" was 2.5 cm cm^{-3} in the top 15 cm but decreased to 0.2 cm cm^{-3} at 150 to 180 cm depths (Fig. 8.2). McCloud (10) observed that peanut root weight in the top 15 cm of soil was 37% of total crop dry matter at 21 days, but about 1.5% at harvest. Root weight in the top 15 cm of soil reached a maximum by 78 days after planting. This coincided with development of a full fruit load.

The seasonal progression of peanut rooting depth can be estimated from soil water extraction profiles throughout the season. Figure 8.3 shows the seasonal progression of soil water uptake by "Florunner" roots in a deep Lake fine sand in 1976 based on the presumption that water extraction indicates root presence (L. C. Hammond, W. K. Robertson, K. J. Boote, L. H. Allen, Jr., and R. J. Varnell, unpublished data). The

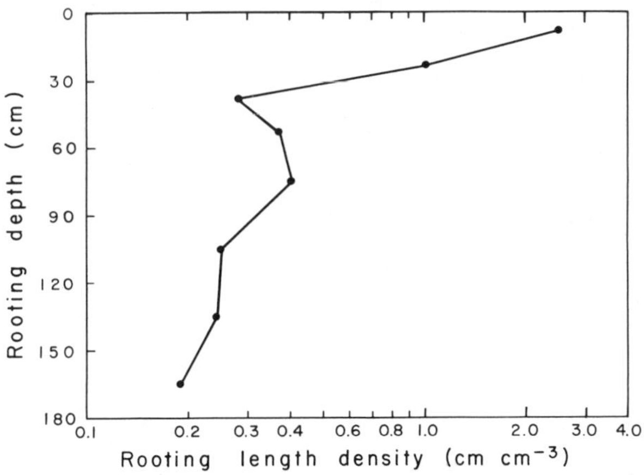

Figure 8.2. Rooting length density of 138-day old Florunner peanut with respect to soil depth in Lake fine sand during 1977 season at Gainesville, Florida. [Adapted from Robertson et al., *Agronomy Journal* **72,** 548–550 (1980) by permission of the American Society of Agronomy.]

deepest observed rooting depth was 280 cm at the end of the season using the soil coring technique. Water extraction profiles by Stansell et al. (30) can be similarly interpreted to show that "Florunner" roots extended to 30, 60, 90, and 120 cm depth in a Tifton loamy sand by 30, 42, 72, and 87 days, after planting, respectively. Others have observed rooting depths to 200 cm (29, 31). Stern (32) found roots of a "Valencia" peanut reached 170 cm, but only a few were at depths greater than 130 cm depth in a sandy to clay loam. Research cited by Harris and Bledsoe (33) reported a maximum rooting depth of 136, 150, and 192 cm at 110 days for the cultivars "Small Japan," "Small Spanish," and "Burmese," respectively. In a barrier-free soil, peanuts have the genetic ability for deep rooting and deep water extraction. This gives the peanut a drought tolerance mechanism that postpones desiccation during extended droughts.

Root Growth and Water Relations

During drought stress periods, lower roots continue to grow down into unrestricted moist soil zones even though top growth may appear to stop (34). The taproot elongation rate of nontranspiring peanut seedlings was found by Taylor and Ratliff (27) to be unaffected by changes in soil water potential between −0.19 and −12.5 bars, provided comparisons were made at the same soil strength as measured by needle penetrometer resistance. Elongation rate was decreased as soil strength increased. Thus, peanut roots can grow into relatively dry soil so long as soil strength is low. Peanut root elongation was less severely affected by

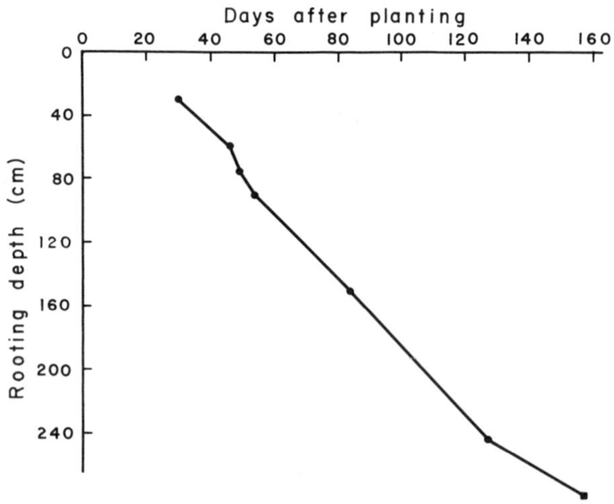

Figure 8.3. Rooting depth of Florunner peanut in a deep Lake fine sand during 1976 season at Gainesville, Florida. Rooting depth based on soil water extraction patterns and actual root observations in soil cores. [Unpublished data of L. C. Hammond, W. K. Robertson, K. J. Boote, L. H. Allen, Jr., and R. J. Varnell.]

***Table 8.3.** Effect of irrigation frequency on water extraction from various soil layers by Virginia Bunch peanuts in Israel during July–August 1960.*[a]

	Water Extracted (mm day^{-1})				
	Irrigation Frequency (days)				
Soil depth (cm)	40	30	21	14	10
0–30	1.57 (28)[b]	2.08 (32)	2.07 (34)	2.65 (37)	3.58 (48)
30–60	1.17 (21)	1.63 (25)	1.77 (29)	2.00 (28)	1.71 (23)
60–90	1.12 (20)	1.30 (20)	1.10 (18)	1.22 (17)	1.12 (15)
90–120	0.90 (16)	0.84 (13)	0.67 (11)	0.71 (10)	0.67 (9)
120–150	0.84 (15)	0.65 (10)	0.49 (8)	0.57 (8)	0.37 (5)
Total	5.60	6.50	6.10	7.15	7.45

[a] After Mantell and Goldin (37).
[b] Expressed as percentage of total water extracted from all depths.

compacted soil than was cotton, and maintained substantial elongation even to 35-bar penetrometer resistance.

Rooting depth is frequently deeper for water stressed compared to irrigated peanuts (28, 35, 36). Peanuts irrigated at 25 or 75% soil water depletion in the 0–30 cm soil zone had roots to 66 and 74 cm depth, respectively, in a sandy loam (28). Robertson et al. (29) observed increased rooting length density in the top 30 cm of irrigated peanuts as compared to nonirrigated peanuts in a dry season. Rooting density was no different at greater depths for irrigated than for nonirrigated peanuts.

Water extraction by roots in various soil layers was calculated from gravimetric soil water measurements during peak rooting and water use periods by Mantell and Goldin (37) (Table 8.3). The crop was grown on clay loam under nearly rain-free conditions in Israel and was irrigated at various frequencies (10, 14, 21, 30, or 40 day intervals). As shown in Table 8.3, peanuts irrigated every 10 days extracted 48% of their water from the upper 30 cm. As the irrigation interval was lengthened, less and less water (on absolute or percentage basis) was extracted from the upper 30 cm, whereas more water (on absolute or percentage basis) was extracted from the 90–150 cm depth. The water uptake pattern for the most frequent irrigation interval probably approximates conditions of peanuts under well-distributed rainfall.

TOP GROWTH AND DEVELOPMENT

Normal Developmental Sequences

Vegetative development of the leaf and stem begins slowly while roots are established. Then, the rate of canopy development increases rapidly to form the familiar sigmoid-shaped curve. The first leaves appear on the main stem and two cotyledonary laterals, but first-order, second-order, and even third-order vegetative branches appear on Virginia botanical type cultivars (*Arachis hypogaea* spp. *hypogaea*). Spanish-Valencia botanical type peanuts (*Arachis hypogaea* spp. *fastigiata*) have four to five main erect vegetative branches which may have specialized reproductive branches at every node. Virginia type peanuts flower longer, mature later, and have fruiting branches formed in a double-alternating sequence on lateral vegetative branches (1, 38).

The flowering date of peanuts is nearly insensitive to photoperiod although short photoperiods increase the reproductive to vegetative ratio (39). Flowering is delayed by lower temperatures (1). Cultivars differ in flowering response. Under normal warm season planting, short-season Spanish and Valencia types flowered 31 days after planting whereas a long-season Virginia type cultivar "Dixie Runner" required 38 days (K. J. Boote, unpublished data).

One of the unique characteristics of peanut is its underground fruits. Flowers are self-fertilized in early morning and soon wilt, yet 5 to 7 days later the base of the fertilized ovary (gynophore) begins to elongate geotropically toward the soil (40). The gynophore is commonly called a "peg." Gynophore elongation is rapid for 1–2 days. By 8–12 days after flower fertilization, the peg is usually firmly anchored into the soil (40). When the peg has elongated about 5 cm into the soil, the ovary tip begins to swell and turns horizontally away from the taproot. The process of gynophore penetration and fruit formation in the soil requires darkness and some soil media resistance (41). Gynophore penetration and fruit formation are enhanced by presence of moisture and Ca in the fruiting zone (22, 42).

After peg penetration into the soil, the fruit expands rapidly; full-sized fruits are present at 14–21 days after peg penetration, and rigid shells by 21–28 days (43; K. J. Boote, unpublished data). The earlier dates apply to Spanish type peanuts and later dates to Virginia types. At 14–21 days, the seeds are still small inside a fleshy fruit. From this stage on, the number of days required for seeds to reach maturity and full size is characteristic of the cultivar. Schenk (43) observed that seeds of "Dixie Spanish" and "Virginia Bunch" filled for about 4 and 7 weeks, respectively, although 6 to 10 weeks, respectively, were required from peg penetration to maximum fruit dry weight. Boote (44) observed approximately 53, 60, and 74 days from peg penetration to maximum fruit dry weight for individual fruits of Spanish-, runner-, and Virginia-market types. "Starr," "Florunner," and "Florigiant" are typical of the Spanish-, runner-, and Virginia-market type cultivars, respectively, in that study.

Growth, development, and dry weight accumulation in vegetative and reproductive parts of "Florunner" peanut are illustrated in Fig. 8.4 from data of McGraw (11), Duncan et al. (45), and unpublished data of R. L. McGraw, D. E. McCloud, and K. J. Boote. Early vegetative growth pro-

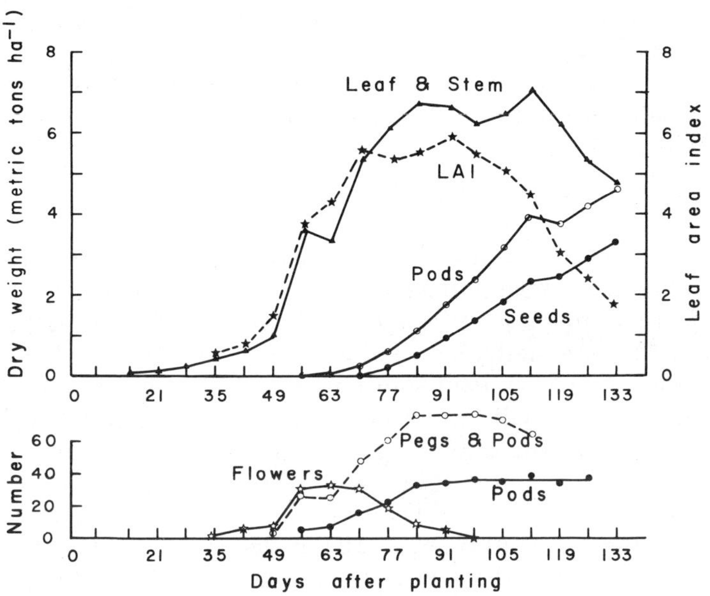

Figure 8.4. Dry matter accumulation and reproductive development of Florunner peanut during 1976 season at Gainesville, Florida. [After McGraw (11)].

ceeded slowly until the leaf area index (*LAI*) exceeded one. For these closely spaced plants (30 × 25 cm spacing), vegetative branches grew out in concentric circles and resulted in 100% ground cover when *LAI* exceeded 3.0 by 56 days after planting on May 5, 1976. The rate of dry matter accumulation became constant after *LAI* of 1.5 was achieved (49 days). Peanuts in 91-cm rows have also been observed to begin a constant rate of dry matter accumulation after 50 days with *LAI* less than 2 (10). *LAI* reached 5.6 by 70 days, remained at this value for about 30 days, then began to drop slowly at first and then more rapidly until *LAI* was near 1.0 at harvest (140 days). New leaf formation slowed drastically during 70 to 100 days. The subsequent decline in *LAI* from 100 days to harvest was primarily from death and loss of lower leaves.

First flowers were observed 35 days after planting and plants had 2–3 pegs per plant at 49 days (11). Peak flowering occurred between 56 and 70 days although the total flowering period spanned 35 to 91 days after planting. The pegging period covered 49–84 days after planting. This relatively long, 7–8 week period of flowering and pegging may be an advantage which allows compensatory fruit set if droughts occur during early flowering and pegging. The pod formation period for Florunner peanuts occurred between 56 and 84 days, at which time the number of expanded fruits stabilized. It appeared that a constant number of fruits were added per day until a full fruit load was achieved. At this time, all the available daily photosynthate was going to the fruits and as a group, the fruits were growing at a constant rate of about 95 kg ha^{-1} day^{-1}. Notice particularly that dry matter increase in the vegetative fraction ceased when a full fruit load was established. From 84 days until harvest can be considered the seed filling period.

Effect of Water Deficit on Top Development

Water deficit reduces peanut crop growth rate (46) and decreases dry matter production of vegetative components (15, 28, 30, 47, 48, 49). Most of these data pertain to samples taken at harvest or from specific growth stages; thus, they are inadequate to document season-long effects of water deficit on the rate of dry matter production. Ochs and Wormer (48) observed vegetative growth rate of potted plants between 31 and 75 days in response to irrigating at 0, 25, 50, 75, and 100% soil water depletion (*SWD*) to resupply water in the amount of 25% of available water. Irrigating at 50, 75, and 100% *SWD* progressively reduced daily top growth rate as compared to supplying water at 25 or 0% *SWD*. Slatyer (46) reported that the rate of dry matter accumulation by peanut was initially reduced when relative turgidity dropped below 90%. The point of reduced growth coincided with decreased transpiration.

Soil water deficiency inhibits leaf expansion and stem elongation through its reduction of relative turgidity (15, 34, 46). Water-stressed peanut plants have shorter main stem and cotyledonary branches (14, 35, 49). Soil water deficit reduces internode length more drastically than it reduces node number (48). Continuous soil water deficit causes fewer and smaller leaves which have smaller and more compact cells (35, 50). Leaves formed during drought retained their xeromorphic structure even after adequate water was supplied.

Water deficits during several stages of growth have been shown to reduce the rate of daily leaf production (15, 48, 51). Differences in *LAI* were reported by Vivekanandan and Gunasena (15) who grew peanuts throughout a life cycle in lysimeters controlled at approximately −330, −550, and −730 millibars of soil water potential. Differences in *LAI* persisted throughout the season with each treatment achieving maximum *LAI* of 6.25, 4.75, and 3.81, respectively. Leaf persistence and leaf area duration were reduced by decreasing soil water potential.

Influence of Water Deficit on Partitioning to Vegetation or Fruits

Vegetative growth is sometimes excessive during wet years or with frequent irrigation. During a dry growing season, irrigation to maintain soil water above −400 millibars potential at 20 cm depth caused main stems to elongate from 34 to 52 cm and cotyledonary branches to extend from 51 to 70 cm (14). Application of a growth retardant (succinic acid, 2,2-

dimethyl-hydrazide) has resulted in yield enhancement in wet years or frequently irrigated plots (14). *LAI* was also increased when soil water potential were maintained near 0 bars (15). High soil water potential failed to increase pod yield; the plants produced excessive vegetative growth rather than pods, and had a low ratio of pods to vegetation (15). Increasing the ratio of pods to vegetative growth under water deficits may be an important mechanism of peanut adaptation to drought.

The degree of partitioning to leaves or fruits during the seed growth phase of Spanish peanut (80–120 days) was shown by Billaz and Ochs (51) to depend on the number of fruits formed earlier, the fruit number potentially being reduced severely by water deficit during pod formation. If pod formation was nearly complete prior to imposing water deficit, then partitioning to fruits was increased by water deficit (resupplying water only at 100% *SWD*) (48). By contrast, water stress during pod formation (re-supplying water at 100% *SWD* from 31 to 90 days) decreased partitioning to fruits compared to other irrigation treatments (48). Billaz and Ochs (51) observed that water deficit during 50–80 days drastically reduced flowering and pod formation and reduced final yield more than did water deficits at any other stage. This treatment resulted in less partitioning to fruits but greater partitioning to leaves during the subsequent 80–120 day seed fill period during which the plants were irrigated. Water deficit from planting to 67 days had yet another effect: the period of rapid fruit growth was delayed by about 10 days and final yield was decreased (52).

Effect of Water Deficit on Flowering

Periods of soil water deficit have delayed flowering by 1–2 days and reduced flowering considerably (28, 35). Severe drought (maintaining soil water at or below 35% of field capacity) delayed formation of floral initials by 7 days and completely inhibited flowering (50). The reduction in flowering varies with the time during the plant life cycle at which water is withheld, the most severe reductions resulting from water deficit during or just prior to peak flowering (47, 49, 51). However, flowering per se is not very important to yield because flower numbers can be 20 times the number of mature pods carried to maturity (49, 53). Although flowering and pegging are reduced by soil water deficit (16, 35, 54), the peanut plant may compensate by producing a later flush of flowers and pegs (51; K. J. Boote, unpublished data) or by maturing a higher percentage of mature fruits from the immature fruits that form (15).

If the water stress is relieved after early pegging, the long flowering and pegging period (8–9 weeks) (10, 11) is sufficient to allow fruit set after water stress is relieved (16, 17, 31). The long flowering and pegging period of peanuts appears to be one of the advantages that indeterminate crops have over determinate crops such as corn (*Zea mays* L.) under water stress situations.

Effect of Soil Water in the Pegging Zone on Fruit Formation

The process of gynophore penetration into the soil requires adequate soil water, whereas both adequate soil water and darkness are required for subsequent development of the peg into an expanded fruit (41). Under dry soil conditions pegs frequently fail to penetrate effectively if the pegging zone is made up of air-dry soil (16, 18, 19). Ono et al. (54) reported that below a certain soil water level in the pegging zone (about one-sixth field capacity), a high percentage of peanut pegs failed to develop into pods, even though the deeper rooting zone contained adequate water for other processes in the plant. They noted that the critical period for adequate soil water for fruit development was the first 30 days after pegging. By the time a given fruit is 30 days old, it is fully expanded with a rigid shell (43) and is sufficiently developed to continue normal growth in dry soil if the roots have adequate moisture (54). Stern (32) also proposed that once a pod has been developed, seed development would continue to maturity even though surface soil water may be inadequate. This is frequently the situation in areas having short periods of inadequate rainfall.

Failure of pegs to penetrate and develop pods in dry pegging zone soil, may result from higher soil temperature (54), greater compaction and soil strength of drier soil (19), lack of turgor if the roots are also in dry soil (16, 23, 34), and lack of Ca uptake by the developing peg or pod (22).

Interaction of Soil Water with Calcium Uptake by Developing Fruits

The application of $CaSO_4$ at early pegging is a common peanut production practice which is based on the need for Ca in the pegging zone for effective fruit development. Slack and Morrill (42) reported fewer fruits, a lower percentage of pegs producing fruits, and less fruit weight as the fruiting zone Ca dropped below 20 ppm. Campbell et al. (55) also observed fewer and smaller peanut fruits, but twice as many pegs if Ca was not present in the fruiting zone.

Effective uptake of Ca from the fruiting zone is enhanced by adequate soil water (Table 8.4). A lack of water in the fruiting zone was reported by Skelton and Shear (22) to cause more unfilled pods (pops), fewer double-loculed pods, and lower Ca content of the hull and seed. Peanuts grown under drought conditions often have lower Ca levels in seeds and hulls (20, 21). The typical symptoms of Ca deficiency such as "concealed damage" and "hollow heart" were more prevalent following droughts

Table 8.4. Effect of soil water in the fruiting zone on peanut fruit formation and the calcium concentrations of the hull and seed.[a]

Treatment[b]	Gynophores per Plant	Dead Gynophores per Plant	Gynophores Developing into Fruits (%)	Ca Concentration (ppm)	
				Hull	Seed
Water	17	2	41	627	150
None	20	9	19	43	7

[a] After Skelton and Shear (22).
[b] Fruiting media of all treatments contained Ca. Water, when added, was added every 2 days.

during the pod addition and formation stage (21). The term "concealed damage" includes poor embryo or plumule development. Peanuts showing these symptoms had lower Ca levels in seeds and hulls than those lacking Ca deficiency symptoms (21).

Cox et al. (20) showed that application of $CaSO_4$ to the pegging zone enhanced Ca uptake by fruits. Even during dry years there was a marked $CaSO_4$ effect on yield and grade (20). Adequate water in the fruiting zone increases the availability of soil Ca to the developing fruits. During wet years excess water causes leaching of Ca from the fruiting zone and reduced responses to applied $CaSO_4$ application. The rapid leachability of soluble Ca from powdered $CaSO_4$ (56) dictates that gypsum should be applied at relatively high rates just prior to pegging to minimize soil Ca leaching from rainfall and irrigation.

The unique Ca requirement of developing underground fruits of peanut is related to their subterranean nature. In aerial fruits of other crops, Ca is passively supplied to fruits in the transpiration stream. By contrast, the underground fruits of peanuts do not transpire significantly, and therefore do not readily acquire Ca from the roots via the xylem transpiration stream (22, 57). Movement of ^{45}Ca from rooting media to fruits was increased greatly if fruits were lifted from the soil and allowed to transpire (22). Apparently the phloem supplies insufficient Ca to the fruits. The lack of xylem flow to peanut fruits was also demonstrated by Weirsum (58) when water soluble dyes did not flow from the roots to the fruits unless fruits were exposed to a drying atmosphere. The dye did move from fruits to nearby stem nodes, thus indicating fruit uptake of water. Xylem water movement from fruits to leaves is consistent with water potential differences that would exist under evapotranspiring conditions (23). Since fruits do take up water, adequate soil water may enhance fruit uptake of Ca from the soil by mass flow. Bledsoe et al. (57) and Slack and Morrill (42) conclusively demonstrated Ca uptake by developing peanut fruits. Their data can even be interpreted to suggest that there is some Ca movement from fruits to the rest of the plant.

Larger-fruited Virginia type cultivars require higher Ca in the fruiting zone (42), show greater response to $CaSO_4$ applications (59), and more frequently have calcium-related quality problems resulting from soil water deficit (30, 60). The increased sensitivity of the larger-seeded cultivars to Ca may be related to a smaller absorbing surface area relative to fruit volume. Some peanut cultivars have increased pod pubescence; however, pods of these lines have not been tested for efficiency of Ca uptake.

Effects of Water Deficit on Pod Number, Quality, and Viability of Harvested Seed

The major yield adaptation to water deficit is caused by a decreased number of fruits set, while the weight per fruit and weight per seed is scarcely affected (15, 28, 51). This is especially true for water deficit periods occurring prior to the end of pegging and pod formation. The quality aspects, shelling percentage, shriveled seeds, and damaged seeds are more greatly affected if droughts occur during the seed growth period (Fig. 8.1) and cause smaller and younger fruits and seeds to stop growing. The termination of growth of seeds in young fruits causes an increased percentage of other kernels (shrivels plus damaged) at harvest.

Peanut quality at harvest is reduced by water deficits during the growing season. The shelling percentage or percent sound mature kernels is frequently 1 to 4 units higher for irrigated peanuts as shown in Table 8.5 (6, 7, 9, 14, 15, 17, 20, 30, 49, 60, 61, 62). Irrigation has been shown to increase the weight per seed (7, 14, 17, 60) and percentage of extra large kernels (14, 30) while decreasing the percentage of damaged or shriveled kernels (9, 30). Irrigation had no consistent effects on oil or protein concentration of the seed (9, 36).

In addition to reduced market quality, soil water deficit may reduce seed germination (Table 8.5) and thereby reduce the quality of seed to be used for the next year's planting (20, 60). Pallas et al. (60) observed significantly lower germination of "Florigiant," "Florunner," and "Tifspan" seed when the seed-producing plants encountered water stress sufficient to cause wilting without overnight recovery. Viability of the larger-seeded "Florigiant" was more susceptible to drought than "Florunner" or "Tifspan." "Florigiant" seed germination was also reduced when irrigation was withheld until the soil water potential of 0–30 cm zone reached -15 bars. Cox et al. (20) demonstrated that drought-depressed germination of "Florigiant" seed was associated with insufficient Ca level in the seed (Table 8.5). The critical Ca level in the seed for good germination was reported to be 420 ppm. Seed germination was reduced 17% for every 100 ppm Ca below 420. In absence of $CaSO_4$ application, irrigation of the seed-producing crop increased next season's

Table 8.5. *Influence of water deficit on quality and viability of harvested seed*

Investigators	Cultivar	Irrigation level[a]	Sound Mature Kernels (%)	Extra Large Kernels (%)	Other Kernels (%)	Germination of Sound Mature Kernels (%)	Dark Plumule (%)	Ca Level in Seed (ppm)
Pallas et al. (60)	Florigiant	−0.2 bar	66a*	33a	3c	93	—	—
Stansell et al. (30)		−0.6 bar	64a	35a	4bc	89	—	—
		−15 bars	55b	25b	6b	82	—	—
		Overnight wilt	43b	19c	8a	68	—	—
	Florunner	−0.2 bar	76a	—	3b	96	—	—
Pallas et al. (60)		−0.6 bar	75a	—	4b	96	—	—
Stansell et al. (30)		−15 bars	69b	—	6b	92	—	—
		Overnight wilt	58c	—	10a	91	—	—
	Florigiant	−1–2 bar[b]	71	—	—	86	4	498
Cox et al. (20)		None[b]	69	—	—	77	10	411
	Argentine	−1 bar	63a	—	4	—	—	—
Matlock et al. (9)		None	51b	—	12	—	—	—

[a] Irrigated to maintain soil water potential (*SWP*) in upper 30 cm at or above given *SWP*.
[b] Averaged over + or − $CaSO_4$ applications.
* Column means within a cultivar followed by the same letter are not significantly different at the 0.05 probability level according to Duncan's New Multiple Range Test.

seed germination from 64 to 80% while increasing Ca uptake in the seed and reducing dark plumule (symptom of Ca deficiency). $CaSO_4$ application by itself increased germination from 64 to 90%. Ca application was more efficient than irrigation in increasing percent sound mature kernels, seed Ca, and viability of next generation seed.

PHYSIOLOGICAL RESPONSES TO DROUGHT

Photosynthesis and Stomatal Control

Very little research has been conducted on the relationship of photosynthesis of peanut to leaf or soil water potentials. Bhagsari et al. (23) observed decreases in net photosynthesis (P_n) of potted peanut plants during a drying cycle which were closely associated with decreases in relative leaf water content (*RLWC*) and transpiration (Fig. 8.5). On the fifth day of water stress, P_n was 8–13% of the watered controls while *RLWC* and water potential were 39% and −30 bars, respectively. Concurrent measurement of transpiration and diffusive resistance showed that the main effect of water stress on P_n was exerted through stomatal closure. Increases in stomatal resistance began when *RLWC* dropped below 80%. Bhagsari et al. (23) compared peanut and soybean and concluded that the relationship of P_n and diffusive resistance to *RLWC* was similar for the two crops. They proposed that any supposed superior drought tolerance of peanut compared to other crops must be due to some characteristic other than leaf P_n. Similarly, photosynthesis of peanut, soybean, and small grains all began to decline just before the first sign of wilting (decline began at 98% of maximum leaf fresh weight), although Iyama and Murata (63) suggested that peanut had a better capacity to maintain leaf blade water content against a soil water deficit.

Stomatal diffusive resistance of field-grown peanuts was measured by Allen et al. (34) on several days during a 21 day drying cycle (Fig. 8.6). Rainfall shelters were used to induce drought. Under field conditions, peanuts adapted gradually to soil water deficit by progressively greater closure of their stomata during the day (increased diffusive resistance). Slight stomatal closure occurred on day 7 and became significant on day 10 when diffusive resistance during the day averaged 2.00 and 1. 24 sec cm^{-1} for induced-drought and irrigated leaves, respectively.

On the day 10, most of the available soil water had been depleted from the upper 75 cm of soil. A comparison of diffusive resistance of dry and irrigated peanuts as a function of photosynthetic photon flux density indicated small differences at low irradiance, but larger differences above 100 nanoeinsteins sec^{-1} cm^{-2} (Fig. 8.7).

By the day 18, most available water was depleted from the upper 120 cm and stomatal diffusive resistance of dry and irrigated leaves during

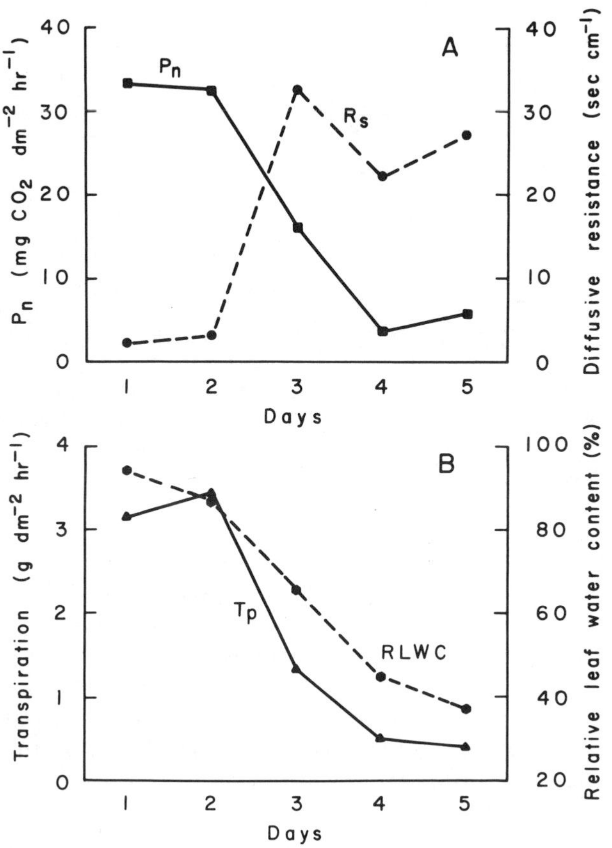

Figure 8.5. Daily changes in (A) net photosynthesis (P_n) and stomatal resistence (R_s); and (B) transpiration (T_p) and relative leaf water content (*RLWC*) of potted Florunner peanut plants during a drying cycle. Measurements made at 200 nanoeinsteins sec^{-1} cm^{-2} of photosynthetically active radiation. [After Bhagsari et al. (23)].

the day averaged 4.75 and 1.47 sec cm^{-1}, respectively. By this time, wilting was very obvious on induced-drought plots during midday sun; peanut leaflets folded together in a manner that minimized radiant energy interception and loss of water vapor. Although wilting, leaflet folding, and stomatal closure (resistance near 10 sec cm^{-1}) occurred in midday sun, stomatal opening to a diffusive resistance of 2 sec cm^{-1} occurred after 1 hr of cloud cover (Fig. 8.6). Diffusive resistance was also relatively low (about 3 sec cm^{-1}) in early morning hours.

Pallas et al. (64) reported similar gradual increases in leaf diffusive resistance as drought duration was extended while the leaf diffusive resistance of irrigated peanuts remained consistently low. One day after irrigation of drought-stressed peanuts, the leaf diffusive resistance returned to near normal. The rapid recovery indicated minimal residual

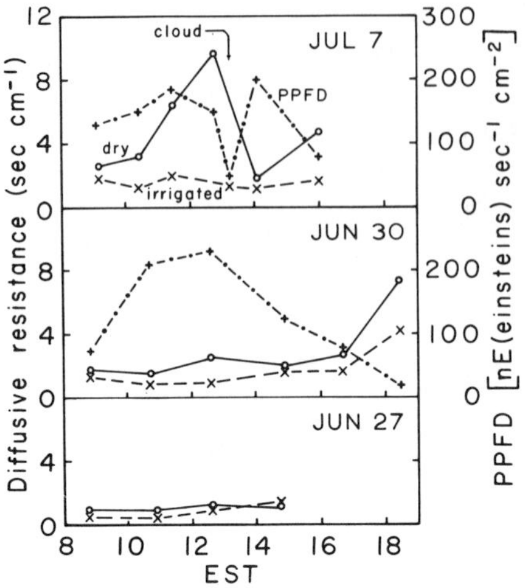

Figure 8.6. Stomatal diffusive resistance for induced-drought (O) and irrigated (X) Florunner peanuts on day 7 (June 21), day 10 (June 30) and day 18 (July 7) of a drying cycle at Gainesville, Florida, in 1975. Average photosynthetic photon flux density (*PPFD*) (+) was measured during periods when stomatal diffusive resistance was measured. [From Allen et al. (34), with permission]

plant resistance either as root resistance or residual stomatal resistance. Rapid recovery may be an important adaptive response of peanuts to water deficit conditions.

Leaf water potential of frequently irrigated peanuts has not been found to be less than −12 bars (64) or −13 bars (34). By contrast, water-stressed peanut plants had leaf water potentials of −30 to −40 bars (64), −30 bars (23), and as low as −60 to −83 bars (34). Based on the Chardakov densimetric dye technique, Gautreau (65) reported leaf water potentials

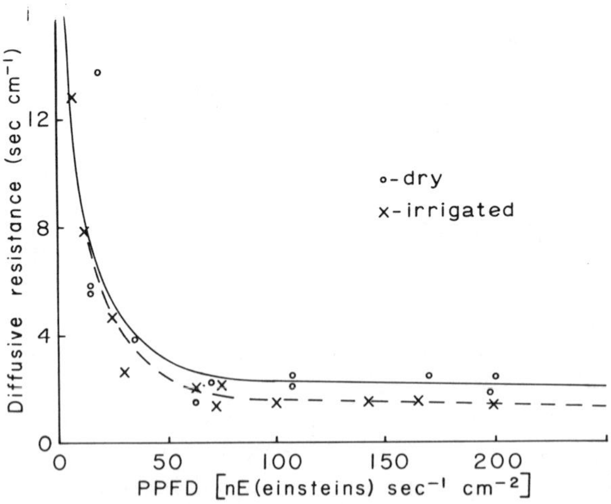

Figure 8.7. Stomatal diffusive resistance of induced-drought (O) and irrigated (X) Florunner peanuts as a function of photosynthetic photon flux density (PPFD) during the afternoon of day 10 (June 30) of a drying cycle at Gainesville, Florida, in 1975. [From Allen et al. (34), with permission.]

of near −12 and −17 bars at midday for 80- to 90-day-old peanuts grown under high and low water treatments, respectively, during the dry season in Senegal. Water potential was more negative on drought-stressed plots, decreased throughout the morning hours, and decreased during the crop life cycle.

Tsuno (66) measured short-term photosynthetic responses of outdoor peanut canopies to different evaporative demand conditions (by varying relative humidity) and reported that during an early growth stage (plant age not stated), relative humidity increases had very little effect on leaf photosynthesis. At a later growth stage (36 days later), increasing the relative humidity caused photosynthesis to increase from 8 to 12 mg CO_2 dm^{-2} hr^{-1}. This suggested that water supply from the roots to leaves was insufficient during the later growth stage. The sugar content of peanut roots and the ratio of root weight to leaf area declined between the two stages. Tsuno (66) suggested that both were related to the ability of roots to supply water to the foliage.

Chen and Chang (67) found peanut photosynthesis responded equally well to soil water regimes between 40 and 80% of field capacity. The response was similar at all growth stages except for some reduction in photosynthesis at 40% of field capacity at a late reproductive growth stage. Wormer and Ochs (68) observed that potted peanut plants depleted up to two-thirds of the available soil water before the stomata began to close.

Diurnal trends in peanut photosynthesis and transpiration are closely correlated (69, 70). The photosynthetic and transpiration rhythm seemed to be correlated with endogenous changes in stomatal aperture (69, 70). Vivekanandan and Gunasena (15) observed diurnal patterns in stomatal opening at different soil water potentials. Peanut required at least 2 hr after sunrise to achieve maximum stomatal opening. Under well-irrigated conditions, significant midday closure of stomata was not observed although minimum relative turgidity was observed between 1200–1300 hours (15, 34).

Leaf Display

In absence of water stress, peanut leaflets tend to be oriented at right angles to the direction of incident radiation (diaphotonasty) and follow the sun in this position. Under severe water stress, peanut leaflets have been observed to orient parallel to the incident radiation (parahelionasty) during hours of high solar radiation (23, 34, 50). The vertical folding of the compound leaflets minimizes exposure of leaflets to direct beam solar radiation and keeps leaf temperature to a minimum when stomata are nearly closed and the leaf is least able to lose heat by latent heat of evaporation. The three reports proposed that this was a drought avoidance adaptation because it permits a reduction in leaf heat load and evapotranspiration. At night, peanut leaflets undergo nyctinastic movement, folding together and pointing up, oriented as during midday water stress. Begg and Torssell (71) observed nyctinastic and parahelionastic leaf movements in *Stylosanthes humilis* and suggested a common mechanism involving turgor changes in the pulvini. Leaflet unfolding in light by *Albizzia julibrissin* has been shown to involve active K transport into the ventral pulvinule motor cells, whereas closing in the dark involves an efflux of K from the ventral cell membranes (72). Ilina (50) observed a layer of transparent, aquiferous cells (also called motor cells) on the lower side of peanut leaflets and along the leaf vein. He suggested that a decrease in water volume of these aquiferous cells during water stress caused the vertical folding of the peanut leaflets.

Response of N_2 Fixation to Water Deficit

Nitrogen fixation of a Spanish type peanut was shown by Reddy and Tanner (73) to be increased by irrigation; moreover, Lenka and Misra (28) observed fewer nodules per plant (240 vs. 553) when peanuts were irrigated at 75% soil water depletion (SWD) compared to 25% *SWD*.

Screening Peanut Cultivars for Drought Tolerance

Drought-tolerant peanut cultivars are needed in the semiarid zone of West Africa (8, 74). Investigations at the Research Institute for Oils and Oilseeds at Bombey have (1) defined physiological differences associated with drought resistance in known varieties, (2) set up three rapid tests to screen germplasm, and (3) developed more precise tests to evaluate drought tolerance of hybrid progeny. Tests to screen germplasm include (1) germination at low osmotic potential, (2) relative rate of early growth, and (3) heat resistance of young plants. Progenies are further screened by comparisons of relative transpiration rate and water potential of leaflets (65).

Early (90 day) and late (120 day) varieties have been developed which have drought tolerance and yield better than other cultivars in dry years, yet have equal yield in normal years (8, 74). Cultivar "C-55-437," has produced yields of 1920, 1275, and 485 kg ha^{-1} with only 296, 242, and 156 mm of rainfall, respectively, in Senegal and Nigeria. In zones normally receiving 400–750 mm, a 120 day cultivar, "59-127," gave higher yield potential in wet seasons, yet had good yield in drier years.

Gautreau (65) measured leaf water potential of peanut cultivars for 5 consecutive days and separated them into three groups having low, intermediate, and high water potential which corresponded inversely to the yield performance (low yielding cultivars having higher water potentials and higher yielding, drought-tolerant cultivars having lower, more negative, water potentials). Presumably, drought-tolerant cultivars with more negative water potential would have stomata closing at a more negative water potential and would have a steeper water potential gradient driving root uptake of water. Varieties lacking the low water potential trait (tolerance) may also be adapted to dry areas, but are adapted because of low *LAI* or short season growth habit (avoidance).

Accumulation of free proline in leaves is an adaptative response associated with drought stress. Mekkri et al. (75) surveyed 10 peanut cultivars for level of free proline accumulation in leaves and found that under stress conditions the drought tolerant cultivar "C-55-437" (see 8, 65) accumulated more proline and fixed more $^{14}CO_2$ under stress than the other cultivars. Proline accumulation was poorly related to relative water content.

APPLIED ASPECTS OF IRRIGATION

Critical Water Demand Periods

The most critical period for adequate soil water appears to be the period of pod formation and pod addition (20, 51, 61, 76, 77). Phenologically, the period starts at beginning pod (first fruit swelling) and ends when a full fruit load is established and vegetative growth has slowed down (Figs. 8.1 and 8.4). For Florunner, this period spans 50–85 days and includes much of flowering and pegging phases as well as some early seed growth; thus, it is not inconsistent with other reports (49, 78) that flowering and pegging stages are the most sensitive stages to water deficit. This 50–85 day period (see Fig. 8.4) is characterized by continued leaf and stem growth as well as pegging and pod formation. Leaf area index reaches a maximum and evapotranspiration is high at this stage. Water deficits during this period reduce turgor for gynophore penetration and fruit expansion, reduce Ca uptake by developing fruits, and most importantly, reduce photosynthesis which determines the number of fruits to be carried. The next most critical period for water deficit is probably the first half of the seed fill period, 85–105 days for Florunner, although Pallas et al. (64) found pod yield equally reduced for droughts from 71–105 days or 105–145 days for Florunner. Their results emphasize the critical need for water during the entire seed fill period.

Water deficit during early vegetative growth has a minimal effect on peanut growth and yield (49, 50, 76). Likewise, less frequent irrigation, or irrigating at a greater SWD has its least detrimental effects prior to pegging and pod formation (76, 77). During the latter half of seed fill, water deficits are also less critical, because fruits have been formed and evapotranspiration has declined somewhat. Excessive and too frequent irrigation or rainfall during fruit growth may be detrimental to yield because of increased incidence of *Cercospora* leafspot (79) and pod loss on heavier soils (14, 30, 37). Decreased yields, probably due to weakening of peg attachments and loss of detached pods, have been observed on clay soils with frequent irrigations (37) and on loamy sand soils irrigated to maintain greater than -0.2 to -0.4 bar soil water potential (14, 30).

Seasonal Evapotranspiration

The seasonal trend of daily water use or evapotranspiration (E_t) of peanuts has been measured under controlled lysimetric conditions by Kassam et al. (80), Stansell et al. (30), and Vivekanandan and Gunasena (15). In each of these experiments, rainfall was excluded and water applications to peanuts were carefully controlled to prevent or allow measurement of percolation. Daily evapotranspiration was calculated by soil water inventory.

Daily evapotranspiration is low during early vegetative growth but increases as the leaf canopy develops and closes (15, 30, 80, 81). Kassam et al. (80) observed that peak E_t occurred shortly before peak *LAI* was achieved (Fig. 8.8). The maximum daily E_t was 0.53 to 0.60 cm day^{-1} for Spanish-type peanuts and was variously observed to occur at 67–77 days (80), 55–80 days (15), and 60–80 days (30) after planting. Maximum daily E_t for "Florunner" and "Florigiant," later maturing cultivars, occurred 10 and 20 days later, respectively, than "Tifspan," a Spanish type cultivar (30). Under high evaporative demand conditions in Israel, daily

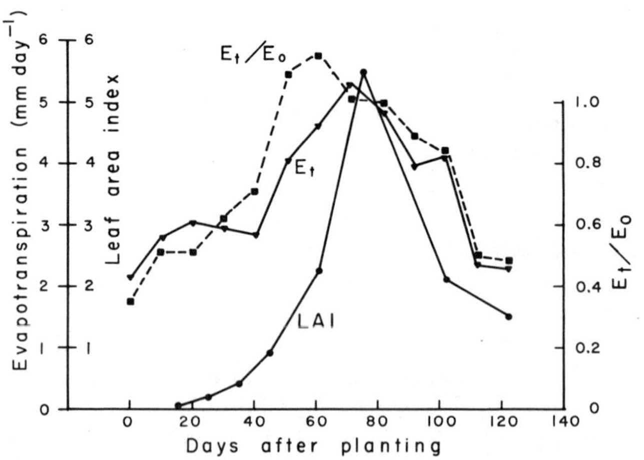

Figure 8.8. Seasonal progression of evapotranspiration (E_t), evapotranspiration/potential evapotranspiration (E_t/E_0), and leaf area index (*LAI*) of Spanish-type peanuts. [After Kassam et al. (80). Copyright 1975 by *Tropical Agriculture (Trinidad)*]

evapotranspiration of Improved Virginia Bunch averaged 0.69 cm day^{-1} during July (53–83 days after planting) (37). Peak E_t/E_0 (evapotranspiration/potential evapotranspiration ratio) for Virginia Bunch peanuts was observed during 70–110 days after planting (82). This was near the period of peak flowering and pegging when there was complete ground cover.

Factors Influencing Daily and Seasonal Evapotranspiration

Daily E_t is reduced by soil water deficit. The maximum daily E_t values for peanuts grown in wet (−330 mbars), intermediate (−550 mbars), and dry (−730 mbars) soil water potential treatments were 0.61, 0.48, and 0.38 cm day^{-1}, respectively (15).

The diurnal pattern of transpiration for nonstressed peanuts is a normal curve with peak height controlled by diurnal radiant energy load and evaporative demand (46). With increasing soil water deficit, the daily transpiration curve was reduced and became progressively skewed to a peak earlier in the morning. This observation of greater transpiration in early morning hours is consistent with nighttime recovery of relative turgidity to allow stomatal opening during early morning hours even during water deficit periods (34).

Daily and seasonal E_t is greatly influenced by percentage ground cover and leaf area development. In absence of any plant ground cover, average evaporation from the soil surface was reported to be 30–40% of pan evaporation (80, 82). Soil surface water content was not defined in these reports.

The E_t/E_0 ratio was found to increase in a nearly linear manner from 0.3 to near 1.0 as the percentage peanut ground cover increased from 0 to 100% (Fig. 8.9) (82) and as *LAI* increased during peanut growth (80). After establishment of full ground cover, the E_t/E_0 ratio remained fairly stable under adequate soil water supply until leaf area began to decline. The ratio of E_t/E_0 during full ground cover was estimated at about 0.8–0.9 under arid conditions in India and Israel (6, 37, 82), but higher (about 1.0) during a rainy, humid period in Samaru (80). The ratio of E_t/E_0 during the season is increased by more frequent irrigations (6, 7, 37, 82).

Certain cultural conditions such as row spacing, row direction, planting density, and fertility levels have small but important effects on E_t. ChinChoy et al. (83) observed that peanuts grown in north–south 30 cm rows in Oklahoma had lower E_t and retained more water in the soil profile than peanuts grown in 90 cm rows. E_t of narrow-row plots averaged 0.18 cm day^{-1} and the wide-row plots averaged 0.22 cm day^{-1}. Not only did narrow rows have lower E_t, but their pod yield was nearly 20% higher. The net result was a sizable increase in water use efficiency (kg of pod yield per ha per mm of water) (83). The water conserving effect of narrow rows was possibly related to more complete shading of

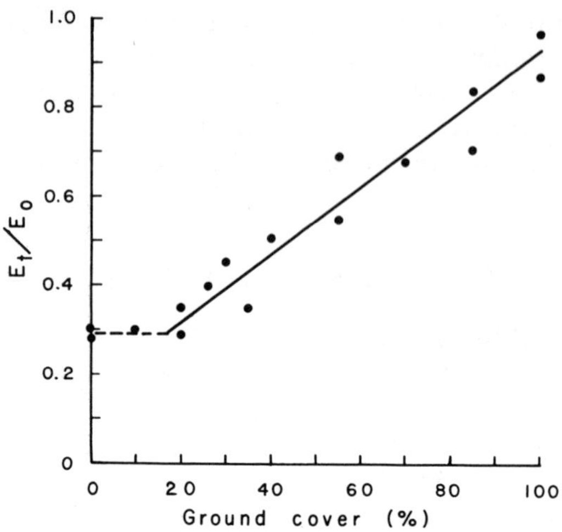

Figure 8.9. Relationship of evapotranspiration/potential evapotranspiration (E_t/E_0) to percentage ground cover of irrigated peanuts in Israel. [After Goldberg et al. (82).]

the soil, improved water infiltration, and avoiding the turbulence caused by the hedge-row effect of wider row spacings. Davidson et al. (81) reported substantial yield response to narrow row spacing for both irrigated and nonirrigated Spanish peanuts.

North–south 30 cm rows had lower E_t than east–west 30 cm rows (83). Applications of N and P tended to increase E_t and water use efficiency (84).

Irrigation Management

A common irrigation management strategy is to irrigate when the soil water in a given layer (usually 0–30 cm) reaches a specific level of soil water depletion (*SWD*) as summarized in Table 8.6. Generally, yields were greater where irrigation was scheduled at fairly low levels of soil water depletion, but irrigating at 25% *SWD* resulted in the lowest water use efficiency (*WUE*) and greatest water application (28, 36). Pallas and Stansell (85) similarly observed an inverse relationship between WUE and yield. Reddi and Reddy (76) irrigated in relation to three phases of growth to try to improve *WUE* values. Irrigation scheduling at 50% *SWD* was least detrimental during early growth and resulted in equal yield and greater *WUE* if followed by irrigation at 25% *SWD* at later stages (Table 8.6). Although differences are not clear-cut, Reddi and Reddy (76) obtained highest *WUE* values where soil water levels were maintained above 25% *SWD* in the later growth phases: "pegging to pod formation" and "pod formation to harvest." Irrigating cultivars or fields

Table 8.6. Peanut yield, water use, and water use efficiency in response to irrigating at different soil water depletions (SWD)

Investigators	Cultivar	Irrigation at *SWD* (%)	Pod Yield ($kg\ ha^{-1}$)	Water Use (cm)	Water Use Efficiency ($kg\ ha^{-1}\ cm^{-1}$)
Lenka and Misra (28)	AK12-24	25	2794	83	34
		50	2653	71	37
		75	2249	62	36
	TMV2	25	4455	77	58
Narasimham et al. (36)		50	4314	66	66
		75	3396	48	70
	AH-1192	50/50/50[a]	3170	43	75
Reddi and Reddy (76)		25/50/50	3430	45	76
		50/25/50	3530	46	77
		50/50/25	3750	47	80
		25/25/50	3490	46	77
		50/25/25	4140	45	91
		25/50/25	4000	49	82
		25/25/25	4680	51	92
		Once/week	3360	47	77

[a] Irrigated at defined *SWD* during three stages: up to pegging, pegging to pod formation, pod formation to harvest, respectively.

with low yield potentials does not significantly increase their yields (6, 7, 62).

Another irrigation approach is to apply irrigation water sufficient to supply a given fraction of the cumulative pan evaporation (*CPE*) since last irrigation (77, 82). During full canopy development, irrigation at 0.9 was best. Irrigation at 0.6 was not detrimental during early growth and caused an increase in *WUE* if followed by irrigation at 0.9 at later stages. Irrigation at 90% of *CPE* works well for closed canopies in arid regions because E_t/E_0 under such conditions is nearly 0.90.

A bookkeeping method was suggested by Stansell (86) in which available water holding capacity of the top 60 cm was considered the total water account. Since previous research has shown that water withdrawals of more than 50% of available soil water decreased yield, the peanut crop was only allowed to withdraw one-half of the total available water before irrigation was scheduled. By recording evapotranspiration as a debit and rainfall and irrigation as credits, a running account could be maintained without measuring soil water. Stansell developed his estimate of evapotranspiration with lysimeters. This approach can be extended to arid, low rainfall areas by determining the rooting depth and the atmospheric demand for water and calculating water applications needed to maintain the soil–root profile above 50% available soil water. Assuming the atmospheric demand is fairly constant, irrigations can be scheduled on a simple frequency interval (37, 82).

A system using tensiometers to estimate soil water potential at a specified depth has also been useful (14, 30, 31, 60). The specific depth will vary according to soil type and soil depth available for rooting. Optimum economic yield and water use appears to result from irrigation scheduling when soil water potential reaches −600 mbars at midday in the top 15 or 30 cm of soil (Table 8.7). Irrigating at a higher soil water potential may be desirable during pegging and pod formation.

Peanut yield response to crop water use is shown for controlled lysimetric conditions (Fig. 8.10) and for short-season cultivars under variable field conditions (Fig. 8.11). Both figures show that highest yield of short season or full season peanuts occurred when seasonal water use was 60 cm or more. The yield response was gradual between 40 and 60 cm, but yield rapidly dropped to zero as water use declined from 40 to 10 cm. Not only is 60 cm of water required, but the distribution must be good for optimum yield. Up to 50% of the yield variability was attributable to factors other than water use. Perhaps future irrigation scheduling will take a crop modeling approach as suggested by Hiler and

Table 8.7. Peanut yield, water use, and water use efficiency in response to irrigating at different soil water potentials (SWP)

Researchers	Cultivar	Irrigation at *SWP* (bars)	Pod Yield (kg ha^{-1})	Water Use (cm)	Water Use Efficiency (kg ha^{-1} cm^{-1})
Stansell et al.	Florunner	−0.2	5080	59	86
(30)		−0.6	4728	49	96
Pallas et al.		−15	3341	32	104
(60)		Overnight wilt	2375	22	108
	Florunner	−0.2	3820	46	83
Hammond et		−1.0	3000	40	75
al. (31)		None (wilted)	2260	33	68
	Florunner	−0.3	5120	—	—
Gorbet and		−0.6	5260	—	—
Rhoads (14)		Midday wilt	4840	—	—
		None	4690	—	—
	Tifspan	−0.2	4543	52	87
Stansell et al.		−0.6	4407	44	102
(30)		−15	3125	31	101
Pallas et al.		Overnight wilt	2491	22	113
(60)					

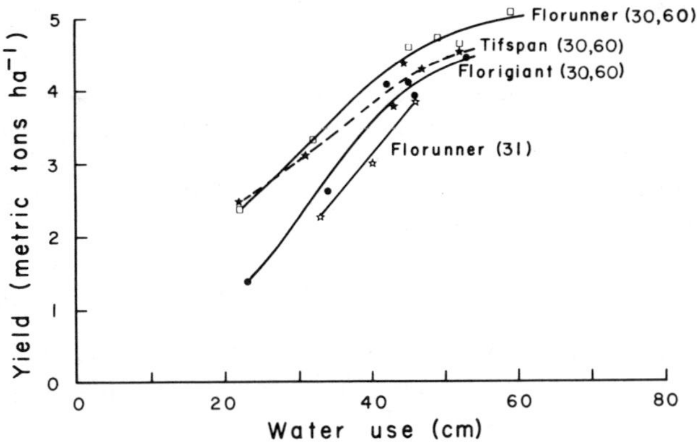

Figure 8.10. Peanut yield response to water use in controlled experiments. [After Stansell et al. (30), Pallas et al. (60), and Hammond et al. (31).]

Clark (87) which will account for crop susceptibility factors, diseases, insects, rooting depth, and soil characteristics in addition to real time weather data on rainfall, humidity, temperature, and solar radiation.

Optimum water management appears to be scheduling irrigation to maintain less than 50% *SWD* in the top 30 cm during early growth and possibly irrigating at 25% *SWD* during pod formation and filling. If soil water potential (*SWP*) is measured in the top 15 or 30 cm of soil, irrigation should be scheduled to maintain *SWP* above −0.6 bars although irrigating at a higher *SWP* may be desirable during long dry hot periods, especially during the sensitive growth stages of pegging, pod formation, and early pod fill.

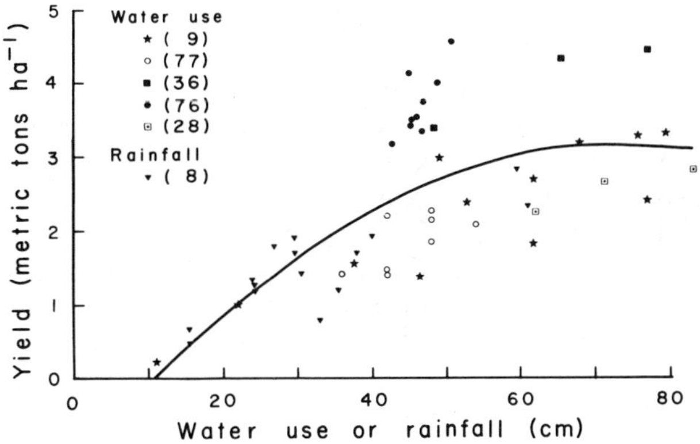

Figure 8.11. Yield response of short-season peanuts to water use or rainfall in a variety of field experiments. [After Bockelee-Morvan et al. (8), Lenka and Misra (28), Matlock et al. (9), Narasimham et al. (36), Reddi and Reddy (76), and Subramanian et al. (77).]

REFERENCES

1. W. C. Gregory, M. P. Gregory, A. Krapovickas, B. W. Smith, and J. A. Yarbrough, Structures and Genetic Resources of Peanuts, in *Peanuts—Culture and Uses*, American Peanut Research and Education Association, Stillwater, Okla., 1973, pp. 47–133.
2. H. Walter, E. Harnickell, and D. Mueller-Dombois, *Climate-Diagram Maps*, Springer-Verlag, New York, 1975.
3. R. O. Hammons, Early History and Origin of the Peanut, in *Peanuts—Culture and Uses*, American Peanut Research and Education Association, Stillwater, Okla., 1973, pp. 17–45.
4. *FAO Production Yearbook*, Vol. 33, FAO, United Nations, Rome, 1980.
5. *Agricultural Statistics 1980*, U.S. Dept. Agric., U.S. Govt. Printing Office, Washington, D.C.
6. S. S. Cheema, H. Kundra, and J. N. Kaul, *J. Res. Punjab Agric. Univ.* **11(4)**, 380 (1974).
7. S. S. Cheema, K. S. Minhas, H. P. Tripathi, and H. C. Kundra, *J. Res. Punjab Agric. Univ.* **14(1)**, 9 (1977).
8. A. Bockelee-Morvan, J. Gautreau, J. C. Mortreuil, and O. Roussel, *Oleagineux*, **29(6)**, 309 (1974).
9. R. S. Matlock, J. E. Garton, and J. F. Stone, *Oklahoma Agric. Exp. Stn. Bull.* B-580 (1961).
10. D. E. McCloud, *Soil Crop Sci. Soc. Fla. Proc.* **33**, 24 (1974).
11. R. L. McGraw, *Yield Dynamics of Florunner Peanuts* (Arachis hypogaea L.), M. S. Thesis, Univ. of Florida, Gainesville, 1977.
12. K. J. Boote, *Peanut Sci.* **9(1)**, in press (1982).
13. J. H. Williams, G. L. Hildebrand, and J. R. Tattersfield, *Rhodesian J. Agric. Res.* **16**, 193 (1978).
14. D. W. Gorbet and F. M. Rhoads, *Agron. J.* **67**, 373 (1975).
15. A. S. Vivekanandan and H. P. M. Gunasena, *Beitr. Trop. Landwirtsch. Veterinarmed.* **14(4)**, 369 (1976).
16. K. J. Boote, R. J. Varnell, and W. G. Duncan, *Soil and Crop Sci. Soc. Fla. Proc.* **35**, 47 (1976).
17. R. J. Varnell, H. Mwandemere, W. K. Robertson, and K. J. Boote, *Soil and Crop Sci. Soc. Fla. Proc.* **35**, 56 (1976).
18. F. R. Cox, *Diss. Abstr.* **22**, 3326 (1962).
19. C. V. Underwood, H. M. Taylor, and C. S. Hoveland, *Agron. J.* **63**, 953 (1971).
20. F. R. Cox, G. A. Sullivan, and C. K. Martin, *Peanut Sci.* **3(2)**, 81 (1976).
21. P. Gillier, *Oleagineux* **32(7)**, 79 (1969).
22. B. J. Skelton and G. M. Shear, *Agron. J.* **63**, 409 (1971).
23. A. S. Bhagsari, R. H. Brown, and J. S. Schepers, *Crop Sci.* **16**, 712 (1976).
24. H. N. An, *Light Intensity Effects on Metabolism, Growth, and Yield Components of Peanuts*, Ph.D. Thesis, University of Florida, Gainesville, 1978. *Diss. Abstr. Intl.* 39-3083-B.
25. A. C. Mixon, *Agron. J.* **63**, 336 (1971).
26. J. A. Yarbrough, *Am. J. Bot.* **36**, 758 (1949).
27. H. M. Taylor and L. F. Ratliff, *Soil Sci.* **108(2)**, 113 (1969).
28. D. Lenka and P. K. Misra, *Ind. J. Agron.* **18(4)**, 492 (1973).
29. W. K. Robertson, L. C. Hammond, J. T. Johnson, and K. J. Boote, *Agron. J.* **72**, 548 (1980).

30. J. R. Stansell, J. L. Shepherd, J. E. Pallas, R. R. Bruce, N. A. Minton, D. K. Bell, and L. W. Morgan, *Peanut Sci.* **3(1)**, 44 (1976).
31. L. C. Hammond, K. J. Boote, R. J. Varnell, and W. K. Robertson, *Am. Peanut Res. Educ. Assoc. Proc.* **10,** 73 (1978).
32. W. R. Stern, *Aust. J. Exp. Agric. Anim. Husb.* **8(34)**, 594 (1968).
33. H. C. Harris and R. W. Bledsoe, Physiology and mineral nutrition, in *The Peanut—The Unpredictable Legume*, Symposium sponsored and published by the National Fertilizer Association, Washington, D.C., 1951, pp. 89–121.
34. L. H. Allen, Jr., K. J. Boote, and L. C. Hammond, *Soil and Crop Sci. Soc. Fla. Proc.* **35,** 42 (1976).
35. H. Lin, C. C. Chen, and C. Y. Lin, *J. Agric. Assoc. China* **43,** 40 (1963).
36. R. L. Narasimham, I. V. Subba Rao, and M. Singa Rao, *Ind. J. Agric. Sci.* **47(11)**, 573 (1977).
37. A. Mantell and E. Goldin, *Isr. J. Agric. Res.* **14(4)**, 203 (1964).
38. R. O. Hammons, Genetics of Arachis Hypogaea, in *Peanuts—Culture and Uses*, American Peanut Research and Education Association, Stillwater, Okla., 1973, pp. 135–173.
39. J. C. Wynne, D. A. Emery, and R. J. Downs, *Crop Sci.* **13,** 511 (1973).
40. B. W. Smith, Am. J. Bot. **37(10)**, 802 (1950).
41. E. Zamski and M. Ziv, *Ann. Bot.* **40,** 631 (1976).
42. T. E. Slack and L. G. Morrill, *Soil Sci. Soc. Am. Proc.* **36(1)**, 87 (1972).
43. R. U. Schenk, *Georgia Agric. Exp. Stn. Tech. Bull. N.S.* **22** (1961).
44. K. J. Boote, *Am. Peanut Res. Educ. Assoc. Proc.* **9,** 39 (1977).
45. W. G. Duncan, D. E. McCloud, R. L. McGraw, and K. J. Boote, *Crop Sci.* **18,** 1015 (1978).
46. R. O. Slatyer, *Aust. J. Agr. Res.* **6,** 365 (1955).
47. P. Fourrier and P. Prevot, *Oleagineux* **13,** 805 (1958).
48. R. Ochs and T. M. Wormer, *Oleagineux* **14,** 281 (1959).
49. K. C. Su, T. R. Chen, S. C. Hsu, and M. R. Tseng, *J. Agric. Assoc. China* **45,** 31 (1964).
50. A. I. Il'ina, *Soviet Plant Physiol.* **5,** 253 (1958).
51. R. Billaz and R. Ochs, *Oleagineux* **16,** 605 (1961).
52. R. Billaz, *Oleagineux* **17,** 35 (1962).
53. B. W. Smith, *Am. J. Bot.* **41(8)**, 607 (1954).
54. Y. Ono, K. Nakayama, and M. Kubota, *Proc. Crop Sci. Soc. Jpn.* **43(2)**, 247 (1974).
55. L. C. Campbell, M. H. Miller, and J. F. Loneragan, *Aust. J. Plant. Physiol.* **2,** 481 (1975).
56. R. J. Jones, D. A. Ashley, and M. E. Walker, *Peanut Sci.* **3(2)**, 78 (1976).
57. R. W. Bledsoe, C. L. Comar, and H. C. Harris, *Science* **109,** 329 (1949).
58. L. K. Wiersum, *Plant Soil* **3,** 160 (1951).
59. M. E. Walker, *Commun. Soil Sci. Plant Anal.* **6,** 299 (1975).
60. J. E. Pallas, Jr., J. R. Stansell, and R. R. Bruce, *Agron. J.* **69,** 381 (1977).
61. R. S. Sandhu, J. S. Saini, and T. Singh, *J. Res. Punjab Agric. Univ.* **9(4)**, 535 (1972).
62. J. S. Saini and R. S. Sandhu, *J. Res. Punjab Agric. Univ.* **10(2)**, 179 (1973).
63. J. Iyama and Y. Murata, *Proc. Crop Sci. Soc. Jpn.* **29,** 350 (1961).
64. J. E. Pallas, Jr., J. R. Stansell, and T. J. Koske, *Agron. J.* **71,** 853 (1979).
65. J. Gautreau, *Oleagineux* **32(7)**, 323 (1978).
66. Y. Tsuno, *Proc. Crop Sci. Soc. Jpn.* **44(1)**, 44 (1975).

67. C. Y. Chen and H. S. Chang, *J. Agric. Assoc. China* **88,** 1 (1974).
68. T. M. Wormer and R. Ochs, *Oleagineux* **14,** 571 (1959).
69. J. E. Pallas, Jr. and Y. B. Samish, *Crop Sci.* **14,** 478 (1974).
70. J. E. Pallas, Jr., Y. B. Samish, and C. M. Willmer, *Plant Physiol.* **53,** 907 (1974).
71. J. E. Begg and B. W. R. Torssell, in R. L. Bieleski, A. R. Ferguson, and M. M. Cresswell, Eds., *Mechanisms of Regulation of Plant Growth,* Bull. 12, R. Soc. N. Z., Wellington, 1974, pp. 277–283.
72. R. L. Satter and A. R. Galston, *Plant Physiol.* **48,** 740 (1971).
73. V. M. Reddy and J. W. Tanner, *Peanut Sci.* **7(2)**, 114 (1980).
74. P. Gillier, *Oleagineux* **33(1)**, 25 (1978).
75. A. A. Mehkri, V. R. Sashidhar, M. Udaykumar, and K. S. Krishna Sastry, *Indian J. Plant Physiol.* **20(1)**, 50 (1977).
76. G. H. S. Reddi and M. N. Reddy, *Mysore J. Agric. Sci.* **11(1)**, 22 (1977).
77. S. Subramanian, S. D. Sundarsingh, K. P. Ramaswamy, S. P. Packiaraj, and K. Rajagopalan, *Madras Agric. J.* **62(9)**, 587 (1975).
78. B. Klepper, Water Relations of Peanut Plants, in *Peanuts—Culture and Uses,* American Peanut Research and Education Association, Stillwater, Okla., 1973, pp. 265–269.
79. P. A. Backman, E. W. Rochester, and J. M. Hammond, *Am. Peanut Res. Educ. Assoc. Proc.* **10,** 68 (1978).
80. A. H. Kassam, J. N. Kowal, and C. Harkness, *Trop. Agric.* (*Trinidad*), **52(2)**, 105 (1975).
81. J. M. Davidson, J. E. Garton, R. S. Matlock, D. Schwab, J. F. Stone, and L. D. Tripp, Irrigation and Water Use, in *Peanuts—Culture and Uses,* American Peanut Research and Education Association, Stillwater, Okla., 1973, pp. 361–382.
82. S. D. Goldberg, B. Gornat, and D. Sadan, *Soil Sci.* **104(4)**, 289 (1967).
83. E. W. ChinChoy, J. F. Stone, and J. E. Garton, *J. Soil Sci. Soc. Am.* **41(2)**, 428 (1977).
84. S. Bhan and D. K. Misra, *Indian J. Agron.* **15(3)**, 258 (1970).
85. J. E. Pallas, Jr. and J. R. Stansell, *Oleagineux* **33(5)**, 235 (1978).
86. J. R. Stansell, *Southeastern Peanut Farmer* **13(3)**, 12 (1975).
87. E. A. Hiler and R. N. Clark, *Trans. ASAE* **14(4)**, 757 (1971).

9

POTATO

Gale E. Kleinkopf

INTRODUCTION

The potato, *Solanum tuberosum* L., is an annual herbaceous dicot belonging to the Solanaceae family and closely related to other important agricultural crops such as tobacco, tomato, eggplant, and pepper.

The potato originated in South America near Peru or Bolivia (1) but was introduced to the European continent by the Spanish and subsequently distributed so widely throughout Western Europe that it was accepted as a staple food in Europe by the late 1700's. The potato was introduced into America from Ireland, thus the origin of the term "Irish" or white potato. From the first established settlement at Londonderry, New Hampshire, potato production expanded into all areas of the United States. During the early years commercial potato production was centered primarily in New England and the mid-Atlantic states. As new lands opened to the west, potato production was expanded and potatoes are now grown commercially in every state.

In dollar value the potato ranks eighth of all field crops and first for crops grown directly for food in the United States. According to the

Table 9.1. 1979 Potato Statistics[a]

State	Hectares Harvested (× 1000)	Yield (Tons/ha)	Production (× 1000 tons)
California	6.9	40.7	700
Colorado	16.0	31.9	1,260
Connecticut	0.8	24.2	46
Idaho	133.6	29.4	9,702
Indiana	1.7	28.0	118
Maine	46.6	27.5	3,162
Massachusetts	1.4	24.2	82
Michigan	12.9	27.5	880
Minnesota	27.5	20.9	1,421
Montana	3.0	26.4	198
Nebraska	2.3	28.6	163
Nevada	6.0	36.3	544
New York	18.3	31.3	1,418
North Dakota	46.1	17.6	2,006
Ohio	4.1	26.4	264
Oregon	25.5	42.9	2,784
Pennsylvania	9.7	27.5	660
Rhode Island	1.6	25.3	98
South Dakota	2.6	20.3	132
Utah	2.1	27.5	143
Vermont	0.3	23.1	16
Washington	41.7	52.2	5,381
Wisconsin	21.8	34.6	1,871
Wyoming	2.1	24.2	125
Total 25 Fall States	434.9	30.9	33,179
Total United States	517.7	29.9	38,241

[a] From USDA Crop Reporting Board (2).

USDA Crop Reporting Board (2) the 1979 potato crop amounted to 517,000 hectares harvested with an average yield of 29 tons per hectare. Table 9.1 lists the potato producing states with their production totals. Idaho, Washington, Maine, and Oregon are the four major potato growing states. Potatoes are grown in many dryland areas, however; irrigated potatoes make up about 50% of the total acreage planted in the United States. This chapter deals with water relations based on field grown, irrigated potatoes with minor reference to dryland conditions.

Today the majority of our potatoes are processed with only 45% of the total potato production consumed as fresh product. The 56 kg annual per capita consumption in 1977 includes 18 kg frozen, 7 kg chips, 4 kg dehydrated, and 1 kg canned. Although the major cultivars grown in America today number well over 100 three or four cultivars make up about 70% of the total production. "Russet Burbank," a late season white, elongated type, is the most widely produced potato cultivar, with production centered in Idaho, Washington, and Oregon. The next most widely produced is "Katahdin," a late season, oval, smooth skinned, white type grown principally in Maine and New England states. Next is "Kennebec," a round, large, white skinned late season cultivar grown mostly for chipping. The last major cultivar produced is the red-skinned type, "Red Pontiac" (3). Other cultivars are grown in lesser quantity throughout the United States with cultivar selection governed mainly by processor and consumer demand and only to a lesser extent by environmental restraints. Potatoes grow best under conditions of uniform and high water availability, warm days, and cool nights. For detailed reviews of potato growth physiology see Harris (4), Burton (5), Moorby and Milthrope (6), and Moorby (7).

POTATO GROWTH STAGES

Based on morphology of the shoot and tuber development, the growth of potato cultivars can be divided into four distinct stages, although Sparks (9) described nine stages of development of the "Russet Burbank" cultivar based on nodal development of the shoot and the number of inflorescences. With reference to water use and tuber development, however, the four-stage description is adequate.

Stage 1: Early Vegetative

Most potato production begins with the planting of tubers or pieces of tubers called seed pieces. This is true vegetative propagation. The flowers and fruits are of interest primarily to the breeder and are not important commercially. Growth stage 1 includes early plant development from planting to the initiation of tubers. This stage varies from 30 to 60 days depending on potato cultivar and environmental conditions. After planting one or more bud or "eye" will sprout and adventitious roots quickly develop as the sprout emerges. The seed piece serves as a source of food for use by the young plant during early development. The stem will develop approximately six nodes during this stage. Rhizome formation and elongation is initiated during this stage and root distribution is quite extensive within the soil 20–30 cm from the seed piece but tubers are not present.

Stage 2: Tuberization

From the start of tuber formation, the majority of tubers are initiated over a 2 week period, but there may be additional tuber initiations into midseason. Stem elongation continues with 10 or more nodes discernible on the stem and floral buds in evidence. Leaf area index (*LAI*) is between 1 and 2. Root development also continues and maximal penetration and density of roots is reached during this stage.

Stage 3: Tuber Bulking

The growth stage including the period of tuber size increase to near maturity is called tuber bulking. Tuber enlargement is nearly linear for the 30–60 days of bulking. Stem growth continues with some branching and flowers appear on main and branched stems. Depending on variety, 15–30 nodes may develop. Growth habit becomes more prostrate due to vine lodging and elongation. *LAI* reaches a maximum (3.5–6.0) and then declines to near 1 as older leaves are abscised. In indeterminant cultivars but not in early or determinant cultivars new leaves replace those lost.

Stage 4: Maturity

Representing the final 10–24 days of growth, maturity is characterized by senescence of the shoot, concomitant with a decline in leaf, stem, and root dry weight. Tubers reach maximum development primarily from translocation of nutrients out of the shoot although early in this stage some photosynthetic activity remains. Early cultivars reach maturity 90–100 days after planting while late season cultivars may take 150 or more days.

GROWTH AND DEVELOPMENT

The potato is normally branched but may be single stemmed. Diagrammatic representation of a potato plant is shown in Fig. 9.1. Each seed

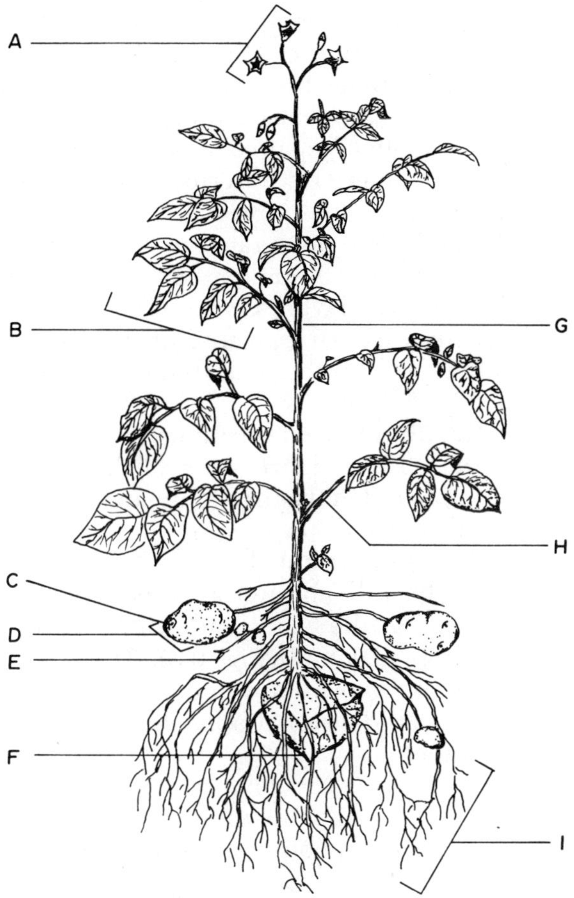

Figure 9.1. Diagrammatic illustration of a potato plant, *Solanum tuberosum L.:* (A) inflorescence, (B) leaf, (C) bud "eye," (D) tuber, (E) rhizome, (F) seed piece, (G) stem, (H) petiole, (I) fibrous root system.

piece is capable of producing more than one individual stem. In the "Russet Burbank" cultivar, stem growth will continue with two or three periods of floral development. The first inflorescence occurs terminally on the main stem with subsequent inflorescences occurring terminally on secondary branches of the main stem (9). It is not uncommon for stem growth to exceed 2 m on vining indeterminant potatoes such as "Russet Burbank."

ROOT GROWTH AND DISTRIBUTION

The root system of the potato is fibrous and all roots are adventitious. Although "Russet Burbank" does not have an extensive rooting system and is considered shallow rooted compared to other field crops such as cereal grains or alfalfa (10) rooting depth may be as great as 120 cm in muck soils (11). Rooting profiles show the majority of the root mass to be confined to the plow layer (about 30 cm under normal tillage practices) and the effective rooting depth of potato plants is considered to be 40–60 cm (12). Lesczynski and Tanner (13) described the seasonal root distribution patterns of field grown "Russet Burbank" potatoes (Fig. 9.2). Two weeks after emergence roots had penetrated to 20 cm. During continued seasonal root expansion and development 15% of the roots had penetrated below the 30 cm tillage layer. These deeper roots may account for expansion of water supply during periods of stress even though they only constitute a small amount of the total root mass, root density being maximal in the 10 to 20 cm depth.

In another study (12), 40 different locations in southern Idaho were examined for rooting depth and distribution of the "Russet Burbank" cultivar. These authors concluded that maximum density and penetration of roots was limited to the surface 30 cm layer. Rooting depth was limited to the tillage layer (35 cm) in all but three medium textured soil sites and in all sites with sandy soil. The fibrous root system is established early in the development of the plant and within seven weeks after planting root dry matter is maximal (Fig. 9.3). This high density root mass is capable of efficient removal of water and nutrients from the soil even though the total available water supply may be decreased because of shallow root penetration.

Fulton (14) compared several vegetable crops for root extension and soil water levels required for maximum yields. Potatoes were more sensitive to soil water stress than any other crop studied probably because

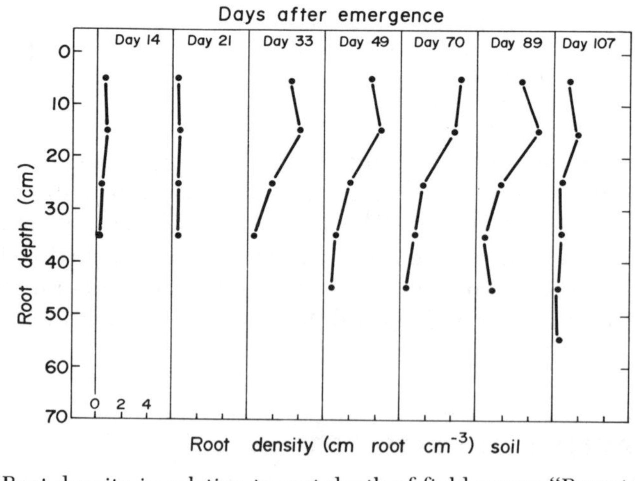

Figure 9.2. Root density in relation to root depth of field grown "Russet Burbank" potatoes. [After Lesczynski and Tanner (13).]

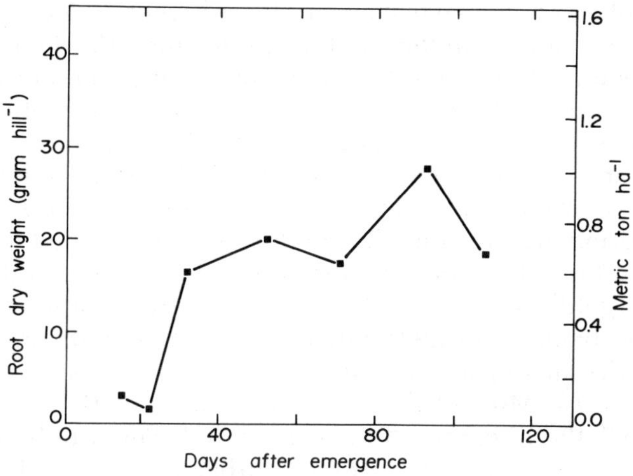

Figure 9.3. Root dry matter accumulation in the 0–30 cm profile of field grown "Russet Burbank" potatoes. Roots below the 0–30 cm depth contributed a maximum of 15% of the total root mass at all sampling periods. [After Lesczynski and Tanner (13).]

of this limited rooting depth. If soil water potential decreased to −0.5 bar, potato yield was adversely affected whereas much lower soil water potentials were required before yields were decreased for other crops.

SHOOT AND TUBER GROWTH AND DEVELOPMENT

LAI increases with time and added nitrogen in "Russet Burbank," a late season cultivar, and "Norgold," an early season cultivar (Fig. 9.4). As

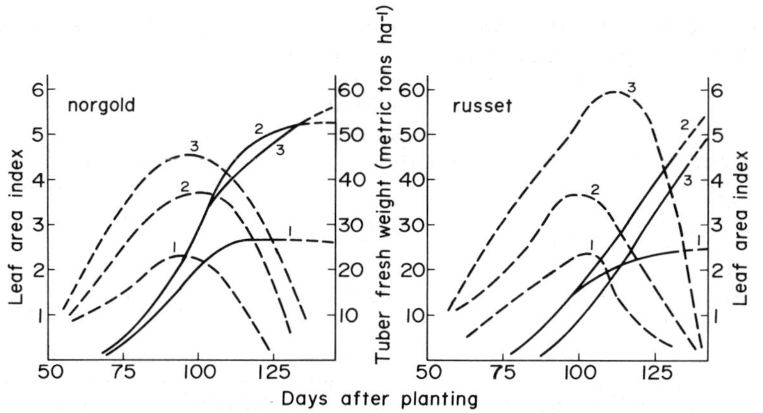

Figure 9.4. Seasonal profile of leaf area index (dashed lines) and tuber fresh weight (solid lines) of "Norgold" and "Russet Burbank" potatoes (1 = low, 2 = recommended, and 3 = excess nitrogen application). [After Kleinkopf and Dwelle (16).]

shown in Table 9.2 maximal leaf area index is higher under well-irrigated treatments than under limited water application (16). Cultivar differences have been noted by Epstein and Grant (15) who reported that leaf area for the "Russet Burbank" is about 1.5 times that of "Katahdin" during midseason. Excessive nitrogen fertilization can cause undesirable vine growth and concomitant delay of tuber bulking. This delay of tuber enlargement shortens the growing season in areas where frost or pathogens may terminate the crop. The end result is decreased tuber yields. The length of the tuber bulking period in the early season determinant cultivar "Norgold" is less sensitive to high nitrogen applications than is "Russet Burbank," an indeterminant late season cultivar.

Seasonal dry matter accumulation of "Russet Burbank" potatoes is shown in Fig. 9.5. Most of the dry matter accumulation occurs during growth stage 3, the linear tuber bulking period. Tuber bulking rates can be 600–800 kg ha^{-1} day^{-1} in "Russet Burbank" and 800–1200 kg ha^{-1} day^{-1} in "Norgold" (15). Primary leaf and stem production occur early in stages 1 and 2 with maximum accumulation 100 days after planting. The continued seasonal increase in dry matter results from increasing tuber size. During growth stage 4 decreasing shoot and root dry matter are due to translocation of nutrients from these parts to the storage organs.

Patterns of Water Usage

Nelson and Hwang (8) reported water usage by potato plants at different growth stages (Fig. 9.6). Transpiration increased during growth stage 1 and 2, peaked early in stage 3, and remained relatively constant during tuber bulking and declined near maturity. These data show that water utilization by potatoes remained uniformly high during growth stage 3,

Table 9.2. Effect of suboptimal water application on yield, quality, and leaf area index (LAI) of "Russet Burbank" potatoes[a]

Total Water Applied (cm)	Maximum *LAI*	Dry Matter Accumulation (Metric tons ha^{-1})	Yield Fresh Weight (Metric tons ha^{-1})	US #1 (%)
60	6.2	14.3	43.0	77
55	6.0	13.2	42.0	71
50	5.7	12.8	37.9	65
45	5.1	13.9	37.0	58
40	4.5	11.1	39.8	55
35	4.3	11.1	33.9	55

[a] After Kleinkopf and Dwelle (16). Water applied is 100% maximum seasonal E_t at 60 cm to 58% maximum seasonal E_t at 35 cm; *LAI* is maximum leaf area attained in each treatment; dry matter accumulation in total above and below ground components; yield is final season harvestable tubers only; and US #1 is percent of final yield in grade.

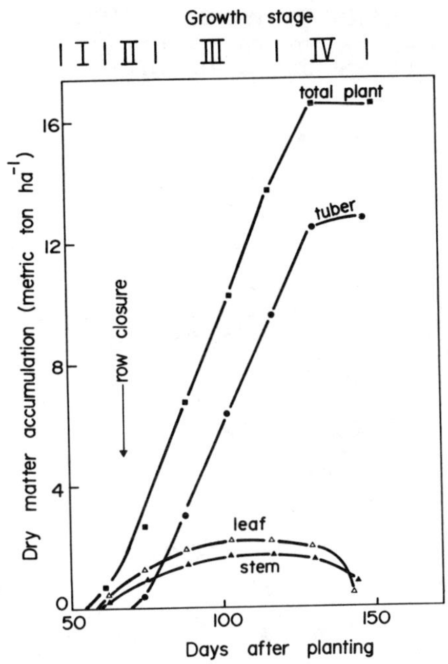

Figure 9.5. Seasonal dry matter accumulation (total, leaf, stem, tuber) of "Russet Burbank" potatoes. [After Kleinkopf and Dwelle (16).]

the stage at which soil water deficits have the greatest effect on tuber quality.

Water Stress Effects on Tuber Quality and Yield

During the rapid vegetative development of the potato (stage 2 and early stage 3) the yield of tubers is very sensitive to changes in soil water

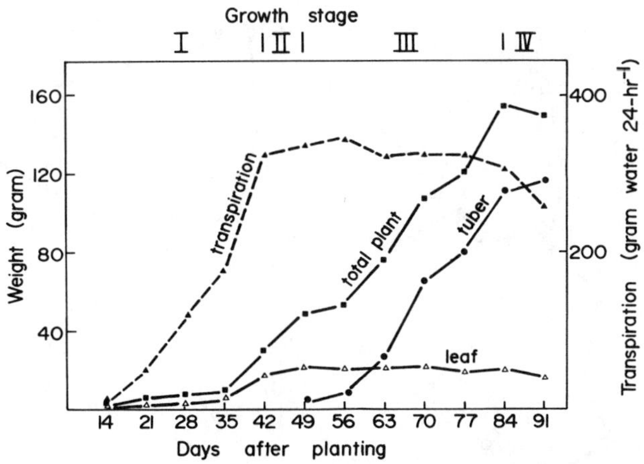

Figure 9.6. Seasonal transpiration and dry matter accumulation (total, leaf, tuber) in "Norland" potatoes in controlled environment chambers. [After Nelson and Hwang (8).]

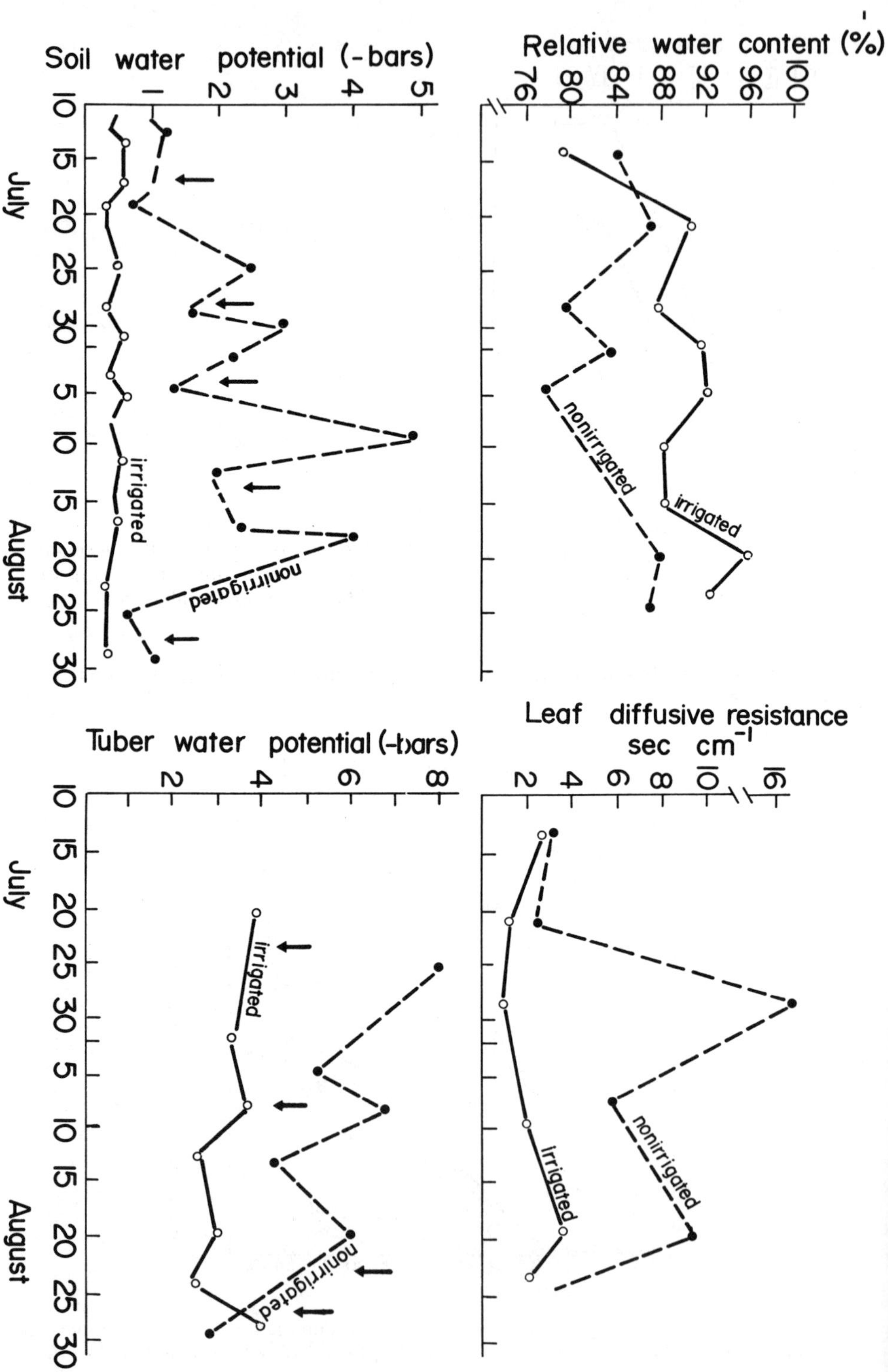
Relative water content (%)
Leaf diffusive resistance sec cm^{-1}
Soil water potential (-bars)
Tuber water potential (-bars)
irrigated
nonirrigated
July
August

potential. Periods of moderate water stress can result in decreased tuber yield or more importantly tuber quality (18–20). Struchtmeyer (21) showed supplemental irrigation to normal dryland production increased yield and tuber number of "Katahdin" potatoes in Maine. Prince and Blood (22), working with "Kennebec" potatoes, showed tuber yields and quality were increased by maintaining 50% available soil water through all irrigations. Table 9.2 shows the changes in yield and quality of field grown "Russet Burbank" potatoes with decreasing water application. Changes in tuber quality were evident with minimum decreases in applied water before total dry matter accumulation declined.

Prior to tuber set water stress affects vegetative growth (23), but after tuber set periods of water stress affect tuber yield and quality (24). During the early tuber set period (growth stage 2) tuber quality is affected more by plant water stress than is total yield (25–28). Increased incidence of second growth, bottleneck tubers, sugar end, and misshapen tubers can result from insufficient water during growth stages 2 and 3. Kleinkopf (29) was able to show that although the total water application given two treatments were equal the percentage of high quality tubers was decreased 30% by allowing the soil water to be depleted to 40% twice during growth stage 2. Other environmental conditions such as high nitrogen application and high temperatures in combination with water stress can also cause tuber quality to decline (20). Not all potato cultivars are as sensitive to water stress as "Russet Burbank," however. Early season cultivars and cultivars with round, less elongated shape appear to resist irregular growth patterns during periods of water stress.

Mechanisms and Critical Levels for Water Stress Effects

Despite the extensive evidence that potatoes are sensitive to water stress, little is known about the various mechanisms by which water deficits affect plant yield. Tuber growth is effectively stopped during periods of mild water deficit. A sudden increase in soil water level permits a continuation of growth although this "regrowth" can result in decreased tuber quality.

The amount of water stress required to cause growth restrictions affecting tuber quality is also relatively unknown. The work of Epstein and Grant (15) showed a high correlation between tuber water potential, leaf water potential and soil water potential (Fig. 9.7). But Ackerson et al. (30) did not find a close relation between leaf water potential, osmotic

Figure 9.7. The effect of soil water stress and precipitation on soil water potential, tuber water potential, relative water content, and leaf diffusive resistance of "Russet Burbank" potato plants. Arrows represents rainfall at Presque Isle, Maine, 1970. [After Epstein and Grant (15).]

potential, or soil water potential. The tuber is botanically a specialized stem structure with water movement into and out of the tuber reflecting the water balance of the plant. The tuber may act as a water storage reservoir of the plant such that during periods of water stress, declining water potential in the leaves causes water to move from the roots, stems, and tubers into the leaves. This is enough to stop tuber growth and affect tuber quality upon resumption of favorable growth conditions.

Water Stress and Leaf Growth

Gandar and Tanner (31) measured the relative growth rates of potato leaves during periods of decreasing soil water (Fig. 9.8). As leaf water potential declined below −2 bars, leaf growth decreased and was effectively stopped at leaf water potentials of −5 bars. Both leaf and tuber growth stopped when leaf water potential declined below −7 bars (32). In a comparison of stressed and unstressed potatoes, stressed plants missing only a single irrigation showed a large decrease in leaf length change compared to adequately watered controls. Even after rewatering the stressed plants were unable to recover the growth restrictions caused by the irrigation treatment (Fig. 9.9).

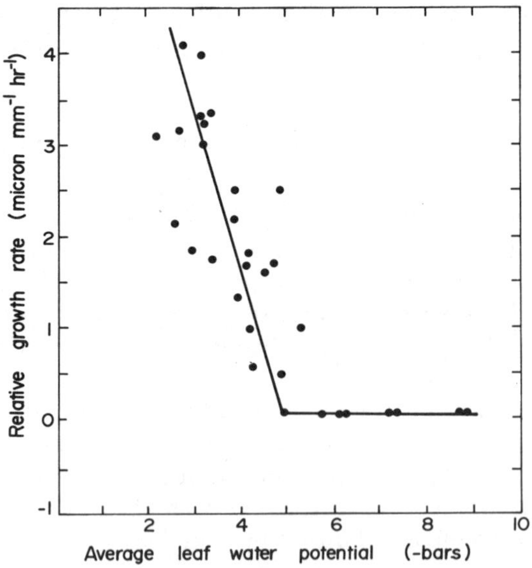

Figure 9.8. Relative leaf growth rate versus leaf water potential of potato. [After Gandar and Tanner (30).]

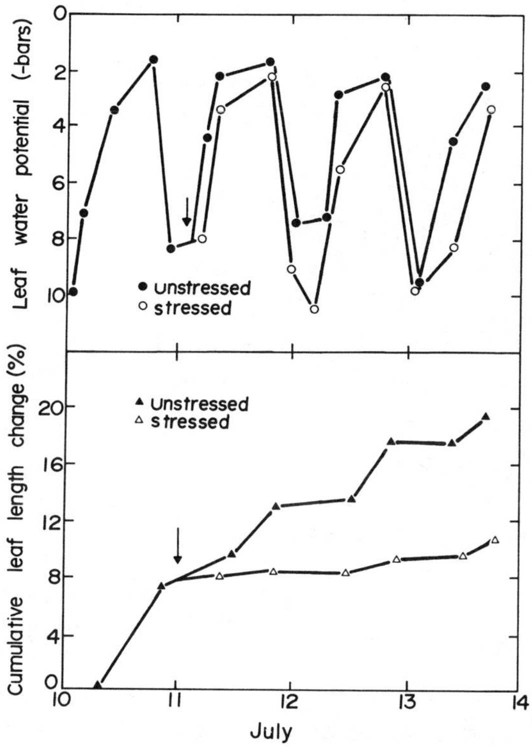

Figure 9.9. The relation between leaf water potential and cumulative leaf length change (%) of diurnal cycles of stressed and unstressed potato plants. The arrow indicates onset of stress period. [After Gandar and Tanner (30).]

PHYSIOLOGICAL RESPONSE TO WATER STRESS

Stomatal Responses

During periods of water stress, photosynthesis and transpiration can be reduced if leaf water potentials are lowered to a point where guard cells lose turgidity and close. Epstein and Grant (15) showed differences in leaf diffusive resistance between irrigated and nonirrigated potatoes (Fig. 10.7). Stomata of well-watered potatoes stay open all day and possibly at night (33) even under mild stress. Potato stomata have been observed to close during the day (15, 34). When droughted potatoes are rewatered, Epstein and Grant (15) observed that stomata did not reopen immediately even though the leaf water content recovered. For some cultivars, the recovery time was 48 hr after moderate stress.

Stomatal density of the abaxial surface of potato leaves is about 3 times the stomatal density of the adaxial surface (16, 120 and 5115 stomata per

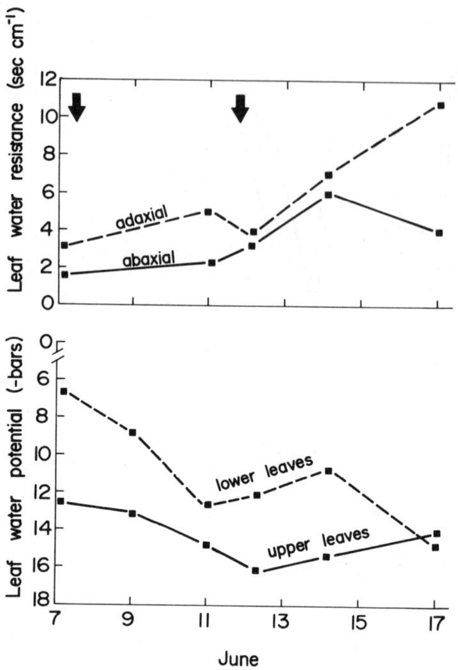

Figure 9.10. Leaf resistance (adaxial and abaxial) and leaf water potential of fully expanded upper and lower leaves of well-watered "Viking" potatoes at Lubbock, Texas, 1976. Arrows represent time of irrigation. [After Ackerson et al. (32).]

cm^2 respectively) (35). Ackerson et al. (30) have shown that the adaxial surface is usually the surface of greatest resistance (Fig. 9.10) and is more responsive to decreasing leaf water potential (Fig. 9.11) than the abaxial leaf surface. The stomata of the adaxial leaf surface begin closing at about −8 bars leaf water potential compared to −12 bars for the abaxial surface (Fig. 9.11).

Shepard (35) found stomatal closure at leaf relative water contents of

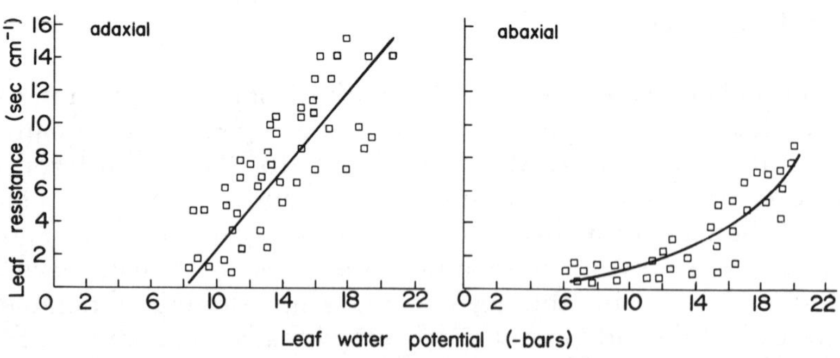

Figure 9.11. Relationship between adaxial and abaxial stomatal resistance and leaf water potential of potato. Values represent upper and lower leaves at radiation levels sufficient (≃1000 $\mu E\ m^{-2}\ sec^{-1}$) to saturate the opening response at Lubbuck, Texas, 1976. [After Ackerson et al. (32).]

85% based on internal conductance measurements. Similar responses have been shown by Stegman and Nelson (37). Epstein and Grant (15) examined the water relations of potatoes under field conditions and showed that leaf resistance increased with decreasing soil water potential (Fig. 9.7). Moorby et al. (34) reported that leaf resistances decreased with leaf age, but did not observe any effects of increasing leaf age on leaf water potential. However, Ackerson et al. (30) found that upper leaves (younger leaves) of potato developed more negative leaf water potentials than lower leaves (older leaves) in irrigated (Fig. 9.10) and nonirrigated (not shown) potatoes.

Photosynthetic Responses

During periods of water stress, yields can be reduced if stomatal closure affects the photosynthate supply or translocation process. Ackerson et al. (30) reported a decrease in photosynthetic rate with a decline in leaf water potential. Similar results have been reported by Moorby et al. (34) (Fig. 9.12).

Developmental stage is also an important factor affecting photosynthetic rates, however. Moorby (7) reported an increase in the photosynthetic rate at about the time of tuber initiation (growth stage 2). The net assimilation rate then declined to a minimum value at maximum leaf area production and finally increased at late growth stage 3. The increase in the late season photosynthetic rate was probably due to an increased demand for photosynthate by the tubers (38–40). The maximum photosynthetic rate in potatoes responded to changes in water supply or temperature, but appeared to be controlled by the sink size of the de-

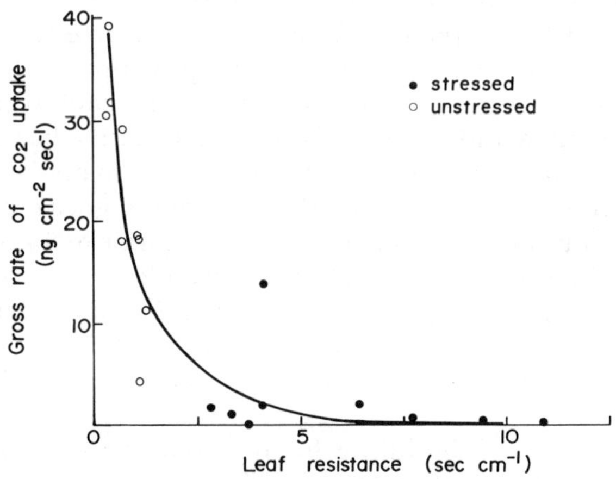

Figure 9.12. Rate of carbon dioxide uptake versus leaf resistance in stressed and unstressed potato plants. [After Moorby et al. (34).]

veloping tubers (7, 41) rather than by short-term changes in radiation level (39).

Temperature also affected the photosynthetic rate. Increasing temperatures resulted in an increase in the net photosynthetic rate becoming maximum from 20 to 30°C, then declining to zero at 36–38°C (42).

Enzyme Activities

The effects of leaf water status on enzyme activities in the potato leaf have been reported by Ackerman et al. (30) and Moorby et al. (34). Activities of ribulose bisphosphate carboxylase and phosphoenolpyruvate carboxylase in well-watered plants were comparable to those of other C_3 species but when leaf water potential declined -16 to -17 bars the activity of both enzymes was substantially reduced (32). Under 5 days of mild stress, however, Moorby et al. (34) did not see a change in the activities of either ribulose bisphosphate carboxylase or carbonate dehydratase; there was, however, less activity of both enzymes in the older leaves when compared to younger leaves. Other biochemical changes observed were an increase in the concentration of ethanol-soluble sugars in the leaves of droughted plants and a decrease of HCl-soluble polysaccharides (34).

IRRIGATION

From the above discussion it becomes apparent that soil water levels are important determinants of potato yields. Short-term stress may affect tuber quality without affecting total yields; however, net economic returns to the grower will be lower if quality factors are used in price determinations. Because potato yields are so sensitive to water deficits, many growers are relying on irrigation scheduling programs to help make on-farm water management decisions. Irrigation scheduling provides a prediction of the time and amount of the next irrigation based on climatic data. These scheduling programs can be used for many crops including potatoes and are gaining widespread favor among growers.

The scheduling programs make use of computer technology to solve equations developed to relate yield and evapotranspiration (E_t). Many current programs are using the USDA-SEA Computerized Irrigation Scheduling Program (43, 44) or modifications of it (45–48). An estimated 300,000 ha were under some type of an irrigation scheduling service in 1978.

Inputs to irrigation scheduling for potatoes are similar to those for other crops but higher soil water levels must be maintained. The first step in predicting irrigation frequency is to relate soil water use or eva-

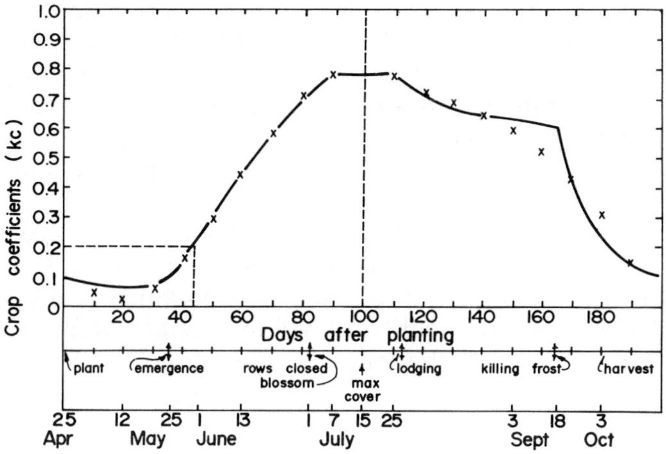

Figure 9.13. Adjusted crop coefficient curve for potatoes based on calculated crop coefficients (K_c, represented as X) at Kimberly, Idaho, 1972. [After Wright (49).]

potranspiration (E_t) to potential evapotranspiration (E_0) estimates. These relationships provide crop coefficients, K_c, (Fig. 9. 13) which adjust for the difference (dependent on stage of development) observed between E_t and E_0 (Fig. 9.14). E_0 estimates and K_c values are used in the computer programs to adjust water requirements of the crop to seasonal demand, precipitation, runoff, and available water holding capacity of the soil.

A seasonal evapotranspiration curve for potatoes is shown in Fig. 9.14.

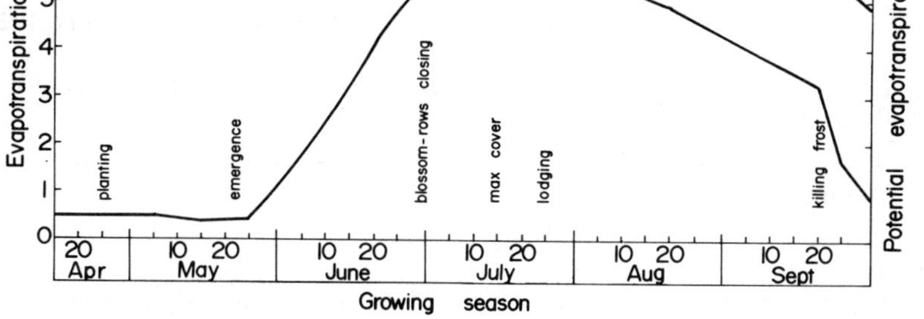

Figure 9.14. Seasonal evapotranspiration (E_t) and potential evapotranspiration (E_0) curves for "Russet Burbank" potatoes during the 1972 growing season at Kimberly, Idaho. [After Wright (49).]

The seasonal evapotranspiration of a well-watered potato crop at Kimberly, Idaho, is approximately 560 mm, which is about half of the potential evapotranspiration, 1166 mm. Use of these predictions in irrigation can result in increased water application efficiency and improved water management. The cost of the irrigation scheduling service can be recovered from decreased water costs or increased yield and quality of the harvested crop.

For some crops mild stress periods during noncritical stages of growth may reduce the total crop water use without loss in grain yield, thus reducing field water costs, but in potatoes suboptimal water application always reduces yield and quality. Development of water stress-tolerant potato cultivars remains a challenge for plant breeders.

REFERENCES

1. F. J. Stevenson. *Econ. Bot.* **5(2)**, 153–171 (1951).
2. USDA Crop Reporting Board, Production Forecasts, Economic and Cooperatives Service, Washington, D.C. (1979).
3. O. Smith, Ed. *Potatoes: Production, Storing, Processing.* AVI Publ., Westport, Connecticut, 1968.
4. P. M. Harris, Ed. The *Potato Crop. The Scientific Basis for Improvement*, Chapman and Hall, London (1978), 730 pp.
5. W. G. Burton, *The Potato: A Survey of Its History and of Factors Influencing Its Yield, Nutritive Value, Quality and Storage*, 2nd ed. H. Venman and N. V. Zoren, Wageningen, The Netherlands, 1966, 382 pp.
6. J. Moorby and F. L. Milthorpe. The Potato (*Solanum tuberosum* L.), in L. T. Evans, Ed., *Crop Physiology: Some Case Histories*, Cambridge Univ. Press, 1975, pp. 225–257.
7. J. Moorby. The Physiology of Growth and Tuber Yield, in *The Potato Crop: The Scientific Basis for Improvement.* Chapman and Hall, London, 1978, pp. 153–194.
8. S. H. Nelson and K. E. Hwang. *Am. Potato J.* **52,** 331–339 (1975).
9. W. C. Sparks, Current Information Series 186. University of Idaho, Moscow, Idaho. 1972.
10. H. C. DeRoo and P. E. Waggoner, *Agron. J.* **53,** 15–17 (1961).
11. J. E. Weaver, *Root Development of Field Crops*, McGraw-Hill, New York, 1926, 291 pp.
12. K. Linford and R. E. McDole, *Proc. 28th Annu. Fertilizer Conf.-Pac. Northwest,* Twin Falls, Idaho, 1977.
13. D. B. Lesczynski and C. B. Tanner, *Am. Potato J.* **53,** 69–78 (1976).
14. J. M. Fulton, *Can. J. Soil Sci.* **50,** 92–94 (1970).
15. E. Epstein and W. J. Grant, *Agron. J.* **65,** 400–04 (1973).
16. G. E. Kleinkopf and R. B. Dwelle. Nitrogen Uptake Efficiency by Potato Varities, in *Proc. 10th Annual Idaho Potato School.* Univ. Idaho Coll. Agric., Moscow, Idaho, 1978.
17. J. D. Ivins and D. M. Bremner. *Outl. Agric.* **4,** 211–217 (1964).
18. W. C. Sparks, *Am. Potato J.* **35,** 430–436 (1958).

19. W. M. Iritani and L. Weller, *Am. Potato J.* **50,** 389–97 (1973).
20. C. G. Painter, D. O. Everson, A. J. Walz, R. R. Romanko, A. Czernik, J. R. Jaeger, N. A. Henninger, and D. A. Gross, *Univ. Idaho Misc. Ser.* **24,** Moscow, Idaho, 1975.
21. R. A. Struchtmeyer, *Am. Potato J.* **38,** 22–24 (1961).
22. A. B. Prince and P. T. Blood, *Am. Potato J.* **39,** 313–319 (1962).
23. J. B. Cavagnaro, B. R. de Lis, and R. M. Tizio, *Potato Res.* **14,** 181–192 (1971).
24. S. T. Jones and W. A. Johnston, *Proc. Am. Soc. Hortic. Sci.* **71,** 440–445 (1958).
25. J. S. Robins and C. E. Domingo, *Agron. J.* **48,** 488–492 (1956).
26. G. L. Corey and V. I. Meyers, *Irrigation of Russet Burbank Potatoes in Idaho,* Univ. Idaho Agric. Exp. Stn. Bull. 246, 1955, 18 pp.
27. R. W. Chase, E. H. Kidder, and R. J. Kunze, *Effect of Plant Environment Changes on Potatoes,* Mich. Agric. Exp. Stn. Res. Rep. 94, 1969, 12 pp.
28. B. R. de Lis, I. Ponce, and R. M. Tizio, *Agron. J.* **56,** 377–381. (1964).
29. G. E. Kleinkopf, Translucent End of Potatoes, in *Proc. 8th Annu. Idaho Potato School,* University of Idaho College of Agriculture, Moscow, Idaho, 1976.
30. R. C. Ackerson, D. R. Krieg, J. D. Miller, and R. G. Stevens, *J. Am. Soc. Hortic. Sci.* **102(5),** 572–575 (1977).
31. P. W. Gandar and C. B. Tanner, *Crop Sci.* **16,** 534–538 (1976).
32. M. D. Campbell, *The Lower Limit of Soil Water Potential for Potato Growth,* Ph.D. Thesis, Washington State Univ., Univ. Microfilms, Ann Arbor, Mich., 1972.
33. J. V. G. Loftfield, *The Behavior of Stomata.* Carnegie Institute Washington Publ. 314, 1921.
34. J. Moorby, R. Munns, and J. Walcott. *Aust. J. Plant Physiol.* **2,** 323–33 (1975).
35. E. C. Miller, *Plant Physiology,* 2nd ed., McGraw-Hill, New York, 1938, p. 422.
36. W. Shepard, Agric. Meteorol. **11,** 213–222 (1972).
37. E. C. Stegman and D. C. Nelson, *N. Dak. Res. Rep.* **44.** N. Dak. State Univ., Fargo, 1973, 15 pp.
38. J. Moorby, *Ann. Bot.* **34,** 297–308 (1970).
39. P. J. M. Sale, *Aust. J. Agric. Res.* **24,** 733–49 (1973).
40. P. J. M. Sale, *Aust. J. Plant Physiol.* **1,** 283–96 (1974).
41. F. L. Milthorpe and J. Moorby, The Growth of the Potato, in *Proc. 3rd Trienn. Conf. Eur. Assoc. Potato Res., Zurich, 1976,* pp. 51–70.
42. W. G. Burton, Response of the Potato Plant and Tuber to Temperature, in A. R. Rees, K. E. Cockshall, D. W. Hand, and R. G. Hurd, Eds., *Crop Processes in Controlled Environments,* Accademic Press, New York, 1972.
43. M. E. Jensen, J. L. Wright, and B. J. Pratt, *Am. Soc. Agric. Eng. Trans.* **14(5)** 954–959 (1971).
44. J. L. Wright and M. E. Jensen, *ASCE Proc., J. Irrig. Drain. Div.* **98,** 193–201 (1972).
45. D. C. Kincaid and D. F. Heerman, USDA-ARS Publ. ARS-NC-12. (1974).
46. D. F. Heerman, H. R. Haise, and R. H. Mickelson, *Am. Soc. Agric. (Eng. Trans.)* **19(2)** 284–293 (1976).
47. P. Tscheschke, J. R. Gilley, T. Thompson, and P. Fischback. Agric. Eng. **59(1)** 45–46 (1978).
48. M. E. Jensen and J. L. Wright, *Trans. ASAE* **21(1),** 82–87 (1978).
49. J. L. Wright, *Agron. Abstr.* **133** (1973).

10

RICE

F. T. Turner and G. N. McCauley

INTRODUCTION

About half of the world's people depend on rice as their principal food source, making rice the most important single food crop in the world (1, 2). The majority of the world's supply of rice is produced and consumed in Mainland China, India, Japan, and Indonesia. In the United States, rice is produced on approximately one million hectares in Arkansas, California, Mississippi, and Texas. The United States produces about 1% of the world's rice but this quantity accounts for at least one-fifth of the total milled rice moving in international trade (3).

In addition to its international importance as a food source, the rice plant has a unique plant–water relationship among the major crop plants. Plants grown in flooded, oxygen-deficient soils develop lysigenous gas spaces or aerenchyma in their roots which are linked with aerenchyma in the leaves and stems through which oxygen diffuses to supply the rice roots with oxygen (4). This adaptation allows rice plants to grow in both flooded (lowland) and nonflooded or rain-fed (upland) conditions. Equally unique are the floating or deep water cultivars which produce rice on stems that elongate as water depth temporarily increases up to 6 m.

Lowland rice is the most productive and requires rather impermeable soils and levees to maintain a water depth of 5 to 20 cm. At least 1 or 2 m of water are required to produce grain yields of 5–10 tons ha^{-1}. Upland rice, grown on arable soils, is totally dependent on amount and distribution of rain. Upland yields, generally about half that of lowland rice, are expected to increase through selections of drought tolerant cultivars (5).

Climate

After years of selection and breeding, rice is grown on every continent except Antarctica, and adapted to altitudes from near sea level on the United States Gulf Coast to 3000 m in Nepal and over a range of latitudes from 49°N in Czechoslovakia to 35°S in Australia (6). In temperate regions, such as the southern United States and Japan, where cultivars weakly sensitive to photoperiod are grown, the rice growing season is limited by low temperatures in early spring and fall. Nishiyama (7) identifies the favorable temperature range for physiological processes of rice plants as approximately 17–32°C. The lower portion of this range probably applies to cool-tolerant rice cultivars of northern Japan and high latitude areas, because day temperatures below 20°C do not support desirable growth and development of most cultivars in the southern United States (8). Likewise, McDonald et al. (9) found that 30°C was the optimum temperature for net photosynthesis of ten rice cultivars from around the world. The adverse effects of high and low temperature on rice production differ with cultivar and plant part. Temperature also indirectly affects such things as pest epidemics and chemical changes in soils through its effect on oxidation–reduction potential of soil.

Rainfall is another climatic factor that influences nonirrigated rice adaptability. In many cases the onset of the monsoon determines the planting time. The seasonal distribution of rainfall greatly influences water availability, for growth and development of rice plants.

Under conditions where temperature and water are not limited, solar radiation has pronounced effects on rice yield. For example, in India and the Philippines rice yields are always higher during the dry season when skies are clear. Stansel (8) has identified a 42 day critical sunlight period for rice grown in Texas as the 21 days prior to and following heading. He points out that decreasing day length during late summer limits the yields of ratoon or second crop rice in Texas.

An example of the effect of climate on rice yield at Beaumont, Texas is evident in Fig. 10.1. The same cultural practices were used for each planting date and water supply was not limited. The yield differences are attributed to the effects of solar radiation and temperature on the growth of the "Brazos" cultivar.

Characteristics and Time Intervals of Growth Stages

The need for increased rice yield has led to the development of improved plant types. The first cultivar approaching the apparent ideal plant type was "Taichung Native 1." Its success in Taiwan has served as a model for the development of more productive cultivars (10) and has spurred the "Green Revolution" for rice in Asia. The new cultivars are shorter (100 cm or less and nonlodging), with erect leaves (efficient receptors of solar radiation), generally photoperiod insensitive, and fertilizer responsive. The length of growing season of 40 cultivars ranged from 90 to approximately 190 days during the dry season at the International Rice Research Institute (IRRI) in the Philippines. Highest yields were from those cultivars with a growing season of 130–140 days (11). The most important cultivars mature in 140–160 days in California, 120–140 in Arkansas and Mississippi, 100–130 in Louisiana, and 100–115 in Texas. To help identify and illustrate the rice plant development stages for photoperiod insensitive cultivars, Stansel (12) outlined three growth stages and their duration as shown in Fig. 10.2. The vegetative stage, reproductive stages, and grain maturation stage all influence the grain yield components: panicle number, grains per panicle and individual grain per panicle, and individual grain weight (12). The growth stage of the rice plant dictates fertilizer, fungicide, and water needs; therefore, it is important to predict the beginning of the various developmental stages. Days from seeding or emergence as a means of predicting the initiation of a growth stage

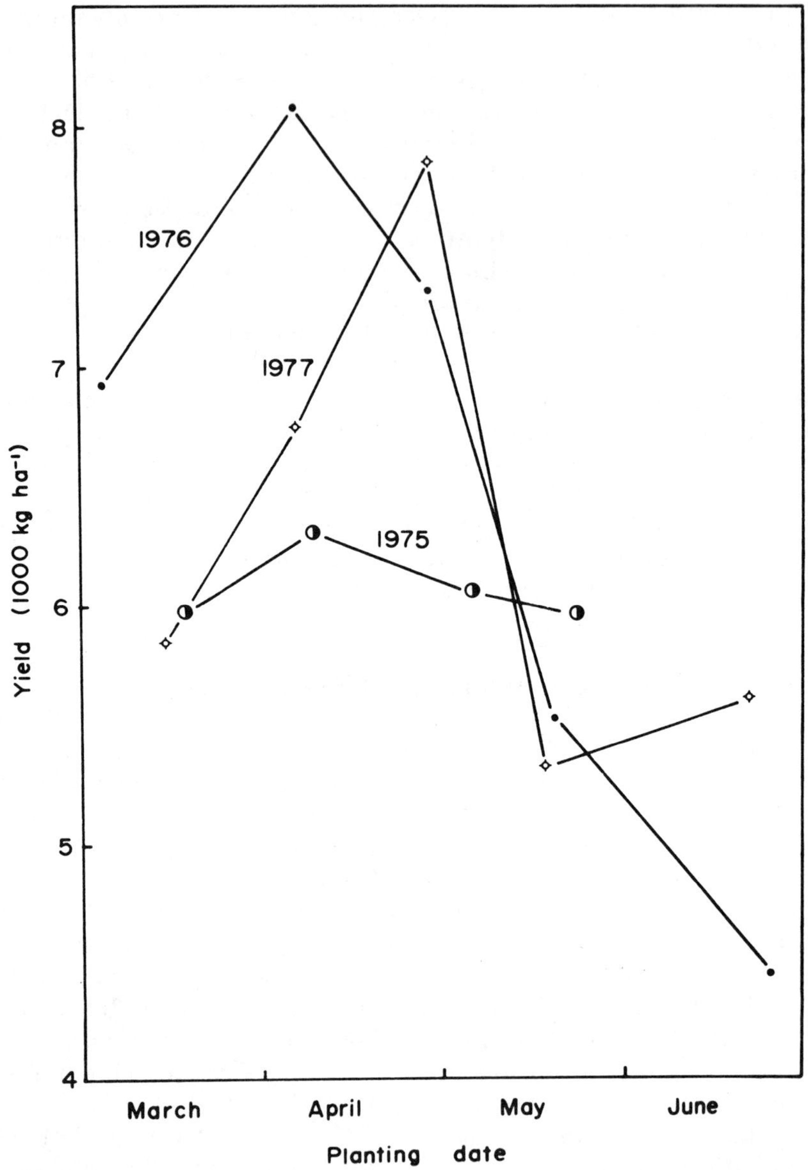

Figure 10.1. The effect of planting date on the rice yield of the 'Brazos' cultivar grown under flooded conditions over a 3 year period at The Texas A&M University Agricultural Research and Extension Center at Beaumont, Texas. [Fred T. Turner, unpublished data.]

has limitations because of variable weather conditions. Therefore, degree days or heat units has been successfully adapted to predict growth stage initiation (12, 13).

Figure 10.2 is representative of direct seeded lowland rice but is also applicable to transplanted rice. Transplanting occurs during the vege-

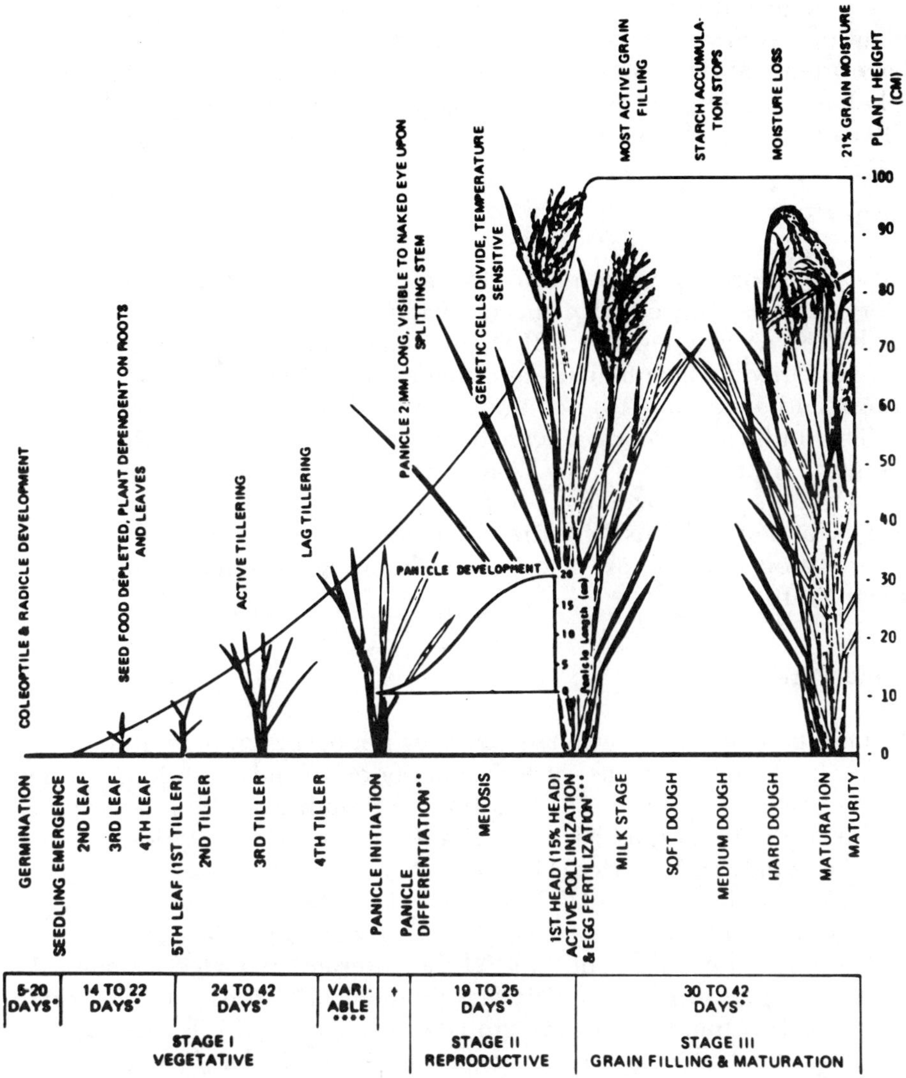

Figure 10.2. Development stages of direct seeded, lowland rice of the photoperiod insensitive, modern plant type (Stansel, 12). *Under warm conditions use the lower number of days and for cool conditions use the larger number of days. **The reproductive stage begins with panicle differentiation, which can be seen only by splitting the stems lengthwise. At this stage 30% of the main culms sampled have a panicle 2 mm or longer. ***Stage III begins when 50% of the florets are pollinated. ****Variable time 0 to 25 days (dependent on cultivar).

tative stage and is followed by a 5–7 day period for recovery from transplanting. After the recovery stage, transplanted and direct seeded rice have identical development stages. However, transplanted rice generally has a lower plant population than direct seeded rice and therefore tillering in transplanted rice is more important than in direct seeded rice.

Crop Use

Rice plants are grown primarily for the grain they produce. Webb and Stermer (14) point out that the most important uses of milled rice are (1) boiled or steamed rice, (2) dry breakfast and baby food cereals, (3) brewers' rice for manufacturing of beer, (4) canned rice for soups, puddings, and mixes, and (5) rice flour as a thickener in sauces and gravies.

The hull, a by-product, constitutes about 20% of the dried grain weight. Disposal of the hulls is sometimes a problem even though varying amounts are utilized as fuel, feed, fertilizer, abrasives, building materials, and absorbents (15). Rice bran and polish comprise about 5–9% of unmilled rice and are used as stock and poultry feeds after oil extraction (16).

Yield Potential

Generally, even when water stress does not occur, grain yield potential under lowland conditions exceeds that under upland conditions. Usually upland rice is grown under less than optimum conditions and typical yields are 0.5–1.5 tons ha^{-1} in Asia, 0.5 tons ha^{-1} in Africa, and 1–4 tons ha^{-1} in Latin America (17). However, under ideal conditions upland rice yields of 5.4, 7.0, and 7.2 tons ha^{-1} have been obtained in Nigeria, Philippines, and Peru, respectively (17). By contrast, the "IR8" cultivar grown under lowland conditions at the International Rice Research Institute in the Philippines (IRRI) has produced 10.3 tons ha^{-1} in a single crop and 16.7 tons ha^{-1} in two crops during 1 year (18). Typical average grain yields for lowland rice vary from 3 to 6 tons ha^{-1} depending on cultivars, management and climate.

ROOT GROWTH AND FUNCTION

Function

Rice is an extraordinary crop in that the primary function of the root differs for the various water management systems. Roots under upland culture must function first to absorb water, second to supply nutrients, and last to support the aboveground portion of the plant. Deep water culture of rice requires roots to change priorities during the season. Early in the season roots penetrate deeply to supply water. Later as the monsoon begins and the water deepens, the roots must serve as an anchor. Also roots, generated at stem nodes suspended in the rising floodwater, absorb water and nutrients. Roots in lowland rice production, except in the Americas, are influenced by transplanting. The roots must recover quickly from transplanting and supply nutrients to the plant. Water absorption is not a problem because seedlings are transplanted into flooded fields. In the Americas all rice is direct sown and grown under lowland conditions. However, rice culture in the southern United States rice belt is similar to upland culture conditions for the first 2–4 weeks of growth. During this nonflood period the seedlings must depend primarily on available soil water though they may be "flushed" (flood irrigated). The seedlings establish a relatively deep root system to supply water and nutrients. After the flood is applied, extensive root development occurs near the surface, primarily to supply nutrients. In California, rice is direct sown in water and a 10 cm flood is maintained throughout the growing season. Roots developing under this continuous flood system form near the soil surface and serve to absorb nutrients and anchor the plant.

Germination and Root Emergence

Water management, seed placement, available oxygen, and temperature all individually affect germination and emergence but the interactions become very complex and entwined. In a given location particular cultural practices are developed that work best for that area. Identifying specific factors affecting germination and interactions with particular locations will be the objective of this section.

The effects of water management on germination per cent and rate are shown in Fig. 10.3 (19). Maximum germination for cultivars "Bala" and "IR24" (lowland cultivar, not shown) were obtained when the soil water was maintained at −33 centibars (cb) potential. Both cultivars reacted similarly. Soil water conditions of −100 cb and flooded (5 cm) suppressed the germination approximately 17–30%, and delayed the germination from 2 to 4 days. Alternate flooding and drying suppressed germination by as much as 37%. Kordan (20) went one step further and described two types of rice emergence depending on water status. On well-drained soils, the coleorhiza emerged first, but under submerged conditions, the coleoptile emerged before the coleorhiza and germination was delayed (21).

Rice will germinate while submerged as long as the oxygen supply is adequate. Oxygen in the soil solution can decrease to 0 ppm within 38 hr after flooding regardless of the initial oxygen concentration of the water (22). This correlates with grower experience that for best seedling

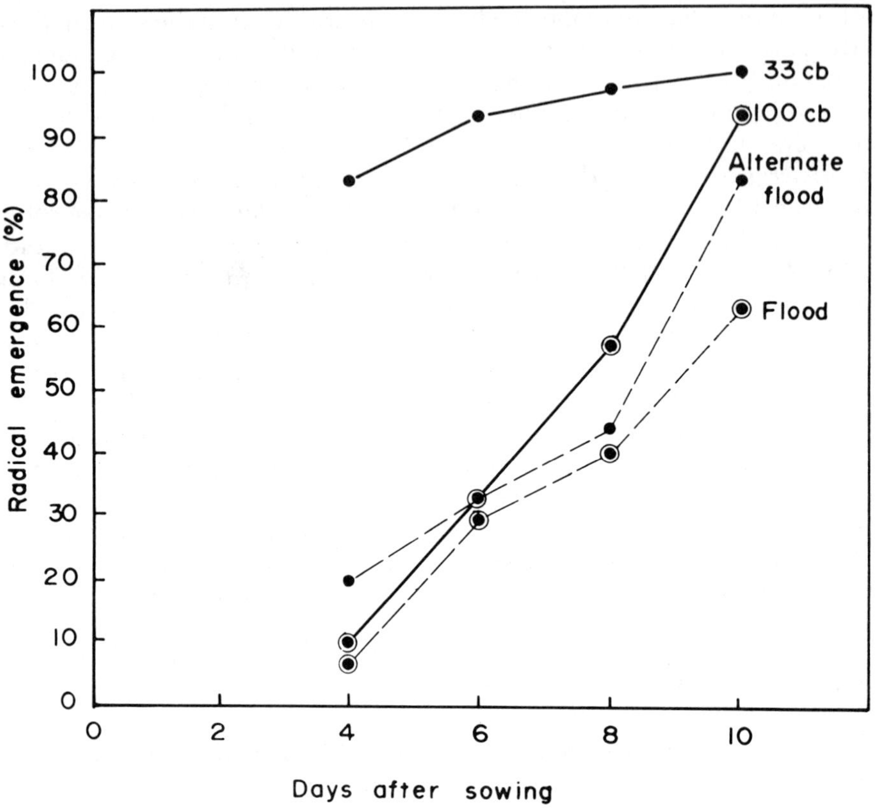

Figure 10.3. Effect of water management on radicle emergence of rice (upland cultivar 'Bala') in relation to time after sowing. [From Maurya and Ghildyal (9).]

establishment, the seedbed should be dry (well aerated) before flooding, and the fields should be sown by plane as soon as flooding occurs. High levels of organic matter in soils increase respiratory activities of the biota and increase the rate of oxygen consumption. Organic matter decomposition in absence of oxygen can lead to hydrogen sulfide formation which inhibits root development. Thus, it is not a good idea to incorporate a green manure crop that will not have time to decompose before flooding the soil. The question then arises, which is more beneficial for seedling establishment of lowland rice, running water or stagnant water? Originally running water was thought to be more beneficial; however, Chapman and Mikkelson (22) found that the dissolved oxygen levels and diurnal patterns under field conditions were similar for running and stagnant water (2 to 18 ppm depending on time of day). Warm water holds less dissolved oxygen than cool water, but even at water temperature around 35°C, the oxygen concentration of flood water was still at air saturation or above for 15–17 hr each day. Air saturation provides sufficient oxygen for germination and root initiation. However, the dissolved oxygen concentrations in stagnant water are insufficient for the normal growth of adventitious roots of submerged germinated rice seedlings in the absence of photosynthetically produced oxygen. [Green algae growth in the presence of germinating rice seedlings increases the net dissolved oxygen of the surrounding water (23).] The relation of the oxygen environment to germination and seedling development in rice is shown in Table 10.1 (24). Note that anaerobic environments (treatments 1 and 2) and submergence have similar effects on seedling development. Kordan (24) has presented evidence that shows (a) the action of inhibition is on growth rather than on the initiation phase of the adventitious root primordia; and (b) this inhibition of adventitious root growth in germinating rice is reversible because adventitious roots rapidly emerge from seedlings transferred to continuously aerated water or normal air environment.

An oxygen deficient environment also results in plants with a total absence of chlorophyll (22) similar to seedlings germinated in the dark. Like dark germinated seedlings, which recover when placed in light, oxygen-deficient seedlings will also recover in a short time if placed in normal air.

Emergence is altered directly by the environmental temperature. Temperatures in the range of 25–30°C are optimum for rice emergence. Shoot initiation is favored by 35°C temperatures, while root initiation is favored by temperatures of 20°C (23). At 30°C secondary root initiation is inhibited more than primary root initiation. Temperatures above 40°C are 100% lethal for rice. Temperatures of 40°C can occur a quarter inch above the dry soil surface in rice where temperatures 10–15°C above air temperature occur. Low temperatures adversely affect germination

Table 10.1. Relation of the oxygen environment to germination and seedling development in rice[a]

Treatment	Germination (%)	Seedling Development[b]		
		Days 3–4	Days 7–9	Days 9–14
1. Submerged, aerated with N_2	89	CE	NARG	ARG
2. Moistened gauze petri dish in anaerobic jar	91	CE	NARG	ARG
3. Submerged	95	CE	NARG	ARG
4. Submerged aerated	97	CE	ARG	—
5. Moistened gauze	100	S-ARG	—	—

[a] From Kordan (24).
[b] CE, coleoptile emergence only; NARG, no adventitious root growth; ARG, adventitious root growth; S-ARG, seminal and adventitious root growth.

of most cultivars. Reports from IRRI (25) show that water temperatures of 13°C and below reduced the percentage germination of "IR667-98" and "Jinheung" (a cool tolerant cultivar) to below 60 and 75%, respectively.

California's water management system of maintaining a continuous seeding to harvest flood would appear to be less than optimum for root emergence. However, California water originates from deep wells and mountain reservoirs and is cool and clear. Oxygen content of water is increased to 50% of air saturation at 20°C as the water is pumped and channeled to the field (23). The seeds laying on the soil surface are submerged in clear water (which allows light to penetrate) and oxygen and temperature are favorable for root development. Submergence may delay root development 2–4 days but the continuous flood system does function satisfactorily in California.

Continual flood during seedling establishment cannot be used in the southern rice belt because warmer water from surface streams with a heavy sediment load results in oxygen and light deficiency. To overcome these conditions seeds are sometime dropped into an established flood. Then after the seeds imbibe water and initiate germination, the floodwater is drained after 3–5 days to allow root system development. Another technique is to drill seed into the soil and flush to wet the seed if necessary.

Conditions for seed germination and root emergence are probably optimum in Asian seedling nurseries. However, the seedlings must withstand a traumatic shock at transplanting as intertwined roots are torn apart and portions of roots are severed. Following transplanting, seedlings must regenerate roots quickly for absorption of nutrients as nutrient reserves are limited during this stage.

Root Depth and Distribution

Information related to rice rooting patterns is limited because of the difficulty of accurate root retrieval and measurement in rice soils. According to Kawata rice plants initiate roots from the upper and lower portion of each node as illustrated by Fig. 10.4 (26, 27). However, after examining several plant types the authors were unable to confirm the root model as described. American long grain cultivars were found to have only one root type (apparently the lower nodal root) emerging from the nodes; however, medium grain types (with *Japonica* parentage) were found to have a root system as described by Kawata. The lower nodal roots are thick and penetrate deeply while the upper nodal roots are thinner and extend upward. Under flooded conditions root mats are formed near the soil surface by the upper nodal roots. Most roots of lowland, transplanted rice are found in the thin surface layer of soil above the plow sole (Table 10.2), but roots have been found as deep as 75 cm (29).

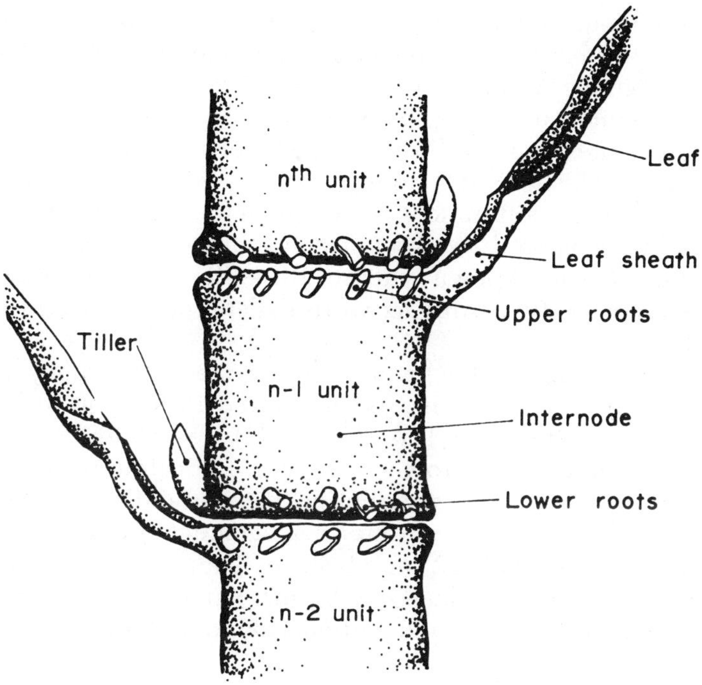

Figure 10.4. Graphic illustration of underground section of rice stem. [From Kawata (26, 27).]

Table 10.2. Soil depth increment containing the majority of roots and depth where deepest roots were found in lowland rice

Reference	Increment Containing Majority of Rice Roots	Depth of Rooting
Kawata (26)	Surface to plow sole	Deep root penetration proportional to depth of plow sole and rate of water percolation
Tei (28)	Shallow surface layer	
Sato (29)	$\frac{2}{3}$ of roots in top 20 cm	2% of total roots penetrated to a depth greater than 50 cm
Guimaraes and Inforzato (30)	95% in the top 35 cm and 35 to 50% in upper 5 cm	Roots to 75 cm less than 4% penetrated deeper than 39 cm
Brown (31)	Root density equal 19 cm cm^{-3} in top cm of soil	Less than 2 cm cm^{-3} at 18 cm depth
Hariuchi (32)	Top 15–25 cm depth	Lack of root penetration below 25 cm depth

Root development under transplanted and direct seeded conditions for lowland rice in Brazil has been compared as shown in Fig. 10.5. Direct seeding of "Iguape Agulha" and "JAC-435" resulted in 45 and 52% of the root dry weight in the top 5 cm of soil, respectively. Transplanting lowered the values to 38 and 34%, respectively. The direct seeded rice produced approximately 40% more root dry weight than transplanted rice with most of the differences occurring in the top 5 cm of soil. Direct seeded and transplanted rice root systems were very similar from 5 to 55 cm depth. It is not known whether the differences observed in root depth between the two cultivars were genetic or because of different soil types.

The effect of three water management systems (lowland, modified, and upland) on root length has been studied in relation to days after planting on four cultivars by Das and Jat (33). They found what appears to be genetic differences in root length (Fig. 10.6) although the cultivars had the same general root distribution pattern and responded in the same general way to the three water management systems. Root length was greatest in cultivar "IR22" and least in "Padma," and greatest under lowland and least under upland conditions. Kar et al. (34), working with "Taichung Native 1," a lowland cultivar, found root length and number

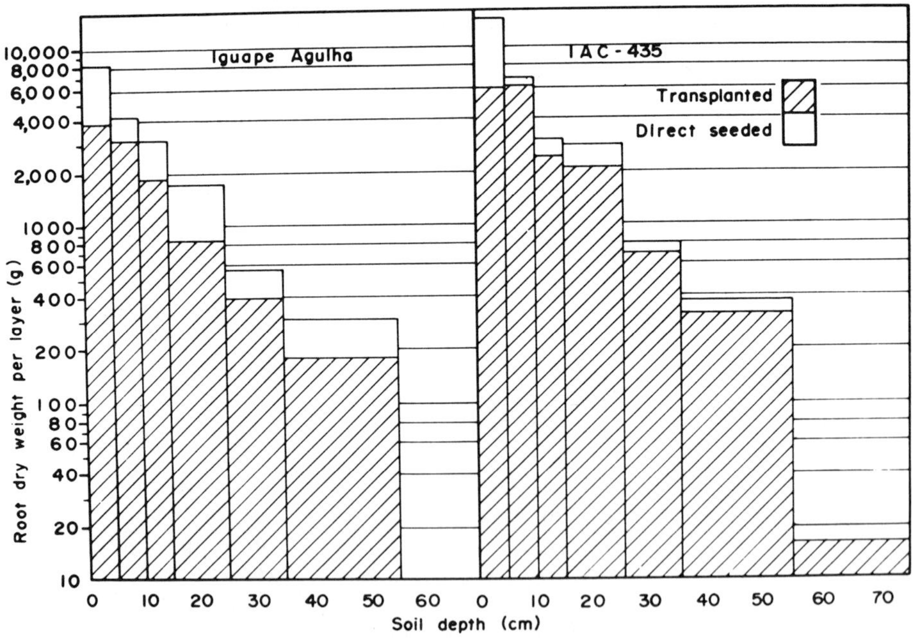

Figure 10.5. Rooting depth and distribution of two rice cultivars when transplanted or direct seeded. 'Iguape Agulha' was on Barro de Telha soil series in 1967–1968 and 'IAC-435' was on Seri Coruputuba soil series in 1968–1969. Surface area of each layer was $0.9m^2$. [From Guimaraes and Inforzato (3).]

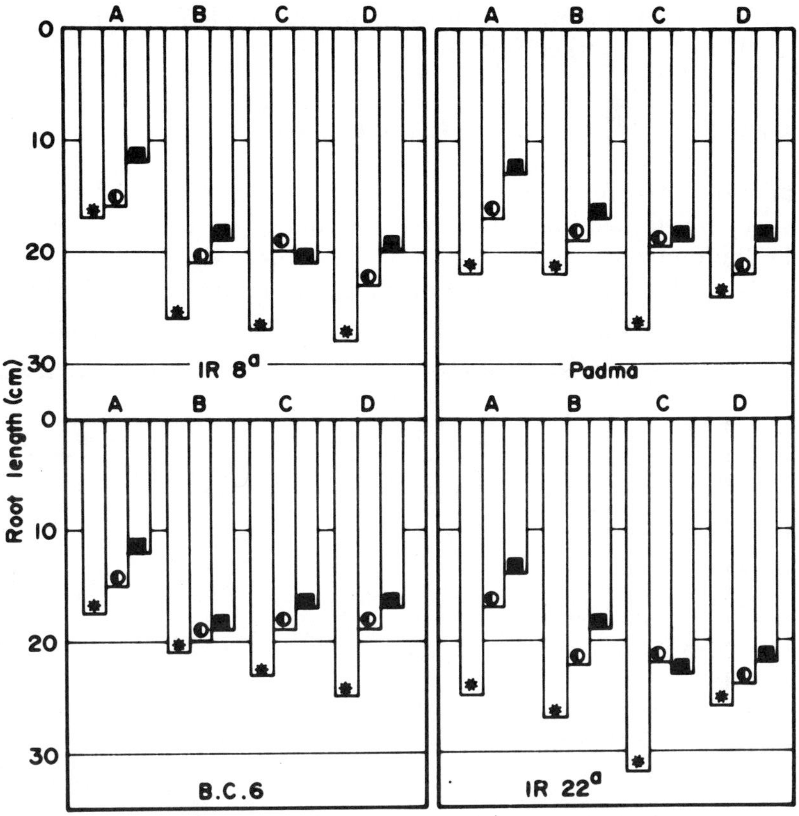

Figure 10.6. Root length of four rice cultivars under three different water management systems. Lowland cultivar = 'IR8'; modified lowland cultivars = 'B.C.6' and 'IR22'; upland cultivar = 'Padma.' Water regime: starburst, flooded; half-filled circle, alternate flooded; filled square, no flood. Days after transplanting: A, 30; B, 60; C, 80; D, 110. [From Das and Jat (33). Adapted from *Agronomy Journal,* Vol. **69**, p. 198, 1977, by permission of the American Society of Agronomy.]

to be greatly diminished by unsaturated conditions when compared to flooded conditions. Maurya and Ghildyal (35) conducted a similar study with six rice cultivars (three upland and three lowland) but reached the contradictory conclusion that nonflood conditions spurred deeper root penetration than flooded conditions. Pradhan et al.'s (36) research with "IR8," a lowland cultivar, supports Maurya and Ghildyal. They found that root length increased with decreasing soil water potential (−100 mb) and root mass at the surface decreased compared to flooded rice (6 cm flood).

Although most agree that in lowland culture, roots are concentrated near the soil surface, contradictions in the cited literature relating water management systems to cultivar root response lead the authors to conclude that more research is needed to clarify and explain root development with varied genetic types (upland vs. lowland), water management systems, and soil conditions.

Root Growth in Relation to Physiological Stage

Marata and Matsushima (37) have shown that root dry weight increases until just prior to heading when root dry matter accumulation levels off and shoot growth continues (Figure 10.7). Maurya and Ghildyal (35) also reported rapid increases in root volume and elongation from 15 to 60

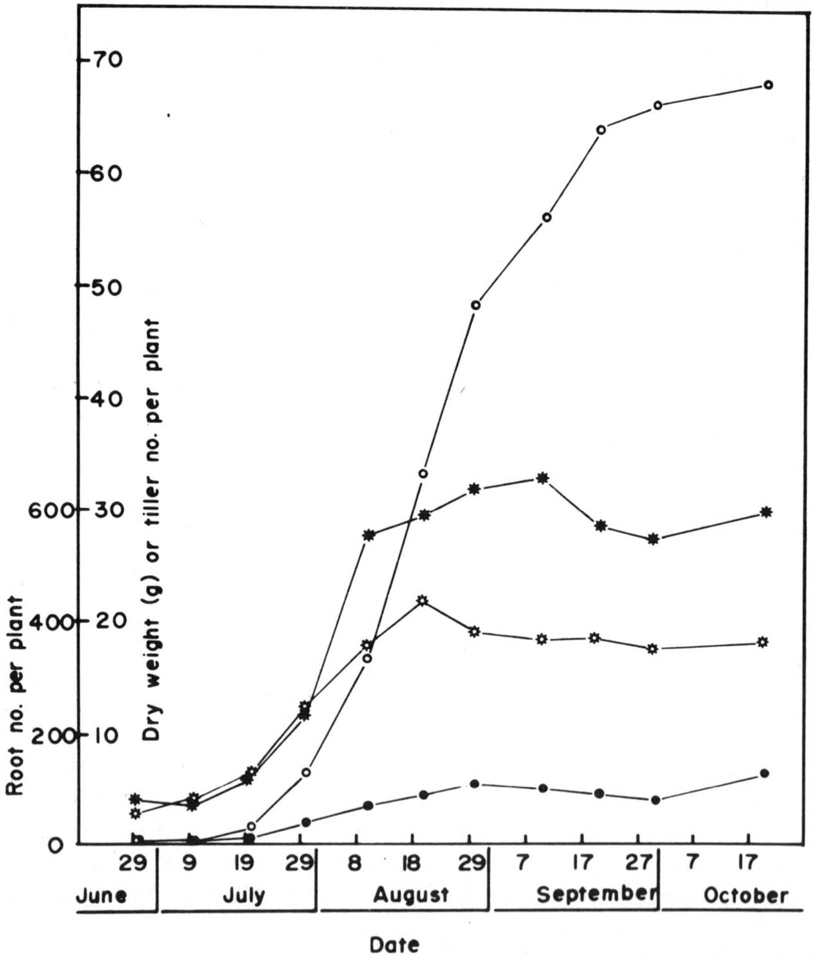

Figure 10.7. Dry weight and number of roots and shoots throughout a growing season as measured in a Japanese rice crop. ○, shoot dry weight; ●, root dry weight; open starburst, tiller no.; filled starburst, root no. [From Murata and Matsushima (37). Adapted from *Crop Physiology*, Cambridge University Press, 1975, page 74.]

days after sowing and a rapid decrease in root elongation and root volume between 60 and 75 days after sowing. This supports the general theory of root pruning during and after the reproductive phase. Carbon-14 studies have confirmed that before heading the lower leaves translocate ^{14}C to the roots while the younger leaves translocate ^{14}C to the young panicle (38).

Studies of shoot–root relationships indicate that rooting pattern should be modeled in relation to the shoot. Fujii (39) and Katayama (40) are both credited with the most accepted solution to the model which is, as the nth leaf develops, the $(n - 3)$th nodal roots develop. The $(n - 3)$th nodal roots are associated with the $(n - 3)$th tiller. Relations such as this indicates: (1) roots increase with tillering, (2) roots reach their maximum at heading, and (3) root growth should be more accentuated in early stage when compared with shoot development.

Figure 10.8 shows that under flooded or unsaturated soil water conditions the shoot–root ratio declines through tillering but increases before reproduction which supports the proposed model. Chellappa and Rao (41) conducted a similar study, using a *Japonica* and an *Indica* cultivar and two crosses, and found a steady root–shoot ratio of approximately 2.8 during the first 30 days which then increased at various rates depending on cultivar (Fig. 10.9).

In the Root Growth and Function section, the authors noted that the root model as described by Kawata (26, 27) may only hold for *Japonica* cultivars and not United States long grain cultivars (*Indica* heritage). Figure 10-9 further illustrates differences in root systems for *Japonica* and *Indica* type rices.

Shoot–root ratios of lines with *Indica* heritage continue to increase up to 90 days after transplanting while the ratios of *Japonica* descendent lines will decrease or level off after heading. These differences indicate that root growth patterns may be more drastically influenced by heritage than by growing conditions.

Root–Water Relations and Dry Matter Accumulation

Under flooded conditions Kar et al. (34) found that roots exhibited accelerated growth in relation to shoot growth during active and maximum tillering (Fig. 10.8). This growth pattern leads to a small decrease in shoot–root ratio during active tillering followed by a gradual increase through the reproductive phase. Shoot and root weights were significantly depressed under unsaturated conditions. The roots in unsaturated soils did not show the same growth burst during the maximum tillering phase as those in the flooded soil. Table 10.3 shows that the results of Sharma et al. (42) agree closely with Kar et al. (35) in that root and shoot dry weights both decline as amount of water decreases from flooded to field capacity. However, the shoot–root ratio showed that root weight had decreased more than shoot weight when plants were maintained at field capacity.

Pradhan et al. (36) conducted a study evaluating six plant characteristics at maximum tillering stage for eight different soil water treatments ranging from 6 cm flood to −100 cb potential (Table 10.4). They found the number of roots decreased steadily as the soil potential decreased down to −100 cb. While the number of roots decreased, the root length increased from 31 to 55 cm at −50 cb. Root dry weight at −100 cb was less than ⅓ the root dry weight obtained at 3 cm flood.

Root development of "IR5" (a lowland cultivar) was more severely restricted by water stress than leaf area index (*LAI*) or tiller number as compared to the upland cultivar "Palawan" which has some tolerance to water stress (43). The ability of upland cultivars to perform better

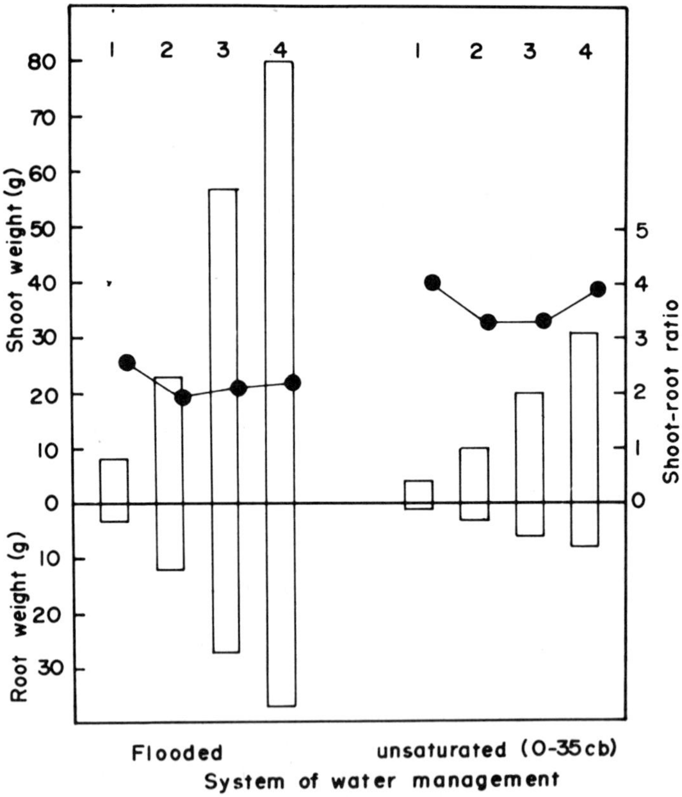

Figure 10.8. Shoot and root weights and shoot–root ratios in relation to the system of water management of rice at different growth phases (growth stage 1, early active tillering; 2, active tillering; 3, maximum tillering; 4, reproduction); ●, shoot–root ratio. [From Kar et al. (34).]

Table 10.3. Development of rice plants under three water regimes 60 days after transplanting[a]

Water Level	Average Height of Plant (cm)	Average Number of Leaves per Plant	Tillers per Plant	Leaf Area per Leaf (cm^2)	Dry Weight of Shoot per Pot (g)	Dry Weight of Root per Pot (g)	Shoot/Root Ratio
Flooding	62.5	13.90	3.7	17.12	23.35	9.00	2.594
Saturation	44.7	14.25	3.2	12.36	10.00	4.93	2.028
Field capacity	52.0	15.40	3.3	9.99	17.50	3.84	4.557

[a] From Sharma et al. (42).

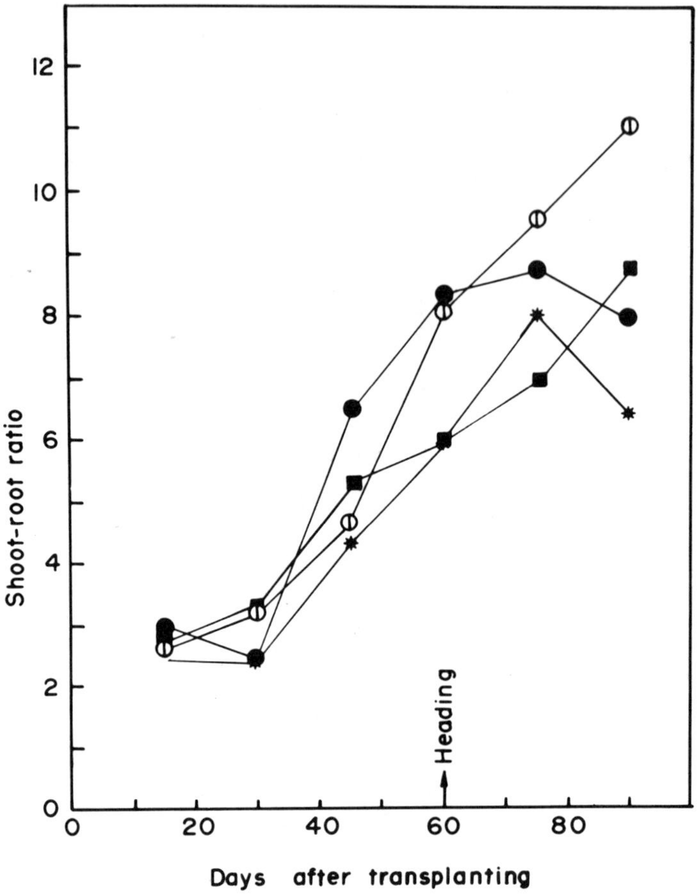

Figure 10.9. Shoot–root ratios of two cultivars of rice and two crosses from transplanting to maturity. ●, 'Norin 1' (*Japonica*); split circle, 'Co 29' (*Indica*); starburst, 'ADT 27' (*Japonica* × *Indica*); filled square, 'Padma' (*Indica* × *Indica*); Chellappa and Rao 41).

Table 10.4. Effect of soil water on several plant characters at maximum tillering[a] (Measurements are on a per pot bases)

Soil Water Treatments	Plant Height (cm)	Number of Tillers (g)	Dry Weight of Shoot (g)	Root Length (cm)	Number of Roots	Dry Weight of Roots (g)
Flooding, 6 ± 1 cm	61.75	49.25	18.75	31.32	532.5	8.25
Flooding, 3 ± 1 cm	59.00	57.25	21.75	33.95	551.5	9.50
Saturation 0 mb	57.50	55.50	19.75	37.45	494.0	7.25
−2 cb	57.00	45.25	17.50	40.95	476.0	7.00
−6 cb	54.12	36.25	13.12	45.20	427.0	5.50
−35 cb	52.75	34.25	9.75	51.30	409.0	4.62
−50 cb	51.25	25.75	8.75	55.12	370.5	3.50
−100 cb	47.25	22.25	6.75	51.50	325.0	3.12

[a] From Pradham et al. (36).

under water stress appears to be related to better root growth during water stress according to work by Krupp et al. (44).

CANOPY GROWTH (LEAF AND CULM)

Dry Matter Accumulation and Physiological Stage

Dry weight changes of the shoot and various plant parts in relation to time and physiological stage are shown in Figs. 10.7 and 10.10. Under optimum growing conditions the lower leaves of early maturing rice (45) begin to senesce between maximum tillering and panicle initiation. Because new leaves continuously develop, *LAI* initially increases with growth, reaches a maximum, then declines as senescence of lower leaves exceeds the rate of new leaf formation (Fig. 10.11). Light transmission ratio (*LTR*) decreases with increased dry weight due to the increase in *LAI*. Ideally, canopy closure, as measured by *LTR*, is about 95% complete at panicle initiation (8). After panicle initiation the stem internodes elongate rapidly, forcing the panicle to emerge or head. After heading the plant soon flowers and panicle weight increases substantially. Dry matter accumulation curves for the late maturing cultivars (150 or more days) have a very definite "dormant" or vegetative lag period between the vegetative and reproductive phases (Fig. 10.2) which the early maturing cultivars do not have (Fig. 10.10).

IRRI scientists compared *LAI* and dry matter production of sorghum and rice (Fig. 10.11). *LAI* of the two crops were very similar during the wet or rainy season when water supply was adequate, except that *LAI* of rice began to decrease prior to heading whereas *LAI* of sorghum decreased after heading. During the dry season maximum *LAI* of sorghum was about one-half that produced during wet season even when supplemental irrigation was provided. It is evident that limited water during the dry season had a greater adverse effect on *LAI* and dry matter production of rice than sorghum. During the wet season rice plants continued to increase in dry matter production after heading but sorghum did not, apparently because rainy weather after heading favored infestation of head mold in sorghum (46).

Submergence Effects on Canopy

Studies at IRRI have demonstrated the pronounced effects of submergence on canopy growth and the survival of lowland rice (47). Plants of the early-short "Peta" cultivar grown for 10 days in porcelain trays containing 5 kg of fertilized soil were submerged in water 35 cm deep. Ninety percent of the plants survived 6 days of submergence but survival dropped to 49, 52, and 25% on day 8, 10, and 12, respectively. As water

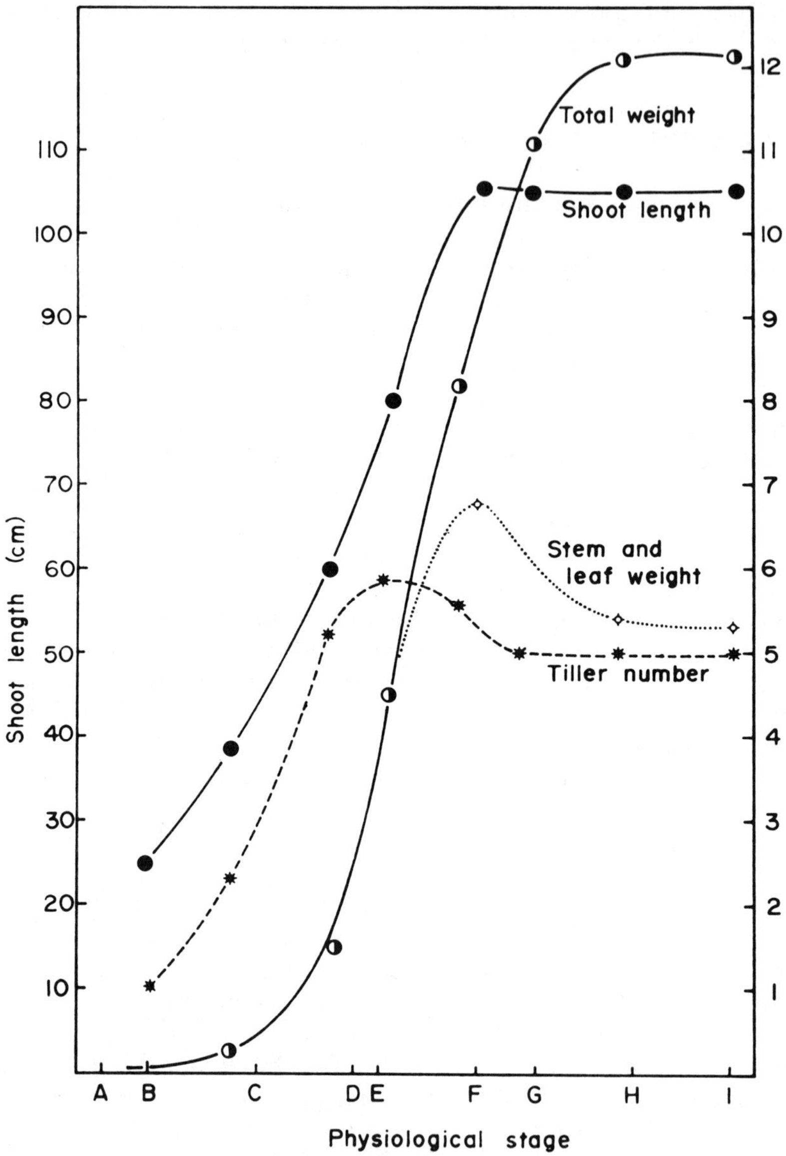

Figure 10.10. Canopy growth of a transplanted, early maturity (110 days) rice plant under lowland culture conditions: (A) transplanting; (B) recovery from transplanting; (C) tillering; (D) panicle initiation; (E) internode elongation; (F) flowering; (G) milk stage; (H) dough stage; (I) incomplete ripening. [From Ishizuka (45). Adapted from *Advances in Agronomy*, Vol. **23**, p. 244, 1971, American Society of Agronomy.]

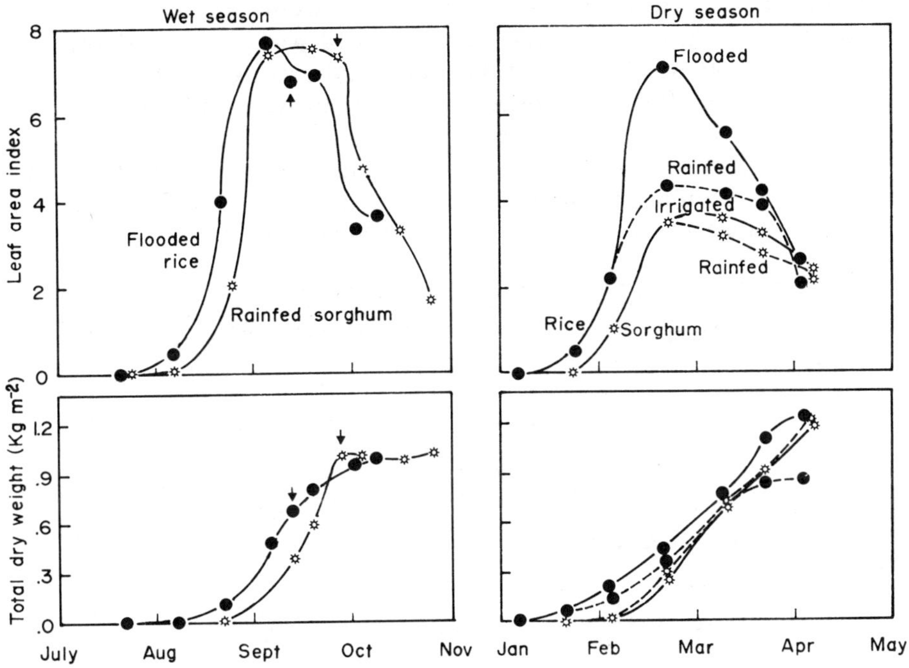

Figure 10.11. Change in *LAI* and aboveground dry weight of lowland rice ('IR154-18') (●) and sorghum ('IS9290') (starburst) grown in the Philippines during the wet and dry seasons of 1971. Heading date indicated by arrow only during wet season. [From IRRI (46).]

temperature increased (from 25 to 30°C) so did the adverse effects of submergence. Turbidity of the water submerging the rice plants reduces light for photosynthesis and affects survival rates. Ninety-one percent of the plants survived 10 days of submergence in clear water, whereas only 52% survived submergence in turbid water which reduced light transmission at soil surface by 56%. Evidence suggested plants receiving low levels of nitrogen fertilizer were more likely to survive submergence than plants receiving high levels of nitrogen because of higher percentage of stored carbohydrates in plants receiving the lesser amount of nitrogen.

Floodwater depth is used as a management tool. Floodwater depths of 10–12 cm apparently provide the best weed control without adversely affecting rice yield while floodwater depths in excess of 25 cm were found to reduce lowland rice yields in Texas (48). Deep water cultivars can tolerate deep floodwater and temporary submergence. They are grown on more than 4 million hectares of land where floodwaters rise up to 90 cm day^{-1} at the beginning of the monsoon (49). They elongate rapidly when submerged. This survival mechanism has spurred IRRI scientists (50) to screen 144 cultivars for the ability to elongate when submerged. All cultivars were grown for 6 weeks in pots containing moist soil and then submerged in 30 cm of water. Water depth was increased to 60 cm the second day, and to 90 cm by the third day. On the seventh day of submergence, 30 of the cultivars had significant increases in height and total internode lengths ranged from 17 to 70 cm.

Figure 10.12 shows that a deep water cultivar must be older than 4 weeks before internodes will elongate significantly upon submergence. While the leaf sheath and blade will elongate when submerged at 2

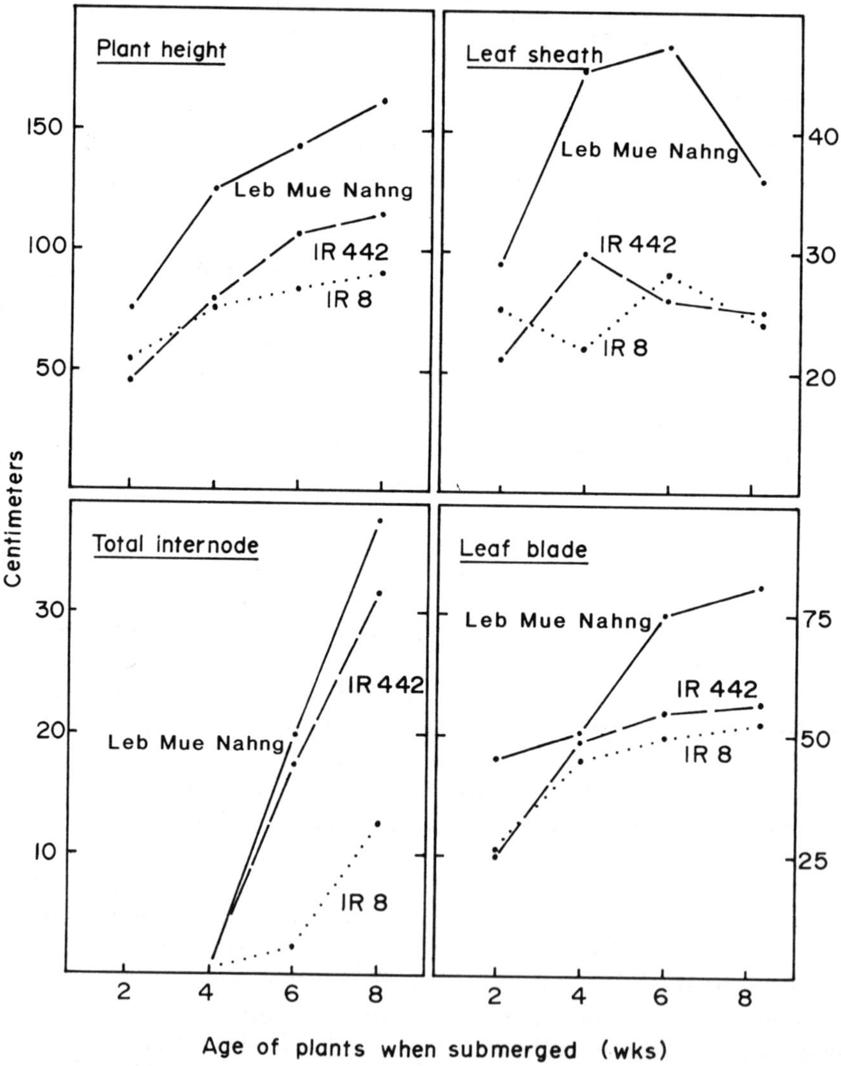

Figure 10.12. Elongation response of three rice cultivars when submerged in 90 cm of water at different plant ages. 'Leb Mue Nahng' is a deep water cultivar and 'IR8' and 'IR442' are lowland cultivars. [From IRRI (50).]

weeks of age, the lowland cultivar "IR8" must be older than 6 weeks before submergence can induce internode elongation, and the elongation response is not usually sufficient to survive submergence (50). The study provides evidence that internode elongation alone cannot account for the rapid increase in plant height and that leaf sheath and blade also contribute to the plant height increase caused by submergence.

Where water depth can be controlled, it is desirable for rice plants to be short statured and strong stemmed as tall plants usually have weak stems which lodge easily causing severe yield reductions. Long, wide, drooping leaves form a flat undesirable layer at the top of the plants which reduces light penetration into the canopy and retains rainwater which encourages lodging when rain and winds occur.

Floodwater Temperature Effects on Canopy

When rice is in the seedling stage in California, irrigation water from streams and wells frequently has a temperature of 18°C (51) and may be as low as 11°C (52). When water of less than 20°C is used for irrigation, the rice plants nearer the water inlet mature slower than the rice away from the inlet, thereby reducing the quality of rice because of nonuniformity at harvest.

Raney (53) reported that in California the mean critical water temperature for normal growth of the "Caloro" cultivar was 20.5°C. Temperatures 3°C lower delayed maturity 30 days beyond the normal 160 days. Mean water temperature of 26.4°C gave highest yields while those above 29°C reduced yield and root development.

A report from IRRI (54) showed that "Jinheung," a cultivar grown in Korea where cool temperature tolerance is necessary, did not develop distinct yellow and orange coloring when subjected to temperatures below 15°C as did "IR667-98." It was also shown that temperature of floodwater affects stem elongation. At floodwater temperatures of 17°C and above, all cultivars tested ("IR667-98", "IR747B2-6", "IR8", "TN1", and "Yokara") had similar increases in height even at 12°C air temperatures but only the cultivar "Yokara" increased in height when both air and water temperature were 12°C during the tillering stage.

Water Quality Effects on Canopy

During low rainfall periods, water levels in streams near the Gulf Coast areas of Louisiana and Texas may become low enough to allow brackish water intrusions from the Gulf. Several workers (51, 55) have noted that water containing more than 600 ppm salt should not be used to irrigate rice. This is especially true during the seedling stage when rice is most sensitive to salt injury. The plant's tolerance to salt water increases with age. Finfrock et al. (56) described good quality rice irrigation water as having less than 525 ppm soluble salts, less than 1 ppm boron, and a sodium adsorption ratio of less than 10. In rice growing areas where evaporation exceeds rainfall, the salinity problem assumes greater importance and cultivars must be tolerant to saline water. Saline tolerant cultivars are available and rice is grown on certain saline soils in Egypt as a means of reclaiming the soil for use by other crops (57).

Floodwater Effects on Mineral Nutrition of Canopy

Anaerobic soil–water systems cause the pH of both acid and alkaline soils to converge toward neutrality (58, 59). The anerobic, neutrally buffered system reduces the oxidized components of the soil (nitrate, manganese, iron, and eventually sulfate if the redox potential becomes low enough), thus influencing the growth and nutrition of rice plants. The main difference between nitrogen behavior in flooded soils compared to aerobic soils is the slower rate of mineralization of organic matter (60) and enhanced rate of denitrification except in the slightly oxidized top cm of soil surface (58). Denitrification may cause losses of 50% or more of the soil nitrogen, especially when rice soils are alternately flooded and dried. Wells and Shockley (61) illustrated the loss of native soil nitrogen from a silt loam soil in Arkansas when they found that plots without fertilizer nitrogen yielded 5690 kg ha^{-1} under continuous flood and only 3700 kg ha^{-1} with intermittent flooding. Increases in phosphorus availability to rice upon flooding a soil (59, 62) have been attributed to the increase in phosphorus solubility as ferric phosphate is reduced to the more soluble ferrous phosphate under the anaerobic system (59, 62). An equally important reason for the increased phosphorus availability following flooding is the increased rate at which phosphorus can diffuse to roots in flooded soils as shown by Turner and Gilliam (63, 64). The report by Barber et al. (65), which suggests that phosphorus diffusion through soil may be the main mechanism by which phosphorus arrives at the plant roots, shows that phosphorus diffusion through soil increases with increased soil water content (66), and that phosphorus uptake increases with increased soil water (67). Submerged soil conditions also have been shown to increase the availability of iron (43), manganese, and silica (59).

Water Stress Effect on Canopy

Because excessive tillers use up water rapidly and increase the probability of water stress later in the growing season (44, 68), upland cultivars have been selected for low tillering ability. The environment also affects tiller number and *LAI*. Under water stress conditions the tiller number and *LAI* were reduced on an unimproved upland cultivar ("Palawan") and a lowland cultivar "IR5" (Fig. 10.13). Water stress during the 45–60 day period after emergence appeared to reduce tillering of the lowland cultivar more than the upland cultivar (44). Wells and Shockey (61) have shown that a 1 week drain period during the early vegetative stage of the "Starbonnet" cultivar produced no adverse effects but additional 1 week drain periods during the vegetative stage resulted in reduced yields.

Panicle number type plants (those whose yield increases are attributed to the increase in panicle number per unit area) and panicle weight types (increased yield results from increased grains per panicle) appear

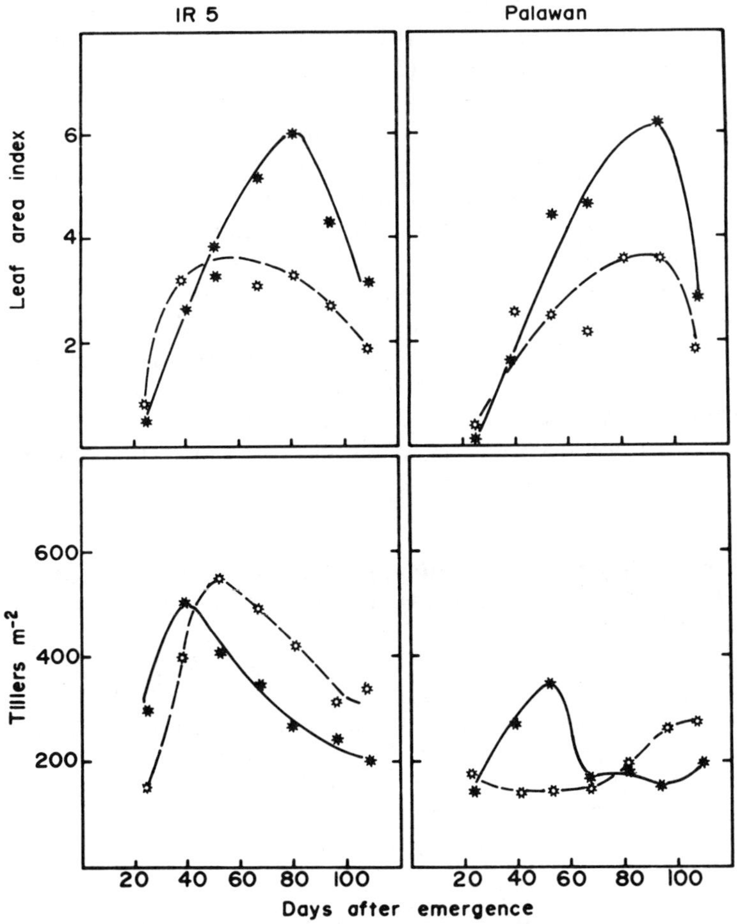

Figure 10.13. *LAI* and tiller number of 'IR5' a lowland variety and 'Palawan' an upland cultivar in response to flooded (filled starburst) and rainfed (open starburst) conditions during the 1970 wet season at IRRI. Severe water stress occurred only during the 45–60 day period after emergence. [From IRRI (44).]

to react differently to water stress (44). The two high tillering panicle number cultivars "IR5" and "IR8" were less sensitive to the growth stage at which water stress occurred than they were to the duration and intensity of water stress. A low tillering, panicle weight-type cultivar, "H4," was sensitive to intensity and duration of water stress but most sensitive to water stress during the reproductive and ripening stages. Thus, the panicle weight-type plants seem to be particularly sensitive to water stress during the reproductive stage.

Cultivar "IR20," grown under dry season field conditions, was subjected to four water stress treatments (69). The no stress (continually flooded), early stress (no irrigation during vegetative growth stage), late stress (no irrigation for the 30 day period from panicle initiation to just after flowering), and extended late stress (no irrigation after panicle initiation) yielded 6.0, 4.5, 2.2, and 0.8 ton ha^{-1}, respectively, indicating that tillering and panicle initiation stages are very sensitive to water stress.

Soil water potentials of -15 cb are sufficient to decrease grain yields (70). Part of the yield reduction associated with water stress may be due to the loss of nitrogen by denitrification under alternately dry and wet conditions. Pradhan et al. (36) found that tiller number decreased when soil water potentials were -2 cb or less. Plant height and shoot dry weight were reduced by absence of a 3 cm deep flood and continued to decrease as soil water decreased to -1000 cb. Sharma et al. (42) observed that as soil water declined the average number of leaves per plant increased but area per leaf decreased by almost 50%, causing shoot dry matter to decrease when measured 60 days after transplanting. Das and Jat (33) compared the effect of flooded, alternate flood, and no flood

Table 10.5. ***Values for transpiration, photosynthesis, and dry matter production are presented with the transpiration/photosynthesis ratio for rice and sorghum***[a]

Regime	Simultaneous Measurement[b] Transpiration (mg H_2O leaf^{-1} hr^{-1})	Photosynthesis (mg CO_2 leaf hr^{-1})	Transpiration/ Photosynthesis Ratio	Dry Matter Production (g pot^{-1})
Rice				
Flooded	—	—	—	10.75
Upland	162	4.1	40	7.55
Sorghum, upland	352	12.8	20	8.45

[a] From Yoshida (72).
[b] Three weeks after sowing.

on shoot dry matter accumulation and revealed that dry weights of shoots were similar except during the 40–100 days after transplanting when the flooded treatment had a shoot dry weight up to 1.3 times higher than the alternate flood or no flood treatments.

An indication of the effect of water stress on photosynthesis of young rice plants was obtained when "IR5" plants were subjected to water stress after 34 days of normal growth in pots. Water stress resulted in leaf resistances of 6, 10, and 26 sec cm^{-1} which restricted CO_2 intake and photosynthesis rates to 42, 18, and 12 mg CO_2 dm^2 h^{-1}, respectively (71). A comparison of photosynthesis and transpiration rates between upland rice and sorghum are shown in Table 10.5.

An additional physiological aspect of water stress is the accumulation of proline as stress increases. Proline content in leaves of intact "IR747B2-6" plants increased from 50 to 7000 μg g^{-1} fresh weight as stress was increased from nil to severe. Similarly as the water potentials of a nutrient solution were increased by addition of polyethylene glycol, the proline content of leaf segments bathed in the solutions with water potentials of approximately 0, −2, −4, −8, and −16 bars were 60, 57, 80, 140, 320 μg g^{-1} fresh weight, respectively (73).

Table 10.6 illustrates the plant changes induced by water stress and summarizes the effects of water stress on canopy growth of rice plants (71).

Table 10.6. Physiological and morphological changes induced by water stress[a]

Feature	Without Stress	With Stress
	Visible Features	
Plant height	Normal	Reduced
Tiller	Normal	Reduced
Leaf area	Normal	Reduced
Leaf rolling	No	Yes
Sterility	Low	High
	Nonvisible Features	
Stomata	Open	Closed
Major path of water loss	Stomatal transpiration	Cuticular transpiration
Leaf water potential	High	Low
Leaf temperature	Low	High
Heat tolerance	Low	High
Photosynthesis	High	Low
Proline content	Low	High

[a] Adapted from IRRI (71).

TRANSPIRATION: REGULATING WATER LOSS

A means of evaluating the water requirement of crops is to compare the number of grams of water transpired per gram of dry matter produced. The water requirement of rice varies with soil, water, climate, cultivar, growth stage, and growth duration of the plant. It ranges from 171 to 438 g g^{-1}, but it is generally around 250–350, which is about twice that of sorghum (72).

Because of the flooded conditions accompanying lowland rice, one might assume that lowland rice is much less efficient in water use than upland rice. But even though transpiration is greater under lowland conditions, so is dry matter production, resulting in comparable water requirements under both upland and lowland culture. This point illustrates the close correlation between amount of water transpired and yield of upland rice and further indicates the sensitivity of rice to water stress. Rice breeders feel that improving the resistance of rice plants to water stress is a promising means of lowering the water requirement and increasing the yield of upland rice. Drought tolerance in rice can be subdivided into two categories: (1) desiccation postponement through such mechanisms as stomatal, cuticular, and root resistance to water flow, and (2) by heat and desiccation tolerance (72). This aspect of rice research is in its infancy.

Stomatal and Cuticular Resistance

IRRI reports (72, 74) show that under normal conditions in solution culture stomatal resistance of 12 rice cultivars ranged from 3.5 to 3.9 sec cm^{-1} (Table 10.7). The addition of polyethylene glycol to the culture solution decreased the solution potential to -6.5 bars and increased stomatal resistance to the range of 6.9 to 12.4 sec cm^{-1} depending on the cultivar.

The drought-resistant cultivars ("Ml-48" and "OS4") exhibited high stomatal resistance while drought-susceptible cultivars ("IR20" and "TN1") had low stomatal resistance values under induced water stress. "IR5" exhibited considerable stomatal resistance for a lowland cultivar but it is regarded as slightly tolerant to drought while "IR20" is most susceptible (75).

Additional insight into factors affecting plant–water relations was gained when workers at IRRI (72, 76) grew "IR5" (a lowland cultivar) in both flooded and upland conditions for 35 days then the upland conditioned plants were subjected to soil water levels of 35 and 70% of field capacity for 3 weeks using a pot culture technique. During the 3 week period diurnal changes in leaf resistance were measured on the second fully developed leaf of the main shoot. A representative day is shown in Fig. 10.14. Stomata opened in the flooded soil at 0600 hours, in the 70% field capacity treatment at 0615 hours and in the 35% field capacity treatment at 0640 hours, when leaf resistance was 2.5–5.0 sec cm^{-1}. Under flooded or lowland conditions leaf stomata remained open till 1500 hours. Under upland conditions (35 and 70% field capacity) water

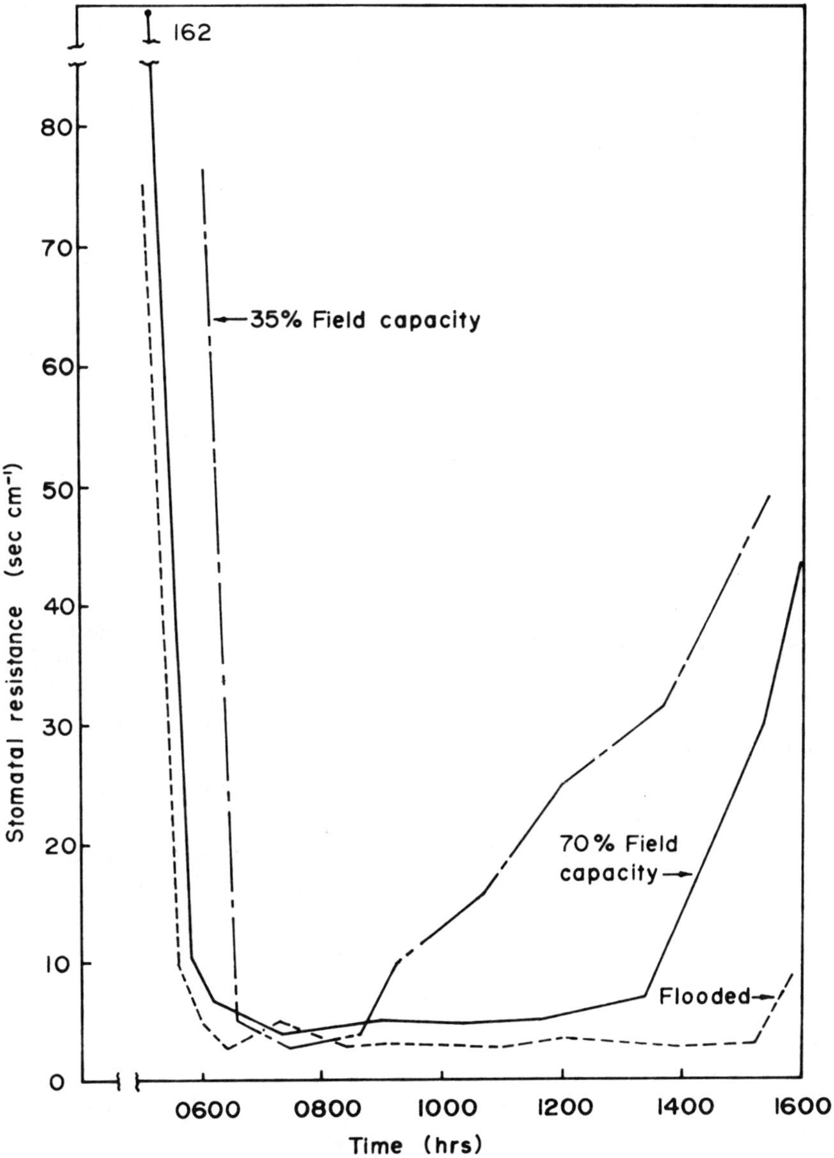

Figure 10.14. Hourly changes in leaf resistance in relation to time and water stress on a clear day at IRRI. (From Yoshida (72).]

Table 10.7. Stomatal resistance in response to water stress of 12 cultivars[a]

Cultivar	Stomatal Resistance (sec cm^{-1}) Without Water Stress	With Water Stress
Jappeni Tunkungo[b]	3.7	12.4
Ml-48[b]	3.8	10.8
IR5[c]	3.6	10.1
OS4[b]	3.7	10.0
Tainan 3[d]	3.7	10.0
Palawan[b]	3.5	9.6
Dular[c]	3.9	9.4
Pl 215936[d]	3.8	9.1
E 425[d]	3.5	8.5
TNl[c]	3.9	8.3
T 141[c]	3.6	7.9
IR20[c]	3.6	6.9
LSD (5)	0.3	2.5

[a] Water stress was induced by exposing root to a solution potential of −6.5 bars (72).
[b] Upland cultivars.
[c] Lowland cultivars.
[d] Thought to be upland cultivars.

stress increased as the day progressed, and the stomata began to close at 0830 and 1315 hours, respectively. Thus, under upland conditions high atmospheric demand for water and a low water supplying capacity of the soil will reduce transpiration and dry matter production by causing stomatal closure.

Yoshida (72) reported large variations (68.3–29.5 sec cm^{-1}) in cuticular resistance of 20 rice cultivars. He suggested that cuticular resistance partially explained cultivar differences in drought tolerance, since upland cultivars tended to have higher cuticular resistance. In contrast he also noted that night time transpiration, which is largely through the cuticle, has been reported to be higher for upland than for lowland rice.

Root Resistance to Water Transport

To study the resistance to water transport in rice plants grown under lowland and upland conditions, Tomar and Ghildyal (77) grew "IR8", a lowland cultivar, in the greenhouse for 45–50 days under flooded and −30 cb soil water potential. Resistance to water flow in plant parts, as measured by a transient flow method, is shown in Fig. 10.15. The resistance to water transport in plants grown at −30 cb was nearly twice as high as those in flooded soil. The greater resistance to water flow in plants grown on nonflooded soil was attributed to higher water transport resistance (16 times) measured in the roots grown at −30 cb soil water potential relative to those grown under flooded conditions. A close look at the anatomy of the flooded and nonflooded roots showed highly developed aerenchyma in flooded roots in contrast to the nonaerenchymatous roots in the nonflooded soil. The aerenchyma develop in flooded

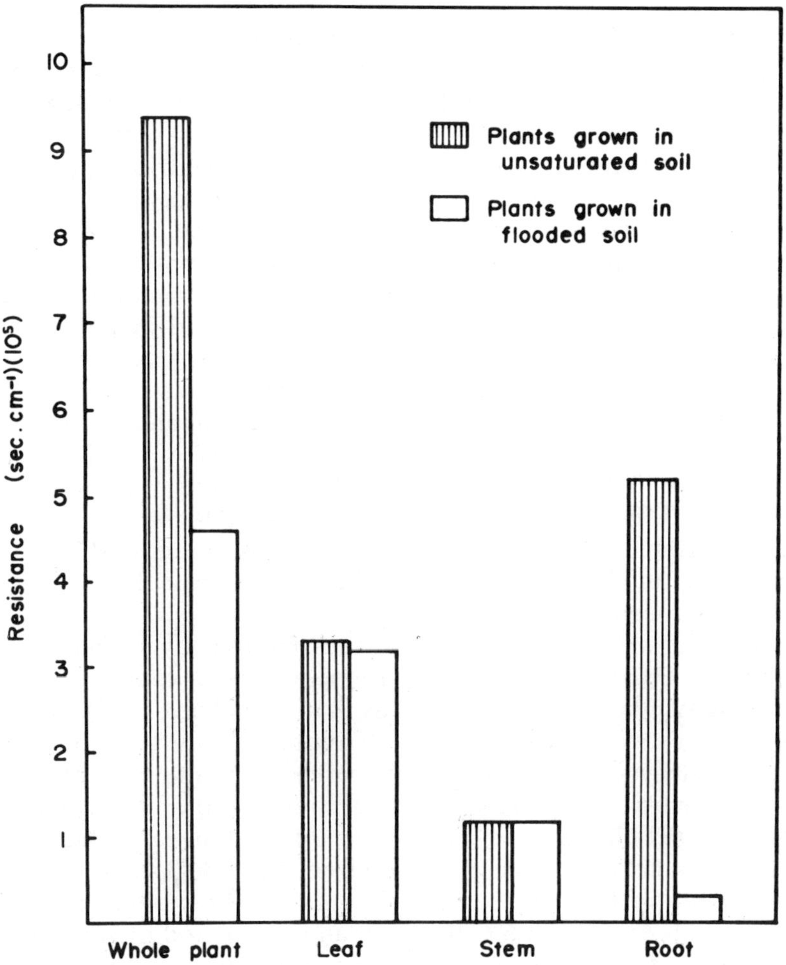

Figure 10.15. Resistance to water transport in different parts of IR8 rice plants grown in saturated and flooded soil. [From Tomar and Ghidyal (77) who measured resistance as $t_{1/2}/0.195d$, where $t_{1/2}$ is the half-time in seconds for recovery and d is half the average leaf thickness. Adapted from *Agronomy Journal*, Vol. **67**, 269–272, 1975, by permission of the American Society of Agronomy.]

roots to allow atmospheric oxygen to diffuse from leaves to roots. They supply needed oxygen not found in flooded soils and apparently allow water to move easily through roots. This study suggests that the roots of rice plants grown under upland culture are more resistant to water transport than those grown under lowland culture, but at present the relationship among root resistant and water transport, transpiration, and drought tolerance is not evident.

Leaf Display

IRRI scientists (78) observed that four out of five leafy cultivars having low light transmission ratios showed severe water stress symptoms, while cultivars exhibiting high light transmission ratios (nonleafy, erect leaf type cultivars) showed no severe water stress symptoms. Thus, it appears that a spreading leaf canopy does not contribute to drought tolerance of rice. Also cultivar differences in leaf dry weight are not correlated to known differences in drought tolerance.

Evidence was provided that leaf rolling in response to water stress could be regarded as a desirable characteristic of drought tolerant cultivars (79). Yoshida (72) pointed out that leaf size and shape are closely related to boundary layer resistance which decreases as windspeed increases and affects water movement in and out of leaves. Rice leaf width ranges from 0.5 to 2.5 cm wide depending on cultivar and environmental conditions. Thus, boundary layer resistances for leaves 0.5–2.0 cm wide will range from 0.92 to 1.83 sec cm^{-1} at wind speeds of 1 cm sec^{-1} and from 0.09 to 0.18 sec cm^{-1} at wind speeds of 100 cm sec^{-1}. Because stomatal resistance of rice leaves without stress averages 3–5 sec cm^{-1}, Yoshida concluded that leaf width will affect the transpiration rate only at low wind speeds.

Leaf angle will influence rate of transpiration by influencing leaf temperature. Provided that stomata are open, droopy leaves will be hotter and have higher transpiration rates than erect leaves (72). When leaf temperature exceeds a critical value of approximately 48°C (80) leaf cell tissue is destroyed, resulting in leaf scorching. White leaf tip, reduced tillering, sterility, and reduced grain filling are other symptoms of excessive heat (81). When rice plants are subjected to water stress (-10.8 bars) the stomatal resistance increased from 2.2 to 29.3 sec cm^{-1} causing leaf temperature to increase (80). The average temperature of water stressed plants was about 5°C higher than nonstressed plants. Leaf temperature varied considerably from one minute to the next as solar radiation varied. Maximum leaf temperature was 48°C. Because leaf temperature increases under severe water stress, heat tolerance may be important in drought tolerant rice cultivars (80).

APPLIED ASPECTS OF IRRIGATION

Rice plants are grown under flooded conditions where possible because rice yields are higher under flooded conditions. Benefits of flood irrigation are attributed to control of nonaquatic weeds (83), enhanced availability of nutrients (43, 64, 83), lack of water stress (72), and disease control (84). De Datta and Beachell (84) pointed out that the incidence of blast and Helminthosporium diseases were generally higher under upland conditions than under flood-irrigated conditions. However, straighthead, a physiological disease associated with flooding and excessive reduction of organic and inorganic soil components under flooded conditions is corrected by draining the flood water (prior to heading) and allowing the soil to aerate and oxidize before reapplying the floodwater.

The practice of alternately flooding and drying a field results in inefficient use of fertilizer and soil nitrogen because it enhances nitrogen loss by denitrification (55, 58, 60, 61). Topdressing nitrogen just prior to flooding results in greater nitrogen efficiency than topdressing into flooded fields (85, F. T. Turner and J. W. Stansel, unpublished data).

Floodwater depth and time of application in reference to the physiological stage of rice have an influence on weeds as well as rice. Water depth of 7.5 cm was sufficient to minimize the population of grasses whereas 15 cm of water was required to curtail growth of broadleaves and sedges in transplanted rice (86). Flooding throughout the growing season helps to control many terrestrial weeds but stimulates aquatic weeds. Oelke and Mueller (87) found that a shallow flood (4 cm) repeatedly produced higher yields than intermediate (8 cm) or deep (18 cm) floods for water seeded lowland rice in California which is flooded from seeding to harvest. Floods deeper than 4 cm reduced the number of tillers per square meter under the California system of rice production.

Ferguson (88) found that flood depth of 5.1–10.2 cm was optimum for "Nova 66" (lowland rice) grown on a silt loam soil in Arkansas. Deeper floods of 15.3–20.3 cm and shallower floods of 2.5–5.1 cm resulted in reduced yields.

Flood depth for optimum yield appears to be about 10.2 cm which is also sufficient for control of most grasses (87, 88). However, a slightly deeper flood is required to control most broadleaf weeds (86).

Critical Water Demand Periods

Researchers around the world have studied the critical water demand period for lowland rice by determining the growth stage most sensitive to water stress or by measuring water consumption during a given growth stage. Using the latter technique Sahu and Rath (89) measured water consumption during each of nine growth stages of transplanted lowland rice under conditions in east–central India. Total water use increased with each successive growth stage, peaking during booting and decreasing thereafter. Matsushima (90) reported that peak transpiration occurred at either the reduction division or flowering stages when lowland rice is grown under Malaysian conditions. He recorded transpiration ratios (water transpired/water evaporated) of 0.5, 1.0, and 1.3 for 7, 3, and 0 weeks before heading, respectively. A few weeks after heading the transpiration ratio had dropped to 1.0 and continued to decrease.

A second but smaller peak in transpiration also occurs during the maximum tillering stage (22, 91, 92). This second peak has been attributed to variability in weather or plant density. Results from most studies designed to measure water consumption during specific growth stages suggest that the early reproductive stage is the most critical water demand period.

Similar conclusions have been drawn by applying water stress at various growth stages and identifying the stage most sensitive to water stress. For example, at IRRI De Datta et al. (94) found that a soil water stress of -75 cb during the vegetative stage reduced grain yields by an average of 1500 kg ha^{-1} whereas stress during the reproductive phase reduced average yields by 2500 kg ha^{-1} or over 50%. These were averages of 48 upland and lowland cultivars or lines grown in water saturated soil except during the growth stage when the 75 cb water stress was applied. Evaluating the effect of water stress at various stages and of various durations, IRRI (95) found water stress at any stage would reduce the yield. However, the duration of these stresses was more

Table 10.8. Effects of irrigation treatments on grain yield, straw yield, and water use for Chianung 242 (dry season)[a]

Treatment[b]	Grain Yield (kg ha^{-1})	Straw Yield (kg ha^{-1})	Total Water Use (mm) (91 days)
1. DCF (10 cm)	7563	5878	683
2. SCF (2.5 cm)	7953	6789	559
3. II	8088	6708	612
4. DS	5250	4994	409
5. DS + F	6966	5535	794
6. AI	7562	6699	606
7. AI + F	8391	7471	728
8. IMS	5561	5795	487

[a] From IRRI (96).

[b] 1, Deep continuous flood; 2, shallow continuous flood; 3, optimum irrigation (season flood with drain at tillering); 4, drained soil; 5, drained soil with flood at panicle initiation; 6, alternate irrigation; 7, alternate irrigation with flood at panicle initiation; 8, irrigation at water stress.

Table 10.9. The effect of water management on the yield and maturity of Starbonnet rice (lowland type on a silt loam soil)[a]

Water Management	Yield ($kg\ ha^{-1}$)	Maturity (days)
1. Continuous flood	8170	109
2. Drain and dry soil at late boot	7990	109
3. Drain and dry soil at flood + 2 weeks	7920	111
4. Drain and dry soil at panicle initiation	7830	109
5. Drain and dry soil at late vegetative	7490	110
6. Delay flood for 6 weeks from emergence	7480	113

[a] From Wells and Shockley (61).

closely correlated to the yield reduction than the stage at which the stress occurred. Water stress during the vegetative stage reduced the tiller number while stress at the reproductive and grain filling stage reduced grain number and weight.

Other work at IRRI (96) designed to evaluate the effect of seven water management systems of rice yield is shown in Table 10.8. Evidence of the need for sufficient water during the early reproductive stage is shown by comparing yields of treatments 4 with 5 and 6 with 7. Treatment 5 and 7 were flooded at panicle differentiation while treatments 4 and 6 were not flooded and had the lowest grain yields. Treatment 8 (irrigation at water stress) shows that their water stress indicator did not indicate water stress. Wells and Shockley (61) working in Arkansas on a light textured rice soil and a lowland rice found that long dry periods early in the season and draining and drying at late vegetative periods have more of an effect on yield than draining and drying at early vegetative or at grain filling (Table 10.9). Draining at panicle initiation may be too late and create water stress during the reproductive period. Maturation period was directly related to water stress while yield was inversely related to water stress.

The beneficial effect of draining has also been shown (89, 96, G. N. McCauley, unpublished data) provided the drain period does not occur during the panicle differentiation to soft dough stage, and provided weed control problems are not encountered. Wells and Shockly (61) investigating the effect of a 1 week drain period (separated by 1 week flood) during the vegetative phase found that up to two drains had no effect on yield though one drain produced a slight yield increase. Three drain periods resulted in marked yield decrease. Each drain resulted in a 1 day delay in maturity. Water stress during the grain filling stage should also be prevented in order to maintain good grain milling quality. Some evidence suggests that if the floodwater is drained early during the grain ripening stage a subsequent irrigation (especially a sprinkle irrigation or any condition that increases moisture in the ripening grain) may result in checking (kernel cracking) which would lower whole grain milling yields and crop value.

Singh and Misra (97) found that water stress (50% depletion of available water) at panicle initiation stage reduced yields approximately 34% compared to the flooded treatment and was the most detrimental of 21 treatments. Grains per panicle and individual grain weight appeared to be the yield factors most adversly affected. Yield reduction because of water stress at stem elongation and tillering stages reduced yields 29 and 19%, respectively. Water stress at plant establishment and at milk stage reduced yields only slightly and had no effect when it occurred during the dough stage. Water stress at plant establishment reduced tiller count but grain yield was not reduced because the remaining tillers produced more grains per panicle and greater 1000 grain weight.

Cruz et al. (98) from Brazil found that the yield components most drastically affected by water stress were seeds per panicle and 1000 grain weight. Sprinkler irrigation at -1 bar potential, just prior to floral differentiation, and flooding thereafter resulted in no yield loss, whereas flooding prior to floral differentiation followed by sprinkler irrigation did result in yield loss. Yields from treatments sprinkle irrigated at -100, -300, and -500 cb potential were not significantly different but all three yielded lower than the continuously flooded treatment.

Matsushima (99), working in Malaysia, measured the effects of water stress in 11 separate periods during the growing season and concluded that water stress had the most serious effect during the reduction–division stage. The second and third sensitive growth stages were flowering and spiklet differentiation, respectively, denoting a critical stress period from about 20 days before to 10 days after flowering. He found the yield limiting factor for water stress in the reproductive phase to be failure of pollination resulting in blanks (unfilled grains) while Singh and Misra (97) found panicle number and 1000 grain weight to be the yield limiting factors resulting from water stress. The apparent differences were probably due to the degree of water stress to which the plants were subjected.

The general opinion from the literature is divided. Some, as mentioned earlier, report that water stress prior to heading reduces yields and delays maturity. Other reports have stated that lowland and upland rice can withstand water stress during the vegetative phase with no yield reduction and even some grain yield stimulation under certain conditions. These apparent contradictions need to be clarified by investigating the undefined parameters.

Consumptive Water Use

Depth of floodwater and irrigation frequency have been shown to have some effect on consumptive water use (evaporation, transpiration plus percolation) during the wet season in Orissa State, India (89). Flooding depths of 10, 7.5, and 7 cm required 1218, 1309, and 1337 mm of water per season, respectively, and irrigation every 4, 6, and 8 days resulted in 1236, 1281, and 1359 mm of water use per season, respectively. The reason for the increase in water use with decreases in depth of floodwater and irrigation frequency is not clear but the data do provide insight on water use of rice plants. Dry season water use values of 1777 mm in 1971 and 1569 mm in 1972 were recorded by Nair et al. (100), using a precision drum culture technique (101) in large Indian fields during the period from transplanting to maturity. Evapotranspiration (E_t) was 821 and 732 mm per season, respectively. Deep percolation averaged 12.01 and 10.47 mm day^{-1}, which was greater than the E_t values 10.4 and 9.15 mm day^{-1} during 1971 and 1972, respectively. The evaporation, 3.03 and 3.26 mm day^{-1} during 1971 and 1972, respectively, was not in proportion with E_t. Possibly evaporative demand during 1972 could have invoked stomatal closure, lowering transpiration rate in spite of the flooded conditions. Table 10.10 serves to illustrate that water use is not always lower during the wet season relative to the dry season (89, 91, 100). The seven water management treatments show the tremendous variation in water consumption that can be expected at one location. The variation was apparently due to inter and intraseasonal climatic variations (dry vs. wet season and dry vs. dry season), water management practices (flooding requires more water than rain-fed), and time from transplanting to maturity.

In summary it appears that in lowland rice culture, evapotranspiration at one location can range from 538 (104) to 821 mm (100), while under upland conditions values of 409 mm have been reported (Table 10.10). Evapotranspiration can increase with increased evaporative demand, increased time from seeding to maturity, and increased water supply. Similarly, total water required to produce rice increases with seepage, runoff, and poor water management systems. Table 10.11 provides a summary of water use around the world. Large variation is shown from country to country but the variation within a state is also as large as the world variation.

Crop Water Consumption Comparison

Water management for rice as discussed in the previous sections tends to indicate that water use for rice would be considerably greater than for other agricultural crops. Table 10.12 (110) compares the E_t per season of rice to 10 other crops grown under the same environmentalconditions in California. Water use efficiency (*WUE*) is another common method of evaluating water management practices and gives relative values that can be compared between crops. Table 10.13 lists *WUE* determinations

Table 10.10. Summary showing variation in consumptive water use of rice at IRRI from seven water treatments during wet and dry seasons for years 1965–1968

Treatment	Reference	Year	Season	Duration (days)	Yield (kg ha^{-1})	Water Use (mm)
1a. 10 cm flood	96	1965	Dry	91	7563	683
1b. 10 cm flood	102	1966	Wet	86	5075	803
2a. 2.5 cm flood	96	1965	Dry	91	7953	559
2b. 2.5 cm flood	102	1966	Wet	86	5110	568
2c. 2.5 cm flood	103	1968	Dry	91	9548	805
3a. Rain-fed	96	1965	Dry	91	5250	409
3b. Rain-fed	102	1966	Wet	86	4600	458
4. Rain-fed, flood panicle	96	1965	Dry	91	6966	794
5. Continuous soil saturation, flood panicle	103	1968	Dry	91	9142	780
6. Continuous soil saturation	103	1968	Dry	91	9025	647
7. Midseason drain	102	1966	Wet	86	5300	635

Table 10.11. Summary of seasonal water use for lowland rice grown at different locations

Location	Reference	Water Use (mm)	
Orissa State, India	89	1236–1359	
India	101	1569–1777	
Lower Egypt	57	1797	
Middle Egypt	57	2095	
Texas	104	538–596	
Eastern Texas (clay)	105	945–1061	
(sandy loam)	106	893–1012	
Western Texas	107	1149–1801	
California	108	914	
(high percolation rate)	109	2530	
Arkansas	108	457	
Louisiana	108	914	
		Wet season	Dry season
Malaysia	91	976	1344
Philippines	96,102,101	5075	7563

Table 10.12. Consumptive water use during the growing season for major field crops [a]

Crop	E_t (mm)
Alfalfa	1067
Barley	406
Beans	483
Corn	635
Grain sorghum	584
Rice	1067
Safflower	610
Sugar beets	762
Tomatoes	711
Wheat	483
Vine seeds	508

[a] From Agric Extension (110).

found in the literature and put on a common base. In general even though the E_t of rice is greater than most crops other than alfalfa, rice is from 2 to 2.5 times more efficient in its water use than wheat. Rice *WUE* data gathered from around the world fall between *WUE* for dryland grain sorghum and grain sorghum which only had one irrigation at the 10 cm joint stage. Table 10.13 further illustrates that even though rice is flooded most of the growing season it does utilize water as efficiently as grain sorghum. *WUE* values do not identify other water losses such as percolation, tailwater, and delivery inefficiencies. With these losses included, rice requires much more water than grain sorghum. However,

Table 10.13. Water use efficiency of rice compared to wheat and grain sorghum

Crop	Reference	E_t (mm)	Yield (kg ha^{-1})	*WUE* [a]
Wheat (dryland)	111	478	1841	3.85
Wheat (one irrigation)	111	599	2231	3.72
Grain sorghum (dryland)	111	338	2856	8.45
Grain sorghum (one irrigation)	111	439	4816	10.97
Rice (rain-fed)	102	458	4600	7.24
Rice (midseason drain)	102	635	9300	8.34
Rice (2.5 cm flood)	102	568	5110	9.00
Rice (10 cm flood)	102	803	5075	6.32
Rice (2.5 cm flood)	103	805	9548	11.86
Rice (10 cm flood)	100	596	5108	8.57

[a] $WUE = DW$ (kg ha^{-1})/E_t (mm).

it appears that with improved delivery systems, good management, and impermeable soils, rice can have *WUE* values equivalent to those of sorghum and other similar crops.

REFERENCES

1. R. Huke, Geography and Climate of Rice, in *Climate and Rice*, IRRI, Los Banos, Philippines, 1976, pp. 31–50.
2. J. H. Martin and W. H. Leonard, Rice, in *Principles of Field Crop Production*, Macmillan, London, 1970, pp. 495–518.
3. J. N. Efferson, Outlook for World Rice Production and Trade, in *Rice, Science and Man*, IRRI, Los Banos, Philippines, 1972, pp. 127–142.
4. S. Mitsui, Dynamic Aspects of Nutrient Uptake, in *The Mineral Nutrition of the Rice Plant*, Johns Hopkins Press, Baltimore, Md., 1964, pp. 53–62.
5. Te-Tzu Chang and S. K. De Datta, Varietal Improvement of Upland Rice in *Major Research in Upland Rice*, IRRI, Los Banos, Philippines, 1975, pp. 91–100.
6. B. S. Vergara, Physiological and Morphological Adaptability of Rice Varieties to Climate, in *Climate and Rice*, IRRI, Los Banos, Philippines, 1976, pp. 67–86.
7. I. Nishiyama, Effects of Temperature on the Vegetative Growth of Rice Plants, in *Climate and Rice*, IRRI, Los Banos, Philippines, 1976, pp. 159–185.
8. J. W. Stansel, Effective Utilization of Sunlight, in *Six Decades of Rice Research in Texas*, Res. Monogr. 4, Texas Agric. Exp. Stn., College Station, Texas, 1975, pp. 43–50.
9. D. J. McDonald, J. W. Stansel, and E. C. Gilmore, *Ind. J. Genetics* **34A,** 1068–1073 (1974).
10. H. I. Oka, *Sabrao J.* **6,** 241–243 (1974).
11. *IRRI Annual Report*, IRRI, Los Banos, Philippines, 1963, p. 41.
12. J. W. Stansel, The Rice Plant: Its Development and Yield, in *Six Decades of Rice Research in Texas*, Res. Monogr. 4, Texas Agric. Exp. Stn., College Station, Texas, 1975, pp. 9–21.
13. D. A. Downey, B. A. Huey, and B. R. Wells, *Arkansas Farm Res.* **26,** 9 (1977).
14. B. D. Webb and R. A. Stermer, Criteria of Rice Quality, in D. F. Houston, Ed., *Rice Chemistry and Technology*, American Association of Cereal Chemistry, St. Paul, Minn., 1972, pp. 102–123.
15. D. F. Houston, Rice Hulls, in D. F. Houston, Ed., *Rice Chemistry and Technology*, American Association of Cereal Chemistry, St. Paul, Minn., 1972, pp. 301–352.
16. D. F. Houston, Rice Brand and Polish, in D. F. Houston, Ed., *Rice Chemistry and Technology*, American Association of Cereal Chemistry, St. Paul, Minn., 1972, pp. 272–300.
17. S. K. De Datta, Upland Rice around the World, in *Upland Rice*, IRRI, Los Banos, Philippines, 1975, pp. 2–11.
18. *IRRI* Annual Report, IRRI, Los Banos Philippines, 1967, p. 149.
19. P. R. Maurya and B. P. Ghildyal, *El Riso* **25(4),** 357–365 (1976).
20. H. A. Kordan, *J. Appl. Ecol.* **11(2),** 685–690 (1974).
21. H. A. Kordan, *New Phytol.* **76(1),** 81–86 (1976).
22. A. L. Chapman and D. S. Mikkelsen, *Crop Sci.* **3,** 392–397 (1963).
23. A. L. Chapman and M. L. Petersen, *Crop Sci.* **2,** 391–395 (1962).

24. H. A. Kordan, *J. Exp. Bot.* **27(100)**, 947–952 (1976).
25. IRRI Annual Report, IRRI, Los Banos, Philippines, 1971, p 58.
26. S. Kawata, Y. Oahashi, K. Yumazaki, and K. Ishihare, *Proc. Crop Sci. Soc. Jpn.* **38**, 434–441 (1969) through T. Yoshida, *Climate and Rice*, IRRI, Los Banos, Philippines, 1976, pp. 265–277.
27. S. Kawata, K. Yamazaki, K. Ishikara, and H. Shibayama, *Proc. Crop Sci. Soc. Jpn.* **32**, 163–180 (1963) through T. Yoshida, *Climate and Rice*, Int. Rice Res. Inst., 1976, pp. 265–277.
28. K. Tei, *Nogyo Oyobi Engei* **8**, 109–118 (1933) through T. Yoshida, *Climate and Rice*, IRRI, Los Banos, Philippines, 1976, pp. 265–277.
29. K. Sato, Ann. Agric. Exp. Stn. Gov. Gen. Chosen **9**, 357–378 (1937) through T. Yoshida, *Climate and Rice*, IRRI, Los Banos, Philippines, 1976, pp. 265–277.
30. G. Guimaraes and R. Inforzato, *Brazantia* **32(5)**, 137–147 (1973).
31. K. W. Brown, F. T. Turner, L. E. Deuel, J. Price, and D. DeMichele, Development of Management Guidelines to Prevent Pollution by Irrigation Return Flow from Rice Fields, Environ. Prot. Tech. Ser., EPA-600/2-78-082, EPA, Ada, Okla., 1978, p. 57.
32. T. Horiuchi, *Res. Bull. Fac. Agric. Gifu Univ.* **39**, 1–7 (1976).
33. D. K. Das and R. L. Jat, *Agron. J.* **69(2)**, 197–200 (1977).
34. S. Kar, S. B. Varade, T. K. Subrumanyam, and B. P. Ghildyal, *El Riso* **23(2)**, 173–179 (1974).
35. P. R. Maurya and B. P. Ghildyal, *El Riso* **24(3)**, 239–244 (1975).
36. S. K. Pradhan, S. B. Varada, and S. Kar, *Plant Soil* **38(3)**, 501–507 (1973).
37. Y. Murata and S. Matsushima, Rice, in L. T. Evans, Ed., *Crop Physiology—Some Case Histories*, Cambridge University Press., 1975, pp. 73–99.
38. A. Tanaka, *Sci. Soil Manure, Jpn.* 29, 327–333 (1958) through T. Yoshida, *Climate and Rice*, IRRI, Los Banos, Philippines, 1976 pp 265–277.
39. Y. Fujii, *Bull. Fac. Agric. Saya Univ.* **23**, 1–117 (1961) through Y. Murata and S. Matsushima, Rice, in L. T. Evans, Ed., *Crop Physiology—Some Case Histories*, Cambridge University Press, 1975, pp. 73–99.
40. T. Katayama, *Yoken-do, Tokyo* (1951) through T. Yoshida, *Climate and Rice*, IRRI, Los Banos, Philippines, 1976, pp. 265–277.
41. M. Chellappa and J. S. Rao, *Madras Agric. J.* **60**, 1797–1798 (1973).
42. B. M. Sharma, M. S. Gangwar, and B. P. Ghildyal, *El Riso* **24(1)**, 31–35 (1975).
43. F. N. Ponnamperuma, Ph.D. Thesis, Cornell University, Ithaca, New York, 1955.
44. H. K. Krupp, W. P. Abilay, and E. J. Alvarez, *Rice Breeding*, IRRI, Los Banos, Philippines, 1972, pp. 663–673.
45. Y. Ishizuka. 1971. *Adv. Agron.* **23**, 241–315.
46. *IRRI Annual Report*, IRRI, Los Banos, Philippines, 1971, pp. 59–60.
47. *IRRI Annual Report*, IRRI, Los Banos, Philippines, 1971, pp. 66–68.
48. N. S. Evatt, Progress Report 2006, Texas Agric. Exp. Stn., Beaumont, Texas, Texas A&M Univ., 1958.
49. B. R. Jackson, A. Yantasast, C. Prechachart, M. A. Chowdhury, and S. M. H. Zaman, Breeding Rice for Deep-Water Areas, in *Rice Breeding*, IRRI, Los Banos, Philippines, 1972, pp. 517–528.
50. *IRRI Annual Report*, IRRI, Los Banos, Philippines, 1974, pp. 139–141.
51. C. R. Adair and K. Engler, The Irrigation and Culture of Rice, in *Water*, Yearbook Agric., 1955, pp. 389–394.

52. F. C. Raney, R. M. Hagan, and D. C. Finfrock, *Calif. Agric.* **11(4)**, 19, 20, 37 (1957).
53. F. C. Raney, *Calif. Rice Res Symp. Proc., Albany, Calif., 1959*, pp. 20–23, through T. H. Johnston and M. D. Miller, *Rice in the United States: Varieties and Production*. Agric. Handbook No. 289. USDA.
54. *IRRI Annual Report*, IRRI, Los Banos, Philippines, 1971, p 58.
55. M. B. Sturgis, Managing Soils for Rice, in *Soils*, U.S. Dept. Agric. Yearb., 1957, pp. 658–665.
56. D. C. Finfrock, F. M. Raney, M. D. Miller, and L. J. Booher, 1960, *Univ. Calif. Div. Agr. Sci. Leaflet* **131**, 1960, 2 pp.
57. H. A. El-Tobgy, *Contemporary Egyptian Agriculture*, 2nd ed., p. 107. Ford Foundation, Cario, Egypt, 1976.
58. W. H. Patrick, Jr., and D. S. Mikkelsen, in R. A. Olson, Ed., *Fertilizer Technology and Use*, 2nd ed., Soil Science Society of America, Madison, Wis., 1971, pp. 187–215.
59. F. N. Ponnamperuma, Dynamic Aspects of Flooded Soils and the Nutrition of the Rice Plant, in *The Mineral Nutrition of the Rice Plant*, Johns Hopkins Press, Baltimore, Md. 1964, pp. 295–328.
60. M. E. Tusneem and W. H. Patrick, Jr., *Bulletin No. 657*, La. State Univ. Agric. Exp. Stn., Baton Rouge, La., 1971.
61. B. R. Wells and P. A. Shockey, *El Riso* **27**, 81–88 (1978).
62. I. C. Mahapatra and W. H. Patrick, Jr., *Soil Sci.* **107**, 281–288 (1969).
63. F. T. Turner and J. W. Gilliam, *Plant Soil* **45**, 353–363 (1976).
64. F. T. Turner and J. W. Gilliam, *Plant Soil* **45**, 365–377 (1976).
65. S. A. Barber, J. M. Walker, and E. H. Vasey, *J. Agric. Food Chem.* **11**, 204–207 (1963).
66. R. E. Phillips, G. A. Place, and D. A. Brown, *SSSP* **32**, 41–44 (1968).
67. J. N. Marais and D. Wiersma, *Agron. J.* **67**, 777–781 (1975).
68. S. K. De Datta, T. Chang, and S. Yoshida, Drought Tolerance in Upland Rice, in *Upland Rice*, IRRI, Los Banos, Philippines, 1975, pp. 101–116.
69. *IRRI Annual Report*, IRRI, Los Banos, Philippines, 1972, pp. 60–62.
70. S. K. De Datta, W. P. Abilay, and G. N. Kalwar, *Water Management in Philippine Irrigation Systems: Research and Operations*, IRRI, Los Banos, Phillipines, 1975, pp. 19–36.
71. *IRRI Annual Report*, IRRI, Los Banos, Philippines, 1973, pp. 52–53.
72. S. Yoshida, Factors That Limit the Growth, in *Major Research in Upland Rice*, IRRI, Los Banos, Phillipines, 1975, pp. 46–71.
73. *IRRI Annual Report*, IRRI, Los Banos, Philippines, 1973, p. 55.
74. *IRRI Annual Report*, IRRI, Los Banos, Philippines, 1972, pp. 213–215.
75. S. K. De Datta, T. T. Chang, and S. Yoshida, Drought Tolerance in Upland Rice, in *Major Research in Upland Rice*, IRRI, Los Banos, Philippines, 1975. p. 110.
76. *IRRI Annual Report*, IRRI, Los Banos, Philippines, 1973, pp. 50–52.
77. V. S. Tomar and B. P. Ghildyal, *Agron. J.* **67**, 269–272 (1975).
78. *IRRI Annual Report*, IRRI, Los Banos, Philippines, 1972, p. 159.
79. *IRRI Annual Report*, IRRI, Los Banos, Philippines, 1973, p. 53.
80. *IRRI Annual Report*, IRRI, Los Banos, Philippines, 1972, p. 214.
81. *IRRI Annual Report*, IRRI, Los Banos, Philippines, 1974, pp. 150–151.
82. C. R. Adair, M. D. Miller, and H. M. Beachell, Rice Improvement and Cultural in the United States, in *Advances in Agronomy*, Vol. 14, Academic Press, New York-London, 1962, pp. 61–108.

83. T. H. Johnston and M. D. Miller, Culture, in *Rice in the United States*: Varieties and Production, U.S. Dep. Agric. Handb. No. 289, USDA Washington, D.C., 1973, pp. 88–134.

84. S. K. De Datta and H. M. Beachell, Varietal Response to Some Factors Affecting Production of Upland Rice, in *Rice Breeding*, IRRI, Los Banos, Philippines, 1972, pp. 685–699.

85. N. S. Evatt, Developing Efficient Systems of Fertilization of Rice, in *Six Decades of Rice Research in Texas*, Res. Monogr. 4, Tex. Agric. Exp. Stn., Texas A&M University, College Station, Tex., 1975, pp. 31–36.

86. *IRRI Annual Report*, IRRI, Los Banos, Philippines, 1968, p. 208.

87. E. A. Oelke and K. E. Mueller, *Agron. J.* **61**, 227–230 (1969).

88. J. A. Ferguson, *Arkansas Farm Res.*, **19**, 4 (1970).

89. B. N. Sahu and D. Rath, *Ind. J. Agron.* **17(3)**, 206–215 (1972).

90. Matsushima, *Proc. Crop. Sci. Soc. Jpn.* **31**, 115–121 (1962), through T. Murakami, Physiological Aspects of Water Control in the Cultivation of Indica Rice Varieties, in *Rice in Asia*, Univ. Tokyo Press, 1975, pp. 252–276.

91. K. Sugimoto, *Proc. Crop Sci. Soc. Jpn.* **39**, 343–349 (1970) through T. Murakani, Physiological Aspects of Water Control in the Cultivarion of Indica Rice Varieties, in *Rice in Asia*, Univ. Tokyo Press, 1975, pp. 292–276.

92. K. Sugimoto, *Proc. Crop Sci. Soc. Jpn.* **39**, 350–354 (1970), through T. Murakami, Physiological Aspects of Water Control in the Cultivation of Indica Rice Varieties, in *Rice in Asia*, University of Tokyo Press, 1975, pp. 252–276.

93. T. Murakami, Physiological Aspects of Water Control in the Cultivation of Indica Rice Varieties, in *Rice in Asia*, University of Tokyo Press, 1975, pp. 252–276.

94. S. K. De Datta, F. G. Faye, and R. N. Mallick, Soil-Water Relations in Upland Rice, in E. Bornemisza and A Alvarado, Eds., *Proceedings of a Seminar on Soil Management in Tropical America*, CIAT, USAID, Cali, Columbia, 1975, pp. 168–185.

95. *IRRI Annual Report*, IRRI, Los Banos, Philippines, 1970, pp. 148–150.

96. *IRRI Annual Report*, IRRI, Los Banos, Philippines, 1965, pp. 222–224.

97. Y. P. Singh and J. P. Misra, *Ind. J. Agron.* **19(1)**, 60–63 (1974).

98. J. C. Cruz, S. S. Brandao, R. M. Del Guidice, P. A. Ferreira, and B. T. Loureiro, *Experiantiae* **19(9)**, 187–209 (1975).

99. S. Matsushima, *Proc. Crop. Sci. Soc. Jpn.* Special Issue 102–109 (1968) through T. Murakami, Physiological Aspects of Water Control in the Cultivation of Indica Rice Varieties, in *Rice in Asia*, University of Tokyo Press, 1978, pp. 252–276.

100. R. R. Nair, P. Suseelan, G. R. Pillai, P. N. Pisharady, and R. Gopalakrishnan, *Ind. J. Agric. Sci.* **43(10)**, 980–981 (1974).

101. *IRRI Annual Report*, IRRI, Los Banos, Philippines, 1963, pp. 141–142.

102. *IRRI Annual Report*, IRRI, Los Banos, Philippines, 1966, pp. 176–178.

103. *IRRI Annual Report*, IRRI, Los Banos, Philippines, 1968, pp. 206–212.

104. K. W. Brown, F. T. Turner, J. C. Thomas, L. E. Deuel, and M. E. Keener, *Agric. Water Mgmt.* **1**, 277–291 (1977).

105. Trinity River Authority, Public Record, Devers, Texas, 1977.

106. Texas Water Development Board, Report 127, 1971, Austin, Texas.

107. U.S. Dep. Interior, Bureau of Reclamation, South Region, File Memorandum, Austin, Tex., 1975.

108. C. R. Adair and K. Engler, Irrigation and Cultural of Rice, in *Water* U.S. Dep. Agric., 1955, pp. 389–394.

109. Agric. Extension, Univ. Calif., Rice Notes, 1975.
110. Agric. Extension, Univ. Calif., Farmers Notes, 1977.
111. F. G. Viets, Jr., Increased Water Use Efficiency by Soil Management, in W. H. Pierre, D. Kirkham, J. Pesek, and R. Shaw, Eds., *Plant Environment and Efficient Water Use*, ASA and SSA, Madison, Wis., 1966, pp. 259–274.

11

Sorghum

D. R. Krieg

INTRODUCTION

General

Sorghum is considered to be one of the five major crops of the world. It currently ranks fourth among the world's cereals following wheat, rice, and maize. Although the statistics are approximate, world production occurs on 40–45 million hectares with an average yield of 1200–1300 kg ha^{-1} (1). About 75% of the world's sorghum production occurs in Asia and Africa where it is consumed principally as food, and grain yields are about 50% of the world average. In much of the world, sorghum is usually produced in environments that are normally too harsh for rice or maize. Sorghum has the inherent ability to outproduce most other cereals under marginal environmental conditions, especially under hot, dry conditions (2). Some of the proposed characteristics of sorghum related to its ability to yield under more demanding environments include a fairly deep, extensive root system allowing full exploration and use of available soil water, a waxy "bloom" covering the stems and leaves reducing cuticular water loss, the ability to maximize water use efficiency through leaf orientation and stomatal regulation, and the ability to utilize pre-anthesis assimilate to fill the grain. These and other developmental and physiological attributes conferring a high degree of water stress resistance to sorghum will be described in some detail.

Origin and Adaptation

Martin (3) and Doggett (4) have provided excellent descriptions of the origin and spread of cultivated sorghum. Briefly drawing from these, it is proposed that sorghum originated in East Africa, probably Ethiopia. The domestication process began between 3000 and 4000 B.C. From Ethiopia, the species spread throughout West Africa. As sea trade between continents developed, sorghum spread throughout the developing world. The movement into the Americas is the most recent extension occurring in conjunction with slave trade from West Africa in the mid-1800's. At present, U.S. sorghum production is primarily in the Great Plains states of Texas, Oklahoma, Kansas, and Nebraska where it is used as a livestock feed or is exported.

Although native to the tropics, many currently used sorghums are well adapted to the temperate regions of the world. Today sorghum is produced on six continents in areas where the growing season can be characterized with 125 frost-free days, average temperatures exceeding 20°C, and in excess of 25 cm of available water. The spread of sorghum into the temperate regions of the world is in part due to the success of the conversion program, a joint venture of the USDA and the Texas Agricultural Experiment Station (5). Sorghum is photoperiod sensitive and is a short day plant in which floral initiation and flowering are hastened by long nights (6–10). In the conversion program, the tall, photoperiod-sensitive, exotic sorghums are converted through a backcrossing technique to short, photoperiod insensitive types that are capable of flowering and producing grain under a rather wide range of environmental conditions. As observed for most organisms, large genotype by environment interactions exist, resulting in large, but localized breeding efforts to produce cultivars adapted to particular environmental niches.

In the 1950s through the efforts of pioneering sorghum researchers in Texas, a system evolved for the mass production of hybrid seed which resulted in a significant increase in yield potential compared to open pollinated cultivars. The major yield component to be increased by hybridization was seed per head (11, 12). Subsequent yield increases have been related to proper cultural management of the hybrids to realize more of their genetic potential. Under favorable production conditions yields of full season hybrids are commonly in the 7000–9000 kg ha^{-1} range. Under dryland (rain-fed) conditions yields are usually considerably lower and highly variable depending on amount and timing of rainfall. The capability to produce a seed head with a fairly large number of seed is observed in the more stress tolerant cultivars under dryland conditions (13).

GROWTH AND DEVELOPMENT

General Growth Characteristics

The growth and development of sorghum is comparable to that of most determinate species in that three major developmental periods exist in the life cycle. Vanderlip and Reeves (14) have thoroughly described the sequential development of a common hybrid (RS610) from emergence through physiological maturity. Vanderlip (15) not only defined the developmental process but has also related this to the mineral nutrition. For the sake of convenience, the terminology adopted by Eastin (16) will be used here. The period from emergence until the apical meristem converts from a vegetative, leaf producing system, to a reproductive apex will be described as growth stage 1 (GS-1). Growth stage 2 (GS-2) will encompass the developmental sequences from panicle initiation until flowering. And growth stage 3 (GS-3) will be the grain filling period from flowering until physiological maturity of the grain.

Growth Stage 1

The time required from planting the seed to emergence of the coleoptile is usually 3 to 10 days depending on seeding depth, soil temperature, and water (17–19). The minimum soil temperature for germination is 8–10°C. During GS-1 total leaf number, total basal tiller number, and a significant portion of the total root system are established. The length of time spent in GS-1 is dependent on both genetic and environmental factors, but usually represents about one-third of the total time required from emergence to physiological maturity. Strong interaction exists between the photoperiod and thermoperiod in determining the length of GS-1 (10, 20, 21). In general, under long day conditions (dark period of less than 11 hr) the thermoperiod exerts the major influence on time spent in GS-1. The thermoperiod–photoperiod response is modified by the daily radiation level, with low levels of radiation increasing leaf numbers by prolonging GS-1 (21).

GS-1 is terminated when the apical meristem stops initiating leaves and begins the panicle differentiation processes.

Growth Stage 2

GS-2 represents the period of exponential increases in leaf area, root extension, and dry matter accumulation. Approximately 65–70% of the total leaf area is developed during this period (Fig. 11.1) which comprises

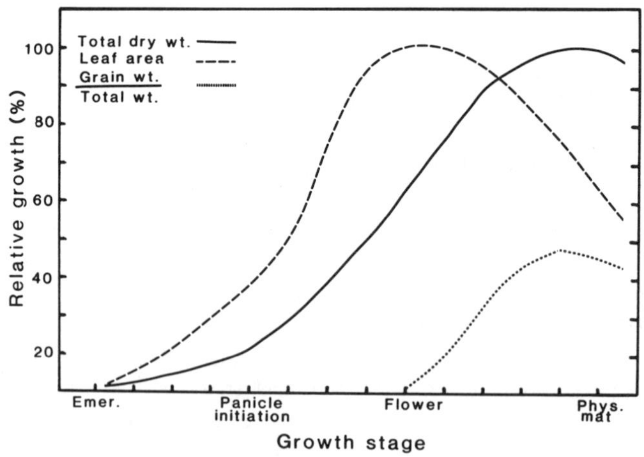

Figure 11.1. Relative growth and development of sorghum as a function of growth stage.

about 33–34% of the total development time. It is during this period that the inflorescence is differentiating and establishing potential seed number and simultaneously building the maximum possible plant size. In general, hybrids spend a slightly greater percentage of their total developmental time in GS-2 compared to their respective parent lines (16), although this is not always the case (22). It is quite obvious that any type of stress which results in disruption of the normal progression of events during this critical growth stage can have very marked effects on yield through both reductions in plant size, especially leaf area, and seed per head.

GS-2 is terminated by extension of the inflorescence from the whorl, attainment of maximum leaf area, and the onset of pollination. At this point in the developmental scheme maximum rooting depth has occurred and 60–70% of the total plant dry weight has been accumulated (23).

Growth Stage 3

The grain filling period represents the last portion of the life cycle and lasts from anthesis to physiological maturity of the grain. This is also a very critical growth stage. Ultimate yield of grain is a function of both the time spent in this period and the rate of dry matter accumulation by the developing grain (16, 24), provided sink size is not limited, such as occurs as a result of stress during GS-2.

The length of the grain filling period for the entire panicle is difficult to ascertain other than by weight accumulation. Pollination begins at the tip of the panicle and proceeds basipetally requiring 4–7 days to complete. Physiological maturity (maximum dry weight) of individual kernels also occurs with a basipetal progression but not at the same rate as pollination. Therefore, the time spent in GS-3 is less for seed at the tip of the panicle compared to those at the base and also varies with position on an individual rachis branch. Physiological maturity of an individual kernel can be ascertained by observing "black layer" formation (25). Seed size and weight follow a pattern similar to the attainment of physiological maturity.

Of the available data concerning time spent in GS-3, no general statement can be made relative to hybrids and parent lines. Considerable variability does exist in the sorghum germplasm as to length of the grain filling period and percentage of total developmental time spent in GS-3 (26). A positive, although not always highly significant, relationship exists between time spent in GS-3 and yield (16). In addition to the time factor, hybrids usually exhibit a higher production rate as compared to their parental lines, both over the entire season and during the grain filling period (16, 27). This response is primarily due to the hybrid exceeding the best parent in total leaf area (27–30).

Leaf area per plant attains a maximum at the beginning of GS-3. Dry matter accumulation in the grain, therefore, occurs during a period of declining leaf area (Fig. 11.1). The relative rate of leaf area decline is dependent on genotype, climate, and nutritional status of the plant (31). Another factor influencing production rate is the area of the top four leaves, since these leaves contribute in excess of 75% of the total dry matter to the developing grain (32–34). Recent evidence (35) indicates that source activity limits realization of potential yield under favorable conditions. Therefore, the effects of stress on photosynthetic activity are even more significant in yield development. The source–sink limitations on grain yield are modified by other plant factors including height, population density, leaf angle, and distribution of photosynthate (36–41).

Specific Growth Processes

Leaf Development

The embryonic seedling possesses four true leaves (42). Subsequent leaves are initiated after emergence with a total of 14–22 leaves being produced in photoperiod insensitive material. There is a positive correlation between time spent in GS-1 and leaf number. However, the leaf plastochron is not constant because the rate of leaf initiation increases with existing leaf area (10). Hybrids tend to initiate leaves at a faster rate than their parent lines, but do not produce greater numbers of leaves (16). The leaves of hybrids are usually longer and wider than midparent values and in most cases the hybrid exceeds the best parent in leaf development characteristics (27–30).

Considerable differences exist in the diurnal growth rates of sorghum leaves, probably reflecting differences in growth environments. Under controlled environment conditions and moderate evaporative demand, leaf growth rate was constant throughout the 24 hr day (43). Leaf expansion rates of approximately 60 cm^2 $plant^{-1}$ day^{-1} were observed which translated into a relative growth rate (ΔLA per unit LA) of 70% per day. In field studies the growth rates are considerably slower (about 20% per day) and the diurnal growth pattern is somewhat different. Under subhumid environmental field conditions, Jordan and Arkin (44) have indicated that maximum leaf extension rates were measured late in the afternoon when turgor was being regained; however, the extension rate was not directly related to the turgor presure. Under semiarid field conditions, Johnson (45) reported that the growth rate of the flag leaf paralleled the change in air temperature with the most rapid growth rate occurring about midday and the minimum rate occurring in the dark.

A study of sorghum leaf growth rates in relation to environment and existing leaf area was just recently completed (Krieg, unpublished). The study was designed to examine the relative contribution to leaf extension rates from turgor pressure and assimilate availability. The diurnal growth patterns for the youngest visibly expanding leaf (L-1) and the oldest expanding leaf (L-2) are depicted in Fig. 11.2 for plants with adequate soil water and those suffering from soil water deficits. The results indicated that the growth pattern is a function of stage of leaf development and stress. Additionally, we observed that the growth pattern was also modified by the amount of existing leaf area. Leaf growth rates were related to turgor pressure but the correlation coeffficients were low. The patterns observed suggested that growth rate was more related to the availability of assimilate when the turgor pressure exceeded a threshold level of approximately 2 bars. Growth during the day probably results from the leaf's own photosynthetic capability, whereas growth in the dark reflects the importation of assimilate from other sources. Leaf expansion of water stressed sorghum plants exhibit not only reduced rates but different diurnal patterns (Fig. 11.2), with most of the daily growth occuring at night in rapidly expanding leaves.

Water stress is one of the most common causes of reduced leaf area and is related to the cell expansion component which is very sensitive (46–48). McCree and Davis (43) have reported that leaf area development of sorghum is not only subject to control by cell expansion but also by cell division. By changing the environment from a warm-humid condition (32°C, 26°C dew point) to a hot-dry condition (41°C, 26°C dew point) in the absence of any soil water stress, a 40% reduction in leaf area was observed. Only half of this reduction could be attributed to cell size; therefore, the other 50% reduction was due to cell number differences. When the hot-dry environment was superimposed on a soil water stress condition an additional 30% reduction in leaf area existed. Essentially

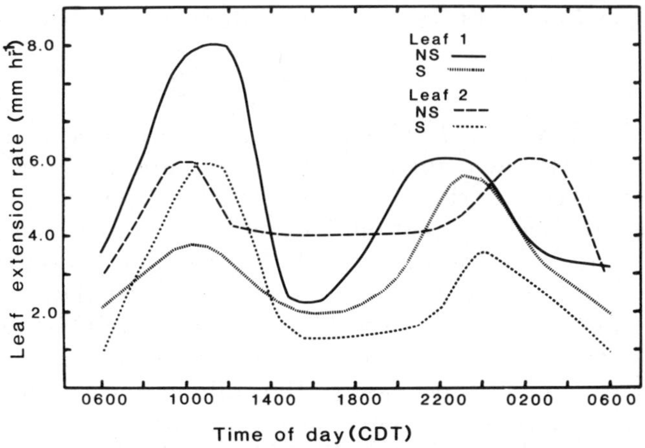

Figure 11.2. Diurnal pattern of extension rates of two different sorghum leaves as a function of water stress. Leaf 1 represents the youngest visible leaf emerging from the whorl; Leaf 2 is approximately 60% fully expanded; NS, nonstressed plants; S, stressed. [From Krieg, (unpublished data).]

all the water stress effect was due to differences in cell number. Stomatal density was not affected when soil water stress was not a factor in reducing leaf area (warm-humid or hot-dry). Imposition of soil water stress on the hot-dry environment resulted in significant increases in stomatal density per unit leaf area, but did not change the ratio of stomata to epidermal cell numbers.

Stout et al. (49) have demonstrated genotypic differences in both rate and duration of leaf growth under water stressed conditions. Reduced rates of expansion are extremely critical during the early stages of plant development when leaf growth is in the exponential phase.

Tillering

Sorghum, like other crops belonging to the Gramineae family, has the ability to produce tillers. Variations in the number of tillers produced by plants determine, in many cases, the differences in yield potential. Downes (50) reported that photoperiod had little effect on tillering, but there was a definite temperature response. If plants at the 4–6 leaf stage were subjected to a few days where the average daily temperature was less than 18°C, they were induced to tiller. Soil fertility level, light intensity, and plant population are also involved in the tillering response (51). Stout et al. (49) indicated that tillers were more sensitive to water stress than was the main stem, especially as related to retention of living leaf area. All these influential factors suggest that tiller initiation and survival are related to available photosynthate. If the growth environment allows more photosynthate to be produced than can be used by primary growth, tillering is initiated. If conditions exist whereby main plant growth can utilize all available photosynthate, then tillering will be arrested, or initiated tillers will die if they are not self-supporting.

Panicle Differentiation

Lee et al. (52) have provided a comprehensive description of the sequence of events in the panicle differentiation process. Briefly, the meristem begins initiating primary panicle branches in an acropetal manner requiring 5–6 days for completion. The total number of primary branches and whorls within the inflorescence is apparently related to the size of the apical meristem at the beginning of the differentiation process. A positive correlation exists between the time spent in GS-1 and the size of the apex. Secondary and tertiary branching of the primary panicle branches also begins at the base of the inflorescence, progressing acropetally. The degree of secondary and tertiary branching decreases from the base of the inflorescence to the tip, resulting in fewer seed per primary branch at the tip (53). Spikelet differentiation begins at the tip and proceeds basipetally. Genotypic differences in the number of grains per primary branch at the tip of the panicle reflect differences in timing between completion of primary branch differentiation and initiation of spikelet differentation, allowing for a greater degree of secondary branching at the tip (52). Yield advantages of hybrids over parent lines are due to seed number increases at the base of the rachis (12). The basis for the greater seed number at the base of the panicle is due to two factors: (1) a larger reproductive apex that allowed for development of larger basal branch primordia, and (2) a 4 day time lapse between termination of acropetal differentiation of primary and secondary branching and the initiation of spikelet differentiation (53). The time differential was utilized by the hybrid to increase basal branch size prior to spikelet initiation. Spikelet differentiation throughout the panicle requires 10–12 days to proceed from the tip to the base. Lee et al. (52) indicate that under favorable field conditions the rate of floret abortion prior to fertilization is very low (less than 1%), except for panicle branches at the base of the rachis.

It is during GS-2 that the sorghum plant is very sensitive to environmental stresses, including water, heat, and light (54). Rather mild water stress, as indicated by leaf water potential differences of only 4–5 bars from the normal nonstressed condition, results in a rather large difference in growth rates of the panicle and leaves (49) and in number of seeds per panicle (55–57). Genotypic differences have been observed in the panicle growth rate response under stress as indicated by time spent in GS-2 (49), which is probably reflected in greater seed number per head found in more stress tolerant cultivars (13). The cause of the developmental response of the panicle to water stress during the GS-2 growth stage is still unresolved. Two possibilities exist: (1) reduced available energy and organic material for the growth of the panicle, and (2) a disruption of the normal hormonal status of the differentiating panicle. Photosynthetic productivity per plant is reduced when (1) leaf area is reduced (43, 58), (2) stomatal resistance is increased (59–66), and (3) biochemical inhibitions are increased (67–70). Hultquist (71) has indicated that genotypic differences exist in the allocation of photosynthate to the expanding inflorescence during the GS-2 stage of development and that possibly lack of adequate organic assimilate was the cause of the seed number reduction. One must consider the total plant during this developmental phase and examine the relative strengths of the various sinks for the limited amount of photosynthate being produced. Rapid expansion of the leaves and roots present rather formidable competition to the expanding panicle for the available assimilate. Sung (72) has demonstrated that the newly expanding leaves represent the major sink for available photosynthate under both nonstressed and stressed conditions. Since soil water stress usually results in increased root activity in comparision to top growth (23) one must assume that the roots continue to be a strong, competitive sink for available assimilate. Therefore, when water stress reduces photosynthetic productivity, the plant must modify its developmental process to ensure that sufficient organic matter will be available to complete the reproductive cycle. It does this through water conservation (stomatal regulation of water loss and reduction in transpiring leaf area), soil water exploration (extension of the root system), and reduction in eventual sink demand (reduced potential seed number). The observed effects of water stress are apparently related to available photosynthate, either directly or indirectly, since a reduction in light intensity at this critical growth stage produces the same effect on seed number (56, 57, 73). When environmental stress reduces seed number it is characterized by abortion largely at the base of the rachis depending on the timing and intensity of stress (56, 74). The role of plant hormones in the panicle differentiation process of sorghum is undefined, especially as related to water stress response. This area of sorghum physiology deserves immediate and intensified efforts.

Seed Development

Just shortly following fertilization, the caryopsis begins increasing in volume due to rapid cell division and expansion. Maximum seed volume is attained 15–20 days after anthesis (75). Starch accumulation is apparent 4–5 days after anthesis. Deposition begins in the crown region where endosperm cells are more mature and progresses downward and outward. Embryo growth is considerably slower than endosperm development. The most rapid period of growth is from 7 to 14 days postanthesis. The first embryonic leaf is initiated at day 10 with radicle development beginning between days 7 and 10 after pollination. The sorghum kernel attains morphological maturity about 14 days prior to attainment of physiological maturity.

Although GS-3 is less sensitive to water stress than GS-2, yield reductions do occur as a result of reduced seed weight, depending on the severity of the stress. Since seed weight is a function of both seed size and degree of filling, stress can affect each process. Although no definitive data exist, one can speculate that water stress would have the same effect on cell development in the caryopsis as it does in the leaf blade. Subsequent effects of stress on seed filling are manifest through reduced photosynthetic activity. Eastin (16) reported that the average grain filling period was reduced by approximately 20% for numerous hybrids grown under dryland conditions in Nebraska whereas yield was reduced by 25%. This indicates that plant assimilation rate became a more critical factor than time as water stress intensified. Dry matter accumulation by the grain is largely a function of current photosynthate produced by the flag leaf and the upper four leaves (32), at least under nonstressed conditions. The panicle itself is capable of assimilating CO_2, both from net assimilation and reassimilation of respiratory CO_2. Differences exist in net assimilation rates due to panicle morphology (76) and degree of water stress. Under severe water stress, when the whole plant photosynthetic activity declines, the relative contribution of the panicle to net assimilation greatly increases (77). It is estimated that under favorable conditions only 10–12% of the final grain weight is due to preanthesis assimilate (78). Under water stress conditions during grain filling large reductions in stem and lower leaf weights are observed, with the material presumably being translocated to the developing grain (31, 79), such as reported for maize (80).

Water stress during grain filling usually results in rapid senescence of the lower leaves, with genotypic differences existing in this characteristic (81). Leaf retention during GS-3 usually does not result in yield increases compared to the more senescent types unless plant populations are very low and the lower leaves are above the light compensation point during much of the daylight hours. Sung (72) has found that at field populations of 200,000 plants ha^{-1}, leaves below the fourth leaf from the top usually import photosynthate for their daily maintenance rather than export it. The photosynthetic rate of surviving leaf area is also reduced because of stress, the degree of reduction depending on stress intensity. The translocation system and sink strength are apparently unaffected until very severe stress exists (72). The grain filling period (GS-3) is less sensitive than GS-2 to water stress in terms of yield reductions (55, 57, 82, 83) presumably because grain filling can draw quite extensively on both organic and nitrogen reserves stored in the vegetative portions of the plant.

Yield Component Compensation

Yield of sorghum is dependent on various components including number of harvested panicles, number of seed per panicle, and weight of the individual seed. Most yield studies have indicated that the seed number component (number of harvested panicles × number of seed per panicle) is more highly correlated with final yield than is the seed weight component (4, 11–13, 84–88). Seed number is a function of genotype, plant population, and management practices. Some significant positive effects on yield can be attained by selecting for larger, heavier seed (89, 90).

Considerable compensation exists among the yield components of sorghum, resulting in yield maintenance within certain limits. If near optimum growth conditions exist during GS-1 and GS-2, low initial plant populations will compensate by increased tillering and increasing the size of the primary panicle, thereby increasing seed number per unit land area. If water stress exists during GS-1 and GS-2 to limit plant population or degree of differentiation of the panicle, then seed number is reduced, and seed weight increases offset the potential yield reductions. However, there is apparently an upper limit to seed size which restricts the degree of compensation attainable. Several reports indicate that only a 15–20% increase in seed weight can occur (16, 57, 91). Therefore, when seed numbers are reduced by more than 15–25%, yield reductions can be expected.

Root Growth and Development

Because of the nature of the root system and difficulties of recovering roots by soil excavation and washing, only limited descriptions of root development are available and these are highly variable. Since the roots are important to the total growth and development of all plants, it is imperative that a better understanding of root development and function be established, particularly with respect to soil water utilization.

In general, the root system of sorghum is comparable to that of most grasses in that a seminal or primary root system produced from growth of the radicle, supports the seedling for the first few weeks of establishment. During this time the adventitious root system is being initiated and developed from buds at the nodes below ground (92, 93). Soon after the plant emerges (within 7–10 days) adventitious roots are initiated at a rate of 0.5–1.0 day^{-1} (94, 95). The initiation process is cyclic, corresponding to initiation at successive nodes, and is referred to as the root plastochron. The total number of adventitious roots per plant is variable, but averages 15–20, depending on the degree of tillering (50). As soon as the adventitious roots become established, the seminal root system usually deteriorates. Early loss of the seminal root system due to mechanical damage or disease can result in death of the seedling or in a stimulation of branching and growth of the existing adventitious roots, thus resulting in greater root volume. During GS-1, linear relationships have been observed between the total root length and leaf area and between root volume and leaf area (95). Some genotypic differences have been observed in gross morphology and organization of the root system (96). Heterosis is also apparent and manifested in faster root growth rates of hybrids compared to parent lines (95). Plants with larger leaf area are also more likely to have larger root volumes and total length of adventitious roots. By the end of GS-1, rooting depths of 120–150 cm have been observed (97) with a lateral spread of 20–45 cm from the main stem depending on spacing and other cultural practices. Hybrids tend to have deeper, more extensive root systems than do parent lines (98) with the greatest concentration of roots in the top 30 cm of the profile (99–102).

During GS-2, extensive branching of the adventitious root system occurs, beginning near the soil surface and continuing to deeper depths with increasing age. Utilizing ^{32}P accumulation as an index of root extension, McClure, and Harvey (98) reported that a sorghum hybrid had 1.5–6 times more root development than its respective parents, with maximum extension occurring about the time of panicle exertion. Maximum dry weight of the root system also occurs about the time of flowering, under both irrigated and dryland conditions (100, 103). Under both dryland and limited irrigation conditions it is very likely that root growth is cyclic in each soil zone, depending on soil water conditions of that zone (104). As the soil dries, fine roots die, and as the soil is rewetted, new root growth replaces the dead tissue. No estimates are available as to the total accumulative root mass under these conditions.

During grain filling (GS-3) the major sink for photosynthate is the developing grain and root growth is probably restricted. Root length density (cm root length cm^{-3} soil) measured during this growth stage range from 4 at the soil surface to 2 at 30–35 cm, to 1 at 60–65 cm, and less than 1 below 85 cm (105). Using ^{32}P to measure depth and lateral growth, McClure and Harvey (98) indicated that hybrids have rather actively growing root systems during grain filling. Their results may be confounded by plant spacing, however, since most root studies indicate greatly diminished root growth during the grain filling period. The uppper 30 cm of soil apparently loses a considerable amount of root activity, whereas the deeper depths gain in activity, indicating a high degree of secondary and tertiary branching of the aging root system. Total root weight declines during this period of development indicating either a redistribution of previously deposited organic matter or increased senescence and death of existing roots due to inadequate supplies of current photosynthate to meet maintenance demands. After physiological maturity of the grain, if favorable soil water conditions exist, significant increases in root growth and weight are observed, especially in the top 30 cm of the soil.

Jordan and Miller (106) have recently provided an excellent review of their sorghum root research efforts with emphasis on genetic differences in rooting patterns. The basic premise of their program is that drought avoidance of sorghum can be enhanced in many agricultural situations by developing a root system that is more efficient in extracting soil water from deep in the profile. The current hypothesis is that root length density (cm root length cm^{-3} soil) of current sorghum cultivars is inadequate at depths below 50 cm to allow sufficient extraction to meet the evaporative demand. Theoretical evidence is presented to imply that increasing the root density to 2 cm root length cm^{-3} soil below 50 cm of depth would allow extraction to match high evaporative demands until the soil dried to approximately -3 bars. Based on their assumptions of soil depth and root dry matter, this increase in root length density would require about 800 kg ha^{-1} of additional dry matter. Whether this can be accomplished without reducing grain yield potential remains to be demonstrated. The identification of more desirable rooting patterns within existing sorghum germplasm is in its infancy, but preliminary results appear promising.

SOIL–PLANT WATER RELATIONS

Consumptive Water Use

Seasonal water use patterns have been described for a number of sorghum production areas representing different soil conditions, water-availability patterns, and environmental demand (107–113). In general, daily water use is largely a function of evaporation from the soil surface until sufficient leaf area is established for transpiration to become a measurable component of the total water use patterns. Evaporation rates from the soil depend on the degree of wetness of the surface since a wet surface will evaporate at a rate equivalent to the atmospheric demand. As the surface dries, the presence of air-filled pore spaces increases the resistance to water flux and the soil evaporation rate declines accordingly. As leaf area develops, the ratio of actual plant water consumption to potential evapotranspiration increases (Fig. 11.3). In a subhumid environment, actual evapotranspiration is equivalent to potential evapotranspiration, derived from the energy budget, when the leaf area index (*LAI*) exceeds 2.7 and the ground cover is in excess of 80% (114–117). When row spacing restricts the degree of ground cover and in environments where the saturation vapor pressure deficit of the atmosphere is quite large (semiarid environments), a rather significant contribution by sensible heat flux from surrounding areas increases the total daily E_t over the potential evaporation, E_0, as defined by the energy budget method. Well-watered sorghum usually has a leaf temperature several degrees below the air temperature, indicating a significant sensible heat

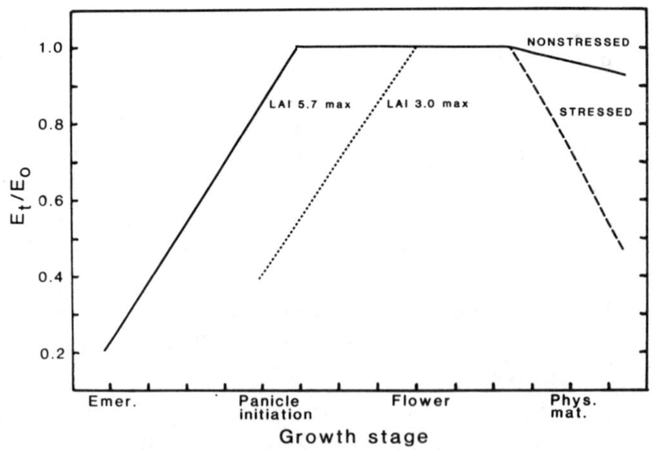

Figure 11.3. Actual evapotranspiration (E_t) as a function of potential evapotranspiration (E_0) as related to growth stage, leaf area index and soil water stress. [Data derived from Kanemasu (113) and Ritchie and Burnett (114); used with permission.]

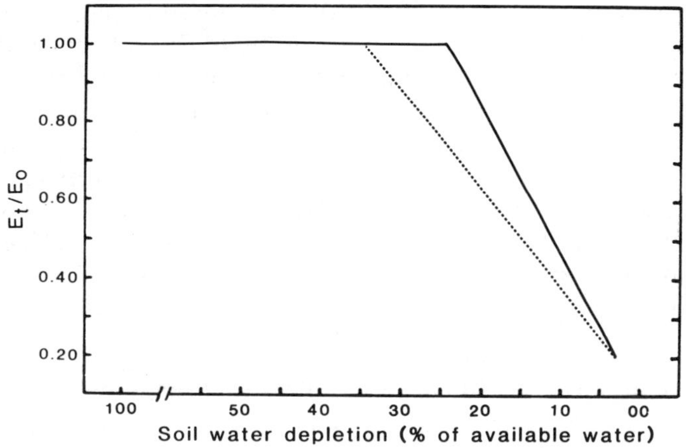

Figure 11.4. Ratio of evapotranspiration to potential evapotranspiration as affected by soil water availability. [From Ritchie et al. (116) and Sumayao et al. (120); used with permission.]

flux contribution to transpiration (60, 118–121). The leaf to air temperature differential is most pronounced in the late afternoon (1600–1800 hours) when the incident radiation is rapidly declining but the air temperature reaches its maximum.

The high rate of plant water use continues well into grain filling with a measurable decline being observed in late grain filling (beginning about the hard dough stage). Much of the reduction in consumptive water use during the late stages of plant development can be related to reduced leaf area due to increasing leaf senescence. It is also possible that the size of the root system is reduced because of root senescence during this period to the extent that absorption cannot maintain the high rates of transpiration. Additionally, the resistance to water transport within the leaf may increase due to aging. Both of the possibilities exist as suggested by decreasing leaf water potentials in old leaves even under nonlimiting soil water conditions (122).

Soil water supply capacity becomes a limiting factor in the $E_t{:}E_0$ relationship when the hydraulic conductivity is less than 10^{-6} to 10^{-7} cm day^{-1} (123). In sandy soils this is equivalent to a matric potential of -1 bar; whereas, in clay soils it is equivalent to a soil water potential of approximately -8 bars. This translates into utilization of 60–70% of the "available" soil water in the effective root zone before the $E_t{:}E_0$ ratio will be altered (116, 124) as indicated in Fig. 11.4. The extent of depletion of the available water in the root zone is a function of root length density. Nakayama and van Bavel (103) indicated that 90% of the root activity (as indicated by ^{32}P uptake) and 80–90% of the soil water depletion occurred within the top 100 cm of a 150 cm soil profile. As previously described, root length density is fairly high in the surface zone (4–5 cm cm^{-3} in the 0–15 cm depth), declining almost linearly to densities of less than 1 cm cm^{-3} at depths of 75–80 cm. Water flux is apparently rather constant throughout the root zone at a rate of approximately 0.003 cm^3 cm root $length^{-1}$ day^{-1} under nonlimiting soil water conditions (105). Genotypic differences have been observed in root density and rooting depth (94, 95, 102, 106). It is not yet apparent whether the genotypic differences in rooting depth and density are related to the plant's ability to utilize "available soil water" more effectively or efficiently. Considerable effort is currently under way to examine these relationships.

Yield and Water Use as Related to Cultural Management

Grain sorghum produces harvestable yields utilizing a wide range of cultural practices dependent on climate, water availability during the growing season, soil fertility, cultivation, and harvesting techniques. A vast amount of literature is available on most aspects of sorghum cultivation as related to yield. Some of the more pertinent literature covers a wide range of environmental conditions (83, 107, 112, 125–138).

A certain amount of water is required to establish the plant and allow adequate vegetative development to support potential yield without actually realizing any grain yield. Although this minimal level is dependent on environmental demand, it is probably in the range of 15–18 cm of available water in the Great Plains (139). When available water during the growing season exceeds this threshold level, grain yield will occur. Since sorghum yield is highly related to the number of seed produced per unit land area, plant population and spacing have been the subject of many studies related to water use efficiency. In general, when water is not a limiting factor throughout the growing season maximum yields are attained with high populations ($\simeq$250,000 plants ha^{-1}) and narrow row spacings (<50 cm between rows), with an additional advantage in using genotypes that utilize the entire growing season. Under rain-fed conditions where some degree of water stress is certain to exist during the growing season, the highest yields are usually attained from production systems where plant populations are reduced to approximately 100,000 plants ha^{-1} with narrow rows or nearly equidistant spacing providing some additional advantage. Of course, the time when the stress occurs in relation to the developmental stage of the plant is extremely critical, as is the intensity and duration of the stress. The panicle differentiation stage is the most critical time. If the plant can proceed through this developmental process with minimal plant stress, seed production will occur with only moderate changes in the seed weight component.

The effect of plant population on evapotranspiration has been recognized as being similar to the effect of degree of ground cover on E_t. Early in the growing season as leaf area is developing, E_t would be greater on a daily basis from a high population, close row spacing, than from a low population or wide row spacing. Several studies in semiarid or arid regions have emphasized the importance of developing ground cover as quickly as possible to minimize evaporation from the soil surface and to reduce sensible heat flux increasing transpiration beyond that attributed to net radiation (114, 121, 127, 140).

The results emphasize the need to modify planting patterns to maximize degree of canopy cover as early as possible within the desired population limits based upon available water supplies.

The relationship between grain yield and the ratio of actual to potential evapotranspiration (E_t:E_0) is depicted in Fig. 11.5. This generalized response suggests that the opportunities to increase water-use efficiency through genetic manipulation or cultural management are somewhat remote. Rather significant grain yield differences are observed among sorghum genotypes of comparable maturity when grown in a common nursery. However, using the most reliable techniques for measuring soil water status, only minor genotypic differences are detected in both rate of water use and total water use once the leaf area index has developed to the point that net radiation is maximized (LAI >3.0). Do these apparent differences in water use efficiency of grain yield reflect genotypic differences in partitioning of total dry matter between grain and vegetative material or do they reflect our inability to accurately measure soil water status, and thus total water use? The genotypic differences in water-use

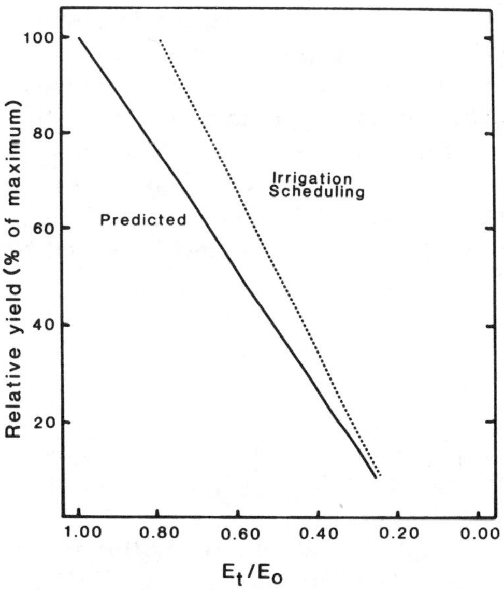

Figure 11.5. Relationship bewteen relative yield and E_t:E_0 ratios. Predicted value derived from data of Hanks (139); used with permission. Dashed line is speculative and represents what might be possible by properly scheduling irrigations to allow plant stress to exist at the least sensitive growth stages.

efficiency of grain yield are even more apparent under conditions where some degree of soil water stress exists.

The growth stage when the stress exists and its duration and intensity are extremely critical in determining the plant response as related to its ability to produce grain. Large genotypic differences exist in the panicle differentiation response to water stress intensity which greatly affects potential yield through effects on seed number. If the stress exists only during the grain filling period, then the seed weight component is affected. Although this developmental stage is less sensitive than the panicle differentiation stage, genotypic differences are observed in grain production water use efficiency. The genotypic differences observed with stress during this growth stage reflect the plants' ability to tolerate the stress via photosynthetic maintenance and the ability to utilize preanthesis assimilate to complete the grain filling process. Other genotypic differences that may play a role in determing water-use efficiency under soil water stress conditions include differences in leaf angle, which alters the photosynthesis : transpiration ratio; rooting depth and density, which result in minor unmeasurable differences at any point in the soil profile but when accumulated over the entire profile represent an appreciable increase in total available water; and degree of stomatal control, which may alter the photosynthesis : transpiration ratio.

From a water management standpoint it may be possible to withhold water at select growth stages and thus reduce $E_t : E_0$ without having any significant effect on yield (Fig. 11.5). The two least sensitive growth periods apparently are (1) during early vegetative growth prior to panicle initiation and (2) during the later stages of grain filling. The consequences of plant stress during the leaf initiation stage on subsequent development are poorly understood at best. Also, maximum stomatal control exists in young leaves and therefore if plant stress does occur transpirational water loss is immediately abated. Severe water stress during late grain filling reduces $E_t : E_0$ without large negative effects on yield. Possibly there are genotypes that can utilize a large amount of existing assimilate for grain filling and not be totally dependent on current assimilate production. However, in many cases where this phenomenum actually occurs, severe lodging often results and harvestable yield is greatly reduced compared to actual yield.

Plant Water Relations

Diurnal and temporal changes in leaf water status of sorghum leaves have been described for a wide range of climatic and thus evaporative demand conditions (53, 59, 62, 117, 141–145).

As evaporative demand increases from dawn, rather large within plant leaf water potential gradients exist among the various leaves (143). During the midday period of maximum evaporative demand, and thus minimal leaf water potentials, only minor within plant gradients are observed in fully illuminated leaves, suggesting only minor resistances exist in the water conducting system in the aboveground portion of the plant. Nonlinear trends have been observed in the leaf water potential transpiration rate response in both expanding and expanded sorghum leaves under nonlimiting soil water conditions (60; Krieg, unpublished). Once a threshold evaporative demand was attained resulting in leaf water potentials of −14 to −15 bars, further increases in transpiration rate were "uncoupled" from leaf water potential suggesting changes in internal plant resistance maintaining leaf water potential nearly constant (Fig. 11.6). As soil water became limiting, a nearly linear decline in the leaf water potential-transpiration rate response curve was noted in expanding leaves prior to the point of stomatal regulation (−18 to −20 bars) and reduction in transpirational water loss. Fully expanded, leaves were capable of maintaining high transpiration rates even as soil water potentials declined and leaf water potentials declined accordingly to −22 to −24 bars.

Under nonlimiting soil water conditions, relatively small genotypic differences are observed in transpirational water flux on a leaf area basis at a given environmental demand (61, 147). The daily leaf water potential required to sustain transpiration is a function of evaporative demand until some critical soil water potential is attained where the rate of supply to the plant begins restricting transpirational water flux. At this point, leaf water potential must decline accordingly to maintain or even steepen

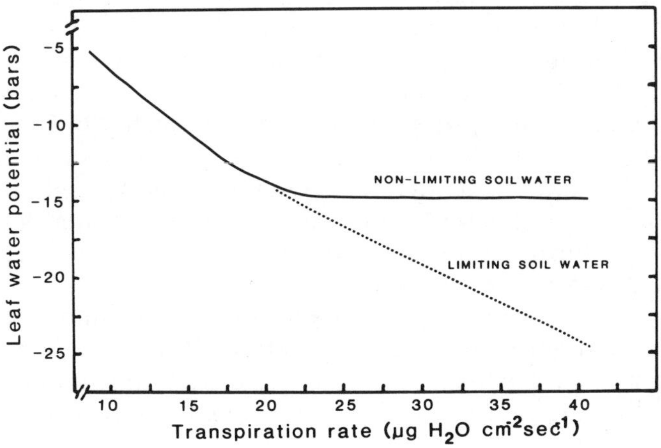

Figure 11.6. Influence of transpiration rate on water potential of fully expanded sorghum leaves subjected to different levels of available soil water. As water potential declined to −18 to −20 bars due to soil water stress, stomatal closure reduced transpiration and expanding sorghum leaves. [Data derived from Ackerson and Krieg (60); used with permission.]

the gradient from the soil to the leaf, or stomatal resistance must increase to reduce transpirational water flux. Reported values of soil water status at which leaf water potential begins to decline vary significantly and are dependent on growth stage, genotype, soil type, and environmental demand. In semiarid environments approximately 40–60% of the available water in the effective soil profile can be depleted before midday leaf water potentials are reduced below the normal −15 to −16 bars (59, 63, 143).

Blum (63, 147) has provided evidence of genotypic differences in both rate of soil water depletion and total water extracted and how this activity related to tissue water relations and drought tolerance. As an example "Shallu" extracted the least amount of total soil water but had the lowest leaf water potential. It is therefore considered very drought susceptible. "Combine Kafir 60" was considered to avoid drought stress by maintaining high leaf water potentials even when leaf diffusion resistance was minimal. "Feterita" extracted the greatest amount of soil water but also had low leaf water potentials and diffusion resistances. Under conditions of high evaporative demand in the field, we have observed no significant differences (<2 bars) in leaf water potential among genotypes under nonlimiting soil water conditions and only minor differences (2–3 bars) among genotypes as soil water potential declined causing a concomitant reduction in leaf water potential (61, Krieg, unpublished).

Many physiological and morphological processes are affected more by turgor pressure than by leaf water potential. Since turgor pressure is the difference between total water potential and osmotic potential, the role of osmotic adjustment in turgor maintenance represents a significant adaptive mechanism to water deficit on a daily and seasonal basis. Several recent studies have examined the role of osmotic adjustment in the total water relations of sorghum as related to turgor maintenance (61, 122, 148, 149, 150). A summary of the observed relationships between leaf water potential and pressure potential is depicted in Fig. 11.7. The data indicate that osmotic adjustment is a function of stress adaptation and of growth stage. Plants that have been preconditioned to soil water stress not only have greater turgor pressures at high water potential but are capable of further osmotic adjustment resulting in turgor maintenance as leaf water potential declines (148).

Sorghum also exhibits a much greater degree of osmotic adjustment capability and turgor maintenance in mature leaves after flowering than prior to flowering. This response may reflect altered source–sink relationships, allowing more of the available photosynthate to be utilized in osmotic adjustment.

Recent evidence indicates significant genotypic differences in the degree of osmotic adjustment and a genotype by growth stage interaction (61). Reductions in leaf growth and physiological responses are not always directly related to cell turgor pressure, suggesting that the total leaf water potential is also of importance.

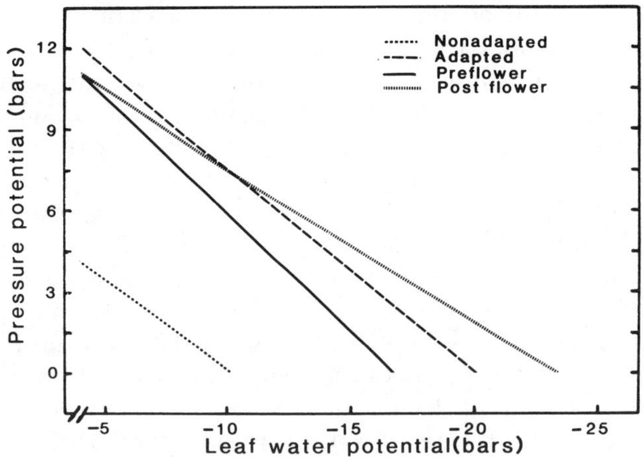

Figure 11.7. Influence of leaf water potential on pressure potential of sorghum leaves as a function of stress adaptation and growth stage. [Data derived from Jones and Turner (148) and Ackerson et al. (61); used with permission.]

Other evidence indicates that stress preconditioning and rate of stress development alters the relationship between water potential components such that plants previously exposed to stress or allowed to develop stress more slowly may have a significantly lower leaf water potential at zero turgor (149). Osmotic adjustment, due to an increase in tissue solute content or an increase in nonosmotic volume, can alter the leaf water potential–osmotic potential relationship such that zero turgor is attained at a lower water potential. Jones and Turner (148) have demonstrated that osmotic adjustment does occur in sorghum due to stress preconditioning. The decreased osmotic potential was related more to increased solute content than to changes in cell elasticity. Differences in the stress response between maize and sorghum were thought to be due to differences in cell elasticity (142). Sorghum had a much lower cell volume change per unit change in water potential than did maize.

Relative Water Content

Drought resistant plants have been described as those having a relatively small change in leaf water content per unit change in leaf water potential. Levitt (151) has stated that dehydration avoidance is related to a relatively small change in water saturation deficit (*WSD*) per unit reduction in leaf water potential, whereas total drought avoidance is more a function of avoidance of soil water stress which causes low leaf water potentials. Several studies, both field and growth chamber, have related relative water content (*RWC*) or water saturation deficit (*WSD*) to leaf water potential and other physiological responses of sorghum. Sanchez-Diaz and Kramer (141) state that sorghum leaf water potential declines more rapidly per unit change in *RWC* than does corn. A sigmoidal relationship between leaf water potential and *RWC* is usually observed. Leaf water potential initially changes rather rapidly per unit change in water content (Fig. 11.8). An inflection point is reached where the rate of decline in *RWC* is much greater per unit change in leaf water potential. This inflection point is usually associated with the attainment of zero turgor pressure. Fereres et al. (122) have indicated that the relationship between leaf water potential and *RWC* changes with leaf age. In older leaves the *RWC* is considerably higher at any leaf water potential as compared with young leaves. Blum (63) has described rather large genotypic differences in the leaf water potential at which the *RWC* begins to decrease exponentially, thus indicating significant differences in leaf water potential at 0 turgor. Jones and Turner (148), using two of the same genotypes that Blum (63) used, found no genotypic differences in the relationship between leaf water potential and *RWC* or between turgor pressure and *RWC*. Zero turgor in Jones and Turner's study was achieved at a *RWC* of 94% irrespective of previous stress history. Blum's data indicated zero turgor at a *RWC* of 80–85% and a leaf water potential of −16 bars for "Hegari" and −10 bars for "RS610." Ritchie and Jordan (117) have indicated that evapotranspiration was constant at a *RWC* in excess of 87% and began to decline rapidly when the *RWC* fell below this value presumably because of increased leaf resistance as turgor pressure approached zero. Pasternak and Wilson (152) indicated that leaf resistance began to increase as the relative water content decreased from 95% or a leaf water potential of −1 bar. At a *RWC* of 80% leaf resistance was 10 sec cm^{-1}. Under high evaporative demand field conditions we routinely measure leaf water potentials in the −15 bar range with *RWC*

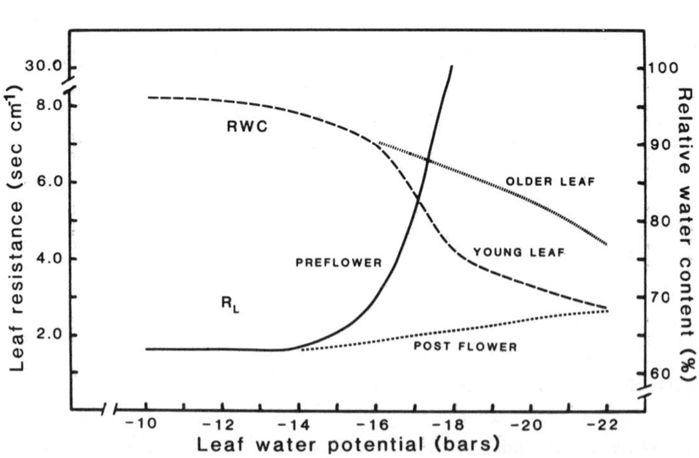

Figure 11.8. Relationships between leaf water potential, leaf resistance (R_L) and relative water content (*RWC*) as a function of growth stage. [Data derived from Ackerson et al. (59), Blum (63), Ferreres et al. (122), and Jones and Turner (148), used with permission.]

around 90% and leaf resistance in the 1.5–2.0 sec cm^{-1} range on fully illuminated, fully expanded leaves (59; Krieg, unpublished). With a soil water stress, leaf water potentials of −22 bars with *RWC* of 85% and leaf resistance of approximately 2.0–2.5 sec cm^{-1} are observed on fully expanded leaves.

Stomatal Regulation

The role of stomatal regulation of transpirational water flux and its relationship to overall drought resistance of sorghum is subject to much debate and to many discrepancies in the current literature. One school of thought is that stomatal regulation and thus water conservation is a very important drought resistance mechanism. Another line of reasoning is that stomatal control of transpiration is one component of total drought resistance and must be coupled with ability to tolerate desiccation. The plants' ability to maintain net assimilation under soil water stress should increase its total drought resistance when defined as the ability to produce an acceptable grain yield under limiting soil water conditions. Considerable evidence exists to indicate that when water supply is relatively high, stomata are open allowing maximum transpiration throughout the daylight hours once the threshold light intensity for opening is exceeded (153). Stomatal opening has been directly related to light intensity (154). Controversy exists as to the tissue water status at which stomatal control of transpiration is exerted. Much of the discrepancy exists due to differences in growth environment, stage of development, root volume : leaf area ratios and thus rate of soil water depletion, and preconditioning or "hardening" the plants by exposure to several cycles of mild stress (155). Significant genotypic differences have also been reported as to the leaf water potential at which stomatal closure occurs (62, 63, 146). Much of the existing discrepancies probably could be resolved if the relationship between turgor pressure and stomatal conductance had been evaluated. Turner (66) has indicated that maximum conductance occurs in sorghum until essentially zero turgor is reached. The relationship between leaf resistance and leaf water potential in Fig. 11.8 represents the response of field grown plants under semiarid conditions at two distinct growth stages. We have observed that during leaf expansion (prior to flowering) increasing stomatal regulation exists over a range of several bars of turgor pressure with some evidence of genotypic differences (61). After flowering leaf conductance was maximum until zero turgor was reached. Genotypic differences were observed in osmotic adjustment capacity, which ranged from 3 to 9 bars, allowing much lower leaf water potentials to exist before zero turgor was reached. These results may explain the previously reported loss of stomatal sensitivity to increasing water stress following flowering in sorghum.

Plant Morphology

A wide range of genetic diversity exists in current sorghum germplasm for leaf angle, blade area, and leaf color, each having an effect on energy interception and dissipation through transpirational water flux. Under high evaporative demand conditions and limiting soil water conditions, leaves of some sorghum genotypes become more erect and tend to roll or fold together to reduce the energy load at the leaf surface. The stomata on the inside of the fold tend to close primarily because of the reduction in light, whereas those on the outer, illuminated surface are still open and transpiring. This response results in a greatly reduced transpiration rate due to reduction in the evaporative leaf surface area and due to the reduced energy load on the leaf surface by changing the angle of incident radiation. The leaf also maintains CO_2 exchange on the exposed surface although at a reduced rate. This adaptive feature is not observed in all sorghum and is definitely more pronounced prior to flowering than after flowering, probably reflecting the degree of osmotic adjustment and turgor maintenance.

Sorghum leaves and stems are also covered with a waxy "bloom," which greatly aids in preventing cuticular water loss and also presents a rather formidable boundary layer resistance reducing transpirational water loss (156) with a differential effect on CO_2 exchange (157) and water use efficiency.

Loss of transpiring leaf area through hastened leaf senescence is another commonly observed response in most sorghum (81), although some are more prone to senesce than others. The leaf senescence response is especially pronounced by stress during grain filling. Much of the dry matter and nitrogen in the lower leaves is readily translocated to the developing grain prior to loss of the leaf (79) allowing for yield maintenance even though net assimilation is greatly reduced.

Another aspect of sorghum's drought tolerance ability is quick recovery after relief of stress. Sanchez-Diaz and Kramer (141) and Glover (158) have demonstrated rapid recovery of stomatal activity after stress is relieved. Following mild stress, growth rates and photosynthetic rates are equal to and sometimes exceed the nonstressed control.

METABOLIC RESPONSES TO WATER DEFICITS

Photosynthesis

Since 90–95% of the dry weight of a green plant is derived from photosynthetic activity, the effect of water stress on photosynthesis has been examined rather extensively as indicated by several recent reviews (159–161).

The photosynthetic process consists of biophysical, biochemical, and physiological components each subject to control by water stress, depending on stage of development and stress intensity. The biophysical component involves the light harvesting and energy conversion processes and the gas exchange process. Results of effects of water stress on the light harvesting process are inconclusive at present. Sullivan and Eastin (162) have indicated that O_2 evolution is affected by the rate of stress induction in sorghum. If plants were allowed to develop stress slowly, no significant effect on O_2 evolution was observed. If stress developed very rapidly, O_2 evolution was severely reduced. Boyer (159, 160) had indicated that significant reductions in electron transport are apparent and are independent of the rate of desiccation. Any effect on the energy conversion functions of chloroplasts would result in reductions in the rate of CO_2 fixation.

Resistances in the CO_2 diffusion pathway include both stomatal and mesophyll resistances. Genotypic differences exist in stomatal density (163) and the stomatal response to light (164), water (61), and temperature (164). Sorghum, being a C_4 plant, has a transpiration : photosynthesis ratio in the range of 300 : 1 (165, 166). Photosynthetic rates range from 30 to 100 mg CO_2 dm^{-2} hr^{-1} depending on genotype, light intensity, and leaf age. Pasternak and Wilson (152) have concluded that genotypic differences in photosynthetic rates were due to differences in leaf conductance properties under glasshouse conditions. Under semiarid field conditions, we have observed a positive correlation between leaf conductance and whole leaf photosynthetic rates (Fig. 11.9) but the correlation coefficient is very low ($r = 0.4$). Rather large genotypic differences are observed in the photosynthetic rates per unit conductance as indicated in Fig 11.10 for a number of common parental lines (68, 69, Krieg, unpublished).

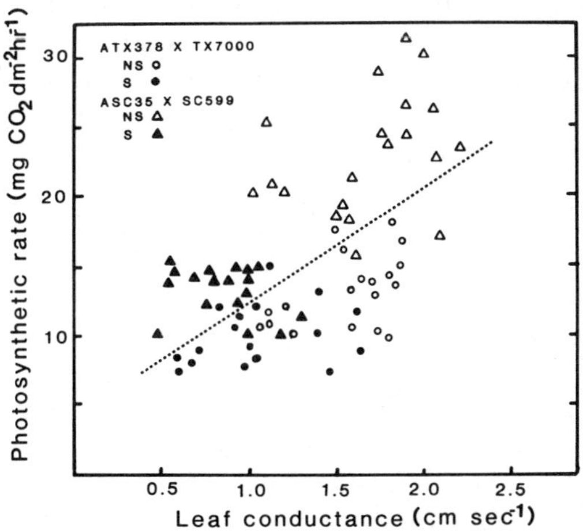

Figure 11.9. Correlation between leaf conductance and photosynthetic rate of two contrasting sorghum genotypes experiencing different levels of water stress. ATX378 × 7000 represented a leaf senescing hybrid, whereas ASC35 × SC599 represents a nonsenescing hybrid. [Previously unpublished data of Hutmacher and Krieg.]

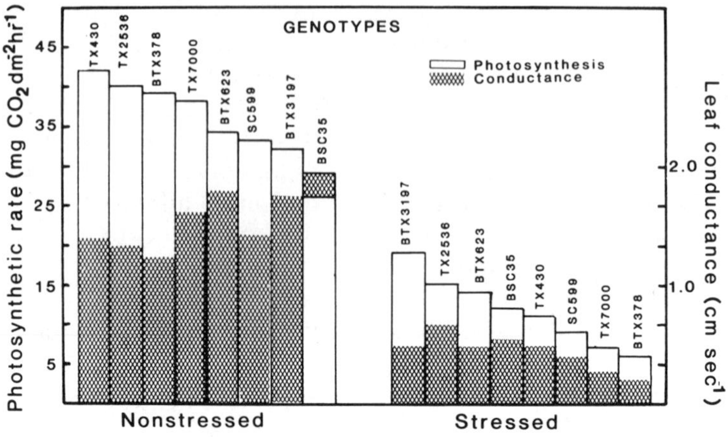

Figure 11.10. Genotypic variability in leaf photosynthesis–conductance responses as a function of plant water status. [Previously unpublished data of Hutmacher and Krieg.]

Much of the previous research, especially controlled environment studies, concerned with the effect of water stress on photosynthesis has concluded that increased leaf resistance with decreasing leaf water potential is the first and major limitation to CO_2 fixation. Although a positive relationship exists between leaf conductance and photosynthetic rate, there is also evidence that nonstomatal factors are involved. In field grown sorghum, under high radiant energy loads, reductions in CO_2 assimilation of stressed plants are not highly correlated to leaf conductance changes (72, 167, Hutmacher and Krieg, unpublished) as depicted for two genotypes in Fig. 11.9 and for several genotypes in Fig. 11.10. This relationship is modified significantly by stage of development.

The biochemical components of photosynthesis include the enzymatic reduction of CO_2 to form organic compounds. Only limited information is available relating carboxylation enzyme activity to photosynthetic activity. Leeton (68) has defined the relationship among phosphoenolpyruvate (PEP) carboxylase, ribulosebisphosphate (RuBP) carboxylase, and whole leaf photosynthetic rates of numerous sorghum lines and hybrids as a function of growth stage under nonstressed conditions. The results indicate that hybrids tend to follow the pattern of the male parent for photosynthetic rate and PEP carboxylase activity. Shearman et al. (168) indicated no significant reduction in PEP carboxylase in water stressed sorghum. Morrissett (69) has evaluated various lines and hybrids of sorghum under both nonstressed and stressed conditions. Both PEP carboxylase and RuBP carboxylase activities were reduced by stress prior to effects on chlorophyll or soluble leaf protein concentrations although no cause and effect relationship was determined. However, RuBP carboxylase activity explained more of the photosynthetic rate response under both nonstressed and stressed conditions than did any of the other biochemical factors evaluated.

Photosynthate Utilization

Source activity (photosynthetic rate) is often regulated by sink activity (use of photosynthate). Sung (72) has described source–sink activities in two sorghum hybrids differing in degree of leaf senescence under water stress. Photosynthetic rates were reduced by increasing water stress prior to any measurable reductions in translocation rates of photosynthate from the source leaf (67). Velocity of translocation in the conducting tissue was unaffected by stress within the range of −22 to −24 bars leaf water potentials. Partitioning of the assimilate leaving the source leaf was unaffected by stress prior to flowering. The expanding leaf tissue continued to be the predominant sink. Although panicle differentiation was affected by stress prior to flowering, resulting in reduced seed per head, no significant reduction in the sink strength of the panicle was observed. Hultquist (71) had previously indicated that significant genotype differences existed in partitioning of assimilate between the root system and the differentiating panicle, which could possibly explain the resultant loss in seed number. During the grain filling stage of development only minor genotypic differences were observed in ^{14}C partitioning under nonstressed conditions (72). In excess of 60% of the daily photosynthate produced was exported to the developing grain. Under water stressed conditions, however, the grain became a stronger sink, with some genotypic differences being observed. Senescent hybrids tended to allocate a larger percentage of the daily assimilate produced by the top four leaves to the grain compared to the nonsenescent types at comparable leaf water potentials (−24 bars). The nonsenescent genotype allocated approximately 16–20% of the production from the top four leaves to the lower leaf tissue, especially those below the seventh leaf, which was reflected in significant seed weight reductions. The results of these experiments and others (35) strongly suggest that the major yield limitation resides in source activity reductions, due both to reduced leaf area and significant photosynthetic rate reductions. The reduction in source activity under stress conditions can be partially offset by modifications within the partitioning system and by elimination of competitive sinks such that the limited photosynthate is preferentially directed to the developing grain. The major effect of stress is on source activity.

Respiration

Dry matter accumulation represents the combined efforts of photosynthetic activity and respiration which provides organic substrates and energy for both growth and maintenance. Approximately 25–30% of the daily carbon assimilation is lost by growth respiration resulting in a conversion efficiency of 70% (169, 170). The growth component is apparently unaffected by water stress, but growth rates are greatly reduced due to reduced CO_2 assimilation capacity (58, 171). Another 1–5% of the dry weight of the plant is consumed daily for maintenance respiration. This value is dependent on species, tissue age, and chemical composition. Wilson et al. (58) have indicated that under severe water stress the maintenance component decreased. The cause of the decrease was not determined. One would suspect that the maintenance cost might increase with mild water stress and decrease with severe prolonged stress, although no definitive data currently exist. Under stress conditions, the maintenance respiration component consumes increasingly greater amounts of the daily assimilation due primarily to reduced photosynthetic activity (171).

Protein Metabolism

Since protein metabolism is so vital to the proper functioning of any biological organism, water stress effects could be manifested through alterations in nitrogen accumulation, reduction, and assimilation into protein. Nitrate reductase activity (NRA), the rate-limiting step in nitrogen assimilation is subject to control by genotype and growth stage (172–174) and has been implicated as being very sensitive to water stress (161). Recent studies concerning NRA as affected by water stress timing and intensity in numerous lines and hybrids indicates that activity is severely depressed by water stress (79, 172). Nitrate reductase activity was independent of NO_3^- supply (as indicated by tissue concentration) which was not greatly affected by water stress. Large genotypic differences existed in NRA under both nonstressed and stressed conditions. However, no significant correlations between NRA and growth rate, yield, or stress tolerance could be ascertained.

The effects of water stress on protein synthesis have been reviewed by others (161). Generally, mild water stress for brief periods results in reductions in protein synthesis, but severe prolonged stress is necessary for major damage to occur. No specific information on protein metabolism as affected by water stress exists for sorghum.

We have recently examined the ability of various sorghum genotypes to utilize both carbon and nitrogen previously assimilated in vegetative tissue for grain development under stress. Although the results are preliminary, it is apparent that more than half of the total grain N can be derived from N assimilated prior to anthesis under stressed conditions as compared to about 30–35% under nonstressed conditions (79). These studies were conducted with plants subjected to rather severe soil water deficits beginning about 20 days prior to anthesis. Significant genotypic differences existed in the ability and degree of remobilization of preanthesis accumulated N and dry matter. Although no data exist for sorghum, one would suspect that even larger amounts of preanthesis dry matter and N would be used if the plant were allowed to develop normally prior to flowering and then stressed during the grain filling stage. This attribute appears to be a very important mechanism allowing cereal crops to mature grain under stress as indicated in several reviews on cereal yield and responses to water stress (80, 175).

Numerous studies have indicated a rather significant increase in free proline concentration in leaves of stressed plants (176–178). In sorghum, free proline increases severalfold in stressed plants, and decreases very rapidly upon relief of stress. Although the change in proline concentration is not highly related to stress tolerance of sorghum (177) possibly it is related to the ability to recover upon relief of stress (176).

Hormonal Relations

Water stress results in major disruptions in the normal hormonal balance of many plant species which result in significant alterations in developmental and physiological process. Only limited information is available concerning the hormonal status of sorghum although considerable effort is under way in both Canada and the United States. Abscissic acid (ABA) is recognized as a stomatal regulator in many plant species (179). In sorghum, stomatal closure apparently precedes measurable increases in ABA concentrations as water stress intensifies (180, 181). In a comparison of stress tolerant and sensitive lines and hybrids, the free ABA concentration was higher in wilted leaves of the tolerant genotypes compared to the drought sensitive ones. Recent evidence (182) indicated that farnesol may function as a stomatal regulator in sorghum much as ABA does in other species. The exogenous application of farnesol resulted in stomatal closure but the inhibition of stomatal opening was less prolonged than that caused by ABA.

Since such limited information currently exists concerning the hormonal regulation of sorghum growth and development, this should prove to be a very fruitful area of research to help elucidate the drought tolerance mechanisms existing in sorghum.

Membrane Integrity

Levitt (15) has indicated that the most probable common denominator in defining plant resistance to various environmental stresses resides in maintenance of membrane integrity. Giles et al. (183) indicated that chloroplast and tonoplast membrane structure was maintained in sorghum under stress as compared to loss of integrity in maize. A simple, but apparently quantitative, test has been developed to measure loss of electrolytes from leaf tissue when subjected to heat or desiccation which is an indication of membrane integrity (184, 185). Using this technique, large differences in apparent heat and drought tolerances of sorghum have been found and generally the two are positively correlated (186). The degree of tolerance, as indicated by this technique, changes with growth stage and culture conditions but relative genotypic differences remain. The simplicity of the technique makes it very suitable to a screening program if significant positive correlations continue to be attained between the response of sorghum cultivars under stressful conditions and their relative ranking using this test.

APPLIED ASPECTS OF WATER MANAGEMENT

Cultural Management

Several agronomic practices have been demonstrated to increase water use efficiency (kg grain produced cm^{-1} water consumed) of grain sorghum. The utility of each practice must be evaluated in terms of the total production scheme. Modification of plant population and spacing are two most commonly used practices as previously discussed. When favorable soil water conditions exist throughout the growing season, high populations (250,000 ha^{-1}) and close row spacings (30 cm) result in highest yields. When the crop is produced under limited available soil water, the optimum stand will be determined by the soil water supply at planting, the potential for rainfall during the growing season, the climatic conditions influencing evapotranspiration, and certain plant factors such as rooting patterns and leaf characteristics. Because of sorghum's ability to compensate among yield components, there is a tendency for yield constancy over a rather wide range of populations under a particular water scheme. Techniques to increase the seed number component (more productive heads or more seed per head) result in yield increases. The seed weight component is somewhat more stable than the seed number component.

Another approach to increasing water use efficiency has been minimizing soil evaporation losses through the use of surface applied barriers early in the season or reflective substances applied to both the soil and the plant during the growing season. Consumptive water use during the early part of the growing season is largely due to evaporation from the soil surface. In the Great Plains of the United States and also in most of the dryland production area of the world, precipitation patterns are such that soil evaporation is maximized. Much of the evaporative losses can be eliminated, making more of the rainfall available for transpirational use, with soil surface barriers such as plastic films or mulches resulting in increased water use efficiency (187). The use of kaolin suspensions as reflectants sprayed on sorghum plants during GS-2, especially immediately prior to head emergence, has resulted in increased yields of 11% primarily due to increasing the number of productive heads per hectare. Accumulative water use was not significantly affected but dry matter production was increased by the kaolin spray resulting in increased water use efficiency (188, 189).

The use of antitranspirants has received only limited attention on sorghum. Fuehring (190) has evaluated phenylmercuric acetate (PMA), atrazine, and "folicote" on field grown sorghum in a semiarid environment. Application prior to the boot stage was most effective in increasing yield under water stress situations with mean yields being increased 5–17% due to the antitranspirant application. No determination of water use efficiency was made. Sig et al. (191) found that PMA applied to sorghum reduced photosynthesis more than transpiration when the concentration was sufficient to have prolonged effects on stomatal conductance. The usefulness of antitranspirants to increase water use efficiency is still questionable on a practical basis.

Crop Improvement Through Plant Breeding

Blum (192) has recently discussed the possibilities and approaches for genetic improvement of drought resistance in sorghum. Basically, two approaches currently exist for developing material capable of performing under soil water deficit conditions. One is based on the premise that the relative ranking of genotypes for yield ability is similar across environments. Selection programs based on this concept are usually conducted under optimal conditions since heritability of yield and yield components is relatively high and selection is more efficient.

The second approach to breeding for yield performance in a stress environment contends that potential yield is irrelevent. Superior cultivars must be developed and selected in the environment in which they are expected to perform. Cultivars developed in such manner show a very narrow environmental adaptability and exhibit extremely large genotype × environment interactions.

A third approach to developing more drought resistant sorghum is evolving which is dependent on a strong interaction among breeders, geneticists, physiologists, and soil scientists. Large differences exist in available sorghum germplasm for almost any characteristic desired. By defining the relationships between developmental, morphological, and physiological characteristics and drought avoidance and tolerance, vast improvement in total drought resistance can probably be achieved. Due to the complexity of drought stress, no single genotype will probably ever be developed possessing all the necessary avoidance and tolerance systems. Therefore, efforts must be made at the local level to define the environment and the stress potential especially with respect to time of occurrence and intensity, and then match this with genotypes possessing characteristics to minimize the adverse effects of the stress. The grain yield potential of more drought resistant genotypes will probably be less than that of more sensitive types irrespective of whether the avoidance or tolerance approach is used. At present one of the best avoidance mechanisms is early maturity, which is associated with lower yield potential. A second approach for some situations is a deeper, more extensive root system to explore more soil volume. Additional root mass requires a considerable amount of carbon. Current indications are that grain yield is limited by source activity even under near optimum conditions; therefore, yield will possibly be reduced even more if additional roots are produced. A third viable approach is through energy shedding by leaf orientation or leaf shedding through hastened senescence. Both approaches reduce CO_2 assimilation on a whole plant basis, thus reducing yield potential. Although stomatal response to water stress is genetically controlled and can be used as a screening technique (193) stomatal control of transpirational water loss also results in reduced CO_2 assimilation and the additional heat load due to reduced latent heat dissipation could result in more damage than benefit.

Plants with high drought tolerance are usually associated with slower growth rate, smaller leaf areas, and lower but more stable metabolic activity. Although these characteristics are desirable under stressful conditions they greatly restrict the plants' ability to respond to more optimal conditions. However, Nour and Weibel (194) have indicated that in the seedling stage the more drought resistant genotypes had greater growth rates than the more sensitive types.

Possibly a combination of avoidance and tolerance mechanisms in one genotype is the most desirable approach. This allows the plant to avoid excessive tissue water deficits and continue growth and development under mild stress conditions and to tolerate desiccation when the stress becomes severe. To some extent sorghum naturally possesses some of each mechanism. Prior to flowering stomatal control of water loss is exerted as tissue water potentials decline due to soil water deficits allowing the expanding leaves and panicle to not be subjected to excessive desiccation. However, once maximum leaf area is attained and flowering has occurred much lower water potentials are required to initiate stomatal closure. Therefore, tolerance to desiccation must become more important during the grain filling period. Other mechanisms such as leaf rolling and energy shedding are also exhibited prior to flowering but are less pronounced during grain filling. Very careful attention must be given to developing germplasm to meet a rather well-defined environmental niche in terms of growing season length and all associated climatic parameters.

In summary, it is recognized that many different avoidance and tolerance mechanisms are present in existing sorghum germplasm. The utilization of each characteristic for particular environments is not well defined, nor is the heritability of each trait. However, efforts are under way throughout the sorghum producing areas of the world to define drought stress potential and to develop relationships between specific avoidance and tolerance mechanisms deemed desirable. The incorporation and utility of the systems in increasing production remain a high priority objective in most research programs.

REFERENCES

1. Food and Agricultural Organization, *Yearbook of Food and Agricultural Statistics*, Vol. 31 (1977).
2. J. H. Martin, *J. Am. Soc. Agron.* **22,** 993–1003 (1930).
3. J. H. Martin, in J. S. Wall and W. M. Ross, Eds. *Sorghum Production and Utilization*, AVI Publ., Westport, Conn., 1970, pp. 1–27.
4. H. Doggett, *Sorghum*, Longmans, Green, New York, 1970.
5. D. T. Rosenow, in N. G. P. Rao and L. R. House, Eds. *Sorghum in the Seventies*, Oxford and IBH Publ., New Delhi, India, 1972, pp. 39–42.
6. W. W. Garner and H. A. Allard, *J. Agric. Res.* **23,** 871–920 (1923).
7. F. R. Miller, J. R. Quinby, and H. J. Cruzado, *Crop Sci.* **8,** 499–503 (1968).
8. F. R. Miller, J. R. Quinby, and H. J. Cruzado, *Crop Sci.* **8,** 675–677 (1968).
9. J. R. Quinby, in N. G. P. Rao and L. R. House, Eds. *Sorghum in the Seventies*, Oxford and IBH Publ., New Delhi, India, 1972, pp. 161–172.
10. J. R. Quinby, J. D. Hesketh, and R. E. Voight, *Crop Sci.* **13,** 243–246 (1973).
11. J. R. Quinby, *Crop Sci.* **3,** 288–291 (1963).
12. A. Blum, *Crop Sci.* **10,** 28–30 (1970).
13. A. Blum, *Exp. Agric.* **9,** 159–167 (1973).
14. R. L. Vanderlip and H. E. Reeves, *Agron. J.* **64,** 13–16 (1972).
15. R. L. Vanderlip, *Coop. Ext. Serv., Kansas St. Univ. Bull.* **C-44** (1972).
16. J. D. Eastin, in N. G. P. Rao and L. R. House, Eds. *Sorghum in the Seventies*, Oxford and IBH Publ., New Delhi, India, 1972, pp. 214–296.
17. W. F. Evans and F. C. Stickler, *Agron. J.* **53,** 366–372 (1961).
18. M. J. Pinthus and I. Rosenblum, *Crop Sci.* **1,** 293–296 (1961).
19. P. M. Saint-Clair, *Can. J. Plant Sci.* **56,** 2124 (1976).
20. J. L. Caddel and D. E. Weibel, *Agron. J.* **63,** 799–803 (1971).
21. J. D. Hesketh, S. S. Chase, and D. K. Nanda, *Crop Sci.* **9,** 460–463 (1969).
22. J. R. Quinby and G. H. Liang, *Can. J. Genet.* **11,** 275–280 (1969).
23. B. K. Kaigama, I. D. Teare, L. R. Stone, and W. L. Powers, *Crop Sci.* **17,** 555–559 (1977).
24. M. D. Clegg, J. D. Eastin, J. W. Maranville, and C. Y. Sullivan, in Research in the Physiology of Yield and Management of Sorghum in Relation to Genetic Improvement. Ann. Rep. No. 4. Univ. of Neb. Lincoln, Neb., 1970, pp. 101–105.

25. J. D. Eastin, J. H. Hultquist, and C. Y. Sullivan, *Crop Sci.* **13**, 175–178 (1973).
26. J. R. Quinby, *Crop Sci.* **12**, 690–691 (1973).
27. P. T. Gibson and K. F. Shertz, *Crop Sci.* **17**, 387–391 (1977).
28. J. R. Quinby, *Crop Sci.* **10**, 251–254 (1970).
29. J. S. Kirby and R. E. Atkins, *Crop Sci.* **8**, 335–339 (1968).
30. G. H. Liang, C. C. Chu, N. S. Reddi, S. S. Lin, and A. D. Dayton. *Agron. J.* **65**, 456–459 (1973).
31. L. W. Balko, M. S. Thesis, Texas Tech University, Lubbock, Tex., 1975, 99 pp.
32. K. S. Fischer and G. L. Wilson, *Aust. J. Agric. Res.* **22**, 39–57 (1971).
33. K. S. Fischer and G. L. Wilson, *Aust. J. Agric. Res.* **26**, 25–30 (1975).
34. K. S. Fischer, G. L. Wilson, and I. Duthie, *Aust. J. Agric. Res.* **27**, 235–242 (1976).
35. R. C. Muchow and G. L. Wilson, *Aust. J. Agric. Res.* **27**, 489–500 (1976).
36. P. R. Goldsworghy, *J. Agric. Sci.* **74**, 523–531 (1970).
37. P. R. Goldsworghy, *J. Agric. Sci.* **75**, 109–122 (1970).
38. P. R. Goldsworghy, *J. Agric. Sci.* **75**, 123–131 (1970).
39. K. J. McCree and M. E. Keener, *Crop Sci.* **14**, 584–587 (1974).
40. K. S. Fischer and G. L. Wilson, *Aust. J. Agric. Res.* **26**, 31–41 (1975).
41. M. D. Clegg, in N. G. P. Rao and L. R. House, Eds. *Sorghum in the Seventies*. Oxford and IBH Publ., New Delhi, India, 1972, pp. 279–310.
42. L. E. Clark, *Crop Sci.* **10**, 307–309 (1970).
43. K. J. McCree and D. S. Davis, *Crop Sci.* **14**, 751–755 (1974).
44. W. R. Jordan and G. F. Arkin, *Proc. 9th Biennial Grain Sorghum Res. Util. Conf. Lubbock, Tex.*, pp. 59–60 (1975).
45. W. C. Johnson, *Agron. J.* **59**, 41–44 (1967).
46. T. C. Hsaio, *Annu. Rev. Plant Physiol.* **24**, 519–570 (1973).
47. T. C. Hsaio and E. Acevedo, *Agric. Meteorol.* **14**, 59–84 (1974).
48. T. C. Hsaio, E. Acèvedo, I. Fereres, and D. W. Henderson, *Phil. Trans. R. Soc. London, Ser. B* **273**, 479–500 (1976).
49. D. G. Stout, T. Kannagara, and G. M. Simpson, *Can. J. Plant Sci.* **58**, 225–234 (1978).
50. R. W. Downes, *Aust. J. Agric. Res.* **19**, 59–64 (1968).
51. A. Blum, *Agron. J.* **59**, 400–403 (1967).
52. K. W. Lee, R. C. Lommasson, and J. D. Eastin, *Crop Sci.* **14**, 80–84 (1974).
53. A. Blum, *Crop Sci.* **17**, 880–882 (1977).
54. R. W. Downes, *Aust. J. Agric. Res.* **23**, 585–594 (1972).
55. R. B. Lewis, E. A. Hiler, and W. R. Jordan, *Agron. J.* **66**, 589–590 (1974).
56. J. M. Bennett, M. S. Thesis, Texas Tech University, Lubbock, Tex., 1975, 74 pp.
57. D. R. Krieg, *Proc. 30th Annu. Corn Sorghum Ind. Res. Conf.* **30**, 13–24 (1975).
58. D. R. Wilson, C. H. M. van Bavel, and K. J. McCree, *Crop Sci.* **20**, 153–159 (1980).
59. R. C. Ackerson, D. R. Krieg, T. D. Miller, and R. E. Zartman, *Crop Sci.* **17**, 76–80 (1977).
60. R. C. Ackerson and D. R. Krieg, *Plant Physiol.* **60**, 850–853 (1979).
61. R. C. Ackerson, D. R. Krieg, and F. J. M. Sung, *Crop Sci.* **20**, 10–14 (1980).
62. A. Blum and C. Y. Sullivan, *Isr. J. Bot.* **23**, 14–19 (1974).
63. A. Blum, *Crop Sci.* **14**, 691–692 (1974).
64. K. J. McCree, *Crop Sci.* **14**, 273–278 (1974).
65. N. C. Turner, *Plant Physiol.* **53**, 360–365.

66. N. C. Turner, in R. L. Bieleski, A. R. Ferguson, and M. M. Creswell, Eds. *Mechanism of Regulation of Plant Growth*, R. Soc. N.Z., 1974, pp. 423–432.
67. F. J. M. Sung and D. R. Krieg, *Plant Physiol.* **64,** 852–856 (1980).
68. B. S. Leeton, M. S. Thesis, Texas Tech University, Lubbock, Tex., 1978, 55 pp.
69. B. L. Morrisset, M.S. Thesis, Texas Tech University, Lubbock, Tex., 1979, 103 pp.
70. N. V. Norcio and C. Y. Sullivan, *Plant Physiol.* **57,** (5 Suppl.), 44 (1976).
71. J. H. Hultquist, Ph.D. Diss., University of Nebraska, Lincoln, Neb., 1973, Diss. Abstr. 73-25.
72. F. J. M. Sung, Ph.D. Diss., Texas Tech University, Lubbock, Tex., 1978, 50 pp.
73. G. E. Pepper and G. M. Prine, *Crop Sci.* **12,** 590–594 (1972).
74. M. D. Clegg, in *Research in the Physiology of Yield and Management of Sorghum in Relation to Genetic Improvement*, Ann. Rep. No. 5. University of Nebraska, Lincoln, Neb., 1971, pp. 107–108.
75. E. Ma, M.S. Thesis, Texas Tech University, Lubbock, Tex., 1975, 62 pp.
76. J. D. Eastin and C. Y. Sullivan, *Crop Sci.* **9,** 165–166 (1969).
77. D. Pasternak and G. L. Wilson, *Aust. J. Exp. Agric. Anim. Husb.* **16,** 272–275 (1976).
78. K. S. Fischer and G. L. Wilson, *Aust. J. Agric. Res.* **22,** 33–37 (1971).
79. S. W. Shau, M.S. Thesis, Texas Tech University, Lubbock, Tex., 1979, 49 pp.
80. J. S. Boyer and H. G. McPherson, *Adv. Agron.* **28,** 1–22 (1976).
81. D. G. Stout and G. M. Simpson, *Can. J. Plant Sci.* **58,** 213–224 (1978).
82. S. Inuyuma, J. T. Musick, and D. A. Dusick, *Proc. Crop Sci. Soc., Jpn.* **45,** 298–307 (1976).
83. J. Shipley and C. Reiger, *Texas Agric. Exp. Stn. Bull.* MP-1202 (1975).
84. G. M. Beil and R. E. Atkins, *Crop Sci.* **7,** 225–228 (1967).
85. A. Blum, *Agron. J.* **59,** 400–403 (1967).
86. A. E. Kambal and O. J. Webster, *Crop Sci.* **6,** 513–516 (1966).
87. F. C. Stickler and A. W. Pauli, *Agron. J.* **53,** 99–102 (1961).
88. F. C. Stickler and S. Wearden, *Agron. J.* **57,** 564–567 (1965).
89. N. R. Malm, *Crop Sci.* **8,** 295–298 (1968).
90. R. L. Voight, C. O. Gardner, and O. J. Webster, *Crop Sci.* **6,** 582–586 (1966).
91. M. Tufail, M. D. Clegg, and J. D. Eastin, in *Research in the Physiology of Yield and Management of Sorghum as Related to Genetic Improvement*, Annu. Rep. No. 4. University of Nebraska, Lincoln, Neb., 1970, pp. 130–132.
92. L. L. Evetts and D. C. Burnside, *Weed Sci.* **21,** 289–291 (1973).
93. C. Hackett, *Aust. J. Biol. Sci.* **26,** 1211–1214 (1973).
94. A. Blum, G. F. Arkin, and W. R. Jordan, *Crop Sci.* **17,** 149–153 (1977).
95. A. Blum, W. R. Jordan, and G. F. Arkin, *Crop Sci.* **17,** 153–157 (1977).
96. W. R. Jordan, F. R. Miller, and D. E. Morris, *Crop Sci.* **19,** 468–472 (1979).
97. I. D. Teare, E. T. Kanemasu, W. L. Powers, and H. S. Jacobs, *Agron. J.* **65,** 207–211 (1973).
98. J. W. McClure and C. Harvey, *Agron. J.* **54,** 457–459 (1962).
99. M. E. Bloodworth, C. A. Burleson, and W. R. Cowley, *Agron. J.* **50,** 317–320 (1958).
100. W. C. Mayaki, L. R. Stone, and I. D. Teare, *Agron. J.* **68,** 532–534 (1976).
101. T. L. Lavy and J. D. Eastin, *Agron. J.* **61,** 677–680 (1969).
102. R. E. Zartman and R. T. Woyewodzic, *Agron. J.* **71,** 325–328 (1979).
103. F. S. Nakayama and C. H. M. van Bavel, *Agron. J.* **55,** 271–274 (1963).

104. W. R. Jordan, M. McCrary, and F. R. Miller, *Agron. J.* **71,** 803–806 (1979).
105. G. J. Burch, R. C. G. Smith, and W. K. Mason, *Aust. J. Plant Physiol.* **5,** 169–177 (1978).
106. W. R. Jordan and F. R. Miller, in N. C. Turner and P. J. Kramer, *Adaptation of Plants to Water and High Temperature Stress*, Wiley-Interscience, New York, 1980, pp. 383–400.
107. T. J. Brown and W. D. Schrader, *Agron. J.* **51,** 339–343 (1959).
108. R. O. Hanks, H. R. Gardner, and R. L. Florian, *Agron. J.* **61,** 30–34 (1969).
109. O. R. Jones, *Soil Sci. Soc. Am. Proc.* **39,** 98–103 (1975).
110. J. T. Musick, D. W. Grimes, and G. M. Herron, *Agron. J.* **55,** 295–298 (1963).
111. K. S. Pharande, N. K. Umrani, and S. P. Kale, *Indian J. Agric. Sci.* **43,** 701–703 (1973).
112. K. B. Porter, M. E. Jensen, and W. H. Slatter, *Agron. J.* **52,** 431–434 (1960).
113. E. T. Kanemasu, *Cont. No. 1643-3,* Agric. Exp. Stn. Kansas State University, Manhattan, Kan., 1977.
114. J. T. Ritchie and F. Burnett, *Agron. J.* **63,** 56–62 (1971).
115. J. T. Ritchie, *Agron. J.* **63,** 51–55 (1971).
116. J. T. Ritchie, E. Burnett, and R. C. Henderson, *Agron. J.* **64,** 168–173 (1972).
117. J. T. Ritchie and W. R. Jordan, *Agron. J.* **64,** 173–176 (1972).
118. C. H. M. van Bavel and W. L. Ehler, *Agron. J.* **59,** 243–246 (1967).
119. C. H. M. van Bavel and W. L. Ehler, *Agron. J.* **60,** 84–86 (1968).
120. C. R. Sumayao, E. T. Kanemasu, and T. Hodges, *Agric. Meteorol.* **18,** 401–408 (1977).
121. R. J. Hanks, L. H. Allen, and H. G. Gardner, *Agron. J.* **62,** 520–527 (1971).
122. E. Fereres, D. Acevedo, D. W. Henderson, and T. C. Hsaio, *Physiol. Plant.* **44,** 261–267 (1978).
123. D. C. Reicosky and J. T. Ritchie, *Soil Sci. Soc. Am. J.* **2,** 293–297 (1976).
124. W. L. Ehler and C. H. M. van Bavel, *Agron. J.* **59,** 243–246 (1967).
125. F. Ajmad, M.S. Thesis, University of California, Davis, 1975, 65 pp.
126. A. Blum, *Agron. J.* **62,** 333–336 (1970).
127. A. Blum and M. Naveh, *Agron. J.* **68,** 111–116 (1976).
128. J. J. Bond, T. S. Army, and O. R. Lehman, *Agron. J.* **56,** 3–6 (1964).
129. D. W. Grimes and J. T. Musick, *Agron. J.* **52,** 647–650 (1960).
130. A. Karachi and Y. Rudich, *Agron. J.* **58,** 602–604 (1966).
131. N. W. Kramer and W. M. Ross, in J. W. Wall and W. M. Ross, Eds. *Sorghum Production and Utilization*, AVI Publ., Westport, Conn., 1970, pp. 167–199.
132. M. Singh, B. A. Krantz, and G. B. Baird, in N. G. P. Rao and L. R. House, Eds., *Sorghum in the Seventies*, Oxford & IBH Publ., New Delhi, India, 1972, pp. 302–332.
133. C. E. Nelson, *Agron. J.* **44,** 303–305 (1952).
134. C. G. Painter and R. W. Leamer, *Agron. J.* **45,** 261–264 (1953).
135. B. M. Reddy and E. Ravindranath, *Ind. Sci. Cong. Assoc. Proc.*, p. 62 (1975).
136. J. C. Stickler, A. W. Pauli, H. H. Londe, H. D. Wilkins, and J. L. Mings, *Crop Sci.* **1,** 297–300 (1961).
137. N. H. Welch, E. Burnett, and H. V. Eck, *Agron. J.* **58,** 160–163 (1966).
138. M. D. Witt, R. L. Vanderlip, and I. D. Bark, *Trans. Kansas Acad. Sci.* **75,** 29–40 (1972).
139. R. J. Hanks, *Agron. J.* **66,** 660–665 (1974).
140. F. W. ChinChoy and E. T. Kanemasu, *Agron. J.* **66,** 98–100 (1974).

141. M. F. Sanchez-Diaz and P. J. Kramer, *Plant Physiol.* **48,** 613–616 (1971).
142. M. F. Sanchez-Diaz and P. J. Kramer, *J. Exp. Bot.* **24,** 511–515 (1973).
143. I. D. Teare and E. T. Kanemasu, *New Phytol.* **71,** 805–810 (1972).
144. N. C. Turner and J. E. Begg, *Plant Physiol.* **51,** 31–36 (1973).
145. N. C. Turner, J. E. Begg, H. M. Rawson, S. D. English, and A. B. Hearn, *Aust. J. Plant Physiol.* **5,** 179–194 (1978).
146. R. G. Henzell, K. J. McCree, C. H. M. van Bavel, and K. F. Schertz, *Crop Sci.* **16,** 660–665 (1976).
147. A Blum, *Crop Sci.* **14,** 361–364 (1974).
148. M. M. Jones and N. C. Turner, *Plant Physiol.* **61,** 122–126 (1978).
149. M. M. Jones and H. M. Rawson, *Physiol. Plant.* **45,** 103–111 (1979).
150. N. C. Turner and M. M. Jones, in N. C. Turner and P. J. Kramer, Eds., *Adaptation of Plants to Water and High Temperature Stress,* Wiley-Interscience, New York, 1980 pp. 87–104.
151. J. Levitt, *Responses of Plants to Environmental Stresses,* Academic Press, New York, 1972.
152. D. Pasternak and G. L. Wilson, *New Phytol.* **73,** 847–850 (1974).
153. R. W. Downes, *Aust. J. Biol. Sci.* **23,** 775–782 (1970).
154. D. Pasternak and G. L. Wilson, *Aust. J. Agric. Res.* **24,** 527–532 (1973).
155. K. J. McCree, *Crop Sci.* **14,** 273–278 (1974).
156. A. Blum, *Isr. J. Bot.* **24,** 50 (1975).
157. N. J. Chatterton, W. W. Hanna, J. B. Powell, and D. R. Lee, *Can. J. Plant Sci.* **55,** 641–643 (1975).
158. J. Glover, *J. Agric. Sci.* **53,** 412–416 (1959).
159. J. S. Boyer, *Phil. Trans. R. Soc. London, Ser. B* **273,** 501–512 (1976).
160. J. S. Boyer, in T. T. Kozlowski, Ed., *Water Deficits and Plant Growth,* Vol. 3, Academic Press, New York, 1976.
161. T. C. Hsaio, *Annu. Rev. Plant Physiol.* **24,** 519–570 (1973).
162. C. Y. Sullivan and J. D. Eastin, *Agric. Meteorol.* **14,** 113–127 (1974).
163. G. H. Liang, A. D. Dayton, C. C. Chu, and A. J. Casady, *Crop Sci.* **15,** 567–570 (1975).
164. D. Pasternak and G. S. Wilson, *New Phytol.* **71,** 683–689 (1972).
165. R. W. Downes, *Planta* **88,** 261–273 (1969).
166. R. W. Downes, in M. D. Hatch, C. S. Osmond, and R. O. Slayter, Eds., *Photosynthesis and Photorespiration.* Wiley-Interscience, New York, 1971, pp. 57–63.
167. R. C. Ackerson, Ph.D. Diss., Texas Tech University, Lubbock, Tex., 1977, 91 pp.
168. L. Shearman, J. D. Eastin, C. Y. Sullivan, and E. J. Kinbacher, *Crop Sci.* **12,** 406–409 (1972).
169. K. J. McCree, *Crop Sci.* **14,** 509–514 (1974).
170. K. J. McCree, in R. H. Burris and C. C. Black, Eds., *CO_2 Metabolism and Plant Productivity,* University Park Press, Baltimore, Md., 1976, pp. 177–184.
171. K. J. McCree and C. H. M. van Bavel, in J. J. Landsberg and C. V. Cutting, Eds., *Environmental Effects on Crop Physiology,* Academic Press, New York, 1977, pp. 199–216.
172. N. Chang, M.S. Thesis, Texas Tech University, Lubbock, Tex., 1975, 44 pp.
173. H. V. Eck, G. C. Wilson, and T. Martinez, *Crop Sci.* **15,** 557–561 (1975).
174. I. D. Teare, R. Manam, and E. T. Kanemasu, *Agron. J.* **66,** 733–736 (1974).

175. L. T. Evans and I. F. Wardlaw, *Adv. Agron.* **28,** 301–355 (1976).
176. A. Blum and A. Ebercon, *Crop Sci.* **16,** 428–431 (1976).
177. R. P. Waldren and I. D. Teare, *Plant Soil* **40,** 689–692 (1974).
178. R. P. Waldren, I. D. Teare, and S. W. Ehler, *Crop Sci.* **14,** 447–450 (1974).
179. T. A. Mansfield, *Phil. Trans. R. Soc. London, Ser. B* **273,** 541–550 (1976).
180. M. F. Beardsell and D. Cohen, *Plant Physiol.* **56,** 207–212 (1975).
181. T. Kannangara, R. C. Durley, and D. G. Stout, *Plant Physiol.* **59,** 93 (1977).
182. R. Fenton, W. J. Davies, and T. A. Mansfield, *J. Exp. Bot.* **28,** 1043–1053 (1977).
183. K. L. Giles, D. Cohen, and M. F. Beardsell, *Plant Physiol.* **57,** 11–14 (1976).
184. C. Y. Sullivan, in N. G. P. Rao and L. R. House, Eds., *Sorghum in the Seventies*, Oxford and IBH Publ., New Delhi, India, 1972, pp. 247–264.
185. C. Y. Sullivan, in K. P. Larson and J. D. Eastin, Eds., *Drought Injury and Resistance in Crops.* CSSA Sp. Publ. No. 2, 1974, pp. 1–18.
186. C. Y. Sullivan and A. Blum, *Proc. 25th Ann. Corn Sorghum Ind. Res. Conf. Am. Seed Trade Assoc.* **25,** 55–66 (1970).
187. R. H. Griffin, B. J. Ott, and J. F. Stone, *Agron. J.* **58,** 449–452 (1966).
188. S. Moreshet, G. Stanhill, and M. Fuchs, *Agron. J.* **69,** 246–250 (1977).
189. G. Stanhill, S. Moreshet, and M. Fuchs, *Agron. J.* **68,** 329–332 (1976).
190. H. D. Fuehring, *Agron. J.* **65,** 348–351 (1973).
191. J. W. Sig, E. T. Kanemasu, and I. D. Teare, *Crop Sci.* **12,** 733–735 (1972).
192. A. Blum, in H. Mussell and R. C. Staples, Eds., *Stress Physiology in Crop Plants*, Wiley-Interscience, New York, 1979, pp. 429–446.
193. R. G. Henzell, K. S. McCree, C. H. M. van Bavel, and K. F. Schertz, *Crop Sci.* **15,** 516–518 (1975).
194. A. E. M. Nour and D. E. Weibel, *Agron. J.* **70,** 217–218 (1978).

12

SOYBEAN

D. A. Ashley

INTRODUCTION

The origin and early history of the cultivated form of soybean [*Glycine max* (L.) Merr.] are unknown; however, it is generally conceded that the soybean is native to Eastern Asia (1). Theories on the specific point of origin vary from eastern Asia (2) to the eastern half of North China (3) to the north and central regions of China proper (4). With respect to the point in time of the first record of soybeans Hymowitz (3) cited evidence referring to soybeans as early as the eleventh century B.C. in the form of pictographs of the Chinese word for soybeans (Shu), in the Book of Odes, and in bronze inscriptions.

The soybean migrated in the early years to South China, Korea, Japan, and other countries of Southeast Asia. Production of soybeans was primarily restricted to China and this region for many years. Piper and Morse (2) gave a rather extensive account of the expansion and distribution of soybeans to other regions of the world, particularily emphasizing the seventeenth, eighteenth, and early nineteenth centuries. They showed that by the early part of the nineteenth century soybeans were produced in many countries throughout the world. In more recent years it is evident that, in addition to Southeast Asia, soybeans can be produced from Canada to Argentina and from Sweden to South Africa.

Adaptation

As demonstrated through the history of world production, the soybean has a fairly wide range of adaptation involving a wide array of climatic and growth conditions. It follows that the areas of adaptation were determined by the way the plants respond to environmental variables and man's ability to genetically modify the plants to accommodate extremes in the environment. Photoperiod, temperature, and available soil water are among the climatic factors which exert a major influence on soybean adaptation.

Photoperiod

Introductions of soybean genotypes into the United States (1880–1920) from various regions of the world differed greatly in their responses to climatic conditions, including variability in flowering and maturity dates (2), suggesting selection of genotypes in their native habitat had involved photoperiodic responses even though photoperiodism would not be identified as such until much later (5, 6). Through the years of more recent soybean development the importance of photoperiodism to the adaptation and successful production of soybeans has been demonstrated repeatedly (7–9).

This variability in photoperiodic response has been used by plant breeders to develop cultivars of twelve maturity groups, 00 to X. Cultivars in each of these maturity groups span latitudinal bands which are rather narrow in the high latitudes and become relatively broad approaching the equator (Fig. 12.1) (10). Hartwig (8) concluded that the wider areas of adaptation at the lower latitudes was due to the fact that change in photoperiod during the period of soybean growth was much less than at the higher latitudes. He presented evidence to show that at latitude 10° the change in photoperiod from May 20 to September 20 was 30 min whereas at 43° it was over 3 hr.

Temperature

With respect to restrictions imposed on the geographic areas for soybean production, there are two growth stages that are most vulnerable to temperature limitations: frost and cold damage during seed germination-seedling stage and late season maturation stage. The minimum temperature for soybean germination is approximately 5°C (10). After seedling emergence soybeans seem to be more tolerant of light frost than most food crops, but they cannot withstand severe cold or frost. Brown (11) found that soybean growth ceased at about 10°C and reached a maximum at 30°C.

In late season heavy frost or a hard feeze will produce visible signs of damage and terminate growth. Plant functions, however, may be greatly impaired or inactived by late season low temperatures even though visible evidence of damage is absent. Sinclair (12) recently showed that plants of "Chippewa 64" soybeans exhibited a drastic drop in photosynthetic activity without visible damage after being exposed to cool night temperatures of 9.5 and 5°C on August 30 and 31, respectively. The photosynthetic activity never recovered to its former level in this cultivar. In the same study a selected line from a "Corsoy" × PI 132,206 cross did not exhibit adverse affects of cool temperatures on August 30 and 31. The selected line matured later and apparently was not vulnerable to this stress. However, on September 12, 13, and 14, temperatures dropped to 4.5, 8.5, and 10°C, respectively, and at this time the selected line exhibited a rapid decrease in photosynthetic activity. He concluded that the decrease in photosynthetic activity associated with cool night temperatures was related to the maturity of the crop as well as potential vulnerability to stress.

Temperature exerts definite influences on soybean plant development and many plant functions such as flowering, pod-set and abortion, photosynthesis, photosynthate translocation, and N_2 fixation (13). The detrimental effects of excessively high temperature on flowering and pod abortion (14, 15) could limit the potential for soybean production in certain areas of the world, particularly where water deficit is an interacting factor, as in the southwest United States.

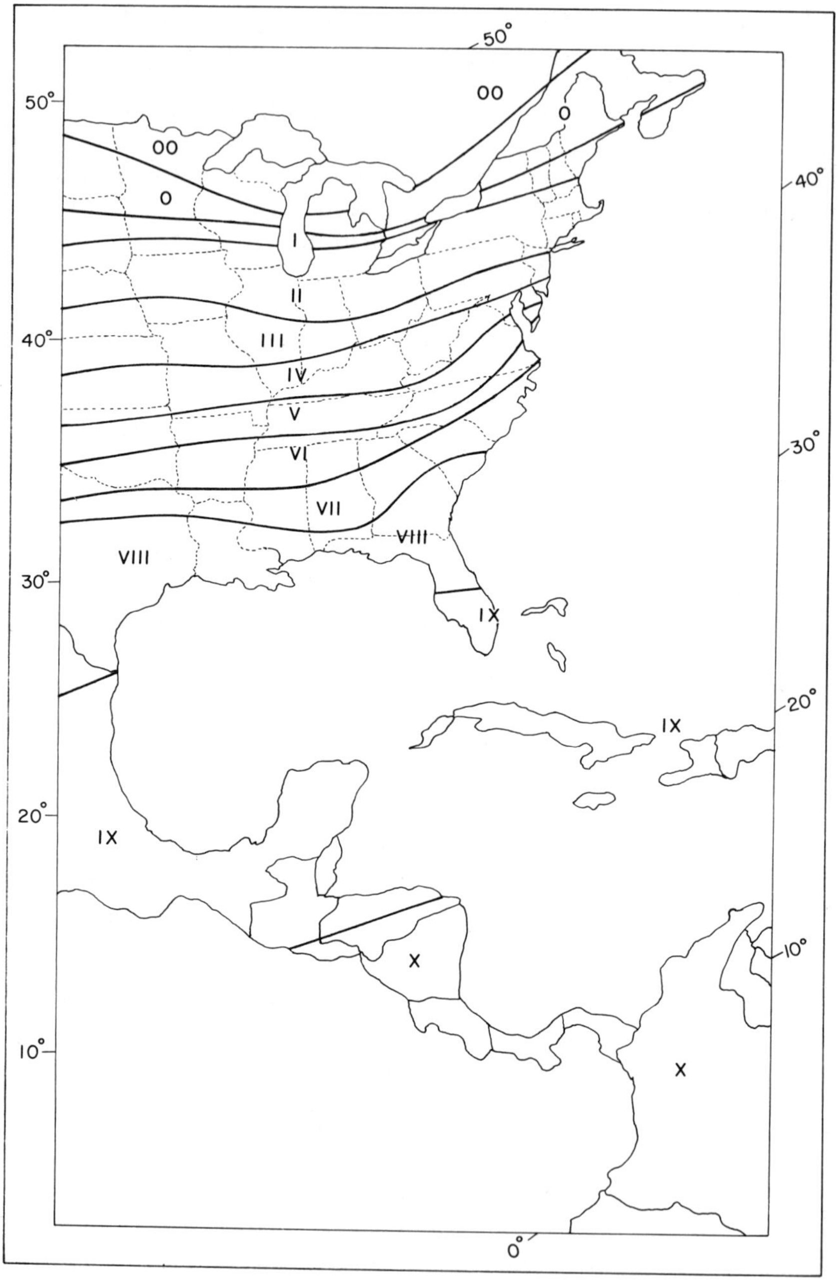

Figure 12.1. Latitudinal adaptation regions for soybean cultivars with maturity groups 00 through X. [From Whigham and Minor (10), p. 79, with permission.]

Water Availability

Availability of water imposes definite limitations on the areas suitable for soybean production throughout the world. In the United States, for example, a map of the major areas of soybean production (16) shows an abrupt drop in soybean acreage west of a line running through eastern Texas, Oklahoma, Nebraska, and the Dakotas. This is primarily due to inadequate rainfall to produce satisfactory soybean yields. If irrigation water is available to supplement rainfall soybeans can be produced in many of these low rainfall areas. Indeed, some irrigation dependant soybeans are produced in western Nebraska, western Kansas, and the high plains of Texas. It has also been demonstrated that soybeans produce well under irrigation in Arizona, California, and in certain parts of Washington and Oregon; however, other agronomic and economic considerations of alternate crops enter as decisive factors in soybean production in these areas.

The specific factors involved in water limitations of soybean production will be discussed in other sections of this chapter.

SEED GERMINATION AND SEEDLING EMERGENCE

Seed germination is the beginning of a complex sequence of events which involves imbibition of water, growth of the embryo, physical rupturing of the seed coat, and emergence of the radicle. Seedling emergence results when continuing events of radicle elongation into soil and hypocotyl elongation combine to force the cotyledons and epicotyl above the soil surface.

Soil water along with temperature and oxygen have been recognized as the most critical environmental factors in seed germination (17) although other factors (18, 19) are also involved. Many studies have been conducted to determine the relationship of soil, seed, and water in the germination and emergence processes. Some of these studies have involved the use of artificial soil conditions in studying the physics of water absorption by seed (17) but the data of Hunter and Erickson (20) have been referred to as more indicative of the soil–seed–water conditions occurring under field conditions. They found that the minimum seed water content for germination of soybeans, corn, rice, and sugar beet was approximately 50.0, 31.5, 26.5, and 31% of dry weight, respectively. They further concluded from data on five soils, ranging in texture from silt loam to clay, that a soil water potential of not less than -6.6, -12.5, -7.9, and -3.5 atm, for the same species respectively, was required for seed germination. Apparently the internal osmotic forces for imbibition of water differ among seeds of these species.

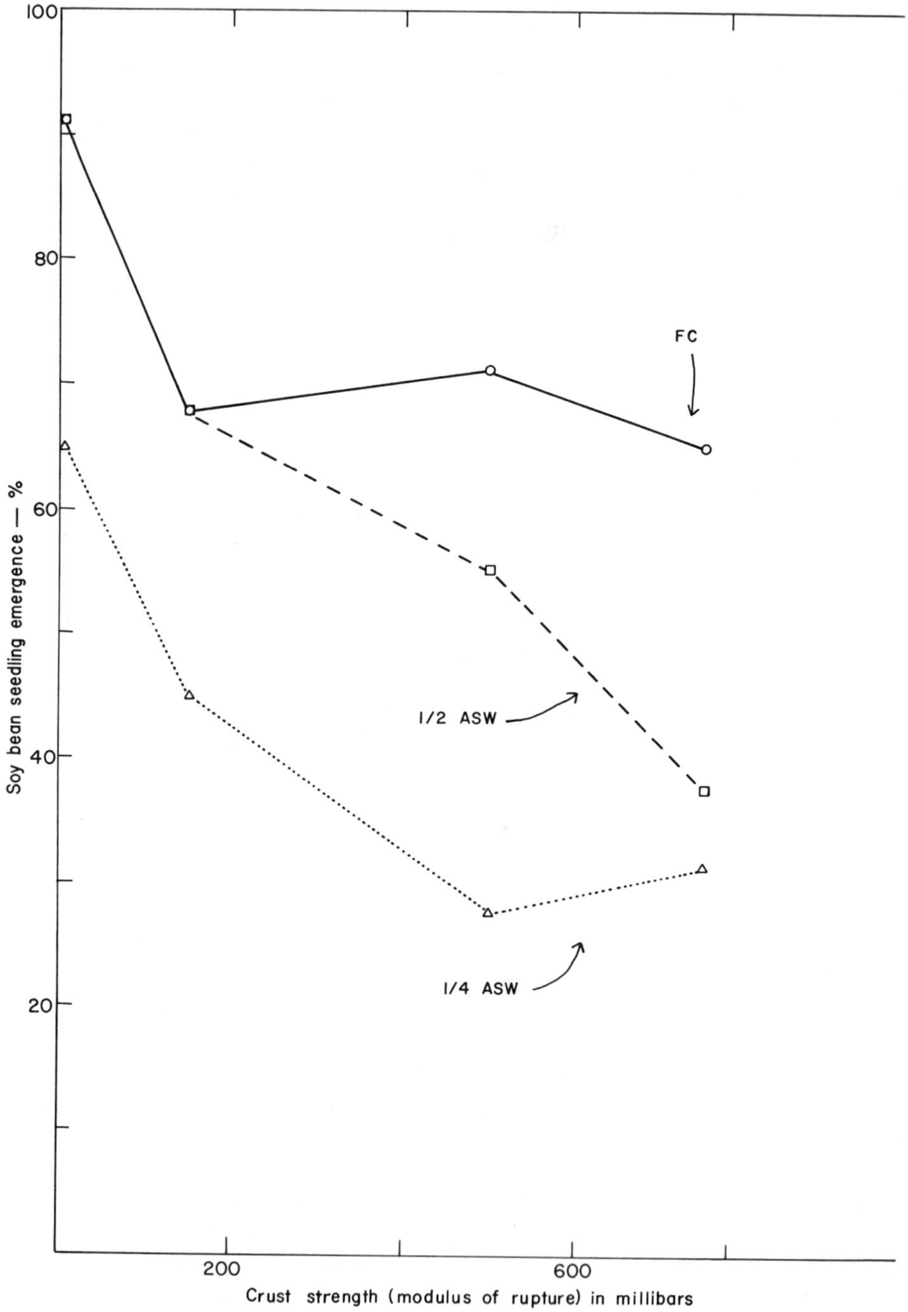

Figure 12.2. Effect of soil crust strength at three soil water levels on soybean seedling emergence. Water levels were field capacity (FC), ½, and ¼ of available soil water (ASW). [After Hanks and Thorp (23).]

Heatherly and Russell (21) pointed out that many previous studies dealt with water requirements for germination and did not measure emergence (20, 22). Their data indicated that a much lower soil water potential (> −1.0 bar) is required for seedling emergence than reported previously for seed germination (20). This concept is supported both directly and indirectly by data from the interaction of soil crust strength, soil water, and seedling emergence. Hanks and Thorpe (19, 23) showed the effect of soil crust strength and soil water content (Fig. 12.2) on seedling emergence in soybean. At a given crust strength seedling emergence was lowest where soil water content was lowest and crust strength increased and emergence decreased as soil water decreased. In summary, soil water requirements for seed germination are better understood than those for emergence, where soil physical properties such as texture, structure and density as well as crust thickness, seed size and seed spacing influence water requirements.

Excessive soil water impedes or stops soybean seedling emergence primarily through restriction in oxygen supply and/or enhanced invasion of pathogenic organisms. Grable and Danielson (24) found that a soil water potential of −0.3 bar resulted in the development of profuse fungal infection on germinating soybean seed which almost completely stopped root growth. Restricted germination was also found in a silt loam when the soil water potential was greater than −0.4 bar (21). In addition to invasion of pathogens at high soil water content, Hunter and Erickson (20) emphasized that infectious fungi can cause severe damage to soybean seed when seed remain in soil with a soil water content too low for germination, yet containing sufficient water to permit seed swelling and fungal growth.

ROOT GROWTH AND FUNCTION

Depth and Distribution

The root system of seedling soybeans grows in a modified taproot fashion (25). In the early stages of development four longitudinal rows of secondary lateral roots develop acropetally from the upper segment of the taproot and soon begin to function as the primary root system (26). These lateral roots grow and branch profusely in the upper soil layer until the plants are approximately 50–60 days old at which time they reach midrow and turn downward sharply. During the same time the "taproot" continues to penetrate to greater depths although it degenerates to a root with a diameter similar to the primary lateral (27). Under soil conditions conducive to deep root penetration in row planted soybean, about five or six of the primary laterals extend downward, eventually reaching about the same maximum depth of penetration as that of the taproot (150–200 cm). These authors also show that the rate of root elongation and depth of penetration varied among the genotypes used.

Vertical distribution of root dry weight and root length in the soil profile has been evaluated by several investigators. Borst and Thatcher (28) reported root penetration of 150 cm of two soybean cultivars with the major portion of the root weight occurring in the upper 60 cm. In a more recent investigation, Mitchell and Russell (26) showed that 90% or more of the roots were located in the upper 15 cm of the soil profile. Data of Raper and Barber (27) from widely spaced plants, with root samples from cylindrical soil shells radiating from the plant center, showed more than 83% of the root weight was contained in the upper 15 cm and more than 90% in the upper 30 cm of the soil profile. They implied through photographic evidence that roots from plants in row plots showed even greater amounts of roots in the upper soil strata than the widely spaced plants. Their data also showed that more than 70% of the root weight was located in the central 15 cm cylinder of soil around the plant. Progressively enlarging cylindrical soil shells around the plant center contained varying amounts of the remaining portion of the root system, but showed a tendency to decrease with distance from the plant. Mayaki et al. (29) presented somewhat similar distribution patterns of root dry weight from a study in which roots were obtained from soil core samples from within the row, one-fourth row, and midrow. Their data showed a greater downward distribution of roots than Raper and Barber (27) in that only about 70% of the root dry weight was in the upper 30 cm of soil.

From the standpoint of depth of rooting, most of the root dry weight of a soybean plant is concentrated in the upper 30 cm layer of soil and exhibits some variability in distribution at depths greater than 30 cm, although exhibiting a tendency to have progressively smaller root dry weight with increasing depth. There is also a tendency to exhibit progressively smaller root dry weight with successively greater radial distances from the center of isolated plants (27) as well as lateral distances from plants in a row (29).

Effect of Physiological Stage on Rooting

Mitchell and Russell (26) concluded that seasonal root development of soybeans could be associated with three general stages of top growth: (1) vegetative top growth with downward tap root penetration and shallow horizontal lateral development, (2) flowering and pod formation accompanied by an increase in roots in the upper soil profile and the beginning of downward penetration of lateral roots, and (3) pod-fill and seed maturity when the major lateral roots elongate and root weight increases in the lower depths as well as in the shallow root zone. The continued development of roots during the period of rapid seed development is in contrast to the report of Borst and Thatcher (28) who concluded that root development diminished during late season. The data of Kaspar et al. (30) support the concept of continued root growth throughout reproductive development. They found in a rhizotron study that the rate of downward root growth in seven cultivars reached a maximum at the R1 to R3 stages of development but continued substantial root growth through the R6 stage. Some evidence indicates that root development and activity (31) during different physiological stages may be different in genotypes exhibiting a determinate rather than an indeterminate growth habit. This is particularily true of rooting during reproductive development since cessation of vegetative development occurs much earlier in determinate than indeterminate genotypes (32).

The rooting patterns described above have been rather widely accepted as being representative of root behavior under soil conditions that are conducive to root penetration. Kaspar et al. (30) felt that this rooting pattern would occur in many soils on which soybeans are grown; however, they recognized that soils showing large changes in their physical and chemical characteristics with depth could cause large differences in root development. Indeed, dense soil layers and "traffic pans" which occur frequently in soils of southeastern United States have been shown to reduce soybean root penetration into subsoil horizons and cause enhanced proliferation of lateral roots above the dense soil layers (33).

Root Water Relations

Soil water extraction patterns have been used by some investigators to indicate depth and activity of soybean roots. Peters and Johnson (34) demonstrated that water was extracted to a depth of 75 cm by the time of early flowering in plastic covered plots receiving no water after plants were 15 cm tall. As the plants advanced in age water extraction by the roots extended to a depth of 125 cm, the maximum depth of sampling. The amount of water removed at all depths generally increased with plant age until the plants neared maturity. They emphasized that water extracted during the latter one-half of the growing season from the lower portion of the soil profile was not replenished even on plots that received abundant rainfall and/or irrigation. A somewhat similar behavior in profile soil water depletion was observed by Stone et al. (35) from both irrigated and nonirrigated soybeans. Depletion of water by roots reached a depth of 150 cm by the time of first bloom with quite similar extraction patterns on irrigated and nonirrigated plots (Fig. 12.3); however, differences in depletion patterns on irrigated and nonirrigated plots were

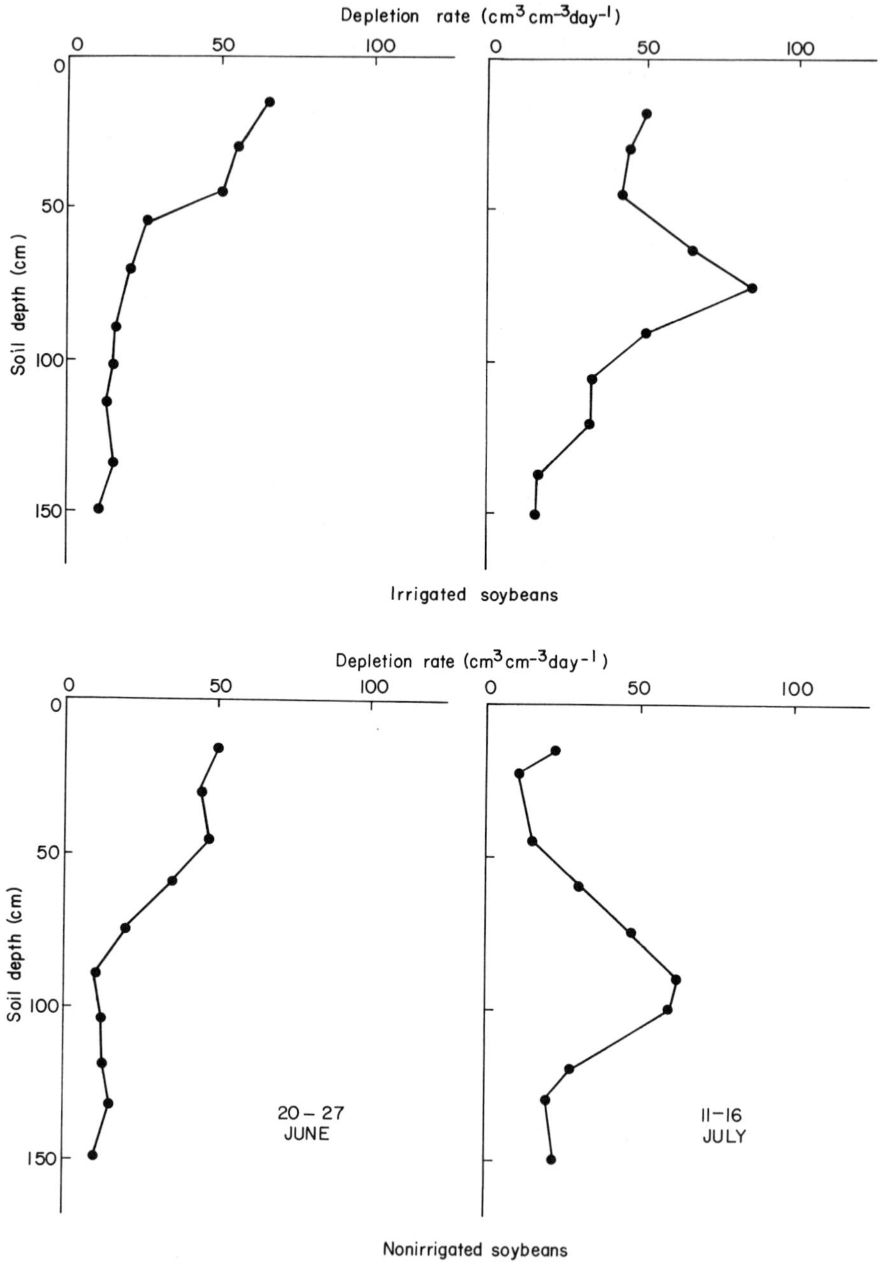

Figure 12.3. Depletion rate of soil water by irrigated and nonirrigated soybeans at two time periods during development. [After Stone et al. (35).]

present by early pod development. Depletion rate from this study and Arya et al. (36) was less at the shallower depths on nonirrigated than on irrigated plots primarily due to water availability. Most of the water removed from the nonirrigated profiles, as the drying cycle advanced, came from the lower half of the root zone.

Total root dry matter (37) and root length (38) in the root zone did not differ for irrigated and nonirrigated soybeans. These studies did show that roots of nonirrigated soybean plants proliferated more laterally and in the lower portion of the soil profile than irrigated plants, but the differences, though significant, were relatively small in relation to the total root mass. Barber (39) also concluded that climatic factors did not influence profile root distribution of soybean roots appreciably, though he recognized that some data to the contrary (40) existed.

Although the total quantity of roots appears not to be drastically affected by seasonal soil water regime in many instances, this does not mean that the rate of soil water removal is not affected by soil water availability. Stone et al. (35) showed that roots of irrigated plants were more effective in depleting water from the upper soil layers per unit root dry weight than roots of nonirrigated plants. They also found that the ability to extract water on a unit root dry weight basis increased with increase in soil depth in both irrigated and nonirrigated plots. This is in agreement with the results of Allmaras et al. (41) and Reicosky et al. (42) from soybean experiments. Taylor and Klepper (43) found that deep corn roots also extracted water more effectively than roots near the soil surface and proposed that the increased effectiveness of deep roots was related to younger roots, wetter soil, and less competition for the available water at these profile positions.

Soil Water Availability and Nitrogen Fixation

Since soybean plants are dependent on symbiotic nitrogen fixation for a large portion of their nitrogen supply, consideration of root zone soil water conditions must include the effects on *Rhizobium* and the N_2 fixation process. It has been known for some time (44) that soil water availability affects nitrogen fixation in soybeans, but only in recent years (45–49) has detailed information on the relationship become available.

Sprent (50) concluded that both nitrogen fixation (C_2H_2 reduction) and respiration of detached soybean nodules almost ceased when the nodule fresh weight dropped below 80% of fully turgid weight and that fixation was positively related to water content between 80 and 100%. In subsequent experiments nitrogen fixation (C_2H_2 reduction) by nodules attached to soybean roots was also reduced in water stressed plants (51) or by osmotically induced root zone water stress (46). Apparently, the vacuolate cells of the outer cortex of the nodule show structural

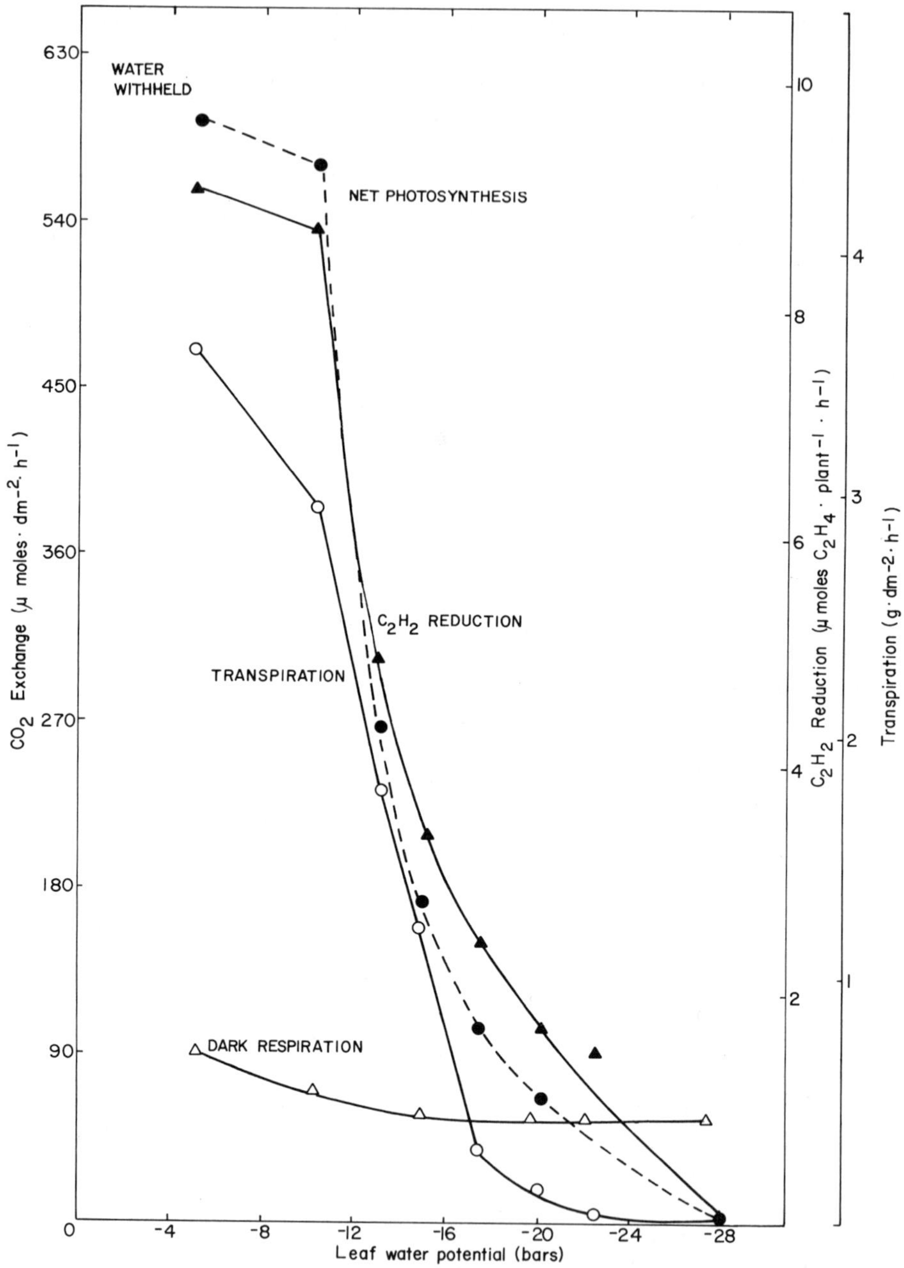

Figure 12.4. Responses of photosynthesis, dark respiration, transpiration, and acetylene reduction to soybean leaf water potential. [From Huang et al. (48), p. 226, with permission.]

changes in cytoplasmic constituents under moderate stress with more severe stress needed to affect the infected and pericycle cells (52).

Pankhurst and Sprent (47) associated the reduction in nodule activity under stress with changes in O_2 diffusion and respiration rates in nodules. Leaf water potential was found to be rather closely associated with the rate of nitrogen fixation (48, 49). In the experiments of Huang et al. (48, 53) a close association of rates of photosynthesis, transpiration, and nitrogen fixation was evident under decreasing leaf water potential, but respiration remained fairly constant at all water stress levels (Fig. 12.4). These responses led to the conclusion that photosynthesis, translocation, or some direct effect on the nodule other than respiration was most likely the cause of the decrease in nitrogen fixation as soil water potential decreased.

CANOPY DEVELOPMENT

General Growth Patterns

Seasonal dry matter production by a soybean canopy exhibits a typical sigmoidal response curve in a plot of plant dry matter versus time. To quantitatively evaluate the seasonal growth pattern of the total system, however, the overall growth curve must be partitioned into plant components based on growth stages (54). The data of Hanway and Weber (55) give a particularly good example because seasonal dry weight production in aerial plant components (Fig. 12.5A) was measured and an estimate of dry weight loss from leaf and petiole drop (Fig. 12.5B) obtained. Following an initial lag in rate of plant growth, a rapid and approximately linear increase in weight occurs. Leaf and petiole growth accounts for a large portion of the increase in weight in the early linear phase of vegetative development. By the time the plants are approximately 60 days of age, accumulation of weight in stems begins to account for a significant portion of total plant weight. Vegetative dry weight reaches a maximum soon after reproductive components begin developing. With advances in plant age reproductive development proceeds in a linear manner and a rapid decline in vegetative dry weight occurs. The decrease in vegetative weight results largely from leaf and petiole drop; however, some weight loss occurs from the stems and pod carpel tissues, presumably due to export of stored metabolites and mineral nutrients.

Genetic variability in the time schedule of weight partitioning into component plant parts has been observed (55) in cultivars of different maturity dates. The genetic effects on weight partitioning was particularly evident when Egli and Leggett (32) compared determinate and indeterminate genotypes. They showed that at initial flowering the determinate genotype had produced 78% of its vegetative material, whereas the indeterminate genotype had produced only 58%. At the same time the stem dry weights of the determinate and indeterminate genotypes were 67 and 30% of the maximum weight, respectively. Neither genotype produced very much vegetative growth after a measurable amount of dry weight had accumulated in pods and seeds.

Soil Water Effects on Canopy Development

General field observation of drought stressed soybeans indicates that the amount of soil water available to the plants throughout all developmental stages exerts a major influence on plant growth. In their review of soybean water relations, Mederski et al. (17) concluded that essentially no information available at that time related periodic field measurements of plant growth rate and seed yield to internal water stress at appropriate intervals throughout the development of the plants. Since then a limited number of studies have contributed information on this topic (29, 56–58). The data of Mayaki et al. (29) are the most descriptive of seasonal soybean development under irrigated and nonirrigated field conditions in that seasonal dry weight accumulation in both roots and aerial plant parts were measured (Fig. 12.6). They found that the total amount of roots was not affected appreciably by irrigation treatment even though greater lateral and deep root development was observed from the nonirrigated plants than from irrigated plants. All aerial components were adversely affected by soil water stress on the nonirrigated plants. Maximum plant height, total plant dry weight, and *LAI* of nonirrigated plants were only 62–64% of that in irrigated plants. Stem weight showed the greatest response to soil water stress with maximum dry weight from nonirrigated plants reaching only 55% of the maximum from irrigated plants. Fruit weight of nonirrigated plants was about 74% of the irrigated plants near the time of physiological maturity. Ashley and Ethridge (56) also showed that changes in stem weight account for a large portion of the vegetative response of three cultivars to water stress. They further point out that branch development was markedly reduced by water stress, and this was responsible for much of the difference in stem weight.

Thus, aerial plant parts of soybeans are reduced in size if subjected to water stress, and the weight of stressed plants may be reduced to less than one-half that of unstressed plants in some cultivars (56) without adversely affecting yield, but the percentage of the total dry weight partitioned to the vegetative aerial plant parts (harvest index) does not differ greatly in stressed and unstressed plants.

The lack of response in total root weight to soil water variables (29, 37) presents an interesting contrast to that of aerial plant parts. At least part of the root system of nonirrigated plants was in moist soil most of the time with direct access to water. Root tip growth probably could

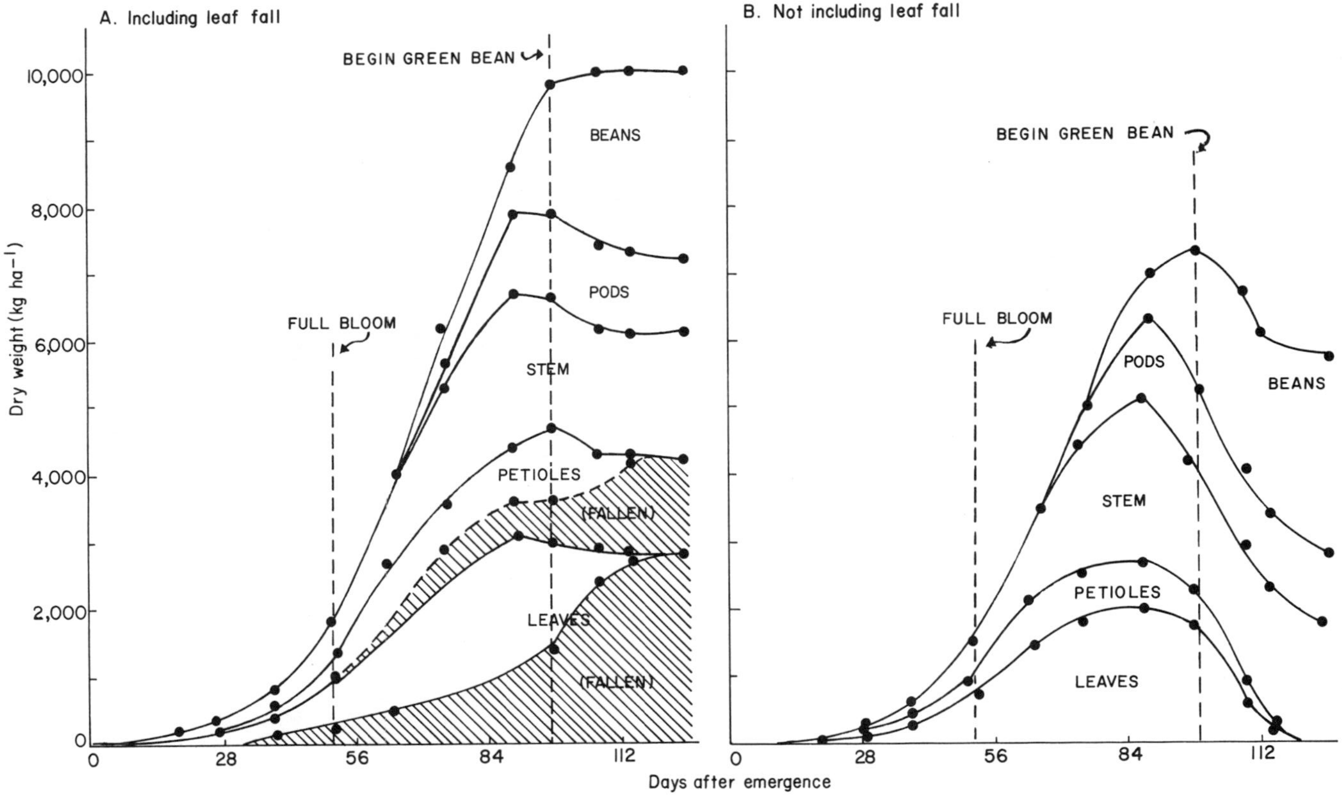

Figure 12.5. Seasonal distribution of dry weight in soybean plant parts, including (A) and not including (B) leaves and petioles abscised from the plant. [From Hanway and Weber (55), p. 265, with permission.]

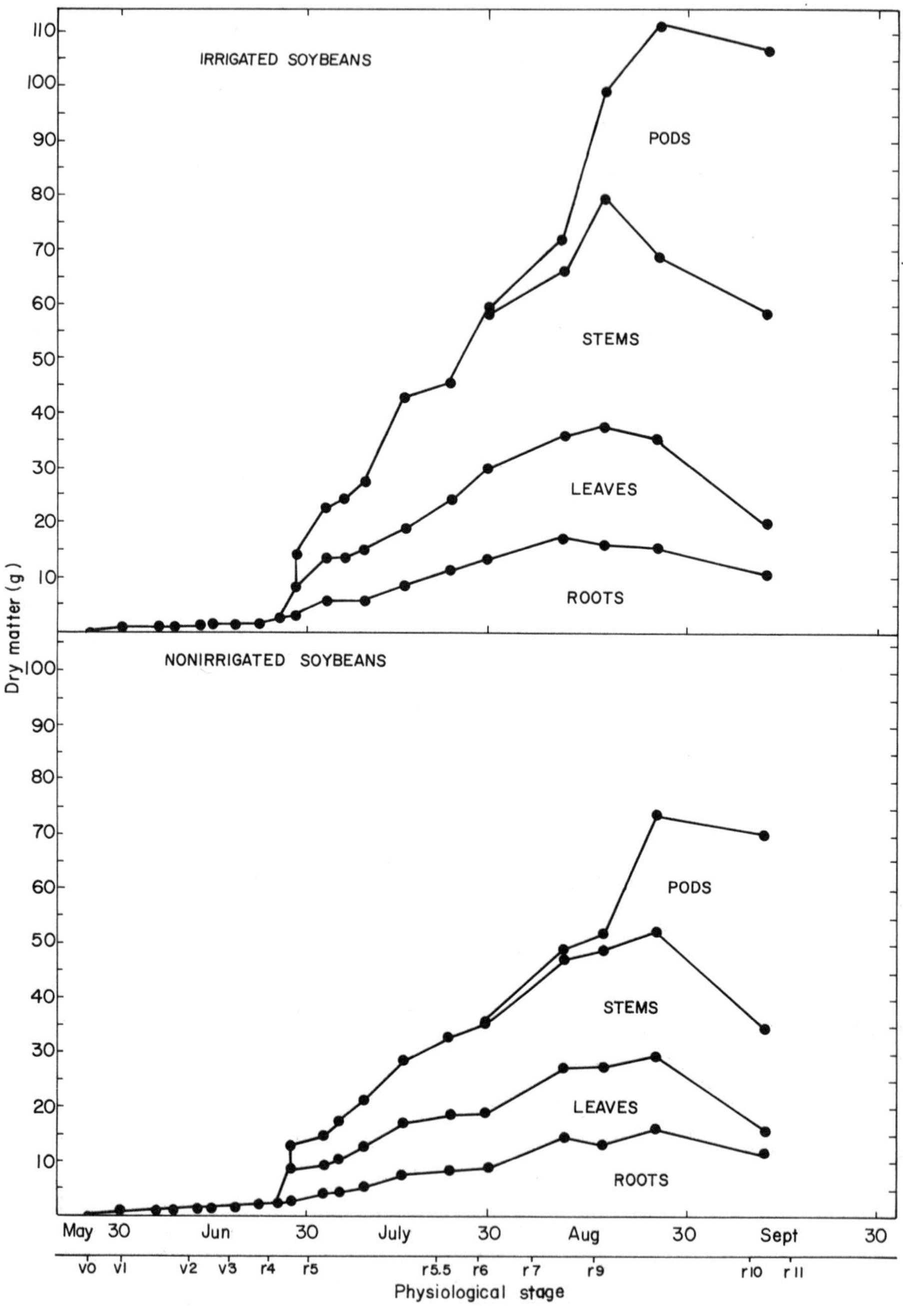

Figure 12.6. Effect of irrigation on the seasonal dry matter accumulation in component parts of two soybean plants. [After Mayaki et al. (29).]

occur independent of xylem water potentials, even though aerial plant parts were under severely restricting water stress. These conditions could permit continued root growth even though water stress severely retarded growth of aerial parts. In support of this concept, Huck and Davis (59) proposed that when abundant water is available shoot growth will consume most of the available carbohydrate until evapotranspiration and low water potential restricts shoot growth. At this point, root growth in regions of moist soil will begin to dominate in use of available carbohydrate. This type of behavior could help cancel the negative effects of water stress on root development as in the Mayaki et al. (29) experiment. It could also result in root generation and growth alternating between surface and deeper soil layers as rainfall causes the surface to be alternately favorable and unfavorable for root development.

Water Relations and Physiological Activity

Leaf Expansion

There is general agreement that turgor pressure is an important determinant of cell elongation. Water status of a plant through its effects on cell turgor exerts a direct effect on plant tissue growth. Boyer (60) reported that leaf enlargement rate of soybeans was dramatically reduced as leaf water potential decreased from -2 to -4 bars. Enlargement rate progressively decreased with a decrease in leaf water potential and approached zero at -12 bars. Other investigations (61–64) have shown that stem elongation and leaf enlargement are reduced by water stress. If a well-watered plant is subjected to water stress, initially the major effect is on cell elongation (65). Prolonged periods of stress apparently result in a reduction in the rate of cell division in meristematic tissue (63, 66).

Dry weight accumulation rates are much less sensitive to water stress than that indicated for leaf expansion (60). During the summer growth period, midday leaf water potentials in the field almost daily exceed that required to markedly reduce leaf expansion rates, even when abundant soil water is available. However, during darkness and periods of low canopy heat load the leaf water potential returns to a value closely approximating the soil water potential (67). Apparently a significant amount of leaf expansion and stem elongation occurs at night which makes it possible for growth of leaves under moderate stress not to be limited appreciably by turgor pressure (68). Nonetheless, prolonged periods of water stress will result in a reduction in leaf area (69–71).

Photosynthesis

It is well established that photosynthesis is decreased by water deficit in soybeans (17, 60, 72) and in other plant species (73). When plants are subjected to intense water stress almost complete cessation of photosynthetic activity occurs (72). Ultimately, plant growth is reduced as a result of restricted supply of photosynthate.

With respect to responses to water stress levels in single soybean leaves, the data of Boyer (60) (Fig. 12.7A) show that photosynthesis was relatively unaffected until leaf water potentials reached about −11 bars in a growth chamber environment. Decreasing the leaf water potential from −11 to −16 bars resulted in a rapid decrease in photosynthesis followed by a gradual decline to a leaf water potential of −40 bars. Subsequent experiments (74) showed similar responses with soybeans and also showed that photosynthesis in corn was more sensitive to water stress than soybeans.

Field grown soybean plants did not show a decrease in photosynthesis rate until leaf water potential reached −15 to −17 bars (76, Fig. 12.7B). This suggests that the field environment conditioned the plants to withstand a somewhat greater water stress than plants in a controlled environment. Proposed factors which contribute to the differences in field and controlled environment plants include (1) short drying cycles and physiological adjustment periods due to restricted soil volume for roots of potted plants, (2) high light intensity and heat load exposure in the field, (3) differences in other atmospheric conditions such as air turbulence and humidity, (4) differences in leaf morphology, and (5) osmotic adjustments of leaf cells (71, 73, 75).

Stomatal closure is the primary factor associated with inhibition of photosynthesis during the initial phases of plant water deficit. Direct evidence of this behavior in soybeans was presented by Boyer (74) in which he showed stomatal diffusive resistance increased beginning at a leaf water potential of −11 bars, the point at which initial effects on photosynthesis were observed. As the leaf water potential decreased stomatal resistance progressively increased. Mesophyll resistance remained constant through −16 bars leaf water potential and only increased slightly at −40 bars. This was interpreted to mean that the influence of stomatal resistance was dominant in the response of soybean leaf photosynthesis to water stress with only minor effects beyond the mesophyll cells surface. Indirect evidence for stomatal closure as a major controlling factor in soybean response to water stress include the fact that transpiration and photosynthesis exhibit very similar responses to increased water stress (53, 74), and the increase in stomatal resistance with a decrease in soil water potential (76).

Respiration

Dark respiration is currently considered to be reduced by a moderate water deficit in soybean (60) and in other species (77). The data of Boyer

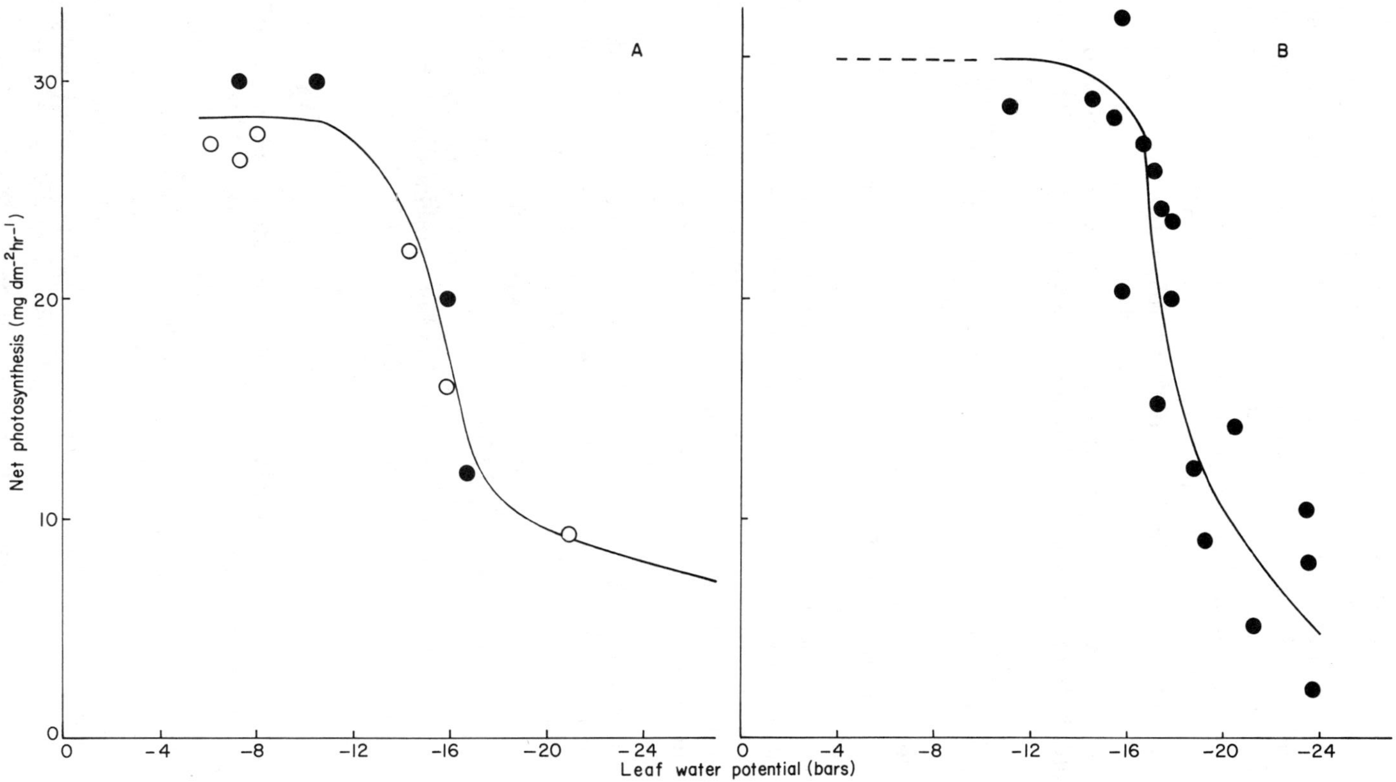

Figure 12.7. Response of photosynthesis to changes in soybean leaf water potential. Net photosynthesis (A) and $^{14}CO_2$ fixed (estimate of gross photosynthesis) (B). [After Boyer (60) and Turner et al. (75).]

(60) show that dark respiration rates of soybean, sunflower, and corn plants progressively decreased as plant water potential decreased from −8 to −18 bars and did not show a species difference in the response. In that study plant respiration remained about the same as leaf water potential decreased from −18 to −40 bars. In work from the same laboratory, soybean plant shoots showed a decrease of 48% in dark respiration rate as leaf water potential decreased from −5 through −28 bars (48). It is significant to note that while dark respiration was reduced by approximately one-half, photosynthesis and transpiration were reduced by more than 95%. Dark respiration in tomato was decreased by leaf water potentials lower than −8 bars and was reduced by 75% at a leaf water potential of −36 bars (77). He also found a different type of response in loblolly pine in that dark respiration decreased as leaf water potential decreased from −8 to −14 bars, increased as leaf water potential decreased from −14 to −30 bars, and again decreased as leaf water potential decreased to −48 bars.

It is difficult to delineate a direct effect of water stress on photorespiration because stomatal closure results in an increase in leaf temperature and a decrease in the supply of metabolites for photorespiration from photosynthesis. Therefore, it is not surprising to find reports that water stress decreases, increases, and has no effect on photorespiration (73). Since photorespiration is increased by increased temperature (78) it is reasonable to assume that soon after water deficit causes stomatal closure the rise in leaf temperature could cause an increase in photorespiration. With continuing water deficit the restricted supply of photosynthetic products, primarily glycolate, probably reduces photorespiratory release of CO_2 from leaves. Recycling of CO_2 liberated from both dark respiration and photorespiration is possibly enhanced when water deficit causes stomatal closure. The net effect of prolonged and intense water deficit would be a decrease in photorespiration which is compatible with the fact that responses in photosynthesis and photorespiration to other variables have shown similarities (79).

Translocation

A number of reports show translocation of assimilates to be reduced by water stress in several plant species (17, 78, 80) although few of the studies have included soybeans as the test plant. The mechanism by which translocation is affected by water stress has been the object of much research and discussion. Most scientists agree that transport of assimilates is by the mass flow mechanism, which, in turn, is dependent on a potential water gradient for its operation. Transport should continue as long as a gradient between source and sink is maintained. Crafts (80) proposed that in a plant subjected to water stress the movement of water into the phloem is reduced and the volume of the assimilate stream and velocity of flow are reduced, although the resulting change in solute concentration can have a mediating effect. Hartt (81) found that low water supply decreased the velocity and mass rate of translocation of ^{14}C-photosynthate. Since ^{14}C-photosynthate transport was slowed more severely than photosynthesis in water stressed plants, she also concluded that water supply had a direct effect on the velocity of translocation through the conducting system.

In contrast, the data of Wardlaw (82) show that translocation to the grain of wheat was affected very little by water stress even when leaves wilted for several days. He suggested that when the growing organs are not subjected to the same stress as other plant parts, as the grain during its development, the velocity of labeled assimilate movement through the conducting tissue is relatively insensitive to water stress. The developing seed on stressed soybean plants may act as a similar sink providing the water stress does not cause excessive fruit abscission, though no data were found to support this hypothesis.

Nelson (83) proposed that water stress affects the movement of assimilates into, but not through, the conducting tissue. A number of studies have provided support to this concept in that export of ^{14}C-photosynthate from a labeled leaf or leaf segment was retarded by water stress (82, 84, 85). The data in Table 12.1 (86) show that more of the ^{14}C-photosynthate was retained in the labeled segment of stressed than unstressed leaves 4 hours after labeling. This type of response implies that phloem loading from photosynthetic tissues is the process limited by water stress.

Assimilate distribution was affected in *Lolium* (Table 12.1) by water

***Table 12.1. Distribution of ^{14}C-labeled assimilates from expanding and mature leaves of* Lolium temulentum *L. in response to water stress*[a]**

Plant Part	% Distribution after 4 hr		
	Control	Mild Stress	Severe Stress
		Expanding leaf	
Leaf 8 (exposed tip)	48.7	59.1	65.8
Leaf 8 (remainder)	38.4	35.2	28.2
Other leaves	4.6	3.2	1.1
Crown and roots	8.3	2.6	5.0
		Mature leaf	
Leaf 6 (blade)	48.1	60.8	66.0
Leaf 6 (sheath)	2.7	3.5	4.6
Other leaves	3.1	4.6	3.9
Crown and roots	46.8	31.2	25.5

[a] Adapted from Wardlaw (82).

stress. The young, expanding leaf (leaf 8) retained approximately 87% of the ^{14}C-photosynthate in control plants, whereas about 94% was retained in leaves of stressed plants. Most of the ^{14}C-photosynthate that moved from the exposed area was recovered in the growing tissue at the base of the same leaf and the quantity was less in stressed plants. In the mature leaf (leaf 6) most of the ^{14}C-photosynthate exported from the source leaf was present in the crown and roots. The amount transported to the crown and roots progressively decreased with increase in water stress. In soybean plants (87) ^{14}C-photosynthate distribution was also affected by water stress. They found that alterations in ^{14}C distribution at leaf water potentials of -15 to -20 bars similar to alterations in dry weight distributions. In water stressed plants a greater portion of the assimilates was transported to the roots than in unstressed plants. This is compatible with previous discussion of root growth in which Mayaki et al. (29) found little effect of water stress on total root weight.

It would appear that water stress effects on translocation varies among species, organs on the same plant, and the state of physiological maturity of the organ or plant (Table 12.1; 73, 80, 86). While translocation of assimilates is undoubtedly affected by water stress, these responses are considered to be secondary in their influence on total plant development to the effects of stress on photosynthesis of the source or meristematic activity and growth of the sink. There is little reason to surmise that the net effect of water stress on growth would be different in soybeans even though the effects of water stress on fruit abscission and its interaction with the translocation phenomenon have not been described.

COMPONENTS OF WATER TRANSPORT

Transpiration occurs as a result of water flow through a dynamic and physically unified soil–plant–atmosphere system (88). Movement of water through the components of this system occurs along gradients of water potential from the soil to the surfaces of leaf cells and then through gradients of vapor pressure to the surrounding atmosphere. Radiant energy balance is the principal driving force in the transpirational processes.

Soil Supply of Water

The rate of water movement from the soil to the root surface is controlled primarily by the soil water potential and hydraulic conductivity (17). When the soil water content is high, the water is held at high potential which results in high capillary conductivity and water supply to the roots. As evapotranspiration depletes soil water the remaining water is held by soil particles with greater cohesive and adhesive forces. This reduces capillary flow of water to the roots through flow restrictions imposed by such factors as water film thickness and path tortuosity in the soil matrix. Many other soil factors influence the capacity of the soil to supply water to plants including the volume of water held per unit volume of soil, depth to water table, soil density, particle size, and organic matter content.

The depth of root penetration and degree of root proliferation have definite effects on soil supply of water to the plant through their influence on the total soil volume effectively explored by roots. Soybeans can effectively utilize water from deeper soil horizons providing root density along with hydraulic conductivity of the soil are favorably associated at or above minimum levels to exploit the soil water reservoir (42). Conversely, a dense soil layer or plow-pan often restricts root penetration and utilization of water in deeper soil horizons (89).

Water Movement into and through Roots

There is a steep gradient in water potential from the soil to the plant root during periods of high evaporative demand in the aerial plant environment (90). This gradient is established diurnally even under conditions where soil water content is relatively high (67, 91). On the other hand, during darkness or when the heat load of the plant is low, the water potential of the soil and plant roots approach equilibrium with the highest leaf water potential occurring just before sunrise. This suggests that leaf water potential of soybeans might be used to accurately estimate soil water potential (67) and emphasizes that sufficient time must elapse for equilibration in water status of the entire plant if leaf water potential is to approach soil water potential.

The relative importance of the various factors causing the differential in soil and root water potential has not been clearly defined. Resistances to radial water flow that contribute to this differential include root–soil surface contact (92), root epidermis and cortex, and root endodermis and stele (43). All three contribute to the total resistance of water entry into the plant with the relative importance of each assuming a different status depending on prevailing conditions. For example, the shrinking of plant roots during periods of low plant water potential (93) would enhance the limitations of root–soil contact as a factor impeding water flow to roots. On the other hand, when soil water content is high the restrictions in water flow across root tissue to the xylem would be a relatively larger component of the total resistance.

Regardless of the specific causes of impedance of water flow from the soil into the roots, there is considerable evidence to support the concept that the soil represents a major portion of the total resistance to flow of water in the soil–plant system and the roots represent the largest liquid

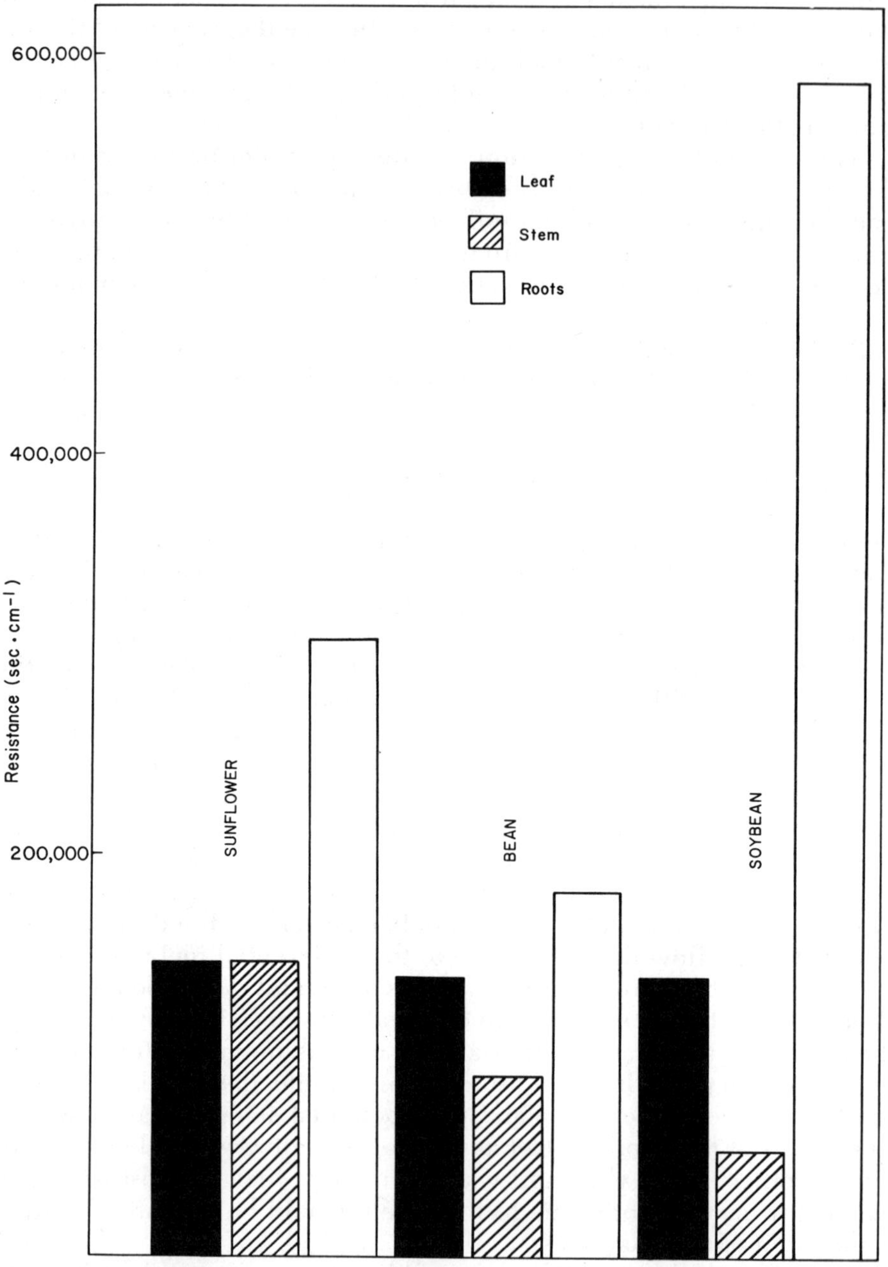

Figure 12.8. Resistance to water transport in soybean, bean and sunflower plants during recovery from water deficit. [From Boyer (94), p. 405, with permission.]

phase resistance to water transport within the plant (90, 91). The work of Boyer (94) supports the conclusion that root tissue was clearly the dominant component of total plant resistance to water flow in soybean, bean, and sunflower (Fig. 12.8). In soybean, leaf resistance was a distant second to root resistance, with stem vascular tissue making only a minor contribution to total plant resistance.

Boyer estimated two components of root resistance: a radial component (the resistance from all the tissue outside the root vascular tissue). The former contributed about 4 times as much to the root resistance as the latter, and the remaining vascular component.

Taylor and Klepper (43) recently reviewed the literature pertaining to axial resistances to water flow in plant roots. These resistances arise in the vascular tissue as a result of frictional forces between the moving water column and vessel walls. They point out that almost no quantitative data exist on the magnitude of axial resistance because the many factors influencing the resistances, including plant species, root length and depth, rate of transpiration, number and activity of adventitious roots, root diameter, and pathogenic organisms in root vascular tissue, make meaningful experimental measurements difficult. They do discuss evidence, however, indicating axial resistances could contribute significantly to root resistance to water flow in soybeans. The data of Taylor and Klepper (95) and Willatt and Taylor (96) were used to infer that substantial axial resistance occurs in soybeans (0.07 bar cm^{-1}) while the resistance in cotton was found to be negligible. The relative importance of axial resistance to flow needs further clarification. It appears that this offers interesting possibilities for genetic improvement in water transport through soybean roots, provided differences in root resistance can be found among soybean genotypes.

Water Movement through Stems and Leaves

Resistance to water movement through the vascular system of soybean stems was found to be small in comparison to other parts of the plant system (94). While water may be withdrawn from the transpiration stream by living cells along the route, these cells develop their own potential deficits subject to diurnal fluctuations but contributing minimally to restrictions in stem water flow (80). No evidence was found indicating that soybean stems exhibit axial resistance of the magnitude proposed for roots by Taylor and Klepper (43).

Leaf resistance to water flow in soybeans was found to be of the same magnitude as in sunflower and bean in the study of Boyer (94); however, it accounted for a much smaller portion of the total plant resistance because of the high root resistance. Leaf resistance to water transport has both liquid and gas phase water transfer components. In most crop plants, liquid flow of water in the leaf involves movement from the vascular elements across surrounding mesophyll tissue to cell surfaces in the stomatal cavity. The majority of this water movement is considered to be passive and to occur through the apoplast system (80) although some transport occurs through cell cytoplasm. In addition, some water moves through the leaf cuticle. Although the cuticle is known to be rather resistant to liquid transfer, values of 10% or more of total water loss have been reported through the cuticle (97).

Vapor phase resistances to water movement out of plants include evaporation of water from mesophyll walls, movement through stomatal openings, and movement across the boundary layer of air into the surrounding atmosphere. The rate of vapor movement from the leaf depends on the vapor pressure gradient between the evaporating surface and the open atmosphere. Vapor phase resistance to movement of water has long been recognized (98) as the major rate limiting step in transpiration under most conditions. Vapor phase movement is particularly important because it is through these events that the plant exercises control of transpiration through stomatal movement.

Transpiration

Seasonal water loss from a soybean crop includes both evaporation from the soil surface and transpiration by plants. During early stages of plant development significant evaporative losses occur from the soil. As the crop canopy develops evaporation from the soil contributes a progressively smaller portion of the total evapotranspiration. The depth of water loss through evaporation from bare soil surfaces is dependent on capillary transport to the surface. Therefore, profile depletion of soil water becomes more extensive when plant roots proliferate in deeper soil horizons and extract water in response to transpirational demands.

Leaf area and its distribution comprise the primary physical plant factors affecting transpiration of field crops. Shaw and Laing (88) presented data from Laing (99) showing that seasonal increases in evapotranspiration from soybeans closely paralleled the increases in leaf area index (Fig. 12.9). In the study of Peters and Johnson (34) doubling plant population and decreasing row spacing by one-half increased transpiration almost twofold during the period from flowering to maturity. These responses led Shibles et al. (13) to emphasize leaf surface area as the primary determinant of transpiration rate in soybeans until a complete ground cover is achieved by leaf development; thereafter, the principal controlling factor is the aerial environment. Radiant energy from the sun is the primary energy source supplying the latent heat requirements for vaporization. Other contributing energy sources come from scattered and reflected radiation from the sky and clouds, as well as sensible heat from the surrounding air, crop, and soil (78). Wind speed and relative humidity are factors also involved in regulating transpirational water loss from plants.

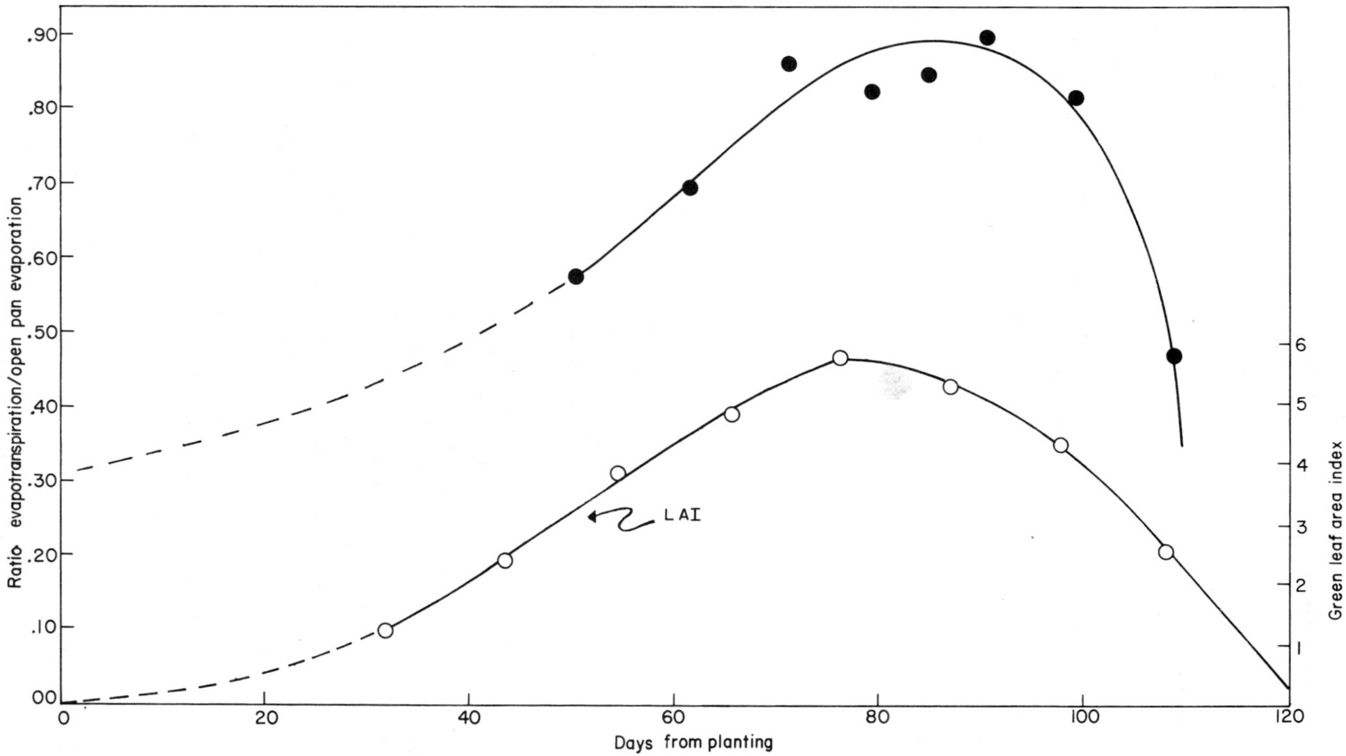

Figure 12.9. Relationship of seasonal changes in soybean leaf area index to the ratio of evapotranspiration to open pan evaporation. [After Shaw and Laing (88).]

Regulation of transpirational loss of water through stomatal action and the responsiveness of stomata to water deficit have been known for many years. It is generally recognized that there is little stomatal response to the development of water deficit until critical values are reached (73). After the critical level is reached stomatal closure occurs rather rapidly with further decrease in leaf water potential. The critical value varies among species (74, 100), the environmental conditions under which the plants have developed (71), and the number of stress cycles to which the plants have been exposed (75, 101). In soybeans the critical leaf water potential for controlled environment plants was in the range of -8 to -11 bars (74, 102) whereas field grown plants may be as low as -15 or -16 bars (102, 103). These findings indicate that the soybean plant exhibits some internal adaptive responses in stomatal closure to environmental conditions and water stress. Osmotic adjustment in leaf cells has been proposed as the main mechanism for these adaptive responses (73) although other factors, such as auxins and hormonal levels and interactions, have been implicated (104).

Morphological factors and physical changes of leaf tissues can affect transpiration. Chia and Brun (69) showed that stomatal frequency varied significantly among soybean gentypes and that water-stressed plants had a greater stomatal frequency and smaller leaf area than nonstressed plants. This type of response could be antagonistic to stress tolerance of soybeans unless the reduction in leaf area more than offset the effect of increased stomatal frequency. Physical factors such as leaf drooping or rolling and changes in leaf angle can alter transpiration by effectively reducing leaf area exposed to radiation and thereby decreasing the heat load of the plant.

Woolley (105) suggested a role for leaf pubescence in reducing transpiration in older leaves. He theorized that leaf hairs on older leaves lose their water and would not contribute to transpiration, yet the hairs would slow air movement at the leaf surface and reduce transpiration. The beneficial effects of leaf hairs on transpiration were not demonstrated under field conditions with a high energy budget, however. Methods of inducing changes in transpiration artificially include compounds that cause stomata to close, film forming polymer materials to impose physical barriers, and reflective materials to reduce net radiation load and leaf temperature (73). These procedures have met with questionable success primarily because the decreased photosynthesis, in many cases, more than offsets the benefits of reduced transpiration.

WATER USE AND YIELD

Estimates of water used in producing soybeans show that more than 500 g of water are required to produce 1 g plant dry matter (106, 107 as cited by 13). Calculations from the data of Doss and Thurlow (108) and Teare et al. (109) show that water required to produce a gram of soybean seed may be 2 or 3 times this value. Obviously, large quantities of water are required to produce acceptable yields of soybeans and data such as those of Teare et al. (109) show that water requirements of soybean are usually as much or more than other summer row crops.

Daily water use rates vary greatly depending on numerous factors including calendar date, leaf area, available soil water, radiant energy, temperature, and windspeed. Evapotranspiration (E_t) rates (108) ranged from 0.05 to 0.84 cm day^{-1} with a daily average for the growing season of 0.25 cm for nonirrigated, 0.38 cm for intermediate, and 0.46 cm for high irrigation levels. Water use rates are low in early season, reach a maximum during reproductive development, and decline as the plants approach maturity. These data substantiate the fact of E_t dependence on available soil water supply and show why the approximate total seasonal water use by these soybeans ranged from 34 (nonirrigated) to 62 (high irrigation) cm. The total amount of water required to produce high yields would probably approach the higher portion of this water use range in late maturing soybeans of the South, whereas less total water is required with a shorter growing season.

Within limits, the soybean plant shows a remarkable capacity to adapt to water stress. The data of Laing (99), as discussed by Shaw and Laing (88), show that short-term water stress imposed during the early flowering stage resulted in flower and pod drop in the lower part of the plant, but the plants compensated for this by subsequently producing more pods at the upper nodes. In a similar manner, partial compensation for flower and pod loss occurred during late flowering and early pod development, although some yield loss occurred as the result of water stress at this time. During the latter part of pod development and throughout grain development the plants' ability to adjust to water stress was less flexible. Changes in seed size constitute the primary mode of yield response to water availability after pod number becomes fixed.

The data of Laing (99) represent one of many reports which show that water stress during reproductive development has the greatest influence on yield (56, 108, 110, 111). Soybeans planted early in the season can be subjected to water stress during the vegetative stages of development without showing significant effects on yield. For example, the data of Ashley and Ethridge (56) show that the size of plants irrigated throughout the season was almost twice that of plants that were irrigated from flowering to maturity; however, yields were approximately the same in both cases. Although prolonged drought during vegetative development can result in plants without sufficient size to support high yields, apparently water deficit of this magnitude during vegetative growth occurs infrequently in most production areas.

In contrast, soybeans planted late in the growing season following small grain or corn harvest often show yield response to water stress during the vegetative stage of development. When the length of the vegetative development period is shortened, as in the double-crop system, vegetative growth must proceed at a rapid rate if plant height and size are to be adequate for acceptable yields.

Excessive water or flooding has been shown to reduce soybean yields. Spooner (111) showed that flooding for 21 days reduced yields by almost 60% and concluded that cultivars differ markedly in their tolerance of excess water. In general, the data available at this time indicate that flooding during early reproductive development was proabably most detrimental (10).

Even though research has repeatedly demonstrated that yields of soybeans can be increased by supplemental irrigation in most years, the economic feasibility of irrigating soybeans is often questionable in humid areas. The major reason for this situation rests largely with the inability of scientists to describe the conditions necessary, including water supply, to consistently produce the super high yields in soybeans. This obviously is a complex issue and involves the "state of the art" in soybean production and not just the issue of water management. Yields from most of the research plots in water requirement studies have been less than 4000 kg ha^{-1} (33, 56, 108, 112). We do not know if the water requirements and responses would be the same where yields approaching 6000 kg ha^{-1} are consistently made.

Yields of this magnitude have been reported from farmers fields in a number of production areas (J. E. Elsner and J. M. Woodruff, personal communication). It appears that the record yields are sporadic in that they are repeated infrequently on the same farm. This means that specific and unknown conditions existed on that farm which permitted the high yields. The data of Cooper (113) offer encouragement in consistantly producing high yields. Through several years of research he has developed genotypes along with a set of management practices specifically for high yield environment. Possibly more research and breeding efforts with maximum yield as the goal are justified. Achieving the goal of repeatedly producing exceptionally high yields predictably will involve judicious water management schemes and further research information.

REFERENCES

1. A. H. Probst and R. W. Judd, Origin, U. S. History and Development, and World Distribution, in B. E. Caldwell, Ed., *Soybeans: Improvement, Production and Uses*, American Society of Agronomy, Madison, WI., 1973, p. 1.
2. C. V. Piper and W. J. Morse, *The Soybean*, McGraw-Hill, New York, 1923, pp. 35–54.
3. T. Hymowitz, *Econ. Bot.* **24**, 408 (1970).
4. T. Nagata, *Proc. Crop Sci. Soc. Jpn.* **28**, 79 (1959).

5. W. W. Garner and H. A. Allard, *J. Agr. Res.* **18,** 353 (1920).
6. W. W. Garner and H. A. Allard, *J. Agr. Res.* **41,** 719 (1930).
7. H. A. Borthwick and M. W. Parker, *Bot. Gaz.* **101,** 341 (1939).
8. E. E. Hartwig, *U.S. Dep. Agric. Cir.* **943,** 13 pp.
9. J. W. Johnson, H. A. Borthwick, and R. C. Leffel, *Bot. Gaz.* **122,** 77 (1960).
10. D. K. Wigham and H. C. Minor, Agronomic Characteristics and Environmental Stress, in A. G. Norman, Ed., *Soybean Physiology, Agronomy, and Utilization,* Academic Press, New York, 1978, p. 77.
11. D. M. Brown, *Agron. J.* **52,** 493 (1960).
12. T. R. Sinclair, *Crop Sci.* **20,** 196 (1980).
13. R. Shibles, I. C. Anderson, and A. H. Gibson, Soybeans, in L. T. Evans, Ed., *Crop Physiology, Some Case Histories,* Cambridge University Press, Cambridge, England, 1975, p. 151.
14. P. H. Van Schaik and A. H. Probst, *Agron. J.* **50,** 192 (1958).
15. J. D. Mann and E. G. Jaworski, *Crop Sci.* **10,** 620 (1970).
16. A. G. Norman, Background, in A. G. Norman, Ed., *Soybean Physiology, Agronomy, and Utilization,* Academic Press, New York, 1978, p. 1.
17. H. J. Mederski, D. L. Jeffers and D. B. Peters, Water and Water Relations, B. E. Caldwell, Ed., *Soybeans: Improvement, Production* and *Uses,* American Society of Agronomy, Madison, Wis., 1973, p. 239.
18. R. M. Hagan, Soil Temperature and Plant Growth, in *Soil Physical Conditions and Plant Growth,* Academic Press, New York, 1952, p. 367.
19. R. J. Hanks and F. C. Thorp, *Soil Sci. Soc. Am. Proc.* **20,** 307 (1956).
20. J. R. Hunter and A. E. Erickson, *Agron. J.* **44,** 107 (1952).
21. L. G. Heatherly and W. J. Russell, *Agron. J.* **71,** 980 (1979).
22. A. R. Grabel and R. E. Danielson, *Soil Sci. Soc. Am. Proc.* **29,** 12 (1965).
23. R. J. Hanks and F. C. Thorp. *Soil. Sci. Soc. Am. Proc.* **21,** 357 (1957).
24. A. R. Grabel and R. E. Danielson, *Soil Sci. Soc. Am. Proc.* **29,** 233 (1965).
25. J. B. Carlson, Morphology, in B. E. Caldwell, Ed., *Soybeans: Improvement, Production* and *Uses,* American Society of Agronomy, Madison, Wis., 1973, p. 17.
26. R. L. Mitchell and W. J. Russell, *Agron. J.* **63,** 313 (1971).
27. C. D. Raper, Jr., and S. A. Barber, *Agron. J.* **62,** 581 (1970).
28. H. L. Borst and L. E. Thatcher, *Ohio Agric. Exp. Stn. Res. Bull.* **494,** 96 p. (1931).
29. W. C. Mayaki, I. D. Teare, and L. R. Stone, *Crop Sci.* **16,** 92 (1976).
30. T. C. Kasper, C. D. Stanley, and H. M. Taylor, *Agron. J.* **70,** 1105 (1978).
31. R. R. Allmaras, W. W. Nelson, and W. B. Voorhees, *Soil Sci. Soc. Am. Proc.* **39,** 764 (1975).
32. D. B. Egli and J. E. Leggett, *Crop Sci.* **13,** 220 (1973).
33. C. K. Martin, D. K. Cassel and E. J. Kamprath, *Agron. J.* **71,** 592 (1979).
34. D. B. Peters and L. C. Johnson, *Agron. J.* **52,** 687 (1960).
35. L. R. Stone, I. D. Teare, C. D. Nickell, and W. C. Mayaki, *Agron. J.* **68,** 677 (1976).
36. L. M. Arya, G. R. Blake, and D. A. Farrell, *Soil Sci. Soc. Am. Proc.* **39,** 437 (1975).
37. W. C. Mayaki, L. R. Stone, and I. D. Teare, *Agron. J.* **68,** 532 (1976).
38. W. K. Robertson, L. C. Hammond, J. T. Johnson, and K. J. Boote, *Agron. J.* **72,** 548 (1980).
39. S. A. Barber, *Agron. J.* **70,** 457 (1978).
40. W. Bohm, *Z. Acker-und Pflanzenbau* **144,** 103 (1977).

41. R. R. Allmaras, W. W. Nelson, and W. B. Voorhees, *Soil Sci. Soc. Am. Proc.* **39,** 771 (1975).
42. D. C. Reicosky, R. J. Millington, A. Klute, and D. B. Peters, *Agron. J.* **64,** 292 (1972).
43. H. M. Taylor and B. Klepper, The Role of Rooting Characteristics in the Supply of Water to Plants, in N. C. Brady, Ed., *Advances in Agronomy*, Academic Press, New York, 1978, p. 99.
44. J. K. Wilson, *J. Am. Soc. Agron.* **23,** 670 (1931).
45. J. I. Sprent, *Plant Soil* **225** (Special Volume) (1971).
46. J. I. Sprent, *New Phytol.* **71,** 443 (1972).
47. C. E. Pankhurst and J. I. Sprent, *J. Exp. Bot.* **25,** 287 (1975).
48. C-Y Huang, J. S. Boyer, and L. N. Vanderhoff, *Plant Physiol.* **56,** 222 (1975).
49. R. P. Patterson, C. D. Raper, and H. D. Gross, *Plant Physiol.* **64,** 551 (1979).
50. J. I. Sprent, *New Phytol.* **70,** 9 (1971).
51. J. I. Sprent, *New Phytol.* **71,** 603 (1972).
52. J. I. Sprent, *New Phytol.* **71,** 451 (1972).
53. C-Y Huang, J. S. Boyer, and L. N. Vanderhoff, *Plant Physiol.* **56,** 228 (1975).
54. L. C. Hammond and D. Kirkham, *Agron. J.* **41,** 23 (1949).
55. J. J. Hanway and C. R. Weber, *Agron. J.* **63,** 227 (1971).
56. D. A. Ashley and W. J. Ethridge, *Agron. J.* **70,** 467 (1978).
57. G. A. Constable and A. B. Hearn, *Aust. J. Plant Physiol.* **5,** 159 (1978).
58. H. D. Scott and J. T. Batchelor, *Agron. J.* **71,** 776 (1979).
59. M. G. Huck and J. M. Davis, Water Requirements and Root Growth, in L. D. Hill, Ed., *Proceedings of the World Soybean Research Conference*, Interstate Printers and Publishers, Danville, Ill., 1976, p. 16.
60. J. S. Boyer, *Plant Physiol.* **46,** 233 (1970).
61. G. Y. Blair, L. A. Richards, and R. B. Campbell, *Soil Sci.* **70,** 431 (1950).
62. M. V. K. Sivakumar and R. H. Shaw, *Agron. J.* **70,** 619 (1978).
63. R. O. Slatyer, *J. Biol. Sci.* **10,** 320 (1957).
64. C. H. Wadleigh and H. G. Gauch, *Plant Physiol.* **23,** 485 (1948).
65. E. N. Martin, *Plant Physiol.* **15,** 449 (1940).
66. R. F. Meyer and J. S. Boyer, *Planta* **108,** 77 (1972).
67. R. A. Brady, W. L. Powers, L. R. Stone, and S. M. Goltz, *Agron. J.* **66,** 795 (1974).
68. W. Wenkert, E. R. Lemon, and T. R. Sinclair, *Agron. J.* **70,** 761 (1978).
69. A. J. Chia and W. A. Brun, *Crop Sci.* **15,** 309 (1975).
70. R. A. Fischer and R. M. Hagan, *Exp. Agric.* **1,** 166 (1965).
71. J. C. Thomas, K. W. Brown, and W. R. Jordan, *Agron. J.* **68,** 706 (1976).
72. S. R. Ghorashy, J. W. Pendleton, D. B. Peters, J. F. Boyer, and J. E. Beverlein, *Agron. J.* **63,** 674 (1971).
73. J. E. Begg and N. C. Turner, in N. C. Brady, Ed., *Advances in Agronomy* Vol. 28, Academic Press, New York, 1976, p. 161.
74. J. S. Boyer, *Plant. Physiol.* **46,** 236 (1970).
75. N. C. Turner, J. E. Begg, H. M. Rawson, S. D. English, and A. B. Hearn, *Aust. J. Plant Physiol.* **5,** 179 (1978).
76. R. A. Brady, S. M. Goltz, W. L. Powers, and E. T. Kanemasu, *Agron. J.* **67,** 97 (1975).
77. H. Brix, *Physiol. Plant.* **15,** 10 (1962).

78. R. O. Slatyer, Physiological Significance of Internal Water Relations to Crop Yields, in J. D. Eastin, F. A. Haskins, C. Y. Sullivan, and C. H. M. Van Bavel Eds, *Physiological Aspects of Crop Yield*, American Society of Agronomy, Madison, Wis., 1969, p. 53.
79. W. L. Ogren and G. Bowes, *Nature (London), New Biol.* **230,** 159 (1971).
80. A. S. Crafts, Water Deficits and Plant Physiological Processes, in T. T. Kozlowski, Ed., *Water Deficits and Plant Growth*, Vol. 2, Academic Press, New York, 1968, pp. 85–133.
81. C. E. Hartt, *Plant Physiol.* **42,** 338 (1967).
82. I. F. Wardlaw, *Aust. J. Biol. Sci.* **20,** 25 (1967).
83. C. D. Nelson, Effect of Climate on the Distribution and Translocation of Assimilates, in L. T. Evans, Ed., *Environmental Control of Plant Growth*, Academic Press, New York, 1963, p. 243.
84. Z. Plaut and L. Reinhold, *Aust. J. Biol. Sci.* **18,** 1143 (1965).
85. E. R. Brevedan and H. F. Hodges, *Plant Physiol.* **52,** 437 (1973).
86. I. F. Wardlaw, *Aust. J. Biol. Sci.* **22,** 1 (1969).
87. J. E. Silvius, R. R. Johnson, and D. B. Peters, *Crop Sci.* **17,** 713 (1977).
88. R. H. Shaw and D. R. Laing, Moisture Stress and Plant Response, in W. H. Pierre, D. Kirkam, J. Pesek, and R. H. Shaw, Eds., *Plant Environment and Efficient Water Use*, American Society of Agronomy, Madison, Wis. 1966, p. 73.
89. R. W. Pearson, Soil Environment and Root Development, in W. H. Pierre, D. Kirkham, J. Pesek, and R. H. Shaw, Eds., *Plant Environment and Efficient Water Use*, American Society of Agronomy, Madison, Wis., 1966 p. 95.
90. P. J. Kramer, *Plant and Soil Water Relationships*, McGraw-Hill, New York, 1969, p. 258.
91. P. J. Kramer, *Am. J. Bot.* **25,** 110 (1938).
92. W. N. Herkelrath, E. E. Miller, and W. R. Gardner, *Soil Sci. Soc. Am. J.* **41,** 1039 (1977).
93. M. G. Huck, B. Klepper, and H. M. Taylor, *Plant Physiol.* **45,** 529 (1970).
94. J. S. Boyer, *Crop Sci.* **11,** 403 (1971).
95. H. M. Taylor and B. Klepper, *Soil Sci.* **120,** 57 (1975).
96. S. T. Willatt and H. M. Taylor, *J. Agric. Sci.* **90,** 205 (1978).
97. J. E. Pallas, Jr. and A. R. Bertrand, *U.S. Dept. Agric. ARS Prod. Res. Rep.* **89** (1966).
98. T. H. Van der Honert, *Discuss. Faraday Soc.* **3,** 146 (1948).
99. D. R. Laing, *The Water Environment of Soybeans*, Ph.D. Thesis, Iowa State University, Ames, Iowa (1966).
100. I. D. Teare and E. T. Kanemasu, *New Phytol.* **71,** 805 (1972).
101. R. A. Fisher, T. C. Hsiao, and R. M. Hagan, *J. Exp. Bot.* **21,** 371 (1970).
102. W. J. Davies, *Crop Sci.* **17,** 735 (1977).
103. R. E. Carlson, N. N. Momen, O. Arjmand and R. H. Shaw, *Agron. J.* **71,** 321 (1979).
104. T. C. Hsiao, *Annu. Rev. Plant Physiol.* **24,** 519 (1973).
105. J. T. Woolley, *Agron. J.* **56,** 569 (1964).
106. A. R. Whitson, *Wisc. Agric. Exp. Stn. Rep.* **190** (1902).
107. I. Kato, *Tokai-Kinki Nat. Agr. Exp. Stn.* 14 pp. (1967).
108. B. D. Doss and D. L. Thurlow, *Agron. J.* **66,** 620 (1974).
109. I. D. Teare, E. T. Kanemasu, W. L. Powers, and H. S. Jacobs, *Agron. J.* **65,** 207 (1973).

110. P. Grissom, W. A. Raney, and P. Hogg, *Miss. State Agric. Exp. Stn. Bull.* **531,** (1955).
111. A. E. Spooner, *Ark. Agric. Exp. Stn. Bull.* **644,** (1961).
112. H. J. Mederski and D. L. Jeffers, *Agron. J.* **65,** 410 (1973).
113. R. L. Cooper, *Proceedings of the Solid Seeded Soybean Systems for Success Conference*, American Soybean Association, Indianapolis, Ind., 1980.

13

SUGARBEET

S. S. Martin

INTRODUCTION

Adaptation

The sugarbeet is one of several forms of cultivated beets, all of which are included in a single species, *Beta vulgaris* L. (Chenopodiaceae). The diversity of plant form within this species is large, as it includes the familiar red or yellow table beet; leafy foliage beets such as chard, eaten as vegetables or salad greens; and fodder beets or mangels, grown for cattle feed. Members of *Beta vulgaris* are closely related to the wild sea-beet, *Beta maritima*, that today is found in coastal habitats along the Mediterranean Sea and northward along the Atlantic coast to the North Sea, as well as inland to the Asiatic steppes and East India (1, 2, 3). The diverse cultivated forms of *B. vulgaris* may have been derived by selection from *B. maritima*, or *B. maritima* and a progenitor of *B. vulgaris* may have evolved from a common maritime ancestor (3). Whatever the origins of the earliest cultivated beets, the sugarbeet is one of few agriculturally important plants developed in comparatively recent times. Only in the early 1800s was the first factory constructed in Germany for the extraction of sucrose ("sugar") from a type of fodder-beet; thus, the genetic development of the modern sugarbeet, which may contain over 20% of its fresh weight as sucrose, has been achieved in less than 200 years. Accounts of the early steps in sugarbeet history and development of the U.S. sugarbeet industry have been given by Coons (4) and Theis (5). Today the sugarbeet is the source of over one-half the sucrose produced in the United States, and of about 40% of world sucrose production (6).

The sugarbeet is a biennial plant; it grows vegetatively in a near-rosette form until induced by a combination of low temperatures and long days (photothermal induction) to elongate its stem ("bolt") and subsequently to flower. In the United States, the sugarbeet most commonly is planted in early spring in temperate climates, grown under irrigation, and the root is harvested at the approach of winter (thus, for sucrose production, the plant is grown as a nonflowering annual). In a few mild-winter areas of the United States, such as the Imperial Valley of southern California and in parts of Arizona, bolting-resistant sugarbeets can be planted in the fall and harvested the following spring or summer. If seed production is sought, the plants must be grown in climates having winters cold enough to provide the necessary induction of flowering, but not so cold that soil freezes and kills the plants (7); alternatively, the first-year-growth roots can be harvested and stored under the proper photothermal conditions, then replanted the following spring for bolting and flowering. Sugarbeets currently are grown commercially in about 16 states, and they have been grown occasionally in about 10 others. Because climate varies widely among the growing areas, sugarbeets have been bred for adaptation to specific combinations of environmental variables. It is to be expected, then, that not all cultivars respond identically to a given stress, including water stress (8).

The fleshy sugarbeet root (which technically includes at its top some hypocotyl tissue and stem) has been selected and is cultivated to be as large as possible. The plant is very slow developing, and must be sparsely planted because of the large size of the mature roots; thus, a long growing season for largest root size is desirable. In the U.S. growing areas, low temperatures, relatively low light intensities, and short days of early- and late-season periods usually limit ultimate root size.

Within the period of vegetative growth of sugarbeets grown for root harvest, it is possible to identify shorter periods of relative dominance of certain developmental phases (Table 13.1). It must be emphasized that these divisions are somewhat arbitrary, as growth is a continuum; other divisions could be made. The sugarbeet seed is small, so emergence is easily inhibited by crusting, and by the low temperatures at which planting often takes place in an attempt to provide a growing season that is as nearly the length of the frost-free period as possible. Also, cool temperatures often favor several seed or seedling diseases. The rate of leaf appearance is quite temperature sensitive (9), so early canopy development usually is slow, with 70 or more days from planting to a leaf area index (*LAI*) of 1.0 not uncommon. As more leaves are available to produce photosynthate, the canopy growth rate continues to accelerate in the more favorable climatic conditions of early summer, and canopy development to *LAI* of 4 or more occurs comparatively rapidly. The fibrous roots and the tap root are developing simultaneously, but at much slower rates of growth than the top, and neither claims a large fraction of the net assimilate produced up to this point. At about 2.5 months after planting, there is a transition from top-dominated growth

Table 13.1. Growth stages of vegetatively grown sugarbeets with typical conditions for spring-planted, fall-harvested plants assumed

Growth Stage	Approximate Number of Days in Stage	Comments
Early establishment		
Germination and emergence	5–10	Very temperature and moisture sensitive
"Seedling"; to first true leaves	10–15	Very temperature and moisture sensitive
Canopy development predominates		
to *LAI* 1	30–50	Root development also is occurring, but at a slower
to *LAI* 4 or greater	15–30	rate than canopy growth
Storage root development predominates	50–70	Canopy growth rate slows, and reaches zero about the time of maximum root growth rate
Preharvest	30	Especially affected by management of water and nitrogen supply

to root-dominated growth. The rate of growth of leaves falls to about zero at the time root growth is at its maximum, then net leaf growth eventually becomes negative because leaf loss through senescence exceeds growth of new leaves. In late season, leaf appearance rate is again affected by cool temperatures. Finally, in about the month preharvest, agronomic practices become particularly important, with low nitrogen availability desirable to reduce growth, and some degree of root dehydration sometimes permitted by withholding late watering. Potential evapotranspiration usually is moderate in the preharvest period, and any apparent gain in root sucrose content (expressed as percentage of root fresh weight) is mainly due to decreased water content. More detailed accounts of sugarbeet development are available elsewhere (9–12).

Crop Use and Yield Potential

In recent years, about 600,000 ha have been planted annually to sugarbeets in the United States (6). About 33% of the planted area is unirrigated, principally in the Red River Valley area in North Dakota and Minnesota, and in Michigan. The major irrigated production is in California, followed distantly by Idaho, Washington, Colorado, Nebraska, and the other producing states. Currently, economic conditions are leading to rapid changes in sugarbeet production and production areas.

Because of higher yield in states with more favorable climates, especially California (the state with greatest area planted and highest yield), the unirrigated sugarbeet areas contribute about 22% of total fresh root production. The U.S. average yield of sugarbeet roots has been around 45 metric tons (MT) per ha; the California average yield usually is about 40% higher, and the California record is 132 MT per ha (9). In most of the U.S. sugarbeet-growing area, economic yields could not be obtained without irrigation.

Its sucrose content of 15–20% of storage root fresh weight clearly is the primary economic value of the sugarbeet, but several useful byproducts also are obtained. The plant tops, which are cut off at the time of harvest, represent about 40% of the total fresh weight of the plant and are relatively nutrient rich. They can be fed to livestock, either fresh or as a cured silage (13), or they can be reincorporated into the soil as a fertilizer. After sucrose is extracted from the storage roots, the pulp remaining is used as a cattle feed, either directly or in a mixture with molasses, which is the syrupy material remaining after repeated crystallizations and removal of sucrose from the sugarbeet extract. Molasses still is relatively sucrose rich (about 50% by weight), so it makes an excellent substrate for fermentation processes such as production of alcohol or glutamic acid (as monosodium glutamate, the flavor enhancer), or for the culture of bakers' yeast (14).

ROOT GROWTH AND FUNCTION

Germination and Emergence

The sugarbeet as originally developed by selection from its wild ancestors produced "multigerm" fruits, that is, a fruit comprised of the fertilized ovaries of several flowers that were enclosed by a common receptacle. The discovery of the gene for "monogerm" fruits (15) was an important advance in sugarbeet breeding. Smith (16) provides a detailed description of sugarbeet floral structure and seed production techniques.

Monogerm sugarbeet seed is small; the seed itself typically weighs about 5 mg, and the entire indehiscent fruit with its corky overlayers averages only about 20 mg (17). Thus, there is little reserve energy supply in the seed. Ideal conditions for seed germination and emergence include a fine, loose planting medium, adequate moisture, and root temperatures about 22–32°C (18). Such ideal conditions almost never exist, of course, under actual field conditions, so potential germination levels seldom are attained. In practice, mechanical impedance to germination and emergence is overcome by shallow planting at about 2 cm depth in a fine, well-prepared seed bed.

Most importantly, the sugarbeet seems to have a higher soil water requirement for its germination than some other crops. At 25°C, sugarbeet does not germinate at a soil water potential of less than about −3.5 bars (19). Low temperatures increase the water requirement; germination at 9°C may be severely diminished even at soil water potentials of −1 bar (20). In addition, both cultivar and degree of fruit maturity affect potential germination (21). After germination, the survival rate of sugarbeet seedlings under light frost conditions seems to be inversely correlated with their sap osmotic pressure (22); that is, the more negative the sap osmotic pressure, the more frost resistance.

Depth and Distribution

The sugarbeet ultimately is deep rooting, but early fibrous root growth is slow. For example, in one field study, roots reached depths of 30, 60, and 90 cm, respectively, about 60, 68, and 74 days after planting (23). These workers also showed that the moisture level of the soil (within defined limits) had no effect on the rate of root extension, as measured by uptake of radioactive tracers at several soil depths. This is but one of many experiments that have throughly disproved an old idea that withholding early irrigation of sugarbeet seedlings would "drive roots deep to water," forming a desirably shaped, tapering taproot.

An interesting comparative study of root development in sugarbeet, potato, and barley was made under field conditions and natural rainfall in England (24). At comparable plant age, sugarbeet extracted water to lesser depth than the other crops. At the same calendar date, when the three crops were of slightly different age, sugarbeet again obtained water more shallowly than the others. For instance, a 3 year average showed that sugarbeet, potato, and barley extracted water on June 15 from the top 44, 50, and 80 cm, respectively. Eventually, all the crops' roots penetrated and extracted water to at least 100 cm, but by the end of the season sugarbeet had extracted the most below 80 cm.

Under the most common U.S. conditions of sugarbeet growth, in which natural soil water is insufficient to meet potential evapotranspiration, fibrous roots continue to go deeper as the season progresses, especially if the water content of upper soil horizons is not kept high by irrigation. Root depths of 150 cm are common, even with occasional irrigation. Under dryland conditions in North Dakota, sugarbeets extracted soil water to at least 183 cm depth, whereas maize on the same soil utilized water only to about 91 cm depth (25). At high plant population density, roots penetrate more rapidly and more water is extracted than at lower density (26).

The most detailed information on fibrous roots has come from studies in vermiculite and nutrient solution cultures (9, 27). Fibrous roots reached 14 g fresh weight in 113 days of growth; the estimated total length of these roots was almost 4 km. As these authors point out, however, little is known about the effectiveness of different parts of the root system in water and nutrient uptake.

Storage Root Growth

The sugarbeet storage root develops by division and enlargement of cells produced by a series of concentric secondary cambia that arise early in the plant's growth. The first two secondary cambia are evident at about 3 weeks after emergence, and six cambial rings are detectable by 6 weeks after emergence (28). Additional rings form slowly, about one a month, to a total of 12 or more. Enlargement of cells produced by these supernumerary cambia proceeds simultaneously, but at different rates depending on location.

An early hypothesis suggested that storage root development and sucrose accumulation occurred only when there was assimilate in excess of needs for top and fibrous root growth, which had priority (29). Subsequent experiments have shown that this is not the case (30–32). Instead it appears that a control mechanism, presumably hormonal, regulates the distribution of assimilate to top growth, fibrous root growth, storage root growth, and sucrose storage. Genotype and some exogenous factors, especially temperature, plant density, and nitrogen fertility, affect the control mechanism and can alter the distribution or partitioning of photo-

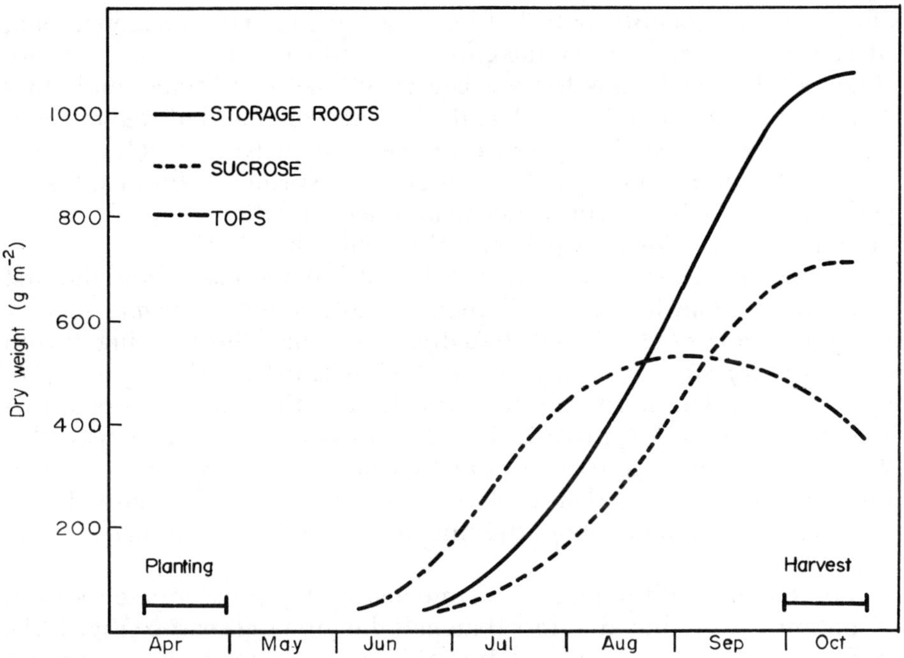

Figure 13.1. Typical dry matter accumulation curves for tops (leaf blades plus petioles), storage roots, and sucrose in spring-planted sugarbeets.

synthetic products. Under typical field conditions, both storage root dry matter and sucrose content (as percentage of dry matter) increase continuously throughout the growing season; their accumulations are described by S-shaped, logistic curves (Fig. 13.1). The rates of dry matter and sucrose increase are maximal in midseason, when warm temperatures and relatively high light intensities act on a canopy of sufficient size for full light interception, allowing a high potential photosynthetic rate. By late season, decreasing light intensities and lower temperatures, especially night temperatures, inevitably lead to reduced growth rates. Nevertheless, storage root dry matter and sucrose continue to accumulate gradually until harvest.

CANOPY DEVELOPMENT

Dry Matter Accumulation

In sugarbeet as in other crops, leaf area, leaf area duration, leaf photosynthetic efficiency, and type of leaf display all are involved in determining how much light energy will be converted to assimilate. There is fundamental genetic control of leaf size, shape, and appearance rate, but each changes throughout the growing season because each is strongly influenced by exogenous factors. Under optimal conditions, including a temperature of about 24°C (30) and adequate water and nitrogen, about four to six leaves usually appear each week. However, the Q_{10} for leaf appearance rate is about 3 (34), so under cool spring conditions, leaves appear very slowly. A typical dry matter accumulation curve for sugarbeet tops (leaf blades plus petioles) illustrates this (Fig. 13.1).

The size of each successive leaf at its maturity increases from the first leaves up to about leaf 20 or 25, then leaves gradually become smaller and often thicker (11, 35, 36). Usually, nearly half the growing season is required for sugarbeet to reach an *LAI* of 1, then leaf accretion proceeds rapidly. The maximum top growth rate (Fig. 13.2) occurs at this time of rapid increase in *LAI*, then growth rate of tops decreases and eventually becomes zero or negative because leaves are senescing and being lost more rapidly than replacements are formed. Root growth rate reaches its maximum at about the time leaf growth rate approaches zero (Fig. 13.2).

Note that most of the plant's organic matter (top plus storage root and its contents) is produced in less than half the growing season (Fig. 13.1). Leaf area present in May and June is not adequate to intercept and utilize fully the long hours of sunlight at that time. It is apparent that any treatment or procedure that can increase the rapidity of early-season growth should be beneficial to ultimate root and sucrose yield. Seedlings can be grown in a protected environment and transplanted to the field

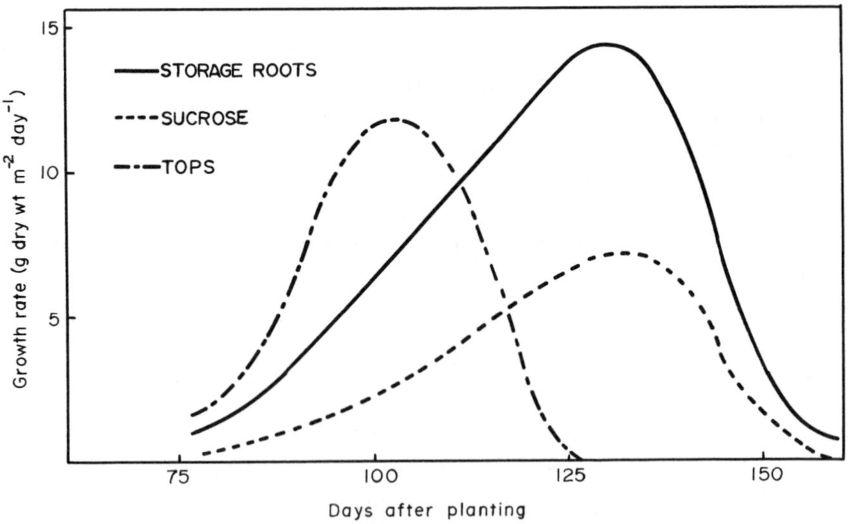

Figure 13.2. Growth or accumulation rates for sugarbeet tops (leaf blades plus petioles), storage roots, and sucrose. [After Storer, Schmehl, Hecker (11).]

after frost danger has passed, and yield benefits have been shown (37, 38), but this is not considered practical on a commercial scale.

Because of leaf crowding and mutual shading on an almost acaulescent plant such as sugarbeet, there logically might be an optimal *LAI* (for each cultivar and environment) above which the energy expenditure necessary to form and maintain the leaf is not equaled or exceeded by the gain in photosynthetic assimilate. In typical experiments, maximum sucrose yield is reached with maximum *LAI* of about 4. Treatments that increase *LAI* above this may result in greater total plant dry weight, but usually give no root or sucrose yield increase, and sometimes a decrease. For example, the total number of leaves produced during the growing season is particularly sensitive to nitrogen nutrition levels. An adequate supply of nitrogen is required for rapid early growth, but if high nitrogen levels are maintained into late season, leaf appearance rates continue high, resulting in increased top weights relative to root or sucrose weights.

The net assimilation rate (*NAR*, net assimilate gained per unit leaf area per unit time) of sugarbeet usually is highest well before *LAI* 1 is reached. Thus, although sugarbeet gains leaves slowly in cool early season conditions, those leaves are efficient. For example, Thorne (39) found that early sugarbeet *NAR* exceeded that of potato and barley. Although sugarbeet *NAR* decreases gradually from the early season maximum, the simultaneously increasing leaf area more than compensates for decreasing *NAR* in midseason, when the maximum total crop growth rate is attained. Watson and Witts (40) compared growth in sugarbeet and its wild relative *B. maritima*, and found that they had similar *NAR* before *LAI* 1, but that sugarbeet maintained a high *NAR* longer, distributed photosynthate differently, and achieved a high *LAI* with fewer but larger leaves. Thorne and Evans (41) compared reciprocal grafts between tops and roots of sugarbeets and spinach beet, another form of *B. vulgaris*. *NAR* and storage root dry weight were greatest for a sugarbeet root with either top, and for a sugarbeet top with either root. They suggested that graft combinations having sugarbeet roots provided better "sinks" for photosynthate and thereby increased photosynthetic rates. More specifically, Rapoport and Loomis (42) studied reciprocal grafts between sugarbeet and chard, and reported that the degree of cell expansion attained was the main difference between the two root types.

Water Relations and Canopy Development

Sugarbeet canopy development is strongly affected by interactions of water supply, the availability of nutrients (especially nitrogen), and temperature. The rate of leaf appearance is affected mostly by nitrogen level and temperature, with mild water stress having little effect (37, 43).

Prolonged or continuous drought, of course, decreases leaf numbers. Final leaf size also is temperature sensitive, but water availability also is a determinant. Terry (44) showed the extreme importance of cell expansion in the growth of young leaf laminae. Laminar size is very important in the early growth states, when the largest possible photosynthetic surface is needed. Hsiao (45) has pointed out that cell expansion is often one of the plant developmental processes most susceptible to water stress. Clearly, because early root development is minimal, the maintenance of high soil water potentials in the upper soil zone is essential to full early-season leaf development in sugarbeet. That this effect still exists in later development was shown by Milford and Lawlor (46), who grew plants under identical conditions for 8 weeks, then found that those subsequently grown at -0.2 bar average soil water potential developed greater leaf area than those grown at -2 bars. Other evidence of reduced leaf size under water stress was obtained by Scott and Bremner (37). However, an exception to this has been reported by Orchard (47). The reduction in leaf size probably is caused by reduced cell expansion (45). It is doubtful that cell multiplication is reduced by very mild, osmotically induced water stress, as has been reported (48).

NAR usually is depressed by water stress. Recovery time to the original *NAR* depends on the severity of the drought. After the original *NAR* is regained following a prolonged drought, there is often a period of compensatory activity in which the *NAR* and relative leaf growth rate of recovering plants exceed those of well-watered controls (47, 49).

Mature sugarbeets are relatively drought resistant because their deep root system allows the utilization of extensive volumes of soil water. However, sugarbeet shows poor control of its internal water balance, frequently wilting under moderate to high evapotranspiration demand even when water is readily available to the roots. Some of the underlying reasons for this will be presented in the next section.

Row Spacing, Plant Density, and Evapotranspiration

Optimal plant density for sugarbeet is determined to maximize sucrose yield, not total plant yield. Some consider an ideal planting design for sugarbeet, which can grow quite large at maturity, to be a series of squares in which plants are about 0.3 m apart. However, practical factors such as the type of irrigation to be employed and the requirement for mechanical planting and cultivation usually cause departure from the ideal. As a result, sugarbeets in the United States usually are planted in single rows spaced about 50–80 cm apart, with planting densities between about 50,000 and 80,000 plants per hectare (50). The effect of plant density on total water use is small, but high densities usually use slightly more water over the entire growing season, probably as a result of higher transpiration loss from the greater early-season leaf area in high-density plantings. For example, in a study in Great Britain, where sugarbeet can be grown successfully without irrigation, plant densities of 22,000 and 134,000 plants per hectare evapotranspired 766 and 841 mm of water per hectare, respectively (26).

Photosynthesis and Water Relations

Stomatal frequencies of 87–153 mm^{-2} have been reported for the upper surfaces of mature sugarbeet leaves (51–54). On lower leaf surfaces, stomata are usually about 15–20% more frequent than on the upper surfaces of the same leaves (51, 52, 54). Photosynthesis in sugarbeet is directly related to stomatal conductance (Fig. 13.3). Although stomatal

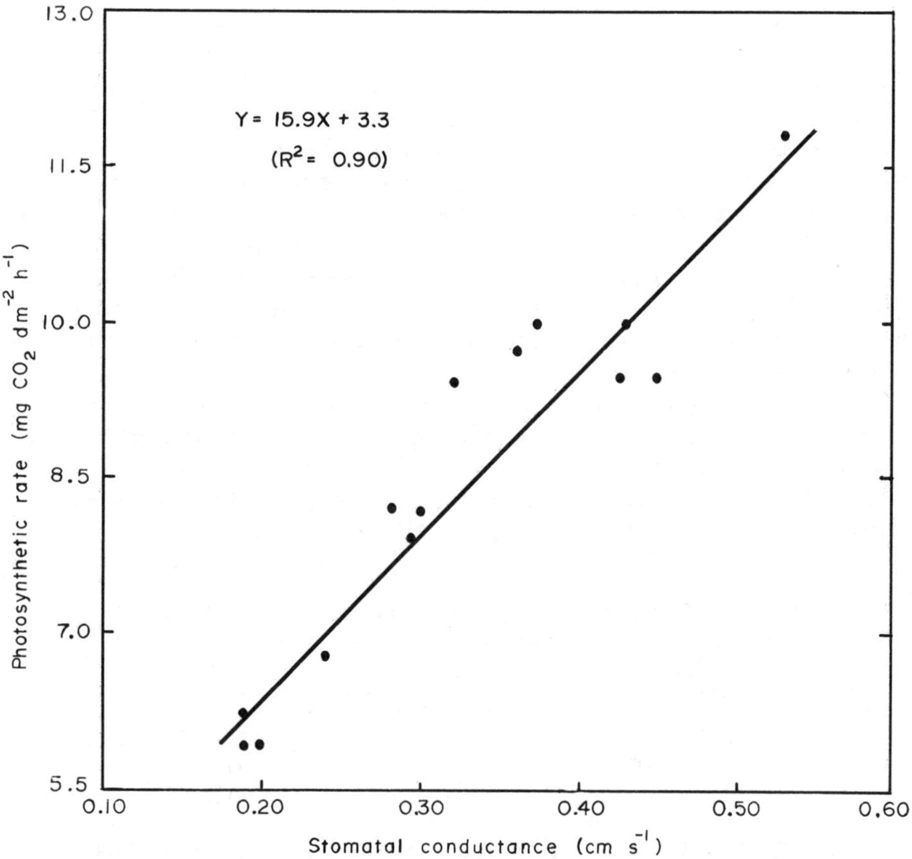

Figure 13.3. The relationship of sugarbeet photosynthetic rate to stomatal conductance. [After Milford and Lawlor (46), combining data for sugarbeets of two ages at two soil water levels, two relative humidities, and two times of day.]

conductance frequently is considered to be directly related to leaf water potential, it actually is affected by many factors, including soil water content, air moisture content (humidity), internal resistance factors, leaf and plant age, and time of day (i.e., solar radiation level and air and canopy temperature). Because most of these factors vary diurnally as well as in longer cycles, the magnitudes of their effects on stomatal conductance vary from hour to hour and from day to day. At times their combined effect can exceed that of ψ_l in determining stomatal opening. For example, in a controlled experiment in which both young and mature sugarbeet plants were grown under different combinations of soil water levels and relative humidities, photosynthetic rate was very closely associated with stomatal conductance, but not with leaf water potential (Table 13.2). In this case, stomatal opening apparently was being controlled by factors other than ψ_l. See Lawlor and Milford (55) for multivariate analyses of the factors affecting leaf water potential and stomatal conductance.

Despite the contribution of other factors, stomatal conductivity (and, therefore, photosynthesis) fundamentally is related to leaf water potential, which, in turn, is largely dependent on soil water potential (46, 55). For example, the data of Table 13.3 show that although ψ_{leaf} and photosynthetic rate decreased during the day at each of two soil moisture levels, the decreases were much smaller in plants growing in the wetter soil. Ultimately, the plants grown in the wetter soil yielded about 8% more root dry matter (and about 10% more sucrose per root) than the plants from the less wet soil (46). Note that leaf water potentials were low even in the plants provided with plentiful soil water; some reasons for this will be considered later.

The importance of the soil water reservoir to maintenance of high photosynthetic rates in field-grown sugarbeets is underscored by a laboratory experiment in which photosynthetic rates were determined on attached leaves of plants under high light intensity, low humidity, and with water freely available from a nutrient solution. Under these con-

Table 13.2. Correlation coefficients among photosynthetic rate (mg CO_2 dm^{-2} h^{-1}), stomatal conductance (cm sec^{-1}), and leaf water potential (bars)[a]

	Stomatal Conductance	Leaf Water Potential
Photosynthetic rate	0.95[b]	0.35 ns[c]
Stomatal conductance	(1.0)	0.57 ns[c]

[a] Calculated from data of Milford and Lawlor (46), Table 2, combining data for sugarbeets of two ages at two soil moisture levels, two relative humidities, and two times of day.
[b] $P < 0.01$.
[c] Not significant.

Table 13.3. Leaf water potential and photosynthetic rate of leaves of mature sugarbeet plants, at two soil water levels and at two times of day[a]

Average Soil Water Potential (bars)	Leaf Water Potential (bars)		Photosynthetic Rate ($mg\ CO_2\ dm^{-2}\ hr^{-1}$)	
	Time of Day			
	A.M.	P.M.	A.M.	P.M.
−0.2	−11.2	−12.2	10.0	8.9
−2.0	−13.2	−18.2	9.7	6.7

[a] After Milford and Lawlor (46).

ditions, net photosynthesis remained constant with time, but if the source of water were removed by detaching the leaf, net photosynthesis began to decline immediately (56).

In midseason, the sugarbeet is efficient in capturing and utilizing the energy of incident solar radiation. Gaastra (57) found a mean of about 5–6% of incident photosynthetically active light energy fixed as net photosynthetic products. However, overall photosynthetic efficiency calculated from about mid-July to harvest is considerably higher than if calculated from about early June to harvest (58). This is because early-season photosynthesis is inefficient, with the plants being too small to intercept light efficiently. Consequently, much light is "wasted," while at the same time water use may be high because of evaporation from bare soil. Even in midseason, after *LAI* has reached 1 or 2, water use efficiency has been found to range from 171 g water per g dry matter accumulated (on cloudy days) to 365 g water per g dry matter accumulated (on clear days) (59). Light saturation for the field-grown crop occurs at about $\frac{2}{3}$ of full sunlight, so on clear days photosynthesis reaches a maximum at less than full sunlight, but evapotranspiration continues to rise with increasing solar radiation and temperature.

TRANSPIRATION

Regulating Water Loss

A major factor affecting stomatal conductance in sugarbeet is leaf water potential, which, in turn, is determined largely by the interaction of soil water potential and evapotranspirational demand (46, 60). Leaf water potentials in sugarbeet seldom exceed −5 bars (55, 60, 61); stomata are fully open under these conditions, with stomatal conductance of about 1 cm sec^{-1} (52). Stomata are fully closed at leaf water potentials of −13 to −18 bars, depending on leaf and plant age (52, 55). Wilting of mature leaves occurs at approximately −12 to −14 bars leaf water potential. If the combination of air temperature, relative humidity, and light intensity is such that evaporative demand is low, soil water potential is the primary factor determining leaf water potential and, therefore, stomatal aperture (46, 55, 60).

If evaporative demand is high, low leaf water potentials and partial or complete stomatal closure result, even if water is freely available in the soil (46, 62, 63). This inability to maintain high leaf water potentials even when water is plentiful in the root zone suggests that there is significant resistance to the flow of water from soil to leaf evaporative surface, but direct experimental data are few. A suggestion of such resistance, probably within the plant, was made by Lawlor and Milford (55), who calculated that hydraulic resistance was greater in mature sugarbeet plants than in young plants under similar conditions. Perhaps the changes occurring in the vascular system as the sugarbeet root enlarges lead to altered resistance to the flow of water.

Midday stomatal aperture reduction or closure begins earlier in the day as soil water is depleted (64). Furthermore, mature sugarbeet leaves, especially, often wilt in midday on days of high insolation. Thus, although sugarbeet has a moderate degree of stomatal control, obviously the control mechanism does not respond quickly enough to prevent wilting under every combination of soil water and evaporative demand. Even a single wilting episode is undesirable physiologically (65), and if wilting is frequent, the loss of photosynthetic capacity in wilted plants might have significant effects on yield. Milford and Lawlor (46) observed that stomatal aperture was increased with little accompanying change in leaf water potential if the moisture content of the air were increased. This suggested that midday mist irrigation might keep stomata more open and the photosynthetic rate higher. Milford (63) found this to be true, as plants that were mist irrigated maintained higher water potentials and higher photosynthetic rates than those trickle irrigated. However, in another experiment, afternoon wilting was prevented by frequent daytime sprinkling, but no statistically significant beneficial effect on root or sucrose yield could be detected (66).

Relation of ψ to Other Physiological Indicators of Water Stress

Plants adapted to saline habitats are, in effect, adapted to low water potentials. Such plants usually accumulate inorganic solutes (often NaCl) in order to maintain turgor and permit growth. The cytoplasm apparently cannot tolerate high concentrations of salts (67); therefore, it is thought that these inorganic ions are actively pumped into the vacuole to keep cytoplasmic levels suitably low. A balance between vacuolar and cytoplasmic osmotic potential may be achieved by the cytoplasmic accumulation of nontoxic organic molecules that have little effect on macromolecules, and thus cause little disruption of cytoplasmic function (68, 69). In sugarbeet, as in many other members of the Chenopodiaceae (a group of plants long recognized to have markedly halophytic tendencies), the organic solute accumulated is betaine, $(CH_3)_3N^+CH_2COO^-$, the *N*-trimethyl derivative of the amino acid glycine. Accumulation of this quaternary ammonium compound appears to be characteristic of the Chenopodiaceae and several other families (70, 71). Different compounds (such as proline) may play a similar role in other families. Over 25% of the sugarbeet root's total nitrogen frequently is in betaine, which is present at concentrations of about 9 to 38 μM per g fresh weight (72, 73). However, although we have no information about the relationship in sugarbeet of leaf water potential and leaf betaine concentrations, other evidence leads to the conclusion that betaine levels are rather stable and unaffected by external conditions (e.g., 73). Thus, although betaine may be important in the overall salt or drought tolerance of sugarbeet, it probably has no involvement in the plant's response to daily or seasonal water stress. On the contrary, it could be argued that the sugarbeet's use of the slowly responding betaine as its major cytoplasmic osmotic regulator (rather than the more rapidly accumulated proline as in some other plants) might be a significant factor in its inability to prevent wilting under daily water stress. Although proline is known to be present in sugarbeet, we do not know how its concentration fluctuates under stress conditions.

APPLIED ASPECTS OF IRRIGATION

Critical Water Demand Periods

Sugarbeet, remaining vegetative throughout its growth as a commercial crop, does not have particular developmental stages at which water demand is critical, in the same sense as in some other crops. This does not mean, however, that water is any the less crucial for sugarbeet growth. Several points of particular importance have already been noted:

1. The importance of wet surface soil for germination and emergence.
2. The necessity for adequate water to maintain leaf turgor, which in turn influences leaf expansion (although most important at the seedling stage, this need continues throughout the life of the sugarbeet, which produces leaves continuously).
3. The midseason requirement for plentiful soil water in order to minimize wilting, and therefore to maximize potential photosynthesis and yield.

Old ideas fade away slowly, and there are two water-related myths in sugarbeet culture that, although they have repeatedly been refuted by scientific experimentation, still may be heard occasionally. The first is that early-season water stress is desirable or necessary in order to produce a long, tapering root rather than surface rooting. The evidence against this is clear (74, 75). The second myth is that late-season wilting should be encouraged, and that this will result in a real net gain in sucrose content. Again, evidence shows that this simply is not the case (74, 76). Preharvest wilting may result in root dehydration, so there may be an apparent gain in sucrose content when it is expressed as percentage of root fresh weight, but this is counterbalanced by the loss in root weight. Loomis and Haddock (76) conclude, "Total sucrose yields are not increased and may be reduced significantly by single, brief (wilting) cycles, and almost invariably are lower by repeated wilting. . . ."

Water Requirement or Water Consumption

Water consumption is strongly dependent on weather conditions, irrigation management, and length of the growth period; to a lesser degree it is affected by plant population density, nitrogen fertility, genotype, disease, and occasionally by other factors. There have been numerous studies of water consumption by sugarbeet under particular local combinations of these factors. The range of evapotranspirational water consumption for sugarbeet in western North America is approximately 550–1050 mm (74, 77), with a tendency for consumption to increase from north to south, presumably with increasing average temperature. A detailed estimate of seasonal water consumption by sugarbeet at a "typical" location (where consumption is at about the middle of the range given above) was made by Jensen (77) (Fig. 13.4). A 188 day growing season is assumed, from planting on April 10 to harvest on October 15. Figure 13.4 illustrates monthly water consumption by sugarbeet, and includes for each month estimates of potential evapotranspiration (E_0) and the mean crop coefficient (K_c), which is the ratio of the plants' transpiration (T_p) to the potential evapotranspiration. The crop coefficients show that through April, May, and most of June, transpiration by the young plants is very small relative to potential evapotranspiration. This is a result of the small leaf area of the widely spaced sugarbeet seedlings. Once canopy cover is complete (about August 1, 112 days after planting), K_c nears 1 because of the high transpiration associated with the large leaf evaporative surface area. Water consumption is at its highest in midseason, and maximum daily evapotranspiration rates of about 8 mm day^{-1} occur at this time. Of 790 mm seasonal consumption, about 79% is used in the last 55% of the growing season.

In climates quite different from those of western North America, water consumption figures also differ considerably. For example, at one lo-

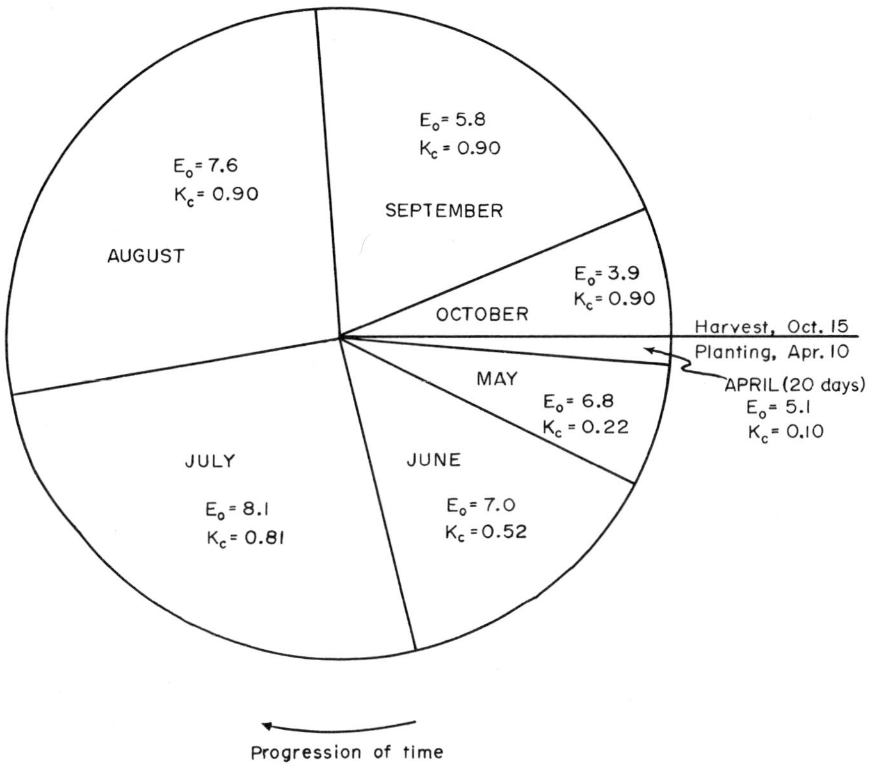

Figure 13.4. Typical monthly sugarbeet water consumption from planting to harvest. The estimated potential evapotranspiration rate (E_0, in mm day^{-1}) and the experimentally determined crop coefficient (K_c, dimensionless) are shown within the sector that represents that month's proportion of the total seasonal water consumption. [After Jensen (77), for Kimberly, Idaho.]

cation in England, the 9 year mean water consumption for a May to October sugarbeet growing season was 347 mm without irrigation, and 408 mm with irrigation (75). Even under these conditions, irrigation resulted in an average sucrose yield increase of 9%.

It long has been evident that growth and yield differences are obtained from applying the same amount of water to sugarbeet in different ways (78). Differences in planting practices, soil type, and requirements for leaching have led to many irrigation management experiments. Jensen (77) has summarized this work and given generalized recommendations for traditional furrow irrigation. After the early-season light irrigations for germination and seedling growth, midseason irrigation intervals of 10–14 days are recommended for the Great Plains and interior western United States. Irrigation intervals usually can be increased after mid-August. During the preharvest period (Table 13.1), water applications usually are managed more to provide good field conditions for harvest than with regard to the plants' physiological requirements.

Furrow irrigation with rather long intervals between water applications obviously cannot maintain optimum water availability in the root zone at all times. Recent advances in mechanization of irrigation have provided the ability to irrigate more frequently, with less water added per application. In most locations, peak daily water consumption rates of about 8 mm day^{-1} usually occur about mid-July. Having chosen either the desired irrigation frequency or the amount of water to be applied per irrigation, the irrigator can estimate the other parameter for mid-season application from the estimated daily consumption, considering anticipated rainfall and temperatures.

The still more recent "high-frequency irrigation" (79) is based on the assumption that if the soil has sufficient water-holding capacity and is at or near full capacity at the start of the season, then high frequency application of less than potential evapotranspiration (less than the daily water consumption) may give satisfactory growth and yield. The deficit between water use and water application must be made up, of course, from the soil reservoir. This technique appears to have potential applicability to sugarbeets (77, 80, 81).

We must recognize that irrigation studies and practices to date have aimed to maximize sucrose yield, with the tacit assumption that water is plentiful and cheap. In other words, maximum sucrose yield has been equated with maximum economic return. Irrigation management affects root yield, sucrose content, and impurities present in the root (8, 82, 83, 84), all of which are involved in determining recoverable sucrose. In the future, it may become imperative to maximize water use efficiency, or some combination of water and fertilizer application (83), if the economics of availability and cost of these inputs continues to change. It is known, for example, that total water use efficiency for sucrose production increases as irrigation rates decrease (85), with a range of at least 50–250 kg sucrose per ha per cm of water applied. In the future, detailed economic studies combined with additional research on water use undoubtedly will be needed.

REFERENCES

1. G. H. Coons, *Proc. Am. Soc. Sugar Beet Technol.* **8,** 142 (1954).
2. M. Zohary, in *Flora Palaestina*, Vol. 1, Israel Acad. Sci. Human., Jerusalem, 1961, pp. 138–189.
3. B. V. Ford-Lloyd and J. T. Williams, *Bot. J. Linn. Soc.* **71,** 89 (1975).
4. G. H. Coons, in *U.S. Dep. Agric. Yearb. 1936*, p. 625 (1936).
5. T. Theis, in R. T. Johnson, J. T. Alexander, G. E. Rush, and G. R. Hawkes, Eds., *Advances in Sugarbeet Production: Principles and Practices*, Iowa State University Press, Ames, 1971, pp. 1–18.
6. *Agricultural Statistics 1978*, U.S. Dep. Agric., U.S. Govt. Printing Office, Washington, D.C.

7. S. Campbell and A. A. Mast, in R. T. Johnson, J. T. Alexander, G. E. Rush, and G. R. Hawkes, Eds., *Advances in Sugarbeet Production: Principles and Practices*, Iowa State University Press, Ames, 1971, pp. 438–450.
8. D. W. James, D. L. Doney, J. C. Theurer, and R. L. Hurst, *Agron. J.* **70,** 525 (1978).
9. G. W. Fick, R. S. Loomis, and W. A. Williams, in L. T. Evans, Ed., *Crop Physiology*, Cambridge University Press, London, 1975, pp. 259–295.
10. P. J. Goodman, *Appl. Ecol.* **5,** 339 (1968).
11. K. R. Storer, W. R. Schmehl, and R. J. Hecker, *Tech. Bull. 118*, Colorado State Univ. Exp. Stn., Ft. Collins, Colorado, 1973.
12. R. S. Loomis, E. Ng, and W. F. Hunt, in R. H. Burris and C. C. Black, Eds., *CO_2 Metabolism and Plant Productivity*, University Park Press, Baltimore, Md., 1976, pp. 269–286.
13. P. B. Smith, in R. A. McGinnis, Ed., *Beet-Sugar Technology*, Beet Sugar Development Foundation, Ft. Collins, Colorado, 1971, pp. 1–24.
14. E. H. Hungerford, in R. A. McGinnis, Ed., *Beet-Sugar Technology*, Beet Sugar Development Foundation, Ft. Collins, Colorado, 1971, pp. 579–587.
15. V. F. Savitsky, *Proc. Am. Soc. Sugar Beet Technol.* **6,** 156 (1950).
16. G. A. Smith, in W. R. Fehr and H. H. Hadley, Eds., *Hybridization of Crop Plants*, American Society of Agronomy, Crop Science Society of America, Madison, Wis. 1980, Chap. 43.
17. V. F. Savitsky and H. Savitsky, *J. Am. Soc. Sugar Beet Technol.* **13,** 621 (1965).
18. J. K. Radke and R. E. Bauer, *Agron. J.* **61,** 860 (1969).
19. J. R. Hunter and A. E. Erickson, *Agron. J.* **44,** 107 (1952).
20. W. R. Akeson, Personal communication to S. S. Martin, 1978.
21. F. W. Snyder, *J. Am. Soc. Sugar Beet Technol.* **18,** 87 (1974).
22. J. W. Cary, *Agron. J.* **67,** 258 (1975).
23. W. E. Larson and W. B. Johnston, *Soil Sci. Am. Proc.* **19,** 275 (1955).
24. M. J. Durrant, B. J. G. Love, A. B. Messem, and A. P. Draycott, *Ann. Appl. Biol.* **74,** 387 (1973).
25. J. T. Moraghan, *Agron. J.* **64,** 759 (1972).
26. A. P. Draycott and M. J. Durrant, *J. Agric. Sci., Camb.* **76,** 277 (1971).
27. G. W. Fick, W. A. Williams, and A. Ulrich, *Crop Sci.* **12,** 108 (1972).
28. G. F. J. Milford, *Ann. Appl. Biol.* **75,** 427 (1973).
29. A. Ulrich, *Agron. J.* **44,** 66 (1952).
30. N. Terry, *J. Exp. Bot.* **19,** 795 (1968).
31. G. F. J. Milford and G. N. Thorne, *Ann. Appl. Biol.* **75,** 419, (1973).
32. R. Wyse, *J. Am. Soc. Sugar Beet Technol.* **20,** 368 (1979).
33. R. S. Loomis and D. J. Nevins, *J. Am. Soc. Sugar Beet Technol.* **12,** 309 (1963).
34. D. J. Watson and E. C. D. Baptiste, *Ann. Bot. N.S.* **2,** 437 (1938).
35. R. F. Follett, W. R. Schmehl, and F. G. Viets, Jr., *J. Am. Soc. Sugar Beet Technol.* **16,** 235 (1970).
36. E. A. Clark and R. S. Loomis, *J. Am. Soc. Sugar Beet Technol.* **20,** 97 (1978).
37. R. K. Scott and P. M. Bremner, *J. Agric. Sci., Camb.* **66,** 379 (1966).
38. M. A. Dillon, B. D. McCaslin, and W. R. Schmehl, *Agron. J.* **64,** 183 (1972).
39. G. N. Thorne, *Ann. Bot.* **24,** 356 (1960).
40. D. J. Watson and K. J. Witts, *Ann. Bot.* **23,** 431 (1959).
41. G. N. Thorne and A. F. Evans, *Ann. Bot.* **28,** 499 (1964).

42. H. Rapoport and R. S. Loomis, Abstr. Papers, 20th Gen. Meet., Am. Soc. Sugar Beet Technol., 1978.
43. A. G. Morton and D. J. Watson, *Ann. Bot.* **12,** 281 (1948).
44. N. Terry, *J. Exp. Bot.* **21,** 477 (1970).
45. T. C. Hsiao, *Annu. Rev. Plant Physiol.* **24,** 519 (1973).
46. G. F. J. Milford and D. W. Lawlor, *Ann. Appl. Biol.* **80,** 93 (1975).
47. B. Orchard, in A. J. Rutter and F. H. Whitehead, Eds., *The Water Relations of Plants*, Blackwell, London, 1963, p. 340.
48. N. Terry, L. J. Waldron, and A. Ulrich, *Planta* **97,** 281 (1971).
49. P. C. Owen and D. J. Watson, *Nature (London)* **177,** 847 (1956).
50. G. E. Nichol, L. M. Burtch, and D. J. Traveller, in R. T. Johnson, J. T. Alexander, G. E. Rush, and G. R. Hawkes, Eds., *Advances in Sugarbeet Production: Principles and Practices*, Iowa State Univ. Press, Ames, 1971, pp. 49–67.
51. E. Artschwager, *J. Agric. Res. Camb.* **33,** 143 (1926).
52. K. W. Brown and J. J. Rosenberg, *Agron. J.* **62,** 20 (1970).
53. E. Ruppel, *Phytopathology* **62,** 1095 (1972).
54. B. Slavik, *Biol. Plant.* **5,** 143 (1963).
55. D. W. Lawlor and G. F. J. Milford, *J. Exp. Bot.* **26,** 657 (1975).
56. D. J. Nevins and R. S. Loomis, *Crop Sci.* **10,** 3 (1970).
57. P. Gaastra, *Meded. Landb. Hogesch., Wageningen* **59(13),** 1 (1959).
58. R. E. Campbell and F. G. Viets, Jr., *Agron. J.* **59,** 349 (1967).
59. K. W. Brown and N. J. Rosenberg, *Agron. J.* **63,** 207 (1971).
60. P. V. Biscoe, *J. Exp. Bot.* **23,** 930 (1972).
61. D. W. Lawlor and G. F. J. Milford, *Ann. Bot.* **37,** 597 (1973).
62. G. Hofstra and J. D. Hesketh, *Planta* **85,** 228 (1969).
63. G. F. J. Milford, *Ann. Appl. Biol.* **80,** 247 (1975).
64. K. W. Brown and N. J. Rosenberg, *Agron. J.* **62,** 4 (1970).
65. C. B. Shah and R. S. Loomis, *Physiol. Plant.* **18,** 240 (1965).
66. R. A. Kohl and J. W. Cary, *J. Am. Soc. Sugar Beet Technol.* **15,** (1969).
67. J. A. Hellebust, *Ann. Rev. Plant Physiol.* **27,** 485 (1976).
68. R. G. Wyn Jones, R. Storey, R. A. Leigh, N. Ahmad, and A. Pollard, in E. Marre and O. Ciferri, Eds., *Regulation of Cell Membrane Activities in Plants*, North-Holland, Amsterdam, 1977, pp. 121–136.
69. J. L. Hall, D. M. R. Harvey, and T. J. Flowers, *Planta* **140,** 59 (1978).
70. R. Storey, N. Ahmad, and R. G. Wyn Jones, *Oecologia* **27,** 319 (1977).
71. R. Storey and R. G. Wyn Jones, *Phytochemistry* **16,** 447 (1977).
72. G. W. Maag, R. J. Hecker, and P. A. Whitaker, *J. Am. Soc. Sugar Beet Technol.* **17,** 154 (1972).
73. G. A. Smith and S. S. Martin, *Crop Sci.* **17,** 469 (1977).
74. M. E. Jensen and L. J. Erie, in R. T. Johnson, J. T. Alexander, G. E. Rush, and G. R. Hawkes, Eds., *Advances in Sugarbeet Production: Principles and Practices*, Iowa State University Press, Ames, 1971, pp. 189–222.
75. A. P. Draycott and A. B. Messem, *J. Agric. Sci. Camb.* **89,** 481 (1977).
76. R. S. Loomis and J. L. Haddock, in R. M. Hagan, H. R. Haise, and T. W. Edminster, Eds., *Irrigation of Agricultural Lands*, American Society of Agronomy, Madison, Wis., 1967, pp. 640–648.

77. M. E. Jensen, Ed., *Consumptive Use of Water and Irrigation Water Requirements*, American Society of Civil Engineers, New York, 1973, 215 pp.

78. P. C. Owen, *J. Agric. Sci. Camb.* **51,** 133 (1958).

79. S. L. Rawlins and P. A. C. Raats, *Science* **188,** 604 (1975).

80. W. O. Pruitt, F. J. Lourence, and S. Von Oettinsen, *Calif. Agric.* **26(10),** 10 (1972).

81. D. W. Fonken, J. C. Steele, and P. E. Fischbach, *Trans. Am. Soc. Agric. Eng.* **17,** 889 (1974).

82. A. P. Draycott and D. J. Webb, *J. Agric. Sci. Camb.* **76,** 261 (1971).

83. M. K. Nicholson, T. Kibreab, R. E. Danielson, and R. A. Young, *J. Am. Soc. Sugar Beet Technol.* **18,** 34 (1974).

84. G. A. Reichman, E. J. Doering, L. C. Benz, and R. F. Follett, *J. Am. Soc. Sugar Beet Technol.* **19,** 275 (1977).

85. D. E. Miller and J. S. Aarstad, *Agron. J.* **68,** 231 (1976).

14

SUGARCANE

G. J. Gascho and S. F. Shih

INTRODUCTION

Modern sugarcane cultivars are trispecies hybrids of *Saccharum officinarum, Saccharum spontaneum,* and *Saccharum robustum*. Cultivars vary widely; each major sugarcane growing area breeds for the particular conditions in their location. In fact, even chromosome numbers vary widely for cultivars of sugarcane.

There are approximately 11 million hectares of land in the world devoted to sugarcane growing. World production is about 90 million metric tons of sucrose per year. The crop is of prime importance to the economy of many countries but the value of the crop can fluctuate widely because of the traditional volatility of the world sugar market.

ADAPTATION

Climatic

Sugarcane is widely adaptable to large ranges of tropical and semitropical climates, soils, and cultural conditions. Geographically, the crop grows from 37°N in southern Spain to 31°S in the Republic of South Africa. Since sugarcane is a tropical crop it only grows well in the tropics or in the subtropics when temperatures are modified by warm bodies of water. For example, growth would not be possible in southern Spain without the influence of the Mediterranean Sea or in South Africa without the Indian Ocean. Survival is possible for all temperatures above freezing but growth essentially ceases at any time the minimum temperatures are below 12°C. The optimum temperature for growth is about 35°C. There is little growth at temperatures below 20°C. The time period of germination (vegetative sprouting) is temperature critical. Clements and Nakata (1) found the absolute lowest temperature for germination was 12°C. The optimum germination temperature is about 30°C. Germination at nonoptimum temperatures, either higher or lower, can result in physiologically weakened plants (2).

The ideal climate would consist of 4–5 months of temperatures of 30–35°C for growth of a 1 year crop followed by a 6–8 week period of cooler temperatures to enhance the maturation (sucrose accumulation) phase. Growth in tropical areas such as Hawaii is usually characterized by two or more periods of growth when the crop is not harvested yearly.

Sugarcane is grown under a wide range of humidity conditions, from areas of rainfall of 1000 cm per year to conditions of extreme drought. Successful culture is possible under irrigation in areas of Iran with as little as 25 cm of rainfall per year (3). Adaptation to conditions of drought has been obtained to some extent through cultivar development. On the other hand, some cultivars of sugarcane can survive for 2–5 months of flood during the period of grand growth.

Solar radiation is directly related to sugarcane growth once the minimum temperature requirements are met and the canopy is completely developed. The tonnage of Florida sugarcane has been accurately predicted from equations (4) considering only degree days and solar radiation for years when freezes are not a factor.

Soils

Soil structure and fertility affect growth and yields of sugarcane but the adaptability of the crop is very wide. High yields are obtainable on all textural classes as well as on organic soils. Likewise, sugarcane is successfully grown where soil pH varies from about 4 to 10 even though there are nutritional problems at the pH extremes. With proper care sugarcane will grow on saline soils (3).

Altitude

Sugarcane is successfully grown at altitudes from below sea level to several thousand meters. The differences in growth and yield potential are directly attributable to changes in temperature, rainfall, humidity, soil type, and solar radiation.

Crop Use

Crystalline sugar is by far the largest product from sugarcane. In addition, molasses as a by-product has a considerable value for food and feed purposes. The spent stalks (bagasse) are used to fire the boilers in the sugarhouse, thus saving on fuel costs. It also is used as a source of pentosans for making the industrial chemical furfural and to make fiber boards. A small portion of the worldwide sugarcane is used to make syrup. Presently there is interest in using the crop for energy purposes because of its relatively high efficiency of production of both biomass (aboveground surface dry matter) and fermentable sugars. The distillation of ethanol from cane juice or from molasses is particularly attractive.

Yield Potential

As high as 22 metric tons of sucrose can be produced on a hectare per year. In Hawaii, record sucrose production has been about 45 metric tons per hectare for crops allowed to grow 2–3 years before harvest (5). The maximum mass of harvestable stalks obtained has been about 225 and 45 metric tons per hectare per year fresh and dry weight, respectively. It is possible to grow up to 65 metric tons per year of total aboveground dry weight. These high potentials are possible because of the high efficiency of conversion of solar energy into organic matter in relation to other plants. An actual field harvest of a 22 month crop in Hawaii showed a conversion of 1.43% of the incident light recorded on a pyrheliometer. Yields approaching the maximums require intensive cultural practices. They have only been obtained with well-managed irrigation or at high altitudes with high uniform rainfall patterns. Unfortunately, average yields are about one-third of the maximum stated here. One of the main reasons for low average yields is the lack of good rainfall distribution or irrigation.

PHYSIOLOGICAL STAGES OF DEVELOPMENT

The growth of sugarcane can be divided into four physiological stages. The time periods required for each stage will vary in relation to climate and genotype. The physiological stages are:

1. Germination and emergence
2. Tillering and canopy establishment
3. Grand growth
4. Maturation (ripening)

For a 12 month crop the time periods involved could be approximately 1, 2, 7, and 12 months (Table 14.1). In areas where cane is grown for more than 1 year before harvest the growth period is extended with cyclic growth rates depending on climate. In a 2 year crop the grand growth period could be extended to 19 months but growth would not be constant during that period unless climate conditions were constant. In areas such as Louisiana where cane is planted and harvested under

***Table 14.1. Physiological stages for 12 month sugarcane crop under ideal climatic conditions*[a]**

1. Germination[b] *and Emergence*	
1.05[c]	Swelling and elongation of vegetative eyes
1.05	Maximum growth of sett roots and ratoon[d] stubble roots
1.1	Emergence of primary stalks
2. Tillering and Canopy Establishment	
2.1	Tillering initiated
2.1	Initiation of shoot roots
2.2	Peak root absorption efficiency
2.3	Rapid tillering and canopy establishment
	Beginning of high tolerance of sugarcane to adverse climatic factors
3. Grand Growth	
3.4	Canopy closure
3.4	End of effective tillering
3.4–5	Primary *LAI* peak
3.4–9	Maximum rate of stalk elongation
3.4–7	Decrease in nonmillable stalk count due to shading and other plant competition
3.8	Threshold for millable stalk count
4. Maturation	
4.10	Maximum biomass
4.10–12	Maximum rate of conversion of carbon to sucrose
4.12	Sucrose peak

[a] Stages are very dependent on climate. In some areas cane is grown for 24–36 months. In general the grand growth phase is increased and the other stages are as above for these longer crops.
[b] Germination refers to vegetative sprouting of either planted stalk cuttings or the ratoon stubble.
[c] The numeral before the decimal point is the physiological stage; the numeral after the decimal point indicates the number of months after planting.
[d] Ratoon crops originate from roots of a harvested crop.

suboptimal climatic (temperature) conditions the germination and emergence stage is lengthened to approximately 5 months.

ROOT GROWTH AND FUNCTION

Rooting is affected by cultivar, crop year (plant cane is the first crop following the planting and ratoons are the subsequent crops growing from stubble after a harvest), temperature, aeration, soil water, acidity (especially high aluminum levels), fertilization, cultivation, soil compression, and wind (6). All contributing factors cannot be adequately discussed in this chapter, however, some studies are summarized below.

Root Growth during Germination and Emergence

Sugarcane is propagated commercially from cuttings of the stalk by vegetative means. Each cutting has one or more buds which are located on nodes of the stalk. The nodes are separated by internodes which contain the sugar. Root primordia are located on a root band which circumvents the stalk at the node. The first roots (often called sett roots) appear and develop during germination (Fig. 14.1). These roots serve the plant until the young shoots produce new roots (shoot roots); then the dependence on the cutting and the original roots gradually ceases. The switch from dependence on sett roots to dependence on shoot roots usually begins during the first 2 months after planting and is complete by 4–5 months (Fig. 14.2). Ratoon stalks originate from shoot roots of the plant crop which serve in the same manner as the sett roots in the plant crop.

Extensive root investigations with sugarcane revealed that the first sett roots started growing within 24 hr of planting and on the third day after planting some were extending at a rate of 10 mm day^{-1} (7). By day 5

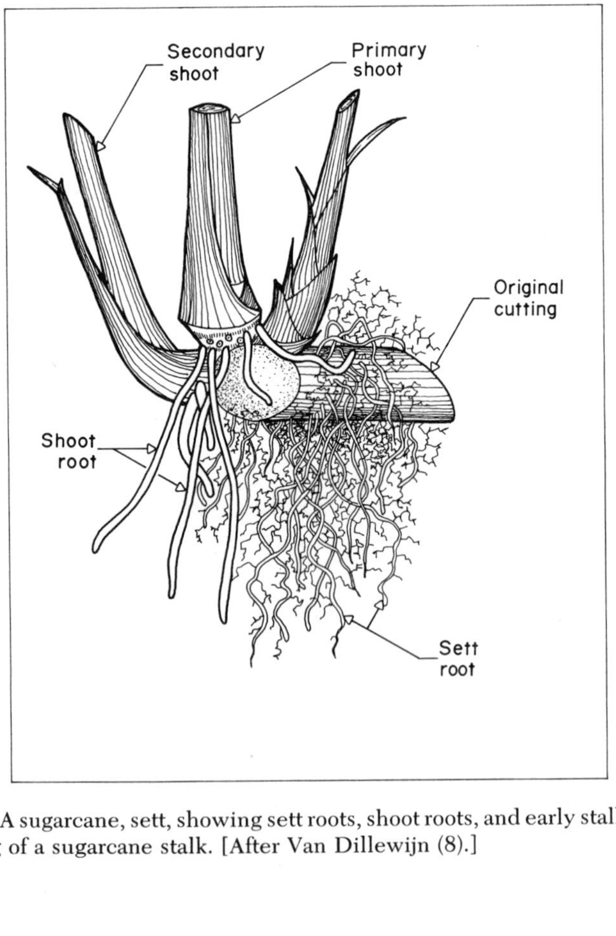

Figure 14.1. A sugarcane, sett, showing sett roots, shoot roots, and early stalk development from a cutting of a sugarcane stalk. [After Van Dillewijn (8).]

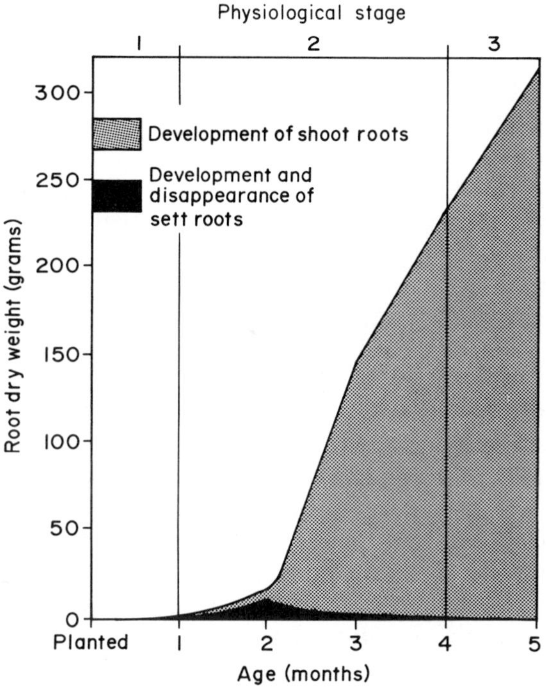

Figure 14.2. Dry weight of sett and shoot roots with age and physiological stage. [After Lee and Weller (11).]

elongation was 20 mm day^{-1}. Sett roots reached a peak rate of growth of 24 mm day^{-1} and quickly developed a branched network of thin subroots. Sett roots ceased growing after 11 days when they averaged 20 cm in length, and their subsequent life was short, none being traceable 2 months after planting. Primary shoot roots started developing 5–7 days after planting, and reached a maximum elongation rate of 75 mm day^{-1}. Maximum root growth is usually at 2–3 months (Fig. 14.2). Glover found the average rate of elongation in light soils was about 40 mm day^{-1}, and in heavier soils was about 28 mm day^{-1}. When there was no physical impedance in the soil, the shape of the root system which developed depended on the water regime at different depths during root development. The branching of roots was always approximately at right angles to the parent root in all soils and turgid root hairs were observed to persist on roots for as long as 4 months.

Sett roots are thin and branched while shoot roots are thicker, more fleshy, and less branched. However, the shoot roots decreased in diameter as they emerge from higher root bands on the shoot. The life of an individual shoot root is also limited but the root system of the plant is continuously renewed. The rate of the shoot root initiation is governed by the rate of tillering. The continuous production of new roots is of great importance since it enables the plant to adjust itself to changing environmental conditions (8).

Root Depth and Distribution

The bulk of the roots of irrigated sugarcane are in the upper 20 cm of the soil. Lee (9) found over 60% of the roots in the 0–20 cm soil layer and the percentage of roots at 20 cm increments decreased markedly with each depth increment (Fig. 14.3). More than 85% of the total root weight was above the 40 cm depth. Calculations of the amount of root organic matter left in the soil by a cane crop range from 2 to 4 metric tons of dry root weight per hectare (10).

Root distribution will vary with age. Lee and Weller (11) found the root percentage in the top 20 cm decreased from 100 to 60% during the 4 months following planting while the percentage increased for the lower depths (Fig. 14.4). After 4 months surface root percentage again increased, which may reflect the perennial nature of sugarcane if the increase was primarily in the upper nodes. However, these time periods cannot be considered to apply in all cases or even as norms; the above data does not reveal the absolute changes in root volume, or activity with age.

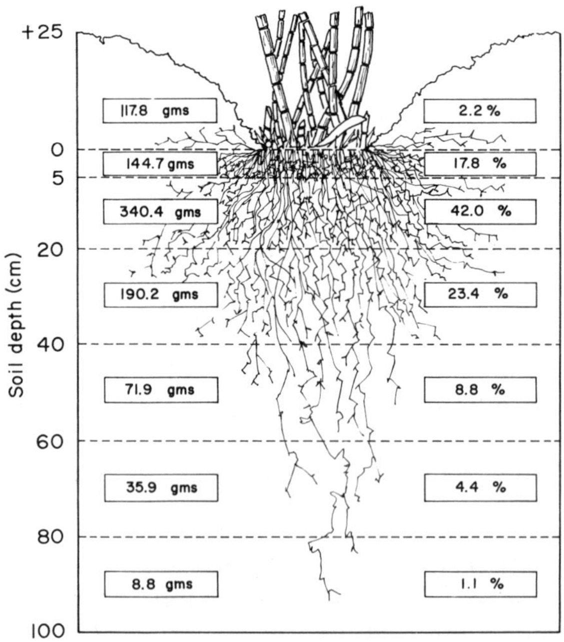

Figure 14.3. Sugarcane root distribution with soil depth during the grand growth stage. [After Lee (9).]

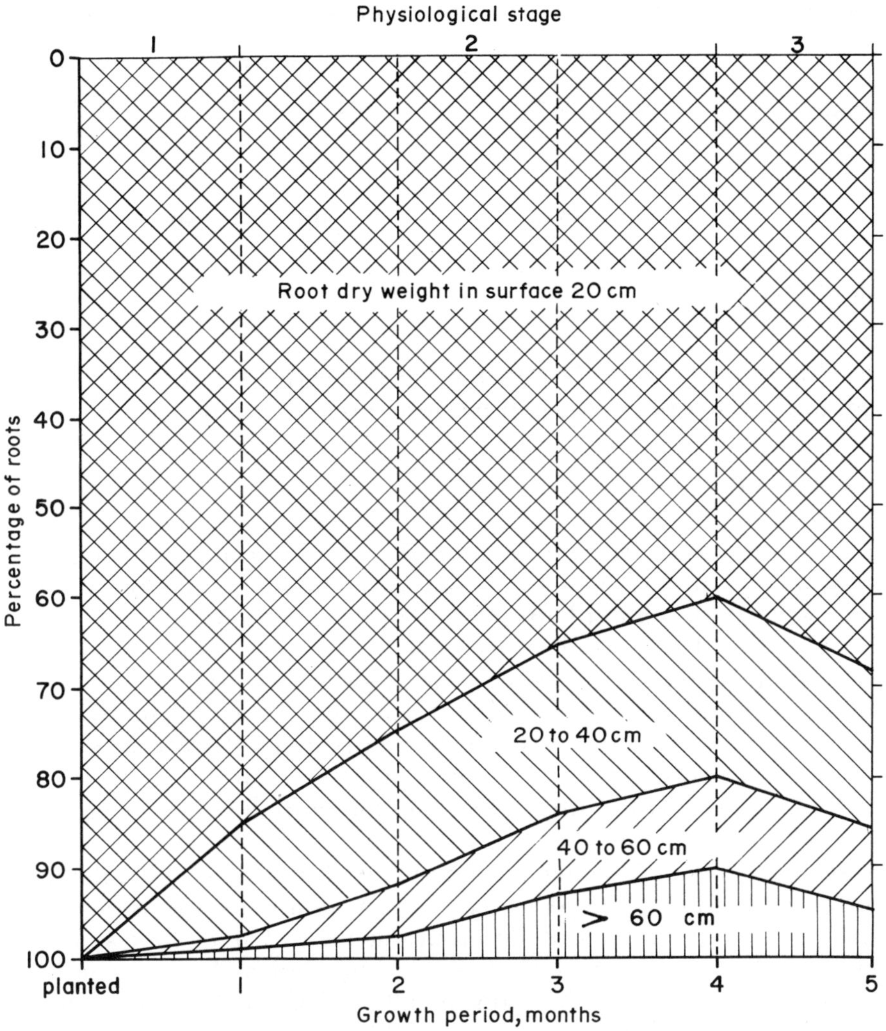

Figure 14.4. The distribution of sugarcane roots with soil depths as affected by age and physiological stage. [After Lee (9).]

Stevenson (12) obtained dry weight of roots four times during the growth of a crop (Table 14.2). He found a maximum dry weight at 11 months. At age 17 months the root system appeared less deep than during the early stages of growth. He also noted differences among cultivars in regard to regeneration of new roots; the noble canes (those of *officinarum* parentage) had longer root vegetative phases than nonnoble canes.

The weight of roots in a volume of soil is not necessarily correlated with the activity of roots in that volume. Evans (13) separated roots into four classes by length and diameter. He found that fibrous surface roots

Table 14.2. Root distribution with time depth for eight sugarcane cultivars[a]

Cane Age (months)	Weight of Roots in Upper 45 cm of Soil (g)	% of Roots in each 15 cm Layer		
		0–15 cm	15–30 cm	30–45 cm
4	32	26	47	27
7	166	37	39	24
11	363	32	40	28
17	160	43	32	25

[a] After Stevenson (12).

spread farther from the plant than other less active types of roots. However, total root surface was still highest very close to the plant.

Root Water Relations

Root growth and distribution is related both to physiological stage and soil water conditions. Srivastava and Ghosh (14) found no root extension below a 30 cm depth until after the monsoon season began. The growth of roots during the tillering stage (premonsoon) is one-tenth of maximum attained during the grand growth stage (postmonsoon).

Root penetration is related to soil water conditions. Yang et al. (15) found that the root density was greater in tile drained plots than untiled plots and the roots in the tile drained plots penetrated to a greater soil depth (70 cm below the surface) than in the untiled plots (50 cm below the soil surface).

Sugarcane has a wide range of genetic variability to drought tolerance. Rostron (16) studied the root system of two cultivars and showed that depth of rooting is an inherited trait in sugarcane. Under good soil water conditions the root systems were similar but when soil water conditions were less than favorable the roots of cultivar NCo 376 extended deeper and were more prolific than those of CB 36-14.

Webster and Eavis (17), while studying the effects of flooding on sugarcane growth, found that the root systems developed under flooded conditions differed greatly from those developed under drained conditions. Drained conditions resulted in fewer roots but a greater proportion of large diameter roots. Under flooded conditions, the root system consisted of a dense mat of small diameter roots. New large diameter roots that would develop under drained conditions were unable to survive. The successive death of primary, secondary, and tertiary root tips probably stimulated laterals of higher order which were thinner and could survive. Nearly all young roots observed during flooding were aerotropic, growing vertically upward to the soil surface. The new roots remained unsuberized and white for at least 30 days after their formation, whereas in unflooded controls the roots quickly suberized.

In the period immediately after flooding there was a surge in root growth and large diameter roots developed. The period coincided with an enhanced growth rate of tillers. The root system was apparently efficiently utilizing nutrients at this time. Recovery in shoot growth may be indirectly related to the stimulation of root growth and nutrient and water uptake in the postflooding period. Although root systems differed, there was no significant difference in the weight of roots in the different treatments.

Rooting depth, distribution, and activity are greatly affected by soil texture. Thompson (18) showed that the sugarcane roots could go as deep as 400 cm in sand. Neutron probe data from a similar soil indicated that water removal was effective to a depth of 180 cm under rain-fed conditions. In contrast, irrigated cane on a clay type soil did not remove water from depths greater than 90 cm while rain-fed cane on the same soil exploited soil water to a depth of at least 120 cm.

Glover (7), working in sand, found an extensive, fine, well-branched root system at all depths down to at least 140 cm. In a disturbed clay there was a well-developed thick primary root system with a poorer secondary root system than in the sand. In the undisturbed clay, primary roots were few in the soil immediately below plot depth until they penetrated the horizon that impeded their growth; then they branched profusely in the lighter subsoil. On an irrigated undisturbed clay, root growth was initially sparse and slow, and frequent light irrigation was necessary to sustain growth. In sand the activity of the root system changed very markedly as the soil wetted and dried. After a winter drought, the first spring rains caused root initials on some basal nodes to grow actively and to form a new superficial root system in the moist surface soil layer. Only when heavier summer rains recharged the soil profile did these roots extend downward.

Root systems can severely limit the yield potential of sugarcane. Since roots reach their peak absorption efficiency about 2 months after planting and decrease in effectiveness thereafter, Hudson (19) concluded that roots damaged by water stress could not support the potential transpiration rate of 0.6 cm day^{-1} in Barbados, West Indies. He correlated poor root systems, as judged by total root surface area, with areas that ratoon poorly.

Eavis (20) studied the effects of flooding on sugarcane growth of two cultivars grown in lysimeters containing either disturbed montmorilinitic or kaolinitic clay. The treatments involved soils that were water logged (water table 15 cm below soil surface) and maintained at field capacity by a soil drainage system 80 cm below soil surface. During a 7 month period the differences in top growth and tiller production between two treatments were observed to be small. During a drought which was imposed at 7 months the plants removed the soil water until it reached the permanent wilting point. The root system of the field capacity treatments consisted of a network of thick primary roots with a relatively small proportion by weight of small diameter laterals. The root system developed in waterlogged conditions consisted of a very

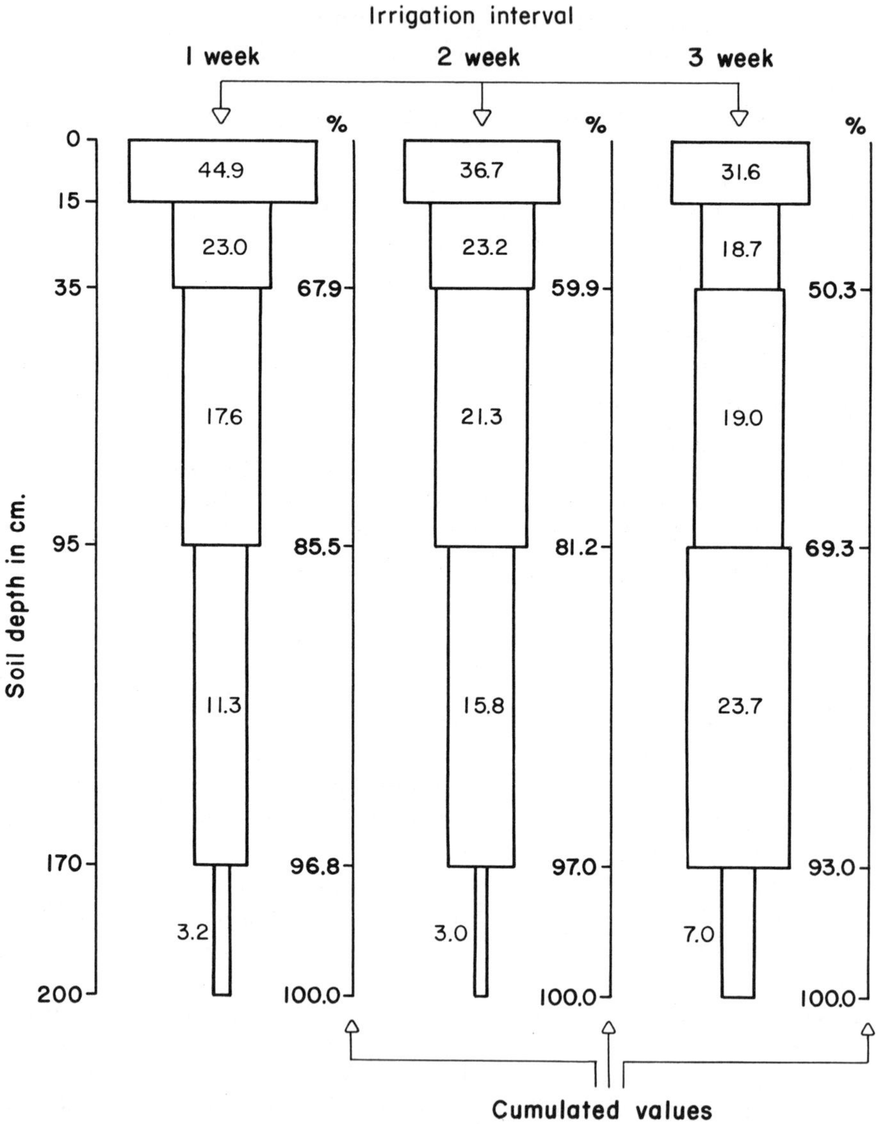

Figure 14.5. Percent root volume in relation to depth in the soil at 12 months of age (maturation stage) as affected by irrigation interval in the grand growth stage. [From Baran et al. (23).]

dense mat of fine laterals with a small proportion of thicker roots. In general total root weight was higher under waterlogged conditions. Waterlogged treatments had 150% greater fibrous to thick root ratios. The proliferation of fine roots seemed to be associated with the wetness of the soil; in other experiments the development of a surface mat of roots under a mulch was also observed to occur only under very wet conditions. Sugarcane was thus able to adapt its root system to accommodate an anaerobic environment since the diffusion path length for entry of oxygen into respiring tissue is less for thinner roots than for thicker roots. Juang and Vehara (21) found that dry weight of sugarcane roots varied from 62 to 77 g (single eye seedling planted) on clay loam soil with a water table between 30 and 80 cm deep, whereas on a well-drained sandy loam it varied from about 48 to 64 g. Their observations on roots under three different water table depths indicated that the increasing rooting depth associated with increasing water table is not related to increasing root weight. This indicates that approximately the same quantity of roots was distributed in a shallower depth of soil because of a high water table. Since fertilizer was thoroughly mixed, it could be anticipated that better utilization of fertilizer in the 80 cm water table treatment allowed roots to extend deeper into the soil.

Gosnell (22) studied the effects of a water table on the growth of sugarcane and found that the majority of roots stopped fairly sharply at 5–10 cm above the water table. The effect of a 25 cm water table on root weight was very marked and yielded only about one-fourth of the root weight per pot of a 50 cm water table. There was no significant reduction in root growth in treatments with water tables below 50 cm.

Baran et al. (23) found that irrigation frequency affected distribution of roots with depth. Figure 14.5 indicates that root volume at harvest (12 months) decreased at depths less than 35 cm when irrigation intervals were extended. At depths of greater than 35 cm, root volume increased when irrigation intervals were extended. Roots proliferated into the areas where there was more available soil water.

CANOPY DEVELOPMENT

The growth of the aboveground portion of the plant is a direct function of the number of stalks per hectare (tillering) and their expansion in both height and diameter. Since cane diameter for a given cultivar does not normally increase to any extent late in the grand growth stage or in the maturation stage, measurements of stalk length correlate well with changes in stalk weights and are commonly used (8).

Growth analysis of world sugarcane crops have revealed similar growth patterns, but with some differences arising from the use of 1 year or longer cropping cycles.

Top Growth

Stalk Development

Total stalk populations were counted monthly for 24 months by Gosnell (24) for both irrigated and nonirrigated sugarcane. In the period from 3 to 7 months (late tillering and early grand growth stages) stalk count was higher for irrigated cane. The maximum stalk number being reached in month 5 when the irrigated populations was 340,000 and the nonirrigated population was 280,000 stalks per hectare (Fig. 14.6). There appeared to be an overproduction of stalks, for up to 50% died and a stable stalk population was achieved after 7 months and remained through the grand growth and maturation stages for both irrigated and nonirrigated treatments. This demonstrated the effect of canopy closure on the final number of stalks per hectare. The decrease in stalk population was mostly due to shading of the shorter stalks (25).

Borden (26) recognized that not all stalks developed into millable cane and partitioned his counts into shoots and millable stalks. Maximum stalk populations were found at 6 months. There was a ratio of about 3 total shoots to one millable stalk. In August (at 9 months) the stalk population was reduced and no small shoots were living. Only about 25% of the large population of shoots (about 200 thousand shoots and stalks per hectare) at 6 months developed into millable stalks of cane. As in Gosnell's data, there was little change in stalk number after 9 months.

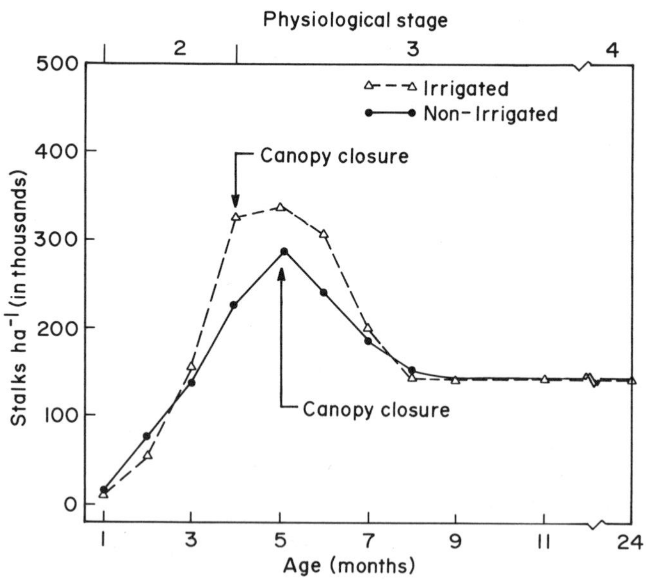

Figure 14.6. Stalk numbers for irrigated and nonirrigated sugarcane, months after planting. [From Gosnell (24).]

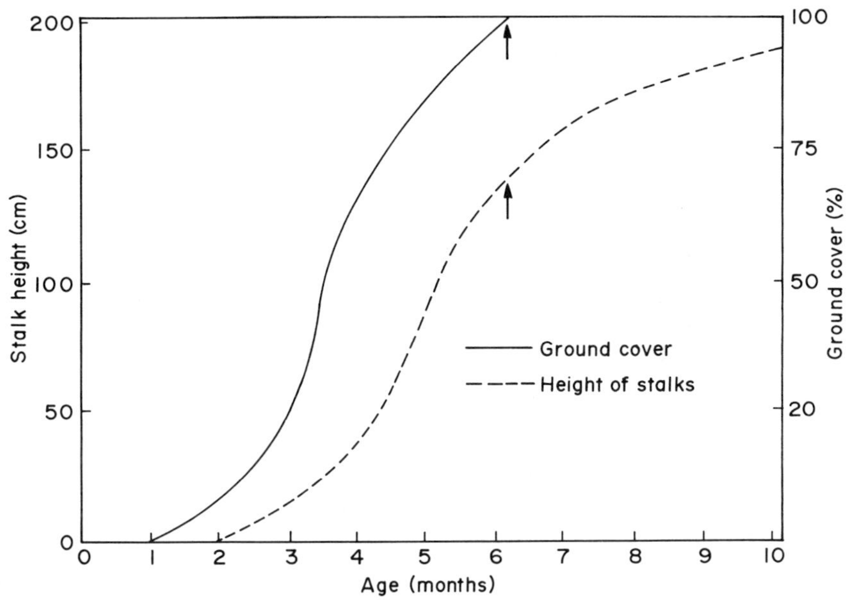

Figure 14.7. Ground cover and height measurements for a plant crop in South Africa, months after planting. Arrows indicate canopy closure. [From Thompson (27).]

The leading Hawaiian sugarcane cultivars grow about 3–4 m per year and can reach a total length of 10–12.5 m (5). Further stalk elongation occurs at 15–17 months in Hawaii because cultivar selection has favored continued crop growth in the second year. Thompson (27) made cummulative height and percentage ground cover measurements as a function of time. Canopy closure occurred in April while the maximum cane height was not attained until about July (Fig. 14.7).

Rostron (16) determined average crop growth rate for plant and ratoon sugarcane over a period of 365 days to be 18 g m^{-2} day^{-1}. This represented a conversion of 2.1% of total incident radiation or 5.0% of the photosynthetically active radiation (*PAR*), assuming that *PAR* is 42% of total radiation, into plant dry matter to give an average total dry matter yield of 65.8 metric tons ha^{-1} $year^{-1}$. About 58–62% of the total plant weight was accounted for by stalks. The average fresh stalk weight yield was 153 metric tons cane ha^{-1} $year^{-1}$ and was 25.4% dry matter.

Leaf Area Index

Leaf area index (*LAI*) is affected little by the rather drastic loss of stalks. In general, the maximum *LAI* is achieved about 6 months from planting (compared to 5 months for maximum stalk number) and then slowly declines. Rate of canopy closure and *LAI* development necessary to absorb 95% of the photosynthetic irradiance are important in sugarcane

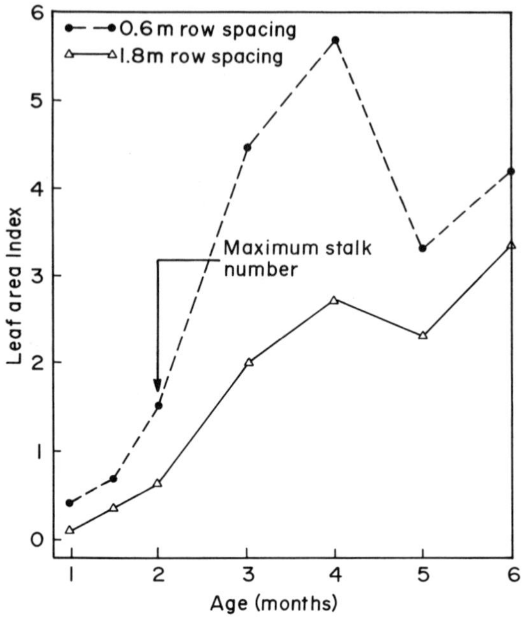

Figure 14.8. Changes in leaf area index at two row spacings as a function of months after germination during a short growing season at Houma, Louisiana. (After Irvine and Matherne (28).]

management because early cover competes with weeds and reduces the evaporation of water from the soil surface. A high *LAI* also provides a larger factory for the production of photosynthate and a higher sugar yield.

The most direct way of increasing the *LAI* early in the season is to space plants closely by reducing the width between rows. Since most commercial sugarcane is grown at row widths of 1.5–1.8 m there may be opportunity to increase *LAI* and yields by narrower row spacing. Irvine and Matherne (28) increased the capture of sunlight by cane and increased sugar yields by increasing *LAI* during the short growing season at Houma, Louisiana, when row spacing was reduced from 1.8 to 0.6 m (Fig. 14.8).

Under the conditions of a longer growing season, *LAI* will not be significantly different by the end of the grand growth stage. Yield responses to row spacings less than 1.5 m under conditions of longer seasons are smaller than for short seasons.

Leaf area can be measured directly in a number of ways, but there are two indirect methods, used for calculating sugarcane leaf area in general, that should be mentioned. One indirect method for determining leaf area requires measuring the length (L) and broadest width (W_b) of the leaf and applying the following formula (30).

$$A = 0.75 W_b L \tag{1}$$

A new method developed by Shih and Gascho (31) is considerably faster. The technique computes the total leaf area of a stalk by measuring stalk length from the ground surface to the top visible dewlap and applying the following formula:

$$Y = e^{[X/(a + bX + cX^2)]} \tag{2}$$

where Y = total leaf area of each stalk; X = length of stalk; and a, b, and c are, coefficients. The values of a, b, and c used in Florida were 1.140, 0.111, and 0.177×10^{-4}, respectively.

Dry Matter Accumulation

Sugarcane may produce up to 65 metric tons of aboveground dry matter per year, of which about 60% is from the stalk. Rostron (16) found total dry matter production (including roots) increased gradually to a maximum of 85.7 metric tons ha^{-1} at 64 weeks. Cumulative dry matter production rate ranged from 16.4 to 19.1 g day^{-1} between 32 and 64 weeks of age and the efficiency of total conversion of solar energy into dry matter ranged from 1.7 to 1.8%. Cane fresh weight yield and sucrose yield, like total dry matter yield, increased steadily up to 64 weeks of age.

Production of dry matter is low during periods of incomplete canopy development, at the beginning of each crop season and again at the end after lodging.

Borden (26) found that the most rapid increases in millable stalk weights occurred between 3 and 9 months in Hawaii. At 9 months, the crop made 63% of the maximum yield of millable cane. One hundred percent of the millable cane was reached at 21 months and the crop lost weight thereafter.

Sucrose Accumulation

Sucrose accumulation in sugarcane storage tissue is usually during the last portion of growth (the maturation stage). During this stage the cane is growing little or not at all. Continuing photosynthesis provides carbohydrates which are eventually stored as sucrose. Gascho and Shih (29) found that sucrose as a percentage of cane increased from 4.5 to 13.7 in the 4 months prior to harvest of a plant cane (i.e., first year cane) of CP 65-357. During that period metric tonnage of fresh stalks per hectare only varied from 95.4 to 99.0 (Fig. 14.9). Thus, sucrose concentration increased about 200% while cane tonnage increased by 2% during the final 4 months of a 12 month crop. Application of a ripener (4 kg ha^{-1} glyphosine) 6 weeks prior to harvest further increased sucrose concentration at the final harvest dates. Sucrose accumulation begins at the bottom of the stalk and progresses upward with maturity to the higher internodes. At full maturity the highest internodes have a sucrose concentration similar to the lowest internodes.

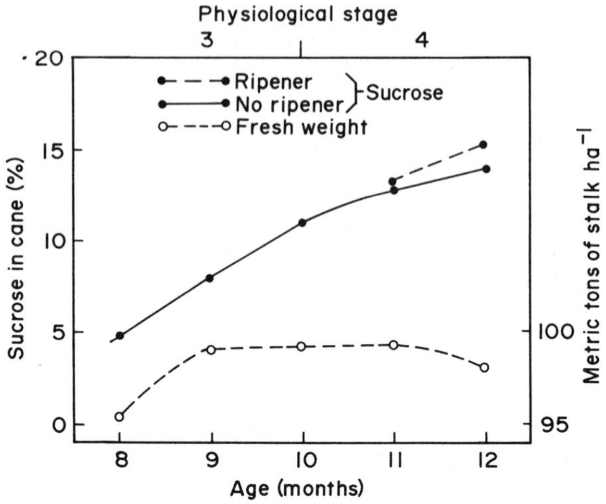

Figure 14.9. Comparison of sucrose accumulation and fresh weight accumulation of cane during the final 4 months of a 12 month crop as a function of months after planting. [After Gascho and Shih (29).]

Water availability has a pronounced effect on sucrose storage. In general, dry weather is preferred during the last 6–8 weeks prior to harvest. The dry weather is detrimental to vegetative growth, thus pushing photosynthates toward sucrose storage. The dry weather also helps concentrate the sucrose by evaporating water from the leaf surface during a time when water supply from the soil is limited. This higher concentration of sucrose in stalks is particularly attractive to growers and millers because of reduced costs for harvesting, hauling, and milling less cane to get an equal amount of sucrose. Caution should be exercised in interpreting data from cane that is harvested after dry conditions to note both the tonnage of sucrose per hectare and the sucrose concentration. A high concentration may not be an increase in total sucrose but only a loss of water.

Water Relations and Canopy Development

Stalk Elongation and Water Quantity

Water relations play an important role in tiller elongation and resultant height of the sugarcane. Webster and Eavis (17) studied the effects of flooding on sugarcane tiller elongation during physiological stage 2 (tillering and canopy establishment) and found that during the period of flooding there were significant reductions in tiller elongation rates. After flooding tiller elongation rates were enhanced in relation to an unflooded treatment. Plants flooded for 1 day at either 1 or 3 months of age caught up with the controls within 2 days. The height of plants flooded at 1 month of age for 4 days and 14 days caught up with the controls 6–10 days after drainage. Plants flooded for 30 days in physiological stage 2 did not survive.

Stalk elongation is positively correlated with the water content of the elongating and meristem tissues (32) and with rainfall for a water deficient area (33). However, there is no generally accepted ideal or optimum soil water content in sugarcane soils. Growth of stalks essentially ceases for plants grown in soils at or below the wilting point (34); however, rate of stalk elongation appears to be constant for all soil water levels above the wilting point.

Chang et al. (35) obtained highly significant correlations between stalk elongation and amount of irrigation water. The most striking elongation of cane stalks took place in June and July (grand growth stage) when the maximum amount of water was required. About 60–70% of cane stalk elongation took place from June to September, the same time that 60–70% of the seasonal available soil water was used.

Gosnell (22) found that a water table at 25 cm depth inhibited germination at planting and ratooning, and caused large reductions in stalk number, height, cane yield, sucrose percent, sucrose yield, and root weight. Water tables at 75, 100, and 125 cm depths gave the best results. A water table at 75 cm gave 25 stalks per pot at harvest compared to a shallow water table of 25 cm which gave 18 stalks per pot. Height was affected in the same way with plant cane height of 283 cm for a 100 cm water table depth and a 223 cm cane height for a 25 cm water table depth. Gascho and Shih (36) found that tolerance to high water tables was cultivar dependent. In their study two cultivars produced as much cane at a constant water table of 30 cm as at water tables of 60 and 90 cm. One cultivar performed best at 60 cm and two other cultivars performed best at a 90 cm water table depth.

Thompson (18) indicated that stalk population, weight, and length were adversely affected when irrigations were applied using a pan evaporation factor of less than 1 for irrigation initiation. Stalk number varied from 182,000 to 139,000, weight per stalk varied from 0.80 to 0.22 kg, and stalk length varied from 214 to 73 cm for pan evaporation factors varying from 1.0 to 0.37, respectively.

Gosnell (24) analyzed the growth of sugarcane on sandy loam soils under irrigated and nonirrigated conditions with a mean annual rainfall of 95 cm. Irrigation treatments were applied on a basis of 1 to 1.25 × class A pan evaporation. He observed significant reductions in stalk elongation rates of cane with age. The elongation rates were 6.1, 2.9, 1.3, 1.4, 2.4, 0.9, and 0.2 cm per week for cane 3–5, 6–8, 9–11, 12–14, 15–17, 18–20, and 21–23 months old, respectively. Hudson (19) studied the relationship between stalk elongation on 3-month-old cane and soil water deficit (cm water). The result showed that stalk elongation decreased with increasing water deficit (Fig. 14.10).

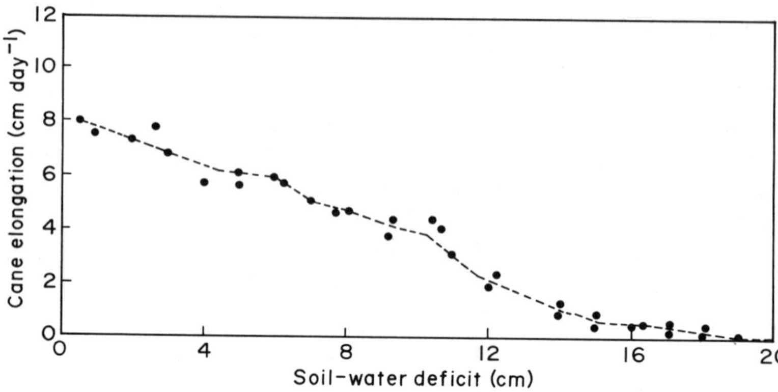

Figure 14.10. Cane elongation in relation to soil water deficit for deep montmorillonite clay soil. [From Hudson (19).]

In Hawaii, Chang et al. (37) related stalk elongation to application of varying amounts of water. Cane elongation rate increased as the amount of water applied increased until the relative water application (E_t/E_p) reached a ratio of 0.9 (Fig. 14.11).

Stalk Elongation and Water Quality

Burr et al. (5) found that stalk elongation ceased by day 5 when the osmotic potential of the root culture solutions was maintained at 2 bars; however, photosynthesis was still 75% of normal. Reduced stalk elon-

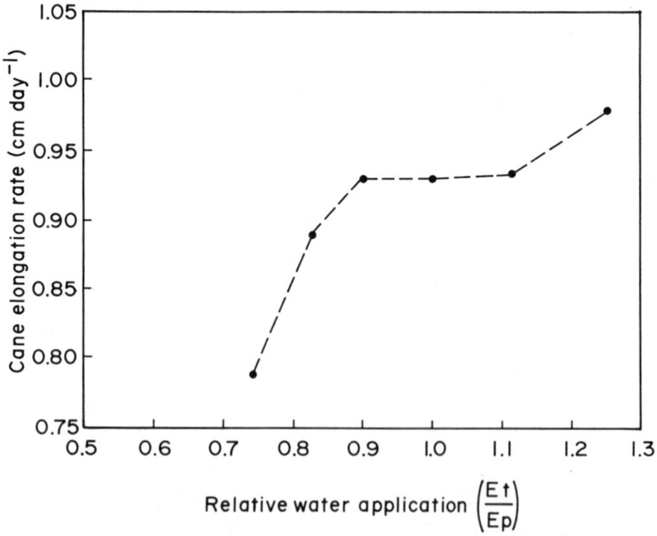

Figure 14.11. Cane elongation and relative water application (E_t/E_p). Water applied included net irrigation plus effective rain. [From Chang et al. (37).]

gation was attributed to the osmotic potential in the root environment. Wilkins and Ateshian (38) reported that production of good quality cane was unlikely at a salinity concentration greater than 2.5 bars.

Mongelard and Minura (39) measured daily stalk elongation of 12 clones whose roots were immersed in nutrient solutions with constant potentials varying from −0.125 to −8.0 bars. They found that after an initial reduction in extension on the first day at all potentials there was a gradual recovery over the succeeding days, the rate of recovery varying with clones. In all clones the rate of recovery was more rapid at solute potentials of −1.5 to −2.0 bars than at lower or higher potentials. Mongelard and Nickell (40) concluded that the reduced growth of plants in media having a solute potential lower than −2 or −3 bars may not be attributed solely to reduced water potentials. There could be two kinds of effects: (1) a short-term effect (not exceeding 24 hr) due to reduced water potential of the substrate, and (2) a long-term effect due to a combination of reduced water potential of the substrate, increased osmotic potential of the vacuole, and ultimately a toxic effect of the salt.

Salinity Effects

Nickell (41) grew sugarcane with the roots in nutrient solution containing sodium chloride at concentrations equivalent to 0.25, 0.5 1, 2, 3, and 4 bars to simulate soil water potentials of drying soil. The conditions were maintained until after tassels appeared. No tasseling occurred at concentrations greater than 2 bars. This was about the same concentration where vegetative growth was reduced. A second experiment, where control plants stayed in the basic nutrient solution (without adding NaCl) and other plants were exposed to nutrient solutions (basic solution plus NaCl) adjusted to 1, 2, 4, 8, and 16 bars with NaCl, showed that harvest total weights and weights per joint decreased in proportion to the amount of salt added. This loss in weight was found to be related to metabolism of sucrose that was stored before treatments were initiated. This weight loss continued for a period of time after the plants were returned to the basic nutrient solution.

Leaf Area Index

Gosnell (24) found a reduction of the *LAI* with maturation (Fig. 14.12). In nonirrigated plots *LAI* varied from 3.1 at 4–6 months to 1.3 at 22–24 months, while corresponding *LAI* values for irrigated plots were 3.8 and 1.8. A secondary peak in *LAI* also occurs in January–May during second year cane production.

Webster and Davis (17) found that the rate of leaf area increase was less in flooded plants than in well-drained moist soils, but was age dependent. At 1 month of age the rate of increase in leaf area was 25% less in plants flooded for 14–30 days compared to nonflooded plants, but at

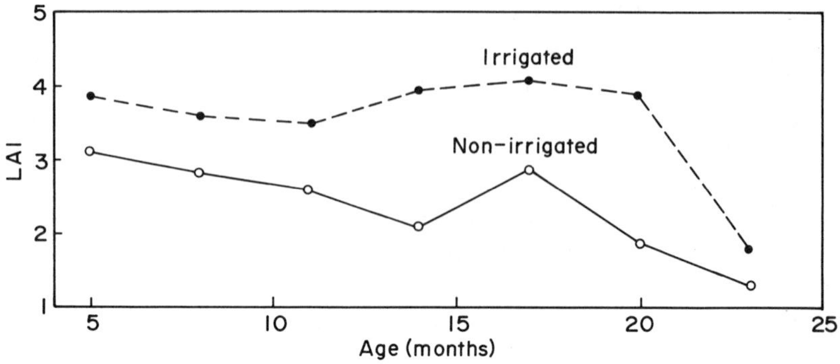

Figure 14.12. Influence of age and irrigation on *LAI* as a function of months after planting. [From Gosnell (24).]

3 months of age flooding did not result in a significant reduction in leaf area when compared to nonflooded plants.

Absorption of water through the leaves of sugarcane was demonstrated by Wadsworth and Das (42). This type of absorption can be fairly important during drought conditions with dew formation at night or trace accounts of precipitation. They showed that water absorbed by the leaf was conveyed from the roots to the soil. Bulliform cells were active in absorption and were also the first cells to dehydrate in dry conditions because of their thin leaf epidermis.

Photosynthesis and Respiration

Assimilation of CO_2 into carbohydrates in sugarcane is via the C_4 pathway and generally increases with increasing canopy cover until closure and is at a maximum during the grand growth stage. The rate of assimilation tends to decrease during the late grand growth stage and during the maturation phase. A sugarcane plant continues to initiate new leaves throughout its life, so that blades of the same rank are of approximately the same age throughout the life of the plant. Hartt and Burr (43) showed that blade four has its highest rate of photosynthesis when taken from a young stalk (Fig. 14.13). The rate decreases steadily until the stalk reaches an age of approximately 15 months, after which the rate remains fairly constant.

Water availability also affects the rate of photosynthesis. Hartt (44) found that some photosynthesis took place in leaves of plants at or below the wilting point because of gains in sugar during the day. However, the amount of photosynthesis was much less than in plants with an adequate water supply. Photosynthesis falls to about 25% of normal when soil water is reduced to the permanent wilting point. When the water potential of the nutrient solution feeding sugarcane plants was decreased from −0.4 to −6.4 bars, the percentage water in the fourth leaf blade decreased 11% and rate of photosynthesis decreased 58% (43).

Respiration may also account for a large part of the reduced growth observed on older cane. Iwaki (45) and Rees (46) stressed the importance of the ratio of nonphotosynthetic to photosynthetic tissue on the net assimilation of perennial crops because of the respiration occurring in a quantity of nonphotosynthetic (old) material. The ratio of stalk dry matter to foliage dry matter increased from 1:1 at 6 months to 10:1 at 24 months. Kamerling (47) found that 9.2 mg CO_2 hr^{-1} were evolved per kg of fresh weight of cane growing in the field at 24 months of age (maturation stage), which indicates that respiration could account for a considerable part of the reduction in growth with age.

Dry Matter Accumulation

The amount of water used by the sugarcane plant to produce a gram of dry substance ranges from 145 to 300 g (5, 37). This range is probably due to differences in cultivars, age, climatic conditions, and experimental techniques. Nickell (41) and Chang et al. (37) found a very close linear relationship between the amount of water utilized and the total dry weight of sugarcane produced.

Sucrose Accumulation

Gosnell (24) studied the effects of time and irrigation on sucrose concentration for a long season crop. The sugarcane was irrigated on a basis

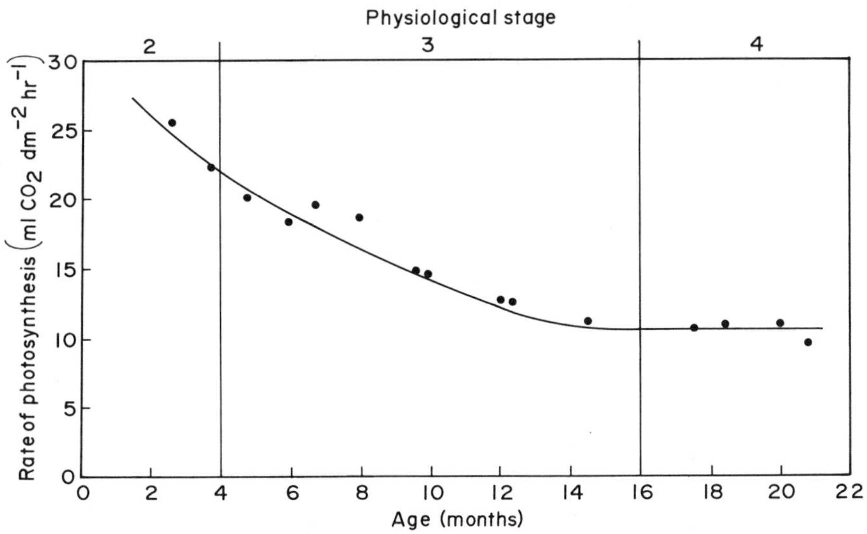

Figure 14.13. Effect of age after germination of stalk on rate of photosynthesis in blade at saturating light intensity. [From Hartt and Burr (43).]

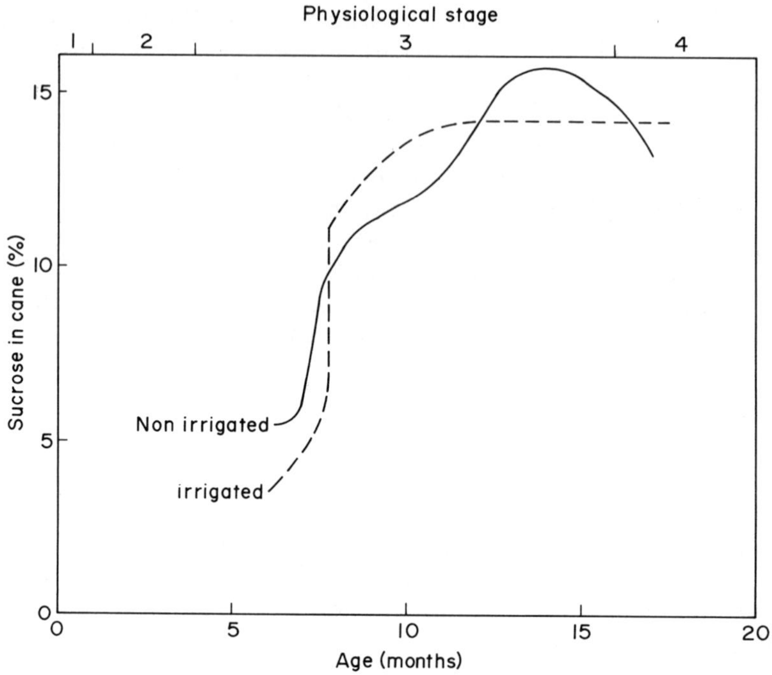

Figure 14.14. The concentration of sucrose in long-season sugarcane as related to age after planting and water treatment. [From Gosnell (24).]

of 1 to 1.25 of class A pan evaporation. At 5 months, the nonirrigated sugarcane had the highest sucrose content, but between 7 and 8 months the percentage of sucrose in the irrigated cane increased rapidly and sucrose content was higher for the irrigated cane from 8 to 12 months (Fig. 14.14) than the nonirrigated cane. After 11 months the percentage

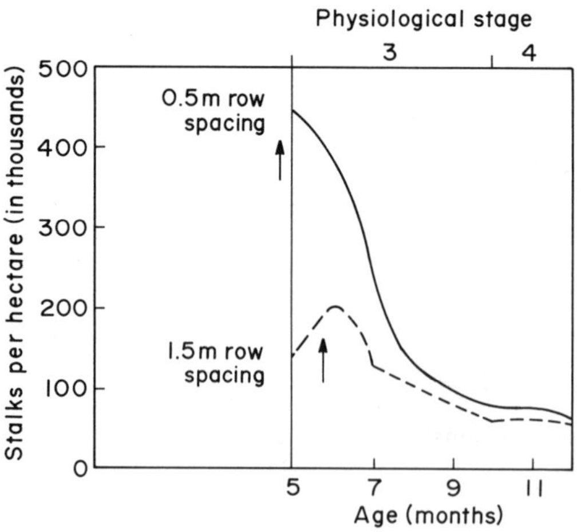

Figure 14.15. Effect of row spacing on stalks per hectare in relation to closure. Arrow indicates canopy age after planting. [From Gascho and Shih (29).]

of sucrose in the irrigated cane became constant while the percentage of sucrose in the nonirrigated cane continued to increase until 14 months when it began to decrease and dropped to less than the irrigated sucrose percentage after 16 months. This is not consistent with sucrose concentration data for a short-season crop (29), which increases rapidly during the last months prior to harvest whether irrigated or not (Fig. 14.9).

Row Spacing

Gascho and Shih (29) determined stalk populations monthly in relation to row spacing from the fourth month after planting through harvest in the twelfth month (Fig. 14.15). Narrow rows produced maximum stalk populations of 440 thousand stalks per hectare at 4 months while 1.5 m rows (standard width in Florida) reached a maximum stalk population of 200 thousand stalks per hectare at 5 months. Canopy closure occurred prior to maximum stalk populations. Narrow rows resulted in significantly more stalks per hectare for most monthly measurements. The differences between the two row spacings decreased rapidly after canopy closure and then remained fairly constant through the late portion of the grand growth and maturation stage and were very similar at harvest. These results were consistent with data from a long-season crop (Fig. 14.16). Stalk populations were much higher for narrow rows up to 5

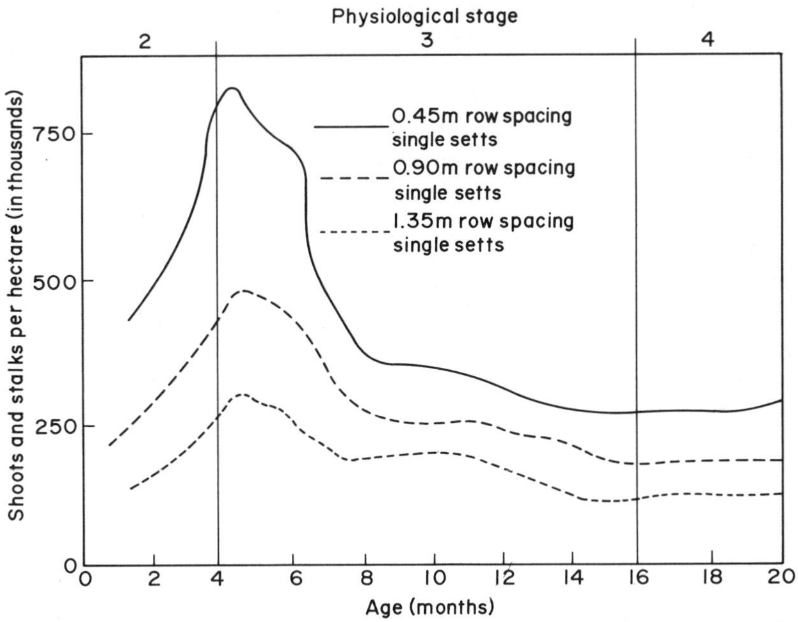

Figure 14.16. Plant populations for three treatments in relation to age after planting. [Thompson and DuToit (25).]

months of age. However, Thompson and DuToit (25) found that the mean stalk weights were lower for the narrow spaced cane resulting in less cane per hectare. The percentage sucrose in the cane was also lower for the narrow rows resulting in less sucrose per hectare.

Thompson (18) reported that the period before canopy closure varies according to cultivar, row width, and the time of the year when the crop starts. Ratoon crops (cultivar NCo 376) started in September, October, and November in rows 1.37 m apart took approximately 160 days to reach canopy closure at two locations. The same cultivar in rows 1.52 m apart at a third location formed a closed canopy in about 120 days when the crop was planted in October or November. However, testing the assumption that canopy closure occurs when E_t first equals E_0, he found in both situations that $E_t = E_0$ after only 90 days. This indicated inaccuracy in the eyeball estimate or in the estimates of E_t or E_0. Cane reached canopy closure after 115 days when crops started in June, July, or August, but after only 70 days when the third ratoon started in September.

TRANSPIRATION

Water conservation under drought conditions is of major importance in obtaining a satisfactory yield of sugarcane. Since drought is common in many sugarcane areas and irrigation is often not possible, it is important to consider reducing transpiration and thereby reducing consumptive water use.

Transpiration occurs in both the leaves and stalks of sugarcane. However, about 90% of transpired water is from the leaves. Stalk transpiration is essentially limited to the nodal region where there is little wax. Lauritzen (48) found that dewaxing cut stalks nearly doubled their transpiration rate.

Meaningful relationships between water potentials and leaf profile and aging have not been established. However, Clements (49) has developed a sheath moisture index as a function of light, plant age, and maximum and minimum temperature. The moisture index reached about 85% during the grand growth stage, then decreases to 73% during the maturity stage. This index could be used as a measure of the crop's well-being and the maturity of cane for harvest.

Stomatal Control

The density and movement of stomata vary with cultivar. Kuijper (50) and Dunlop (51) independently found large differences in stomatal populations among cultivars. The lower epidermis has about twice the number of stomata as the upper epidermis. Kuijper found that the adaxial surface had 115 and the abaxial surface had 253 stomata mm^{-2}. Evans

(52), however, found no relationship between stomatal density and transpiration. The transpiration rate was about the same in the upper and lower leaf surfaces in spite of the twofold difference in stomatal density. Evans concluded that cuticular transpiration was responsible for some water losses in sugarcane, which partially explains why stomatal density differences among cultivars do not account for differences in transpiration rates.

The activity or movement of stomata is greatly affected by external factors of which light, temperature, and humidity are dominant (8). Kuijper (50) found that stomata of sugarcane open in direct sunlight. Weak diffuse light results in a partial closure, the stomata of the lower epidermis closing more than those of the upper. In complete darkness the stomata first close completely and subsequently reach a situation in which they are slightly opened, thus permitting a small degree of diffusion. In order to effect an early maximum opening, direct sunlight is required, and once this occurs, a subsequent decrease in light intensity is of little consequence. This explains the beneficial effect of early morning sunshine on sugarcane.

High humidity, high temperature, and irrigation all favor the opening of stomata. During a sunny day without rain the stomata will open widely until about 10 a.m. and then begin to gradually close corresponding to a normal decrease in air humidity during the remainder of the day. Under conditions of high evaporation, i.e., hot and dry, transpiration may be rapidly reduced by stomatal closure as the water potential at the soil-root interface rapidly decreases (40). Differences in drought tolerance among cultivars are related to differences in stomatal movement. Naidu and Bhagyalakshimi (53) studied the relative turgidity and stomatal movement of four cultivars during a drought. The relative turgidity of two of the drought tolerant cultivars decreased rapidly to about 60% of normal turgidity by the end of the drought. However, nearly 90% of their stomata had closed by day 5 or the drought. In the drought-susceptible cultivars the relative turgidity decreased less rapidly and reached 76–89% of normal turgidity at the end of the drought. The drought tolerant cultivars had the ability to lose turgidity, close stomata, and conserve soil water. They also had the ability to tolerate moderate dehydration.

Cultivars that have high transpiration rates in the early morning hours when water conditions are favorable followed by a rapid closure of stomata and decreased transpiration rate, as the temperature rises and the humidity increases, have been associated with high yield in sugarcane areas with deficient soil water (8).

Other Controls

As mentioned earlier, plant water loss in sugarcane is not entirely regulated by stomata. Cuticular transpiration should also be considered.

However, few data are available to indicate that antitranspirants that induce stomatal closure or create cuticular films are either beneficial or detrimental to sugarcane.

Passive curling of leaves also reduces the radiation receipt by leaves, thereby reducing water loss and increasing water use efficiency to a greater extent than if the leaf remained uncurled. Evans (52) found a reduction of 10–20% in the rate of transpiration when leaves were curled. This rather small effect of curling is understandable because the curling only somewhat protects the area of one surface of the leaf blade (usually the upper surface). Since transpiration from the upper and lower surfaces are approximately equal, the results reported by Evans are reasonable. Sugarcane cultivars differ in their susceptibility to leaf curling. Evans found that some cultivars start curling when their water content has been reduced by 15–20% while other cultivars do not curl or wilt even under extreme drought.

In addition, drought tolerant cultivars will usually have less transpiration area. Evans (52) found that cultivars with low transpiring surface to absorbing surface of root ratios were drought resistant. Because of the tedious nature of studying the absorbing surface of roots, the ratio has not been studied recently. However, the relationship with age is known. Young plants have a high ratio of transpiring surface to root absorbtive surface, which decreases to a minimum during the grand growth stage and increases again rapidly during the maturation stage. Sugarcane plants in the grand growth stage are considered more drought tolerant than plants in the tillering and maturation stages.

Mongelard (54) showed that the differences between a drought tolerant and a drought sensitive cultivar could be due to rooting depth. Other studies also emphasize the importance of rooting depth, a characteristic that is controlled genetically and often limited by the environment.

APPLIED ASPECTS OF IRRIGATION

Water has been called the life blood of sugarcane. Various aspects of water requirements have been and are being studied intensively in different sugarcane growing areas of the world. Potential evapotranspiration of a given area is primarily a function of weather conditions. Solar energy, wind, temperature, and humidity are the four climatic factors affecting evapotranspiration. Mukammal (55) showed that solar energy alone accounts for about 80% of water loss in sugarcane, while wind effects amount to approximately 14%, and the remaining 6% is associated with temperature and humidity. Thus, pan evaporation can prove a useful tool for estimating the evapotranspiration from a crop of growing cane, especially since it is simple and inexpensive to measure.

Predicting When to Irrigate

Most of the studies reviewed have dealt with the prediction of water needed to maintain soil water levels between two predetermined values with little regard to physiological stage of development or other management aspects. Computations of consumptive water use for scheduling and quantifying irrigation have been predominantly based on meteorological considerations. Pan evaporation (E_p), which is commonly gathered from the U.S. Weather Bureau Class A pan, has been widely used as a predictor of sugarcane evapotranspiration (E_t) for estimating the amount of water needed for irrigation. Short-range differences due to seasonal changes have been pointed out but not stressed. Chang et al. (35) reported that 60–70% of the seasonal water consumption of sugarcane occurs in the months of June–September in Taiwan. Shih and Gascho (56) studied the water requirement for related water table depths (Table 14.3) in two seasons (dry and wet) and the grand growth period (April–October). The dry season (November–April), which accounts for about 25% of the annual rainfall in Florida, is a critical period for irrigation. The wet season (May–October) is a critical period for both drainage and supplemental irrigation. About 70% of water usage is in the wet season. The E_t/E_p ratios were 0.67 and 1.22 for the dry and wet seasons, respectively. The monthly pan evaporation and sugarcane evapotranspiration are shown in Fig. 14.17. The monthly rates of E_t/E_p varied from 0.4 in February to 1.3 in August.

The short range differences between E_t and E_p are also influenced by plant age. Ekern (57) found that once canopy closure was attained (at end of 3 months) the monthly E_t and E_p were approximately equal. Chang et al. (37) indicated two distinct segments in the curve formed by plotting E_t and E_p. The first segment of 5 months represents the

Table 14.3. Average pan evaporation (E_p) and evapotranspiration (E_t) for dry, wet, and grand growth periods in three water table depth treatments[a]

			E_t (cm)		
			Water Table Depth		
Period	Crop Cycle	E_p (cm)	90 cm	60 cm	30 cm
Dry	Plant cane	65.61	42.09	52.99	61.79
	Ratoon	67.85	31.70	37.70	43.93
Wet	Plant cane	88.35	112.53	120.64	119.49
	Ratoon	91.58	97.28	102.42	98.42
Grand growth	Plant cane	105.16	117.67	129.96	136.01
	Ratoon	109.66	105.46	115.06	113.16

[a] From Shih and Gascho (56).

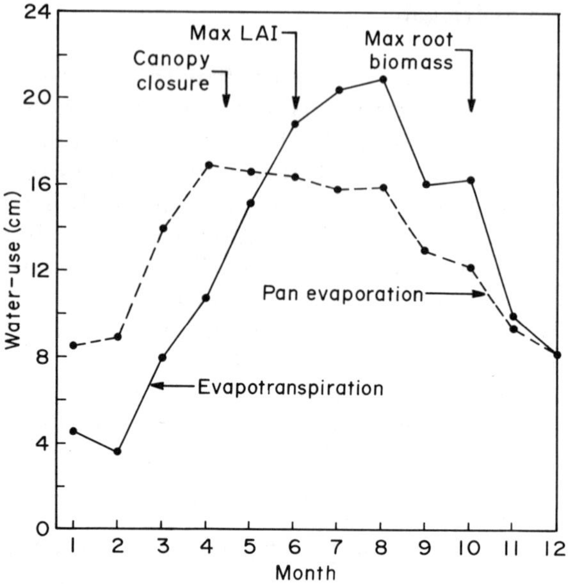

Figure 14.17. Monthly pan evaporation and sugarcane evapotranspiration in Florida. [From Shih and Gascho (56).]

period from germination to the establishment of full canopy when the E_t/E_p ratio increased from 0.4 to 1.01. In the second segment, under a full canopy the E_t/E_p ratio reached a peak of 1.20 at 10 months and then declined gradually to 0.98 at 17 months. The average ratio for all ages of cane was 0.98, and the average for cane with a full canopy was 1.08. Thompson (58) and Shih and Gascho (56) also found this relationship. However, in the grand growth stage, the average E_t/E_p was 1.1. Isobe (59) also reported that a ratio of 1.1 was found during the grand growth stage in Hawaii.

The ratio of E_t/E_p is also different due to crop cycle, i.e., plant cane and ratoon cane. Studies conducted by Moberly (60) in South Africa indicated the decline in the E_t/E_p ratio from 1.0 in the plant cane to 0.8 in the ratoon cane was reflected in the decline in the yield from 166 and 125 metric tons cane ha^{-1} in plant and ratoon canes, respectively. Shih and Gascho (56) also found that E_t/E_p ratios declined from 0.80, 1.33, 1.22, and 1.10 in the plant cane to 0.56, 1.09, 1.01, and 0.86 in ratoon cane for dry, wet, grand growth, and annual periods, respectively (Table 14.3). This difference of water requirement between plant and ratoon cane affected cane yield. There was a 33% reduction in ratoon cane tonnage as compared to plant cane tonnage.

Daily E_t rates were investigated by several researchers. Campbell et al. (61) reported that the daily E_t ranged from 3.8 to 8.9 mm day^{-1} in Hawaii. The peak use of water was 8.1 to 8.6 mm day^{-1} measured during a 1.5 month portion of the grand growth period. Thompson (58) indicated that the daily E_t varied from 2.3 to 6.1 mm day^{-1}. Shih and Gascho (56) showed that the daily E_t varied from 1.3 to 6.8 mm day^{-1} depending on the physiological stage of development and atmospheric demand.

Plant factors could be integrated to develop a useful guide to cane irrigation. Chang et al. (37) reported there was a good linear correlation between the relative water amounts transpired and the logs of the ohms resistance of gypsum blocks. They also found that stalk elongation rate was an index of water stress. Under field practice, irrigation should begin before the cane elongation rate decreases. Robinson (62) has shown that the elongation rate of cane stalks declines as the potential approaches -2 bars at the 30 cm depth.

Soil water has been used to schedule irrigations. Ewart (63) placed the resistance blocks at 23–30 cm below the cane stool and irrigated when the blocks reached 5000 ohms resistance (equivalent to about one-third depletion of available water). Waterhouse and Clements (64) installed porous tensiometer cups at 51 cm below the bottom of the furrow and irrigated when the readings reached 25–35 scale units (approximately -0.25 to -0.35 centibars).

Water Table and Irrigation

Water table depth management is an important factor for determining the need for subsurface irrigation of sugarcane in the Everglades. The depth of water table has an effect not only on the quantity of water used but also on the determination of effective rainfall. Effective rainfall is defined as a rainfall that is temporarily stored in the soil without causing runoff and seepage to the groundwater. Effective rainfall is an important element used to estimate the needs for supplemental irrigation and drainage.

Shih and Gascho (56) indicated that effective rainfall is linearly related to water table depth (Fig. 14.18). Approximately 0.34 cm of rainfall was stored in the soil for each 10 cm water table depth increment. They also reported that the rate of reduction in E_t is significantly different due to water table depth differences in both the dry and wet periods. The rates of reduction in E_t for dry and wet periods, respectively, were 0.30 and 0.24 cm per cm of water table depth. Thus, when a high water table is maintained, not only is more water required to irrigate but more pumping for drainage is also required following a heavy rainfall.

Yield and Irrigation Relations

Due to the variations of location, climate, and cultivar, it is difficult to draw a general relationship between E_t and E_p. Ewart (65) pointed out that predicted water applications derived from pan evaporation data gave no guarantee of optimum yields. Mongelard and Nickell (40) indicated that the 1:1 ratio of E_t/E_p is undoubtedly very helpful in determining when to irrigate but did not give the optimum irrigation efficiency (the ratio of yield to amount of water applied). Yang (66) estimated that 60 cm of irrigation water could result in yields of 90–150 metric tons of cane ha^{-1}. If more than 90 cm of water is applied it would be considered uneconomical under Taiwan conditions.

The relationship between E_t and yield has been studied as another water use index in water resource study. For many crops in particular locations and climates, transpiration has been shown to be linearly associated with dry matter production (67). Water to cane yield ratios as shown in Table 14.4 for Puerto Rico (68), Hawaii (59, 61, 69, 70), South Africa (71), and Florida (56) may be used as a first approximation for defining E_t in relation to cane and sugar yields. Shih and Gascho (56) also indicated that not only the total water requirement for ratoon cane is less as compared to the plant cane, but water use efficiency is also considerably lower.

Critical Water Demand Periods

Good water availability is essential in the tillering stage prior to full canopy development. However, consumptive water usage is fairly low

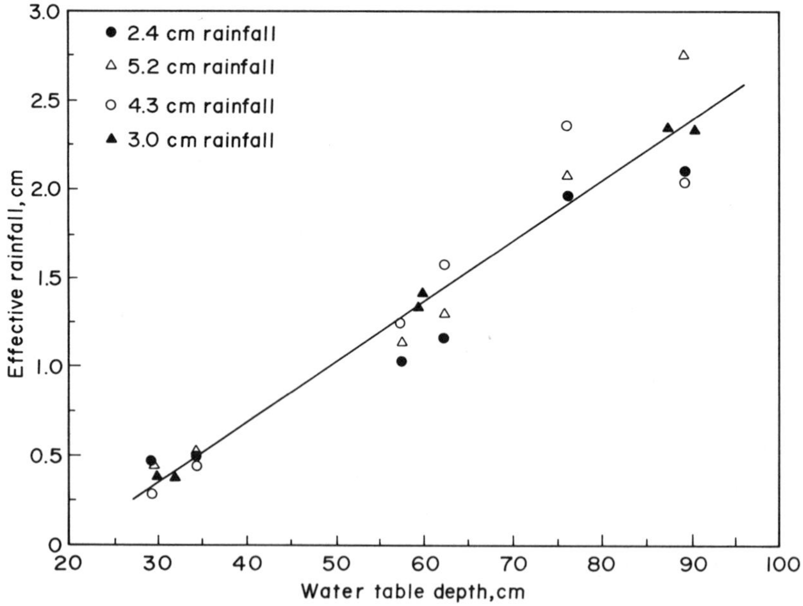

Figure 14.18. Relationship between effective rainfall and water table depth in a Florida organic soil. [From Shih and Gascho (56).]

Table 14.4. The values of water to cane and sugar yields used as a predictor of evapotranspiration (E_t) for estimating the amount of water required for sugarcane production

References	Year	Location	Water to Yield Ratio	
			Cane	Sugar
Fuhriman and Smith (68)	1951	Puerto Rico	71–84	—
Baver (69)	1954	Hawaii		2000
Campbell et al. (61)	1959	Hawaii	94–129	—
Isobe (59)	1968	Hawaii	106–136	846–967
Scott (70)	1971	Hawaii	—	2000
Thompson et al. (71)	1963	South Africa	59–102	—
Shih and Gascho (56)	1980	Florida	92–122	917–1157

prior to full canopy development because of the smaller leaf area index. Adequate water is necessary for vigorous vegetative growth during the grand growth stage. Water stress during this stage will result in lower final yields. Irrigation intervals are commonly reduced during the maturation stage in order to improve the juice quality. The percent sucrose in the juice and in the cane is adversely affected by excess water during the maturation stage. Yang (66) suggested irrigation intervals of 30–35 days for young cane, 20–25 days for cane in grand growth, and 30–35 days for cane being ripened on medium texture soils in Taiwan. Mongelard and Nickell (40) stated: "During the period of vigorous vegetative growth, cane will respond best (in terms of final yield of high quality juice) to soil moisture maintained at relatively low tensions. As harvest time approaches and during the ripening phase, sucrose accumulation is favored by water stress induced by a slowly imposed drought."

REFERENCES

1. H. F. Clements and S. Nakata, *Int. Soc. Sugar Cane Technol. Proc. 12th Congr.* pp. 554–560 (1965).
2. G. J. Gascho, O. C. Ruelke, and S. H. West, *Crop Sci.* **13**, 274–276 (1973).
3. K. A. Sund and H. F. Clements, *Hawaii Agric. Exp. Stn. Res. Bull.* **160** (1974).
4. R. J. Allen, Jr., G. Kidder, and G. J. Gascho, *Proc. Am. Soc. Sugar Cane Technol., N.S.* **7**, 18–22 (1978).
5. G. O. Burr, D. E. Hartt, H. W. Brodie, T. Tanimoto, H. P. Kortschak, D. Takahashi, F. M. Aston, and R. E. Coleman, *Ann. Rev. Plant Physiol.* **8**, 275–308 (1957).
6. R. P. Humbert, *The Growing of Sugarcane*, Elsevier, New York, 1968.
7. J. Glover, *Proc. S. Afr. Sugar Tech. Assoc.* **42**, 123–130 (1968).
8. C. Van Dillewijn, *Botany of Sugarcane*, Chronica Botanica, Waltman, Mass., 1952.
9. H. A. Lee, *Plant Physiol.* **1**, 363–378 (1927).
10. H. A. Lee, *Hawaii Planters' Rec.* **30**, 520–523 (1926).

11. H. A. Lee and D. M. Weller. *Hawaii Sugar. Technol.* pp. 69–72 (1927).
12. G. C. Stevenson, *British West Indies Central Sugarcane Breeding Stn.* Bull. 11 (1936).
13. H. Evans, *Sugarcane Res. Stn. Mauritius* Bull. No. 1W (1937).
14. S. C. Srivastava and A. K. Ghosh, *J. Ind. Soc. Soil Sci.* **18,** 117–119 (1970).
15. S. Yang, P. L. Wang, Y. T. Chang, and T. C. Yang, *Taiwan Sugar* **24,** 378–384 (1977).
16. H. Rostron, *Int. Soc. Sugar Cane Technol. Proc. 15th Congr.* pp. 1001–1010 (1974).
17. P. W. D. Webster and B. W. Eavis, *Int. Soc. Sugar Cane Technol. Proc. 14th Congr.* pp. 708–714 (1971).
18. G. D. Thompson, *S. Afr. Sug. J.* **60(12),** 627–635 (1976).
19. J. C. Hudson, *Int. Soc. Sugar Cane Technol. Proc. 13th Congr.* pp. 484–498 (1968).
20. B. W. Eavis, *Int. Soc. Sugar Cane Technol. Proc. 14th Congr.* pp. 715–721 (1971).
21. T. C. Juang and G. Vehara, *Int. Soc. Sugar Cane Technol. Proc. 14th Congr.* pp. 679–687 (1971).
22. J. M. Gosnell, *Int. Soc. Sugar Cane Technol. Proc. 14th Congr.* pp. 841–849 (1971).
23. R. Baran, D. Basserean, and N. Gillet, *Int. Soc. Sugar Cane Technol. Proc. 15th Congr.* pp. 726–735 (1974).
24. J. M. Gosnell, *Int. Soc. Sugar Cane Technol. Proc. 13th Congr.* pp. 499–513 (1968).
25. G. D. Thompson and J. L. DuToit, *Int. Soc. Sugar Cane Technol. Proc. 12th Congr.* pp. 103–113 (1965).
26. R. J. Borden, *Hawaiian Sugar Planters' Rec.* **52,** 1–51 (1948).
27. G. D. Thompson, *Int. Soc. Sugar Cane Technol. Proc. 12th Congr.* pp. 3–9 (1965).
28. J. E. Irvine and R. J. Matherne, *Sugar Y Azucar* **73(6),** 46 (1978).
29. G. J. Gascho and S. F. Shih, *Am. Soc. Sugar Cane Technol. Proc. N.S.* **8,** 72–76 (1978).
30. J. C. Mongelard, *Int. Soc. Sugar Cane Technol. Proc. 13th Congr.* pp. 643–651 (1968).
31. S. F. Shih and G. J. Gascho, *Agron. J.* **72(2),** 309–313 (1980).
32. H. F. Clements and T. Kubota, *Hawaii Sugar Planters' Rec.* **46,** 17–35 (1942).
33. V. G. Sun and N. P. Chau, *Rep. Taiwan Sugar Exp. Stn. Circ.* **4,** 1–40 (1949).
34. J. A. Swezey and H. A. Wadsworth, *Hawaii Sugar Planters' Rec.* **44,** 49–68 (1940).
35. H. Chang, J. S. Wang, and F. W. Ho, *Int. Soc. Sugar Cane Technol. Proc. 13th Congr.* pp. 652–663 (1968).
36. G. J. Gascho and S. F. Shih, *Soil and Crop Sci. Soc. Fla. Proc.* **38,** 23–27 (1978).
37. Jen-hu Chang, R. B. Campbell, H. W. Brodie, and L. D. Baver, *Int. Soc. Sugar Cane Technol. Proc. 12th Congr.* pp. 10–24 (1965).
38. R. A. Wilkins and K. H. Ateshian, *Int. Soc. Sugar Cane Technol. Proc. 12th Congr.* pp. 54–81 (1965).
39. J. C. Mongelard and L. Minura, *Hawaii Sugar Planters' Assoc. Annu. Rep.* pp. 37–39 (1970).
40. J. C. Mongelard and L. G. Nickell, *Int. Soc. Sugar Cane Technol. Proc. 14th Congr.* pp. 827–840 (1971).
41. L. G. Nickell, *Int. Soc. Sugar Cane Technol. Proc. 13th Congr.* pp. 38–48 (1968).
42. H. A. Wadsworth and U. K. Das, *Hawaii Sugar Planters' Rec.* **34,** 289–299 (1930).
43. C. E. Hartt and G. O. Burr, *Int. Soc. Sugar Cane Technol. Proc. 12th Congr.* pp. 590–609 (1965).
44. C. E. Hartt, *Naturwissenschaften* **21,** 666–667 (1963).
45. H. Iwaki, *Jpn. J. Bot.* **16,** 210 (1958).
46. A. R. Rees, *Ann. Bot.* **27(108),** 615–626 (1963).
47. Z. Kamberling, *Archief. Suikerind Ned-Indie* **12,** 773 (1904) cited in (8).

48. J. I. Lauritzen, Sugar Bull. **21(4)**, 25–31 (1942).
49. H. F. Clements, *Int. Soc. Sugar Cane Technol. Proc. 14th Congr.* pp. 657–672 (1971).
50. J. Kuijper, *Archief Suikerind. Ned-Indie* **23**, 1673–1700 (1915).
51. W. R. Dunlop, *West Indies Bull.* **13**, 314–323 (1913) cited in (8).
52. H. Evans, *Int. Soc. Sugar Cane Technol. Proc. 6th Congr.* pp. 802–808 (1938).
53. K. M. Naidu and K. V. Bhagyalakshimi, *Ind. J. Agric. Sci.* **43(11)**, 1016–1019 (1973).
54. J. C. Mongelard, *Annu. Rep. Mauritius Sugar Ind. Res. Inst.* pp. 92–97 (1967).
55. E. I. Mukammal, *Hydrology Symp. Proc., Nat. Res. Council Canada* No. 2, 95 (1961.)
56. S. F. Shih and G. J. Gascho, *Trans. Am. Soc. Agric. Engr.* **23(4)**, 934–937 (1980).
57. P. C. Ekern, *Int. Soc. Sugar Cane Technol. Proc. 14th Congr.* pp. 805–812 (1971).
58. G. D. Thompson, *S. Afr. Sugar J.* **47**, 478–479 (1968).
59. M. Isobe, *Int. Soc. Sugar Cane Technol. Proc. 13th Congr.* pp. 49–54 (1968).
60. P. K. Moberly, *Int. Soc. Sugar Cane Technol. Proc. 15th Congr.* pp. 694–700 (1974).
61. R. B. Campbell, J. H. Chang, and D. C. Cox, *Int. Soc. Sugar Cane Technol. Proc. 10th Congr.* pp. 637–649 (1959).
62. F. E. Robinson, *Agron. J.* **55**, 481–484 (1963).
63. G. Y. Ewart, *Hawaii Sugar Planters' Rec.* **54**, 257–269 (1953).
64. A. D. Waterhouse and H. F. Clements, *Hawaii Sugar Planters' Rec.* **54**, 271–283 (1953).
65. G. Y. Ewart, *Int. Soc. Sugar Cane Technol. Proc. 12th Congr.* pp. 34–53 (1965).
66. S. Yang, *Taiwan Sugar* **23**, 161–166 (1976).
67. C. T. De Witt, *Verslag van landbouwk. Onderzoek.* **646** 88 pp. (1958).
68. D. K. Fuhriman and R. M. Smith, *Univ. Puerto Rico J. Agric.* **25**, 1–47 (1951).
69. L. D. Baver, *Hawaii Sugar Planters' Rec.* **54**, 163 (1954).
70. P. Scott, *Int. Soc. Sugar Cane Technol. Proc. 14th Congr.* pp. 859–864 (1971).
71. G. D. Thompson, C. H. O. Pearson, and T. G. Cleasby, *S. Afr. Sugar Technol. Assoc. Proc.* pp. 1–8 (1963).

15

WHEAT

M. B. Kirkham and E. T. Kanemasu

INTRODUCTION

Adaptation

Climatic

More wheat is produced annually than any other food or feed crop. It is the world's major source of calories and protein (1). Wheat is grown over a wide range of moisture and temperature conditions. The main wheat regions of the world lie between about latitudes 30° and 55° in the North Temperate Zone and 25° and 40° in the South Temperate Zone, in areas where annual precipitation ranges between 30 and 114 cm (2). The only parts of the world from which it is absent are the hot, low-lying regions of the tropics (3). The wheats of the more humid areas of the world are generally soft and starchy, and those of less humid areas usually are hard. Spring wheat is sown where the winters are dry and cold. Winter wheat grows where plants can survive winter. The greatest amount of the best wheats are produced in countries with cold winters and hot summers. So long as the winter temperature does not fall below −20°C and the air and soil are dry, the exposed plants suffer little. But at lower temperatures, the crop is damaged, unless protected by snow (3) or mulches.

The distribution of wheat and the kind of wheat grown in relation to temperature are largely determined by the length of the frost-free period, the minimum winter temperature, the temperature in relation to the average length of day during the growing season, and the maximum temperatures immediately preceding harvest (2). For most satisfactory growth and development of grain, a cool, moist, growing season, followed by a bright, dry, and warm ripening period of 6–8 weeks, with a mean temperature of 18–19°C, is necessary (3). Maps and tables are available which show the world distribution of wheat (2–7).

Wheat yields are not becoming any more or any less variable because of changing weather (8). The wide separation of the regions of production ensures that large-scale weather phenomena will not wipe out the world supply of food. The frequent, yearly variation in yield suggests that bumper crops and failures will be produced in different parts of the world in a year and that international trade can relieve famine.

Yield Potential (Irrigated and Dryland)

The record wheat yield of 14,100 kg ha^{-1} was obtained with the soft white winter wheat cultivar "Gaines" under irrigated conditions in Kittipas County, Washington, United States, in 1965 (9). The land was recently reclaimed from sagebrush. Sweet corn had been grown for the previous 2 years. Both corn crops received 120 kg N ha^{-1}. No soil tests were made. Organic matter was estimated to be between 2 and 4%, which is higher than normal for the area, but low compared to soils with corn in the midwestern part of the United States. The wheat plot received 130 kg N ha^{-1}, 25 kg P ha^{-1}, and 56 kg K ha^{-1}. The wheat was planted at a rate of 90–95 kg ha^{-1} in October and fall growth was minimal. From April through July, the plot was irrigated 7 times. Water was applied more frequently and in larger amounts than normal. Harvest was in early September, which was late for the region. Environmental conditions in Kittipas County are good for high production. The days are clear, the light intensity is high, and the nights are cool. The average yield of irrigated "Gaines" wheat in the area in 1965 was 6800 kg ha^{-1}.

This record production (14,100 kg ha^{-1}) exceeds the average wheat yields in the United States by 7 times (10), in England by about 3 times (10, 11), and in the Netherlands (9000 kg ha^{-1}) (12) by almost 2 times. Assuming that light is the major limiting factor, Evans (10) estimated that the maximum potential yield of wheat could be about 25,000 kg ha^{-1} if wheat plants were illuminated with 400 cal $cm^{-2}day^{-1}$.

Potential yield of wheat can only be obtained under well-watered conditions. An enormous increase in water use would be required to approach the potential yield however (10). On drylands maximum yields cannot be realized as there is a linear relationship between yield of a cultivar of wheat and the severity of drought (13, 14). Fischer and Maurer (13) give the yield under drought by the following equation:

$$Y = Y_p(1 - SD) \tag{1}$$

in which Y_p = yield without drought of a cultivar; Y = yield with drought of the cultivar; S = drought susceptibility index, which is a dimensionless slope defined by the following equation:

$$S = \frac{\Delta(Y/Y_p)}{\Delta(X/X_p)} \tag{2}$$

The X in Eq. (2) is an environmental index and is defined as either the mean yield of all the cultivars used or the yield of the cultivar common to all experiments. Fischer and Maurer (13) used the latter definition of

X in their work. Therefore, X_p for them was the highest grain yield (under irrigation) of the cultivar which was common to all their experiments, "Yecora 70." Under their conditions in Sonora, Mexico, X_p for "Yecora 70" was about 700 g m^{-2}. The D in Eq. (1) is the drought intensity and defined as $1 - X/X_p$. Equation (1) gives a quantitative relation between the reduction in yield for a given reduction in water. Yield under drought is dependent on both yield potential of a cultivar without drought (Y_p) and the environmental index (X).

Characteristics and Approximate Time Intervals Between Growth Stages

Growth stages have been defined by Feekes (15), Large (16), who reproduced in English the work by Feekes et al. (17), and Tottman et al. (18) for cereals in general, and by Tottman (19) and Waldren and Flowerday (20) for winter wheat in particular. The growth-stage description of Waldren and Flowerday (20, p. 392) will be given because they developed it for farmers and agronomists, while Tottman (19) made his table for a more specialized group, herbicide applicators. In the following growth-stage description, the approximate numbers for the Feekes scale are also given, since it is more universally known than the newer one by Waldren and Flowerday.

Growth stage	Description of wheat
0	Emergence of coleoptile. Usually happens 7 to 14 days after planting. In cold soil, emergence may not occur for up to 10 weeks after planting. (Feekes 1)
1	Crown is visible. Tillers develop. (Feekes 2)
2	Leaf sheaths elongate and form a false stem. Collars are visible. (Feekes 4). A winter wheat plant will usually be in Stage 1 or 2 when interrupted by winter. The stage depends on environmental conditions and planting date. Additional tillering can occur in the spring, if favorable conditions exist.
3	Culm elongation. First internode is visible (jointing). This stage will not occur in winter wheat until after vernalization. Stage usually takes 2 or 3 weeks. (Feekes 6)
4	Tip of flag leaf visible (boot stage). Stage takes 1 to 2 weeks. (Feekes 8)
5	Penduncle elongates. Inflorescence emerges (heading). Stage takes 7 to 10 days. (Feekes 10.1)
6	Flowering (anthesis). Stage takes about 1 week. (Feekes 10.5.1)

7	Anthesis complete. Grain filling begins. Lower leaves turn color. Stage usually takes 2 weeks. (Feekes 10.5.3)
8	Grain is stiff dough. Flag leaf has lost all green color. Stage usually takes about 1 week. (Feekes 11.1)
9	Ripening. Grain is hard, but will not crack. Inflorescence has lost all green color. Uppermost node is still green. Stage usually takes about 1 week. (Feekes 11.2)
10	Maturity. Grain cracks and is easily separated from chaff, but is not dry enough to harvest and store without further drying. No green color is left in the plant. No time is given by Waldren and Flowerday (20). (Feekes 11.3)

A winter wheat crop is in the ground about 10 months (21). The longest stages are 1 and 2, which may last more than 6 months depending on the date of planting.

The above table permits easy measurement of the stage of growth in the field. If field data are not available, however, a biometeorological time scale (22) can be used to estimate stage of growth, as was done by Feyerherm et al. (23), Feyerherm (24, 25), and Hodges and Kanemasu (26).

ROOT GROWTH AND FUNCTION

Germination and Emergence

Wheat seeds from the time of imbibition to emergence of the first leaf are extremely resistant to drought (27, 28). In fact, a slight osmotic stress appears to increase the rate of germination. Hanson (29) said that wheat seeds that had been allowed to imbibe and then redry germinated more rapidly when the treated seeds were under osmotic stress (polyethylene glycol solution of 3 bars) than when they were not stressed. Idris and Aslam (30) showed that soaking wheat seeds in a 0.5% NaCl solution, and then drying before planting, stimulated germination and growth of the seedlings. Salinities of 1.0 and 1.5% NaCl failed to improve germination. Chaudhuri and Wiebe (31) found that soaking of wheat seeds in 1% $CaCl_2 \cdot 2H_2O$ solutions increased percentage of germination.

Gul and Allan (32, 33) in Washington noted that the time required for emergence of 97 wheat genotypes nearly doubled for each decrease of soil water potential of -4 bars. Lindstrom et al. (34), also in Washington, reported that the emergence rate of "McCall" and "Nugaines," soft white winter wheats, progressively decreased with lowering of water potential, lowering of temperature from 25°C, or with increase in planting depth. In their study, threshold water potential for emergence appeared to increase (wetter soil) as temperature increased, which meant that, as the seed-zone temperature was lowered, wheat germinated and emerged in increasingly drier soil. De Jong and Best (35) in Canada found that the minimum temperature for emergence of "Canthatch," a hard red spring wheat, dropped from 1.3 to 0.2°C, as the soil water potential decreased from $-\frac{1}{3}$ to -10 bars. In the study by Lindstrom et al. (34), there was 80% emergence of "McCall" and "Nugaines" at soil water potentials as low as -10 bars. Only slight effects on emergence rate were seen as the soil water potential dropped from -0.4 to -4.0 bars. The lowest water potential at which seedlings emerged was -20.0 bars. Their results suggested that, at soil water potentials above -10 bars, seedling emergence would not be reduced enough to be of concern to farmers. The *rate* of emergence, however, a criterion that influences wheat yields, could be affected by water potentials below -4.0 bars.

Implements used to drill wheat seeds directly into the soil cut slits, the surfaces of which are smeared when the soil is moist. Prebble (36) investigated the penetration of wheat roots into smeared and unsmeared surfaces at 90° (horizontal) and 45°, as a function of soil matric potential. Germination was not affected by smearing. He found, however, that root penetration was reduced by 50% for a tenfold decrease in soil matric potential below -0.1 bar for the horizontal surface, and that the effect of smearing was equivalent to a fivefold decrease in soil matric potential. On a surface of 45°, smearing had little effect.

Some of the changes occurring in the initial stages of germination are stable to dehydration (37). DNA is actively duplicated and transcribed in 3-day-old embryos and is relatively inactive in 1-day-old embryos. Therefore, seeds can be dried early in the germination process without damage to DNA replication and RNA transcription. Carceller and Soriano (38) dried wheat seeds just after germination without deleterious effects on subsequent germination. Viability of seeds may be retained over several cycles of hydration and dehydration (37).

The 97 genotypes that Gul and Allan (32, 33) studied differed in emergence. Helmerick and Pfeifer (39) and Powell and Pfeifer (40), using mannitol solutions to impose drought, also found cultivar differences. These results show that wheat lines might be selected that establish adequate stands and emerge faster under lower soil water potentials than other less resistant lines. Boyer and McPherson (41) suggested that drought tolerance might be improved by selecting seedlings that survive under suboptimal water availability. The International Rice Research Institute, as part of its comprehensive drought research program, routinely screens rice (*Oryza sativa* L.) plants at germination (42). Several studies with wheat germinating under droughty conditions were done in the 1930s and 1940s (43–45). Little work, apparently, has been done recently. But wheat seedlings, whether drought resistant or drought sensitive, may deplete soil water content to the same value. Dolgov and

Vinogradova (46) found that soil water contents at wilting of soils with different wheat cultivars were similar, despite differences in the water potential of the seedlings.

Depth and Distribution

Gregory et al. (47) reported the root depth and distribution of the winter wheat cultivar "Maris Huntsman" grown at Nottingham, England. The crop was planted 30 October 1974, harvested 5 August 1975, and yielded 6.45 tons ha^{-1}. Figure 15.1 shows the distribution of root length with depth at intervals during the growing season. Production of seminal roots was complete by the beginning of March when all plants had six (occasionally seven) axes (roots) which persisted throughout the life of the crop. Nodal axes were produced from mid-February until late May and finally numbered about 20 nodal axes per main stem. (Seminal roots come from the seed and nodal roots come from nodes on the culm, also called the stem.)

Greacen et al. (48) found that wheat plants at flowering had an average of five seminal roots per plant, and 55, 30, and 15 nodal roots per plant at seeding rates of 27, 137, and 595 plants m^{-2}, respectively. A few nodal roots penetrated more deeply than 90 cm at all seeding rates. The largest number of deep nodal roots occurred at the lowest seeding rate (27 plants m^{-2}), although at this low planting rate the numbers were small (13, 5, and 2% of the total number of axes at 30, 60, and 90 cm depths, respectively). At the highest plant density, only one seminal axis per plant had penetrated below 30 cm, while the number of nodal axes at 30 cm depth or greater was negligible. This was surprising to them because the supply of water and nutrients appeared adequate. Another unexpected result was the uniformity of the distribution of axes. At flowering, for the 30 and 60 cm depths, the density of axes was close to 400 m^{-2} or 1 axis per 25 cm^2. At the dough stage, this value increased to about 600 m^{-2}.

Root Growth in Relation to Physiological Stages

For both common and durum wheats grown in Rehovot, Israel, Pinthus (49) observed the highest rate of root formation during the period from spike initiation [approximately stage 3, Waldren and Flowerday (20); stage 6, Feekes (15)] to heading [stage 5, Waldren and Flowerday (20); stage 6, Feekes (15)] and noted that root dry weight increased with length of period.

Figure 15.2 shows data for root dry weight from the study in England by Gregory et al. (47), who, in a later paper (50), also give corresponding soil water contents and soil water potentials. The arrow in the figure indicates anthesis [stage 6, Waldren and Flowerday (20); stage 10.5.1,

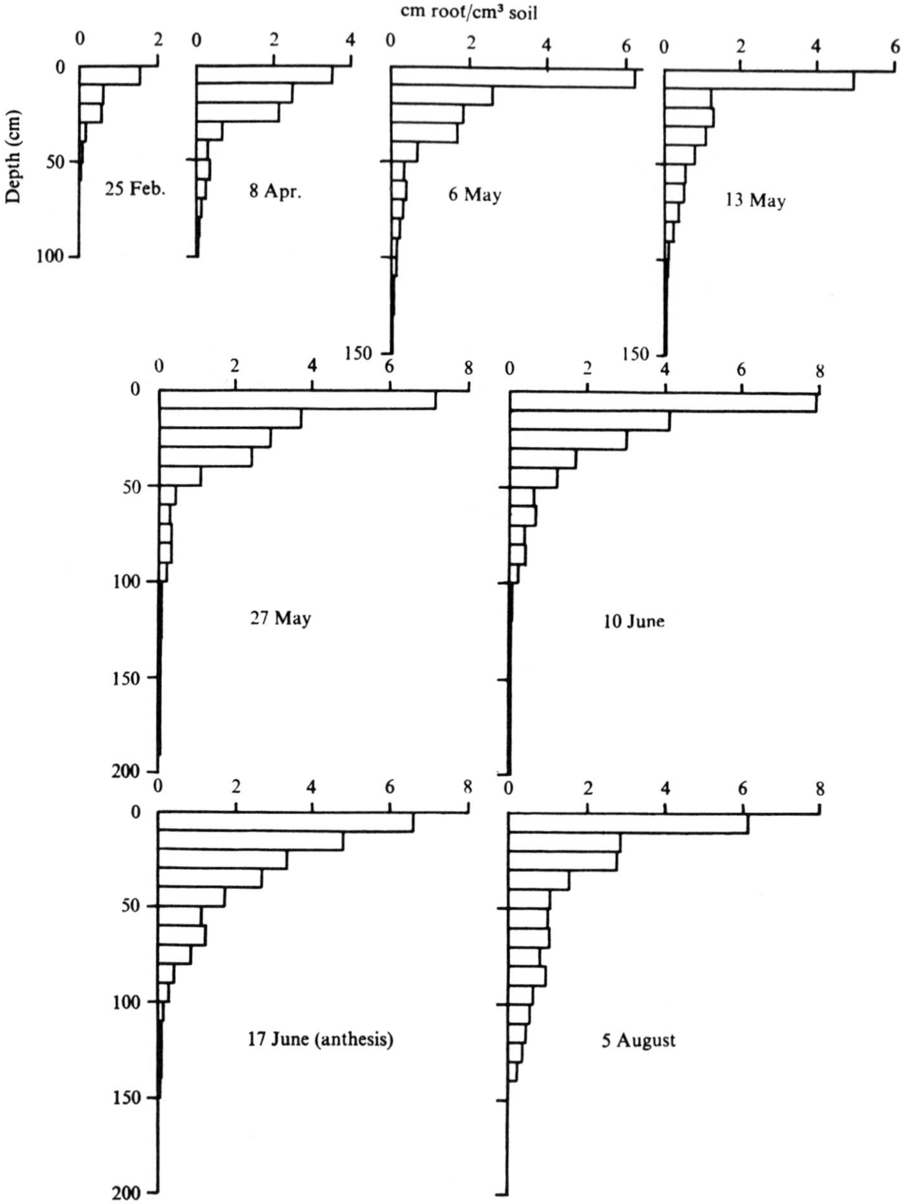

Figure 15.1. Distribution of root length with depth at intervals during the growing season. [From Gregory, McGowan, Biscoe, and Hunter (47), p. 97, with permission.]

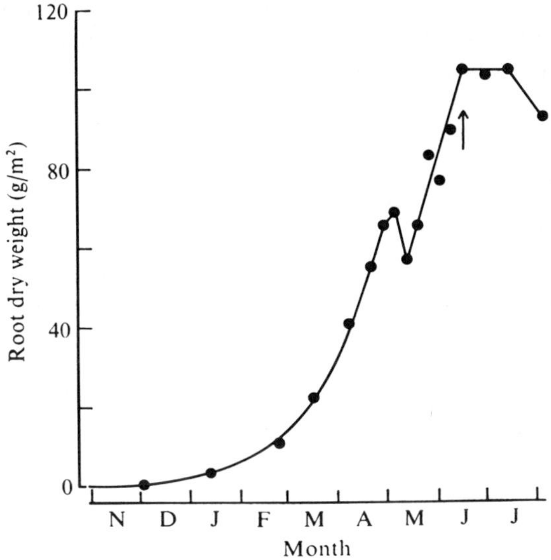

Figure 15.2. Changes in root dry weight with time. The time of anthesis is shown by the arrow. [From Gregory, McGowan, Biscoe, and Hunter (47), p. 96, with permission.)

Feekes (15)]. Root dry weight increased exponentially until the end of March and beginning of April, and then linearly until anthesis (mid-June). From April onward, about 65% of the total root dry weight was in the 0–30 cm layer. After anthesis, even though root growth continued below 80 cm, root weight remained constant at 105 g m^{-2} for 4 weeks and then decreased to 93 g m^{-2} at harvest. Connor (51) also reported that root dry matter stopped increasing at anthesis.

Root Water Relations and Dry Matter Accumulation

Water Conserving Ability

Passioura (52) hypothesized that, when wheat plants are grown on stored water, they should have root systems with high hydraulic resistance, so that they conserve water during early growth. Therefore, more water would be available later at grain-filling time. He felt that high root resistance would reduce water uptake, leaf area, and transpiration while the reduced leaf area would allow some seed production, even if no rain fell. If the soil were wetted by rainfall, adventitious roots would form, root resistance would decrease, water uptake would increase, and the leaves would develop normally. To check his theory, he increased the resistance of roots by forcing a plant, growing with stored water, to rely on one seminal root. Single-rooted plants used less water before anthesis than did normal plants and they produced 2 times more grain. He also observed that the xylem of seminal roots is dominated by one large metaxylem (outer part, and last to be formed, of primary xylem) vessel whose diameter probably determines hydraulic resistance.

Meyer and Alston (53) also altered the number of seminal roots of wheat plants depending on subsoil water (water more than 45 cm below surface of soil). Grain yield was positively related to the amount of water available after anthesis. With wet subsoil, yield increased with increasing number of seminal roots per plant. With drying subsoil, yield decreased. The results showed that wheat yield depended on the interaction of the root geometry and the availability of the subsoil water. The depth of subsoil water storage affects yields. In a 3 year experiment with dryland wheat, Singh and Brar (54) found that deeper storage resulted in greater depletion of stored water and gave higher yield than that obtained when the same amount of water was stored at shallower depths. Some of the shallower depth water would have been evaporated. Deep water would have been transpired.

Meyer and Alston (53) showed that the metaxylem vessel increased in radius along the seminal axis toward the apex from <30 μm near the soil surface to about 45 μm at 1 m depth. Although they found little variation in the number of seminal axes in cultivars, differences occurred in the size of the metaxylem vessels. They concluded that one could control resistance to water flow in the seminal axes by selecting different cultivars.

Rate of Root Extension

Figure 15.3 from the study by Gregory, et al. (47) shows the maximum penetration of the root system of "Maris Huntsman" winter wheat. The mean extension rate during the winter was 6 mm day^{-1} and, during the spring was 18 mm day^{-1}. In their study, by the beginning of April, roots were present to 1 m and reached 1.9 m by the end of May. Root extension beyond 2 m (the limit of sampling) was unlikely because the quantity of roots in the 1.9–2.0 m layer did not increase after the end of May and the roots in the 1.9–2.0 m layer were always unbranched.

Inheritance of Rooting Depth

Heritable differences in root growth of wheat occur, according to Monyo and Whittington (55), Hurd (56), and Townley-Smith and Hurd (57). Robertson, Waines, and Gill (58) obtained greater seedling root number by selecting seeds with higher weights, but MacKey (59) did not. MacKey (59) did find that wheat plants with different top weights could be selected to get variations in root weights.

In Saskatchewan, Canada, Hurd and co-workers (57, 60) produced two high-yielding durum cultivars with extensive root systems, derived from the drought-resistant durum cultivar "Pelissier," which also has an ex-

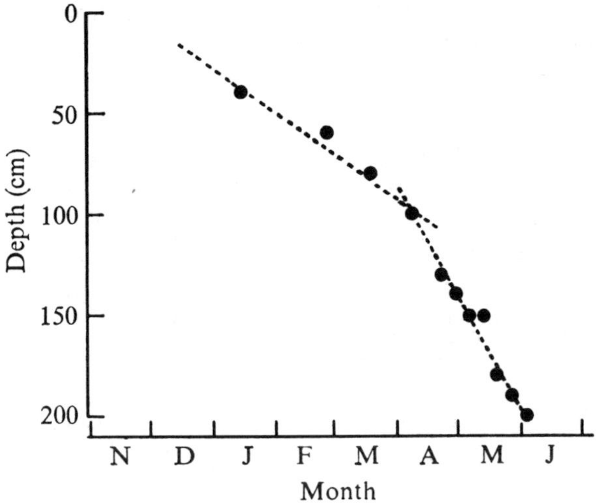

Figure 15.3. The maximum depth of penetration of the root system. The lines represent the mean extension rates during the winter (6 mm day^{-1}) and spring (18 mm day^{-1}). [From Gregory, McGowan, Biscoe, and Hunter (47), p. 96, with permission.]

tensive root system, especially at lower depths. In Zagreb, Yugoslavia, Tomasovic (61) studied 40 winter wheat genotypes, 25 of which showed highly significant differences in number of primary (seminal) roots. The cultivar "Vigorka" had the largest number of primary roots (mean = 5.7) and the cultivar "Centurk" had the smallest number (mean = 3.3). Centurk is drought sensitive and yields poorly under dry conditions (62). Aamodt and Johnston (63) reported that the drought resistant spring wheat cultivars "Milturum" and "Caesium" possess a more highly branched primary root system than the nonresistant spring wheat cultivars "Marquis" and "Reward." Tomasovic (61) concluded that the number of primary roots is an important quantitative characteristic of the wheat root system and it might be used in breeding programs to develop high-yielding genotypes for drylands. Differences in root growth, however, may be brought out only under conditions of water stress (38).

In Victoria, Australia, O'Brien (64) found wheat genotypes varied in their pattern of root development. Length of nodal roots differed after 7–8 weeks of growth. He suggested that root angle of seminal roots could affect the uptake of water and nutrients in the top 10 cm of soil. Wide-angled genotypes would have a greater total seminal root length in the top 10 cm than narrow-angled genotypes and, hence, would take up more water and nutrients per plant. He cited work by Hudig (65) who reported that the deepest root axis was that of the radicle and that the radicle of wide-angled genotypes penetrated more deeply than that of narrow-angled genotypes. The root tips of narrow-angled genotypes were closer together than those of wide-angled genotypes. The relationship between root angle and depth of penetration suggests that simultaneous selection for both is possible.

Root Resistance

Total root resistance is the sum of axial and radial resistances (66). The axial resistance in roots is often considered less than the radial. In other words, it is harder for water to flow from the outside of the root to the xylem in the stele than for water to flow up the xylem (48, 66, 67). The ratio of axial to radial resistance determines the optimum length of a root (68). Even though considered negligible (48, 66, 67), axial resistance can be significant in wheat (52, 69, 70) and in other grass species (71) and soybean (72).

Meyer et al. (70) used a model, which incorporated radial and axial resistances, to describe water uptake by the seminal roots of wheat. They grew wheat that depended on subsoil water extracted from below 45 cm. The plants had one, three, or five seminal roots. Axial resistance was calculated to be the dominant resistance within the soil–root system. The size of the metaxylem vessel in the seminal axes affected water uptake. For example, the flow rate at a subcrown potential of -15 bars would be increased by 30% in a plant with three axes without changing the root length, if the radius of each metaxylem vessel was increased by 3 μm.

Meyer and Ritchie (73) covered a leaf near the bottom of sorghum (*Sorghum bicolor* L. Moench) plants and found that its potential, with no flux through the leaf, was similar to the potential of roots near the crown. So, there was no head loss between the soil surface and the crown; and, knowing the water uptake by the root and the difference in potential between the leaf and the root, they calculated a resistance for the flow of water between these two points, which they called the connection resistance. They were unable to determine the connection resistance of roots that were not near the crown. Connection resistances have not been measured for wheat. Seaton and Landsberg (74), however, used an electrical-analog model to calculate resistance to water movement through the roots of wheat growing in small weighing lysimeters. Calculations were based on hourly measurements of transpiration rate, leaf water potential, and water uptake from three soil layers, determined from measurements of soil water potential at three depths. Total root resistance ranged from 63.9 to 627.3 bars sec^{-1} mm^{-3} (compared to stem resistances which ranged from 24 to 70 bars sec^{-1} mm^{-3}), and was inversely related to flow rate through the main roots. This indicated a constant potential drop (soil minus stem base) of about 10 bars, regardless of soil type. Radial root resistances ranged from 4.6×10^4 to 4.2×10^6 bars sec^{-1} mm^{-1} and were inversely related to the rate of water uptake from the soil. Axial resistances had little effect on the total resistance.

They could not explain the fall in root resistance with increase in transpiration.

Methods need to be developed to determine water potentials of wheat roots that are not near the surface of the soil. Perhaps techniques of So (72), who used a root psychrometer to get root potentials, could be used. Studies should emphasize situations where water potential in the upper layers is higher than that in the lower layers, which occurs after rain or irrigation (68). Most work with wheat roots has been done with upper dry layers and wet lower layers (53, 70, 75–77). Measurements need to be made of water potentials of wheat cultivars that differ in drought resistance. Cultivars with small vessels might have higher connection resistances (73) and, therefore, be better adapted to dry conditions because they use less water.

As stated previously, Meyer and Alston (53) saw that the diameter of root vessels of wheat increased with depth. A metaxylem vessel which changes in diameter from 30 μm at the soil surface to 45 μm at 1 m depth, carries, by Poiseuille's law, 5 times the water at 1 m depth as at the surface (78). Wheat roots absorb water from the top part of the soil first, even if deep roots are in wet soil (75). If large vessels lead to faster water flux, as suggested by Passioura (52) and Meyer et al. (70), why do not wheat roots take up water at depths before they take it up from the top, assuming a uniformly wet soil, since vessels are larger at the bottom of the root system than at the top? One reason is because water potential changes with depth (75). Another reason may involve membrane permeability, and, consequently osmotic adjustment (ability of a plant to lower its osmotic potential). New roots produced near the surface of the soil might have a better capacity to adjust osmotically than deep roots because oxygen is more available in the soil surface. Models of root growth (66, 79) neglect effects of movement of salts between the xylem sap and surrounding tissues because osmotically induced flow has generally been regarded as small (48). Changes in osmotic potential of xylem sap could lead to unexpected water exchange between the xylem sap and surrounding tissues. These exchanges would modify flow velocity along the length of the vessel (78).

Reduction of Soil Water

Gregory et al. (50) measured volumetric soil water content and soil water potential of a winter wheat crop during the 1975 growing season in England. Almost no rain fell between mid-May and mid-July and the soil dried continuously until the potential was less than −20 bars to a depth of 80 cm. Generally, water uptake decreased throughout the period of measurement from a maximum of 2.5×10^{-3} to a minimum of 0.66×10^{-3} ml water cm root^{-1} day^{-1}. A few deep roots were important in taking up water during the dry season. When rain came in mid-July water uptake from the top soil increased and uptake from lower layers decreased. These results indicated that, if water is available in the topsoil, then the plant will preferentially use water higher in the profile. Little production of new roots was observed in the topsoil after the mid-July rain, suggesting that, even after severe soil drying, roots take up water.

GROWTH OF CANOPY

Dry Matter Accumulation and Physiological Stage

Height

In an experiment to determine differences between winter wheat plants on the north and south side of irrigation beds in western Oklahoma, height was measured (80). Five cultivars ("Centurk", "Scout 66", "Tam W 101", "Triumph 64", "Vona") grew on ridges both with and without irrigation. Rainfall during the growing season was 28 cm. Irrigated plants received about 38 cm irrigation water plus rain. Height was measured between March, before culm elongation, and June (harvest). Irrigated plants were the tallest and at harvest averaged about 40 cm taller than plants grown without irrigation water (Fig. 15.4).

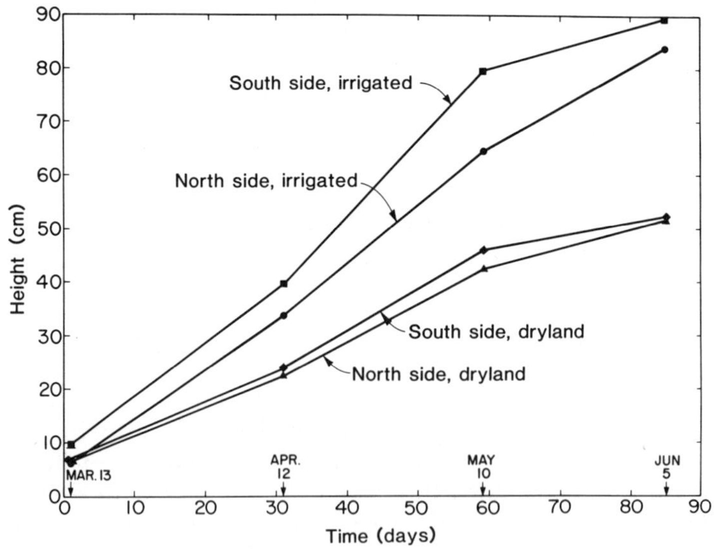

Figure 15.4. Height of irrigated and dryland winter wheat in north- and south-facing rows on east–west ridges. [From Kirkham (80), p. 273.]

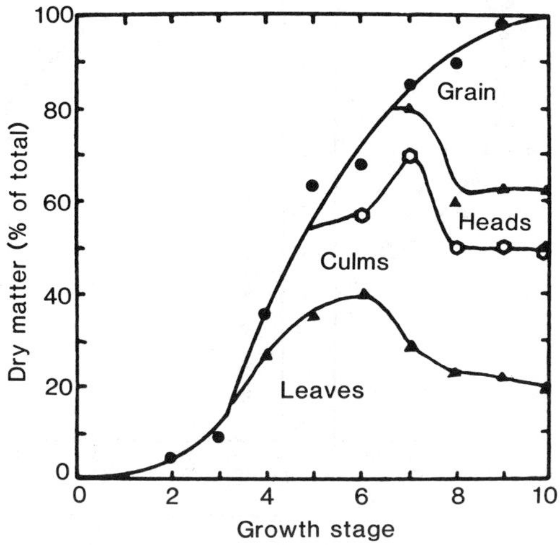

Figure 15.5 Distribution of dry matter in various parts of winter wheat at different growth stages. [From Waldren and Flowerday (20), p. 396, with permission.]

Culm, Leaves, and Grain

Figure 15.5 shows dry matter accumulation as a function of the Waldren and Flowerday (20) growth stages for the winter wheat cultivar "Centurk" grown 3 years near Lincoln, Nebraska (20). Figure 15.6 shows dry matter accumulation, as a function of the Feekes (15) growth stages, for "Centurk" winter wheat grown 3 years at four locations in Kansas (81). Thill et al. (82) report a maximum leaf area index (*LAI*) for "Nugaines" soft white winter wheat grown in Washington varying from about 5 to 8, depending on time of planting. Wheat that was planted early (September) had a higher *LAI* than wheat that was planted late (October). In a dry year, however, Ramsay and McCloud (83) in Florida found that "Holley" winter wheat reached only a maximum *LAI* of two.

Figure 15.7 (84) compares rates of dry matter accumulation for two cultivars of winter wheat, "Tam W 101" and "Vona," grown for 1 year under irrigated or water stressed conditions at Bushland, Texas. For the stress treatment (Fig. 15.7), plants were not irrigated at spikelet differentiation (floral initiation) to late boot heading. The adequate water treatment was not stressed throughout the life cycle. Seeds were planted at three different times: 15 September, early; 12 October, normal; and 7 November 1977, late. With adequate water, plants showed a high accumulation rate of 25 g m^{-2} per day from spring tillering until past flowering. For normal and late plantings, accumulation rates with adequate water averaged 17 g m^{-2} per day from the time of early spring growth to physiological maturity. Accumulation rates for the stressed treatment ranged from 9 to 12 g m^{-2} per day.

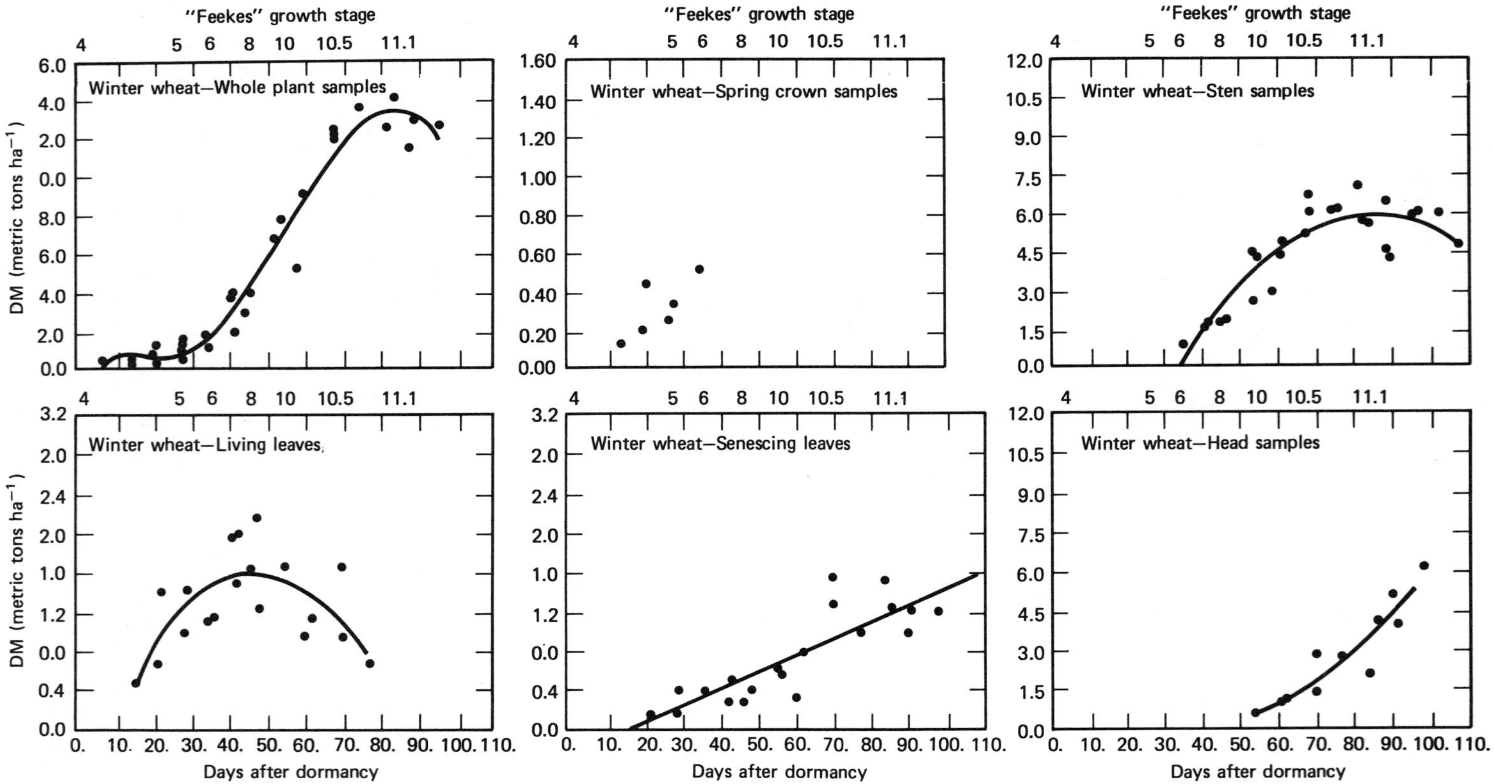

Figure 15.6. Dry matter accumulation by winter wheat plants during the spring. [From Karlen and Whitney (81), p. 283, with permission.]

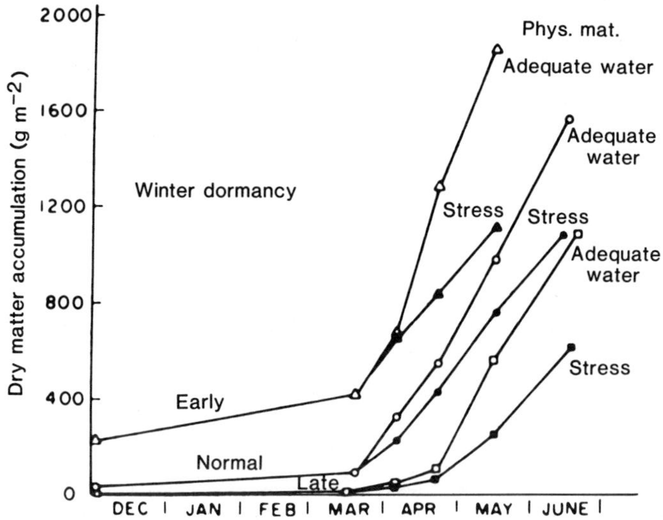

Figure 15.7. Dry matter accumulation above crown level to time of physiological maturity with and without water stress. Planted early (15 Sept.), normal (12 Oct.), and late (7 Nov.). Data are averages of TAM-101 and Vona. [From Musick and Dusek (84), p. 51, with permission.]

Sippola et al. (85), Campbell and Davidson (86), and Karlen and Whitney (81) give data on nutrient uptake (N, P, K, Ca, Mg, Cu, Mn, Zn) of different plant organs (roots, culms, leaves, heads) as a function of stage of development.

Water Relations and Photosynthesis

Frank et al. (87) stressed wheat plants by withholding water. They stressed the fifth leaf at the tillering stage and the flag leaf at the heading, flowering, and grain-filling stages. Recovery of photosynthesis after rewatering was related to recovery of stomatal diffusion resistance at tillering and heading (Fig. 15.8). Photosynthesis never, however, fully recovered to prestress levels, as stomatal resistance did. Photosynthesis of plants stressed at grain filling showed no sign of recovery. Starting 7 days after spike emergence, Johnson et al. (88) withheld water from wheat for up to 7 days. During this time, flag leaf water potentials fell as low as −33 bars and rates of net photosynthesis and transpiration decreased linearly with decreasing flag leaf water potential. Net photosynthesis was zero at leaf water potentials of −31 to −33 bars. Flag leaf transpiration was zero at −28 bars and spike transpiration was zero at −31 bars.

Other than the studies by Frank et al. (87) and Johnson et al. (88) little quantitative data exist for photosynthetic rates of wheat under water

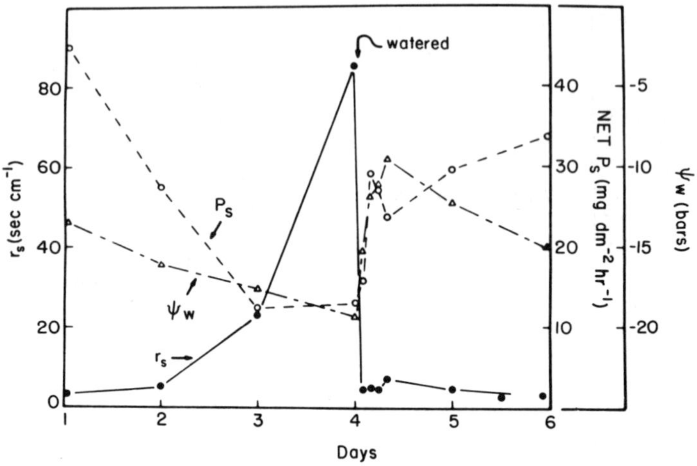

Figure 15.8. Leaf water potential (ψ_w), net photosynthesis (P_s), and stomatal diffusion resistance (r_s) of the flag leaf of wheat at heading when subjected to a soil drying period at 18°C and allowed to recover. Nonstressed plants remained constant for ψ_w, P_s, r_s (like day 1) throughout the 6 day cycle. [From Frank, Power, and Willis (87), p. 778, with permission.]

stress, and corresponding plant water potentials. Sandhu and Horton (89, 90), however, have published information for oats (*Avena sativa* L. "Jaycee"). They stressed oats for 9–11 days at the boot stage, at anthesis through early grain formation stage, and at both stages in combination. Stress at both booting and at anthesis through early grain formation caused the net photosynthetic rate to drop to 20% of its prestress level. Similarly, severe water stress at all stages of development caused about an 80% decline in photosynthetic rate. The plants stressed during early boot stage alone, but adequately watered thereafter, exhibited higher photosynthetic rates and profuse tillering, compared to control plants. But yields were not increased in these plants. The phenomenon needs to be studied to see if the increased photosynthetic rates after release from stress might be manipulated so that *yields* are increased, too.

It is extremely unlikely that the C_4 pathway responsible for the high photosynthetic rate of many tropical species exists in wheat (91). Therefore, it is essential to breed for enhanced photosynthetic efficiency within the C_3 pathway under dryland conditions. Several reports suggest that this may be a fruitful line of research because wheat cultivars do vary in their photosynthetic rate (92–100).

Evapotranspiration

Ryhiner and Matsuda (101) present calculated potential evapotranspiration and measured actual evapotranspiration in relation to crop density

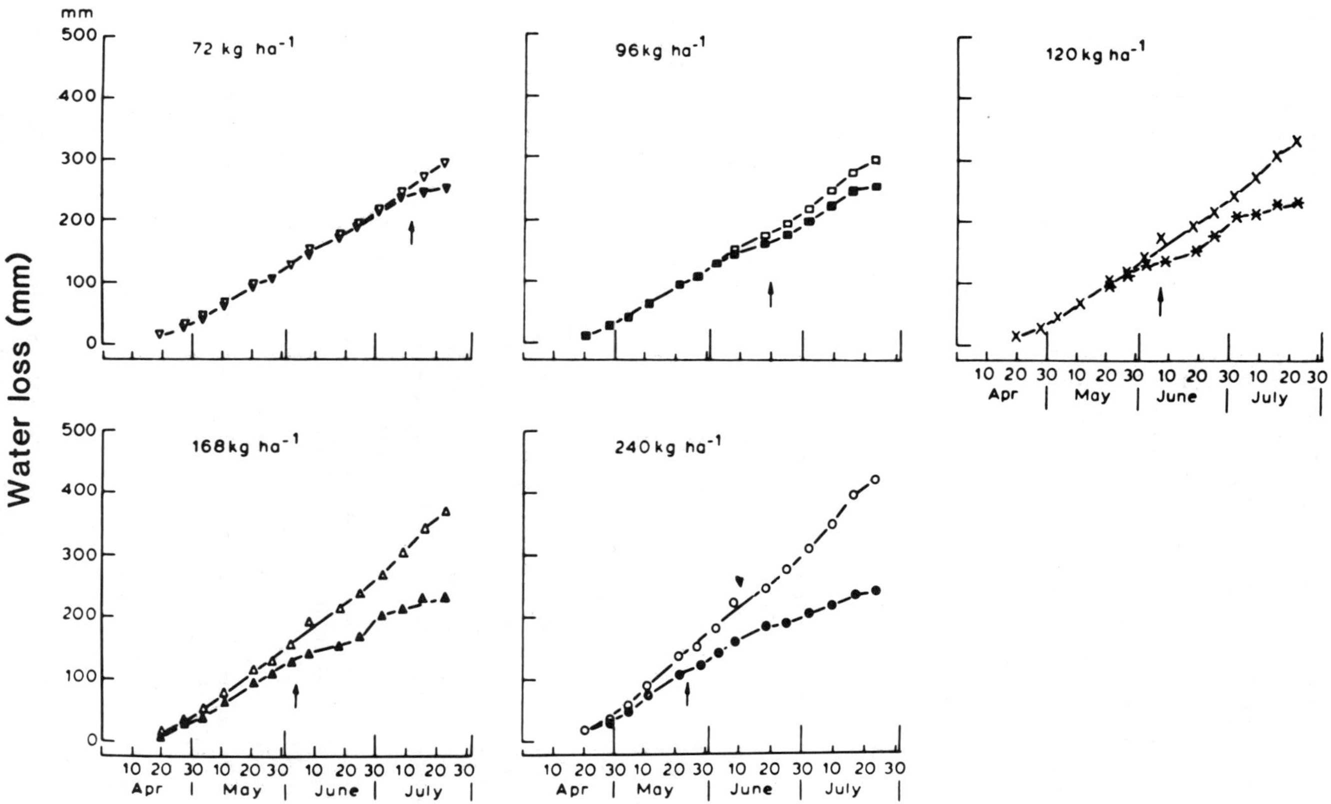

Figure 15.9. Potential (E_p) and actual (E_t) evaporation of winter wheat during the growing season of 1971 for the given sowing densities. Open points represent E_p and closed points represent E_t. Arrow indicates start of serious water stress. [From Ryhiner and Matsuda (101), p. 206, with permission]

for the winter wheat cultivar "Tadovora" grown in the Netherlands in 1970–1971 (Fig. 15.9). With increasing density, stress developed at earlier dates. Potential evapotranspiration increased with increasing density. But actual evapotranspiration was about constant because of limited water supply. They reported that, since severe stress at high plant densities occurred before anthesis, grain production increased slowly with increasing crop density. Under even more severe stress, they predicted that grain production would drop with increasing densities. Early work on wheat growth at different densities is described by Clements et al. (102) and by Russell and Watson (103). Jensen (104) summarizes evapotranspiration data for wheat.

TRANSPIRATION

Regulating Water Loss

Stomata

Stomatal frequency is greater on the adaxial (upper) than on the abaxial (lower) surface of wheat leaves (105, 106). Sheriff (107) reported an average stomatal pore depth of 19 μm; the substomatal cavity averages 28.6 μm deep and 16.3 μm wide. He observed no ectodesmata (strands of cytoplasm that penetrate the cell wall on the outside surface of the plant). Stomatal frequency decreases in relation to position of leaf insertion on the culm, and the top leaf has the highest frequency (87, 105, 108). Barley (*Hordeum vulgare* L.) stomata also decrease in frequency progressively from the flag to the lower leaves, with flag leaves having about twice as many stomata per square millimeter as the basal leaves (109). Salim and Todd (110) found stomatal frequency in barley leaves was nearly twice, and stomatal length was about one-half, that of wheat, when both were grown under well-watered conditions. Barley is more resistant to stress than wheat. Differences in stomata appear to explain partly differences in drought resistance. Consequently, several people have suggested that stomata be studied in programs designed to screen plants for drought resistance (111–114). A selection for a low stomatal frequency, however, does not necessarily result in a reduction in crop water use or improved drought resistance. With barley, Jones (115) found correlated changes in stomatal size (larger) and green leaf area (larger) which had effects opposite to that desired (reducing water use). Jones's results (115, 116) emphasize the need to apply selection pressure on several characters at once to eliminate compensation observed when selection is done for a single character.

O'Toole and Chang (42) of the International Rice Research Institute in the Philippines feel that evaluation of stomatal behavior as an indicator of drought stress or as a tool for screening for drought resistance is problematic because environmental variability makes repeatable screening procedures difficult. However, at CIMMYT (95) in Mexico, stomatal resistance is being used in the drought-screening program. Leaf permeability has a highly significant positive correlation with grain yield of wheat and shows promise as a selection criterion. At CIMMYT, results from the resistance porometer are less encouraging than those from the air flow porometer. Fischer et al. (117) describe how the air flow porometer can be used to take rapid measurements, necessary in any large-scale screening program.

Since transpiration rate usually is directly related to stomatal aperture (118–121), stomatal resistance measurements should be carefully made and studied to find wheat cultivars adapted for dryland conditions. Cultivars known to be drought resistant in the field do have a higher stomatal resistance than those considered to be drought sensitive (122, 123). Cultivars with low transpiration rates might be indirectly identified by finding ones with thick cuticles (42), leaf-rolling capability (42, 124), no trichomes (125), low silicon contents in the glumes (126), or those with high frost resistance (127).

In the study by Frank et al. (87), in which water was withheld from wheat at tillering, heading, flowering, or the grain-filling stage, stomatal closure of stressed plants was affected by both leaf position and age. Closure occurred at −13 bars on the fifth leaf at tillering and at −17 bars on the flag leaf at heading. The result was conservation of water during the vegetative stage.

Morgan (128) in Australia found stomata more open postanthesis than preanthesis in three wheat cultivars grown in soil or a peat and sand mixture under growth chamber conditions. He withheld water preanthesis, at anthesis, or postanthesis. In contrast, in a field study of ten cultivars of spring wheat under two soil water regimes (irrigated or droughted), Jones (129) in England found stomata most open prior to anthesis. Drought was imposed by placing rainwater gutters between the crop rows to intercept part of the rain. Soil water fell to about 50% in the droughted soil. Perhaps Jones's wheat was not stressed to a great extent because Abdel-Samie and Talha (130) found no yield reduction if soil water was kept at 50% or above. Jones did observe a strong negative correlation between stomatal conductance and yield for the period 7–14 days before anthesis. In both studies by Morgan (128) and Jones (129), leaf water potentials decreased with stress. Morgan saw no apparent effect of leaf water potential on stomatal resistance until potentials reached −18 bars. There was increasing stomatal closure as leaf water potentials decreased to about −30 bars. Stomata did not open at lower values. He found the conductance of the adaxial (upper) surface was less variable and more sensitive than that of the abaxial (lower) surface. The upper surface of wheat leaves has a higher conductance than the lower surface (121).

Ehrler et al. (131, 132) in Arizona have used drought-induced stomatal closure, which causes a rise in canopy temperature, to monitor plants for water stress by using infrared thermometers. They demonstrated that crop canopy temperature responded to changes in leaf water potential, and, therefore, felt that temperature measurements, made by remote sensing techniques, could be used to schedule irrigations.

Often no relation is found between stomatal resistance and yield under dryland field conditions (95, 105, 128, 129, 133). Maturity may have an effect. Miskin et al. (112) reported that barley lines with few stomata tended to mature earlier than lines with more stomata. Early maturing lines are often more drought resistant because they avoid late-season droughts (42).

Even though there may be a genetic relation between date of anthesis and stomatal frequency, water stress has an effect per se on time of anthesis regardless of genetic makeup. Angus and Moncur (134) watered wheat until floral initiation and then stressed the plants to −5, or −10, or −15, . . . , −40 bars, and then rewatered. They found that with severe water stress (−25 to −40 bars) there was about 1 day's *delay* in anthesis for each day's delay in watering. With moderate stress (down to −15 bars), dates of anthesis were *ahead* of well-watered controls. They concluded that water stress was an important factor affecting the phasic development of wheat, and, although it was less important an influence than temperature or day length, it should be included in developmental models if they are to predict the timing of wheat within an accuracy of a few days. Workers at CIMMYT (95) in Mexico have observed a highly significant inverse correlation (−0.40**) between anthesis date and yield of droughted wheat.

Leaf Water Potential

Leaf water potential is not uniquely related to stomatal resistance (119, 135–141). In a greenhouse study, Seaton et al. (139) observed that the minimum value of leaf water potential attained each day declined through a drying cycle, but stomatal resistance was not uniquely related to leaf water potential. Initial stomatal closures occurred at leaf water potential values which decreased from −11 to −25 bars as drying progressed. This adaptive mechanism was related to changes in osmotic potential of the leaves. Transpiration rates fell below those of well-watered plants when the average soil water potential in the root zone was reduced to −1 to −5 bars. Under field conditions, Fischer and Kohn (142) found a 40% fall in actual evapotranspiration of wheat relative to potential evapotranspiration at a mean soil water potential of about −10 bars. Millar and Denmead (137) said stomatal closure was induced at the same turgor potential in leaves of wheat at all positions on the stem, but because of a different relation between turgor potential and osmotic potential for each leaf, the corresponding critical water potentials varied from −7 bars for leaves at the bottom of the stem, to −14 bars for the next highest leaves, to −19 bars for top leaves. The different relations between turgor potential and osmotic potential resulted from a systematic decrease of osmotic potential with leaf position from the bottom to the top of the stem.

Osmotic Potential

Some plants wilted at tillering have yielded more grain than plants wilted at later stages of growth. Salter and Goode (143) reported an experiment in Russia by Moliboga in 1927 where "Marquis" spring wheat was allowed to wilt at one of the following stages of growth: tillering, shooting, earing, or when grain was milk-ripe. If the grain yield from the control plants was taken as 100, the yield of plants wilted at tillering was 137, at shooting 50, at earing 34, and at milk-ripeness, 77. Apparently, the plants "hardened" in the early developmental stage and this permitted them to grow and yield better than plants which were always well watered.

The hardening process may be related to osmotic adjustment, which appears to be an important mechanism for adaptation to water-limiting conditions (138, 144–148). The osmotica may be internally generated [for example, proline (149), soluble carbohydrates (150), amino acids (151)], or the plants might take up solutes external to the root. Germinating seeds or very young seedlings have a great capacity for osmotic adjustment (146), which may be one reason why they are resistant to water stress. For example, Shcherbakov (152) grew wheat during the vegetative period in pots of soil at 70 or 30% of total saturation. The growth of the first three leaves of plants grown at 30% saturation was the same as that of the first three leaves of plants grown at 70% saturation. Later-forming leaves of plants at 30% saturation were severely inhibited in growth.

Whiteside (144) reported the osmotic pressure of two cultivars of potted spring wheat of unspecified resistance to drought under field conditions. There was little difference in osmotic pressure between the cultivars. Apparently, no recent work reports the osmotic potential of different cultivars of wheat exposed to drought conditions at tillering. This might be a useful line of research if one wants to identify drought resistant cultivars. The resistant cultivars might be able to lower their osmotic potential more than the susceptible ones. The water-retaining capacity of cut tissue has been used to find, with success, drought resistant cultivars (153). Perhaps the technique works, in part, because there is a difference in the osmotic potential of the cultivars. The drought resistant cultivars probably have a lower osmotic potential than the sensitive ones and the resistant ones are able to retain water longer than the ones that wilt rapidly. Morgan (154) found wheat cultivars varied in their "osmoregulation" (osmotic adjustment). He did not, however, relate differences in osmotic adjustment to differences in drought resistance.

Relative Water Content

Relative water content, also called relative turgidity, is defined as follows: (fresh weight − dry weight) ÷ (turgid weight − dry weight). Connor (51) observed that relative water content remained high (between about 90 and 100%) until 40 days postflowering, at which point noon leaf relative water content fell to about 60%. Morgan (155) saw that the number of dead spikelets increased with decreasing relative water content. Fischer and Kohn (156) said that when dawn leaf relative water contents reached about 75%, leaf senescence and desiccation soon followed. Therefore, if it is impossible to measure photosynthesis, relative water content might be obtained to determine the end of the assimilation period. Fischer (157) found relative water content was directly related to soil water content and suggested that relative water content might also be used to indicate soil water content. Since relative water content is related to transpiration rate (120), it has been used as a method to show differences in drought resistance among cultivars (158–160). But it is a time-consuming method, and, consequently, of limited value.

Drought Resistance and Yield

Water stress at spike initiation causes the greatest reduction in yield (89, 90, 143, 157, 161–172). Fischer (157) found the most sensitive stage was 15 days before anthesis. The effect was mainly on number of grains per ear. This period (15–5 days before ear emergence) corresponded to that of rapid elongation of the ear, anthers (male flower parts), and carpels (female flower parts). Fischer said such an effect of stress has been associated with occurrence of meiosis in pollen mother cells (in the anthers). The stressed plants in his study had abnormal anthers, but normal female parts. Under nonstressed conditions, the male flower parts differentiate before the female flower parts (173).

Grain yield is determined by ear number, grain number per ear, and grain weight. According to Asana and colleagues (174–176), under adequate soil water, ear number has the greatest effect on yield. Under drought, grain number per ear and grain weight are most important. Drought in the vegetative stage reduces the potential number of heads (tillers). Drought at flower formation results in reduced grain number, while drought at later stages results in reduced kernel size because the number of grains has already been established (152, 174–179). Selection is being based on kernel weight as the most important yield component which can help increase wheat production on dryland in Oklahoma (180).

Bingham (181) saw that wilted wheat produced supernumerary florets. [Bonnett (173) defines supernumerary spikes as small spikes which develop at the base of the main spike.] This showed to Bingham that number of florets formed is not limited by the number of initials and may increase until a late stage of ear development. Perhaps it might be possible to find cultivars that produce more of these supernumerary florets than other cultivars when water is lacking. These florets might increase the grain yield.

Skidmore et al. (182) presented stomatal resistance data for six cultivars of winter wheat (Fig. 15.10). "Pronto" is more drought resistant than the other cultivars (Teare, personal communication) and had the highest stomatal resistance in the open field (Fig. 15.10). The study showed that stomatal resistance of high-yielding wheats was lower than that of more drought resistant wheats on soils with the same available water. Many other experiments report that the amount of water evapotranspired is directly related to growth and yield (26, 142, 183–190). The older wheat cultivars are often more tolerant of water stress than newer cultivars. These older cultivars, however, also have a lower yield capacity (191). Therefore, it seems impossible to breed for high yield and drought resistance at the same time. Singh and Tsunoda (99) found that net photosynthetic rate and transpiration rate of wheat were linearly related with an *r* value (correlation coefficient) of 0.98. The literature suggests the following rule; Increasing stomatal resistance conserves water, but productivity is reduced because of reduced photosynthesis.

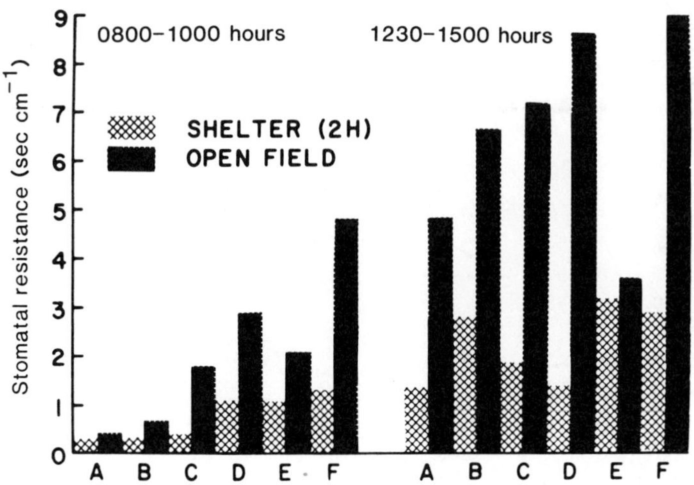

Figure 15.10. Shelter versus open-field stomatal resistance of six wheat varieties for two periods of a day (22 May 1972) having relatively high potential evaporation. Wheat varieties were (A) "Satanta," (B) "Blueboy," (C) "Caprock," (D) "Parker," (E) "Shawnee," and (F) "Pronto." Data presented in order obtained. [From Skidmore, Hagen, Naylor, and Teare (135), p. 502, with permission.]

One exception to this rule appears to be the use of the antitranspirant, phenylmercuric acetate (PMA). Christ (192) studied its effect on transpiration and growth of wheat leaves. He also worked with another transpiration inhibitor, the plant hormone abscisic acid (ABA). Figure 15.11 shows the transpiration rate, and Fig. 15.12 shows the elongation rate, of wheat leaves treated with PMA and ABA. Both treatments reduced transpiration, but only the ABA treatment reduced elongation. Maintaining yields under drought by the use of PMA appears unlikely because PMA is toxic to plants (193, 194).

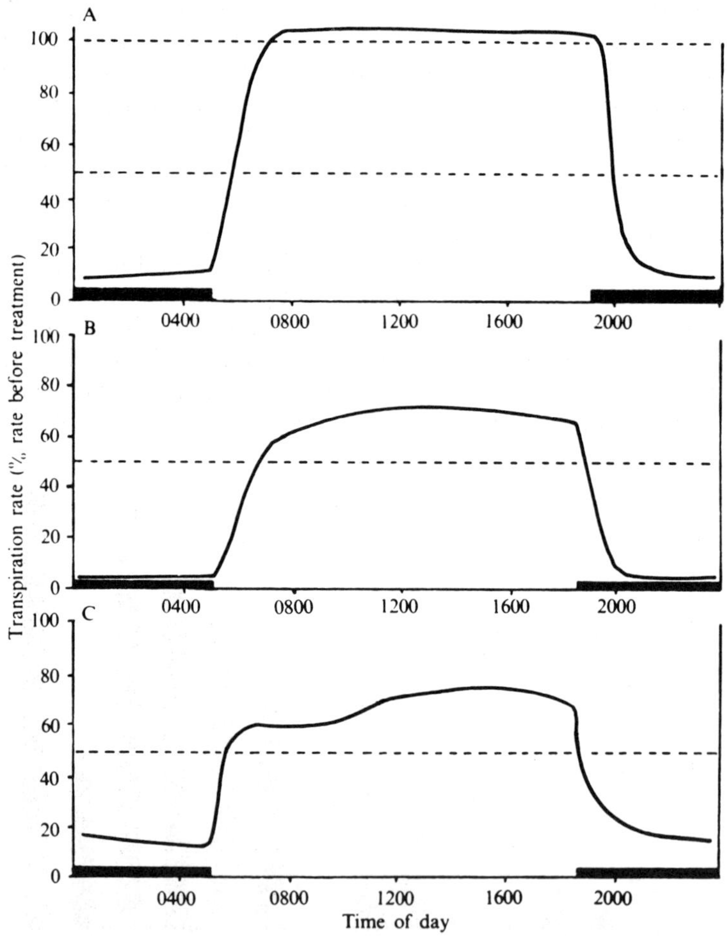

Figure 15.11. Effect of the transpiration inhibitors ABA and PMA on the transpiration rate of wheat leaves. The transpiration rate is recorded as percentage. 100% = transpiration rate of the same leaf (third from base) before treatment. The inhibitors were applied to the leaf during the previous day. (A) Untreated; (B) ABA (100 μg ml^{-1}); (C) PMA (100 μg ml^{-1}). [From Christ (192), p. 616, with permission.]

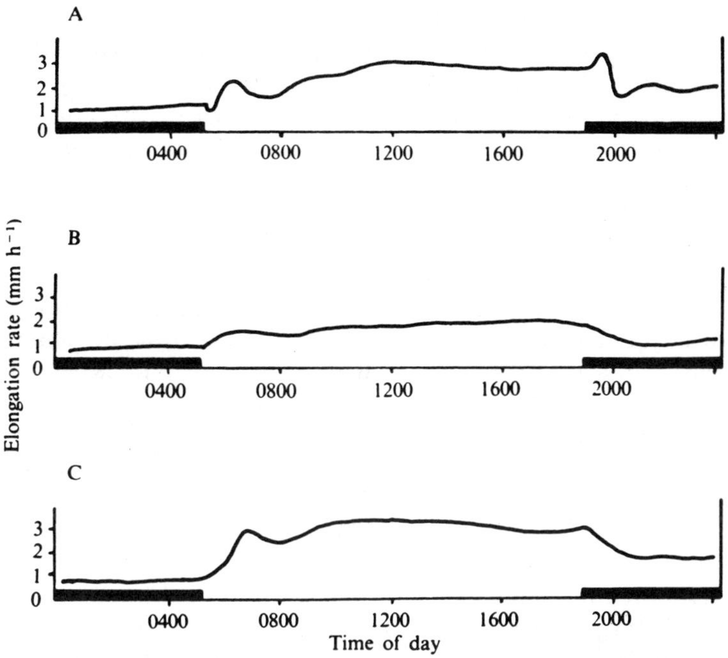

Figure 15.12. Effect of the transpiration inhibitors ABA and PMA on the growth rate of the third wheat leaf, counted from base. The inhibitors were applied to the leaf during the previous day. (A) Untreated; (B) ABA (100 $\mu g\ ml^{-1}$); (C) PMA (100 $\mu g\ ml^{-1}$). [From Christ (192), p. 615, with permission.]

Translocation and Water Stress

There is great debate concerning how much material below the flag leaf is mobilized as a result of water stress. Bidinger et al. (195), Rawson et al. (196), and Thorne (197) feel little material is translocated. Asana and Basu (162), Gallagher et al. (198), Johnson and Moss (199), and Wardlaw (200) believe translocation can still occur after photosynthesis stops and stem reserves may be important under dryland conditions. Thorne (197) reported that under nonstressed conditions, when wheat was supplied with $^{14}CO_2$ at ear emergence, the percentage of radioactivity in the plant at maturity recovered in the grain was 55% when the flag leaf was treated, 5% when the leaf below was treated, and only 2% when the next two lower leaves were treated. Many factors affect these percentages, including respiratory losses, environmental conditions, cultivar, and whether or not cultivars are awned (197, 201–203). Similar figures for water stressed wheat apparently have not been published.

Barley and wheat were treated with $^{14}CO_2$ in a study by Rawson et al. (196). Plants exposed to drought during grain growth did not translocate more stored materials to the grain than those plants frequently watered during this stage. These results contradicted those of McPherson and Boyer (204) who showed that prestressed corn (*Zea mays* L.) plants yielded better when exposed to stress than corn plants not previously stressed. Rawson et al. (196) explained the difference in results by pointing out that corn is a C_4 plant not adapted to semiarid conditions as are wheat and barley, C_3 plants. Also, the stress imposed by McPherson and Boyer was not as severe as that imposed by Rawson et al. The most severely stressed corn plants of McPherson and Boyer had leaf water potentials of −20 bars. Frank et al. (87) found the leaf water potential of well-watered wheat at heading was −17 bars, but the potential fell to −43 bars if the wheat had been water stressed. Wheat was operating *normally* in the experiment by Frank et al. at a potential to which corn was greatly stressed in the experiment by McPherson and Boyer. Wheat develops a hollow stem as it matures, while the stem of corn remains solid. The amount of material that can be translocated by wheat, therefore, would be limited. If, under water stress, material is translocated from stems to grain, wheat cultivars with tall or thick stems would have an adaptive advantage over cultivars with short or thin stems. Recently, McNeal and Berg (205) reported that the amount of solidness of a wheat stem is a highly heritable trait and solid-stemmed plants yielded more than hollow-stemmed plants.

Gallagher et al. (198) studied yield of wheat in England in 1976, the driest year since records were begun in 1916. They assumed grain growth was as follows:

$$\bar{W}_g(f) = \bar{W}_g(i) + R_g D_g \tag{3}$$

where $\bar{W}_g(f)$ is the final mean weight per grain, $\bar{W}_g(i)$ is the initial value of $\bar{W}_g$, R_g is the grain growth rate during the linear phase of growth, and D_g is the duration of the linear phase. Increase in $\bar{W}_g$ was assumed to be linear once $\bar{W}_g$ exceeded 6 mg. With drought and the associated warmer temperatures in 1976, they observed a faster R_g. But this was more than offset by a sharp decrease in D_g so that $\bar{W}_g(f)$ was only 37 mg in 1976, compared to 50 mg in both 1974 and 1975, when more normal amounts of rain fell. They felt that under the dry 1976 conditions, the contribution of translocated materials to grain yield was much greater than in the wetter years of 1974 and 1975. They concluded that the results illustrated the potential importance of translocation of material assimilated before anthesis to grain growth when existing photosynthesis is limited by an adverse environment. They also noted that if cereals are to yield well in regions where drought and high temperatures during grain filling are frequent, cultivars must be bred which not only increase the rate, but also the duration, of grain growth. Such characters would maximize yield.

Bidinger et al. (195) disagreed with Gallagher et al. (198). Bidinger et al. (195) studied irrigated and droughted wheat in Mexico and found little stored preanthesis assimilate was translocated to grain under either treatment. Neither Gallagher et al. (198) nor Bidinger et al. (195) reported soil water potentials or plant water potentials. Perhaps the wheat in England was more severely stressed during the driest year on record than the wheat in Mexico, and this might explain the differences observed.

More work needs to be done concerning translocation of materials in wheat under water stressed conditions. How much material is mobilized from the stem during grain development under water stress? At what plant water potential and soil water potential does translocation stop in wheat? Are these potentials lower than for other crops like corn? Yang and de Jong (206) reported that wheat extracted water from soil at −40 bars. Perhaps wheat can take up water from soil at lower potentials than other crops can. Do cultivars differ in their ability to translocate materials from the stem under water stressed conditions?

Morphological studies should be carried out in association with physiological studies. Segovia and Brown (207) found with peanut (*Arachis hypogaea* L.), tall fescue (*Festuca arundinacea* Schreb.), and soybean (*Glycine max* Merr.) that phloem area in leaf bases or petioles was closely correlated with leaf area. Shading affected this correlation. What effect does water stress have on wheat phloem area? Phloem in wheat has been studied and apparently is stable under a variety of conditions (208–210). Little work has been done under water stressed conditions. Xylem should be studied, too. Under dryland conditions, Denmead and Millar (118) showed that more water flowed through stem tissue than leaf tissue when wheat was grown under dryland conditions than when it was grown under irrigated conditions. Does xylem develop differently under dryland, compared to irrigated conditions? Since models of wheat growth often do not predict grain yield as well as vegetative yield (211), it is important to understand the translocation process.

Leaf Potential and Physiological Indicators of Water Stress

ABA and proline accumulate in water stressed plants (212–215). Bengston et al. (216) give data for both compounds. They stressed 6- to 7-day-old wheat seedlings by exposing them in darkness to air of 35% relative humidity for 2.5–20 hr. This treatment resulted in a water potential of −11 bars in the leaves after 20 hr (Fig. 15.13). During the stress period, both ABA (Fig. 15.14) and proline (Fig. 15.15) accumulated in the leaves. The content of ABA increased up to 6 times the normal level during water stress lasting for 20 hr. In a similar treatment, proline increased about 3 times. After rewatering the leaves, the levels of both ABA and proline rapidly declined and reached, 10 hr later, the levels found in

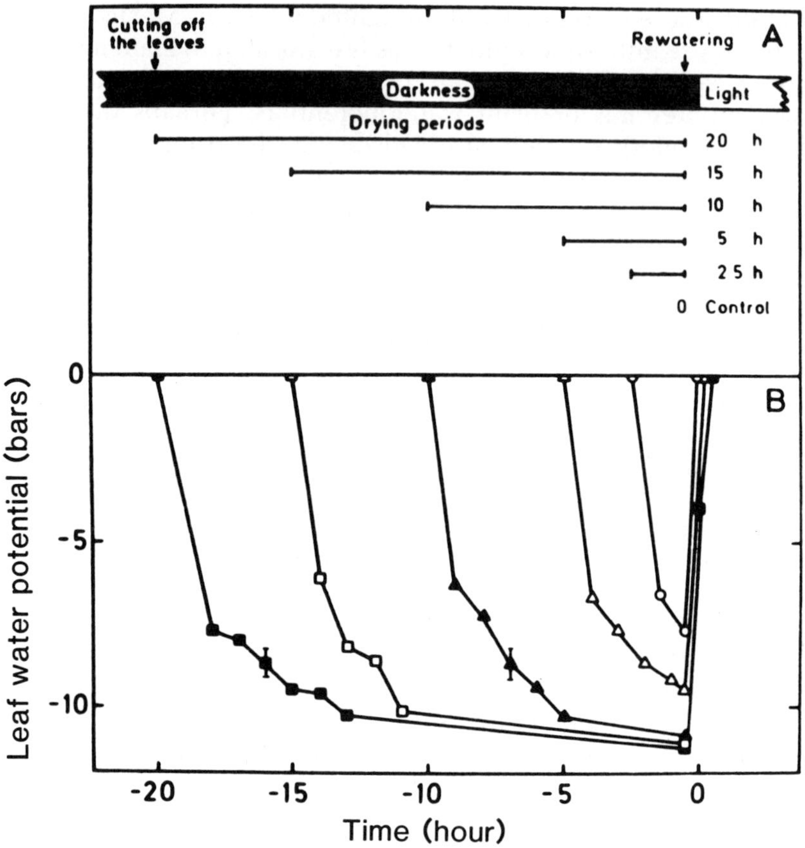

Figure 15.13. The sample treatment (A) and the leaf water potentials of dark grown wheat leaves during water stress (B). (A) Six-day-old leaves were cut just above the seeds. The leaves which were stressed for 20 hr were immediately arranged on wire netting in the climatic chamber. The rest of the leaves were stored at a relative humidity close to 100%, with the cut ends in distilled water, until their drying periods were started. Half an hour before the irradiation, the leaves were allowed to regain full turgor, lying in Petri dishes on filter papers moistened with buffer solution. (B) The symbols represent stress periods of 20 hr (■), 15 hr (□), 10 hr (▲), 5 hr (△) and 2.5 hr (○). The untreated leaves (unstressed) maintained a leaf water potential close to 0 bars (not indicated in the Figure). [From Bengtson, Klockare, Klockare, Larsson, and Sundqvist (216), p. 210, with permission.]

unstressed seedlings. ABA does not appear to affect permeability of roots to water. Pitman and Wellfare (217) found that ABA inhibited salt transport to the shoot by acting on ion transport to the xylem, and not by reducing water flow coupled with salt transport.

The plant hormone ethylene, a gas, also accumulates when plants are stressed (218). Data of Wright (219) give relations between ethylene production and leaf water potential. At a wheat leaf potential of −6.7 bars, 0.23 μl ethylene were produced per 10 g fresh weight after 6.75

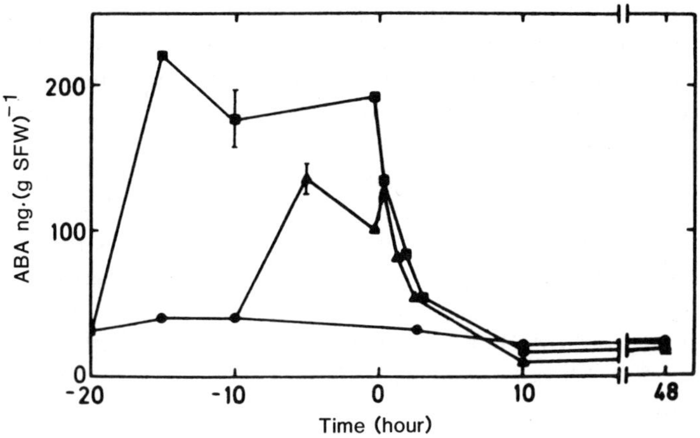

Figure 15.14. The ABA content of dark grown wheat leaves exposed to water stress for different periods of time. Cut leaves which had been exposed to water stress for 20 hr (■), 10 hr (▲), or 0 hr (●) were analyzed for their ABA content during the stress period and after rewatering in light. SFW, saturated fresh weight. [From Bengtson, Klockare, Klockare, Larsson, and Sundqvist (216), p. 210, with permission.]

hr while stressed leaves, which had a leaf water potential of −11 bars, produced 3.30 μl ethylene per 10 g fresh weight during the same time.

Recently, it has been discovered that glycine betaine accumulates in response to salt and water stress (220, 221). Hitz and Hanson (222) reported glycine betaine accumulation for 22 species from the family Gramineae, including wild and cultivated barleys, ryes, and wheats. Water

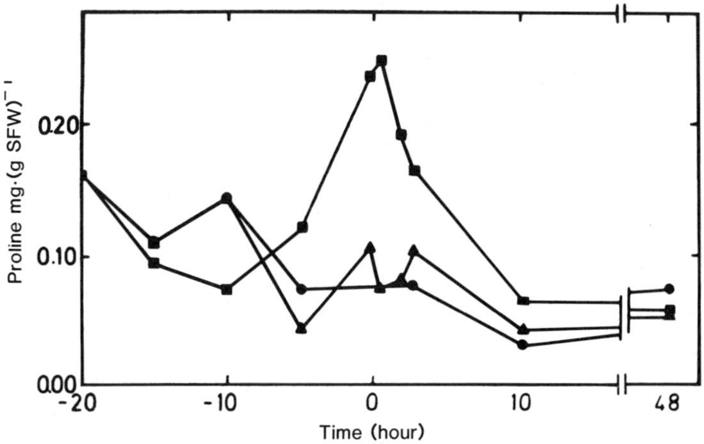

Figure 15.15. The proline content in dark grown wheat leaves exposed to water stress for different periods of time. Cut leaves which had been exposed to water stress for 20 hr (■), 10 hr (▲), or 0 hr (●) were analyzed for their proline content during the stress period and after rewatering in light. SFW, saturated fresh weight. [From Bengtson, Klockare, Klockare, Larsson, and Sundqvist (216), p. 210, with permission.]

stress was imposed by withholding irrigation at the three-leaf stage for 4–6 days. For each species, stress was continued for about 1 day after visible wilting occurred. The content of glycine betaine showed much interspecific variation. All the genotypes except one accumulated glycine betaine during stress. But the only species which accumulated substantial amounts of glycine betaine were from the tribe Hordeae (barleys). They said that, even though glycine betaine has been suggested as an osmoticum during salt stress (217), their results showed that the accumulation of glycine betaine was not of universal importance in the Gramineae. For example, it occurred only to a limited extent in millet (*Pennisetum glaucum* R. Br.), sorghum (*Sorghum bicolor* L. Moench), and upland rice (*Oryza sativa* L.), all of which are adapted to droughty environments.

APPLIED ASPECTS OF IRRIGATION

Critical Water Demand Periods

Figure 15.16 shows the effect of water stress at various growth stages on grain yield reduction (223). A study was done in which uniform amounts of irrigation water were applied to winter wheat at four growth stages. Early or vegetative stage (early), early to late boot (boot), heading to flower (heading), and early grain development (grain fill) (223). With a single water application, yield was 52% of the well-watered treatment if water were applied at the heading stage, but only 25–35% if applied at other stages. With two applications (irrigations at boot plus heading

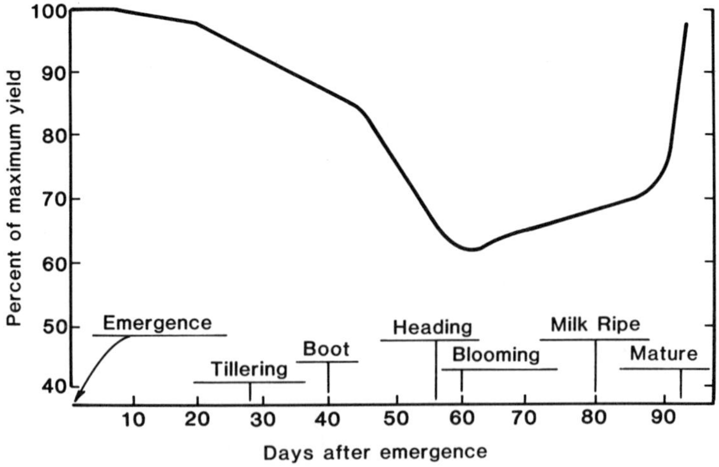

Figure 15.16. Effect of water stress at various growth stages on grain yield reduction in wheat. [From van Brackle (223), p. 23, with permission.]

or early plus heading), yields were about 75% of the maximum, and other combinations with two irrigations were 49–62% of the maximum. With three applications at the early plus boot plus grain fill stages, yield was 60% of maximum. Yield with irrigations at early plus boot plus heading was 75% of maximum. Other combinations (early + heading + grain fill) and (boot + heading + grain fill) produced yields which were 90% of the maximum.

In another study (224), water was applied to wheat at different growth stages. Plots irrigated during grain filling with 4 cm water produced 4800 kg ha^{-1}. Plots irrigated only during the vegetative stage with 5 cm water produced 2900 kg ha^{-1}. All plots received 6 cm irrigation water in the fall after planting. Fall rainfall was 3 cm, and spring rainfall was 17 cm, most of which occurred near heading. Conventional dryland farming (fallow system), used for comparison, produced 1800 kg ha^{-1}.

There is a strong interaction of irrigation and nitrogen on grain yields. In the absence of a late irrigation, Shimshi and Kafkafi (225) found that the optimal nitrogen rate was 120 kg ha^{-1}. But with a late irrigation, the optimal nitrogen rate was 180 kg ha^{-1}. They also saw that lack of nitrogen reduced soil water extraction during the grain-filling stage and that stomatal aperture in the irrigated treatments was larger in nitrogen supplied than in nitrogen deficient wheat.

Hack (226) in the Sudan suggested using a stomatal infiltration technique to time irrigations. He chose the method because erratic rainfall precluded fixed irrigation intervals and a simple procedure was needed to avoid complex instrumentation. The infiltrating liquid was a solution of water and alcohol (isopropanol). Peak scores occurred after rain or irrigation. Low scores indicated that stressed plants needed to be irrigated.

Water Requirement

Larcher (227) and Woolhouse (228) reported that 540 g water are needed to produce 1 g dry matter of wheat (type not specified). Black (229) summarized work of Shantz and Piemeisel (230), who gave water requirements for emmer wheat, two hybrid wheats, 13 cultivars of durum wheat, and 22 cultivars of common wheat. Their average water requirements were, respectively, 517, 534, 542, and 557 g water per g dry matter.

REFERENCES

1. V. A. Johnson, in M. D. Thorne, Ed., *Agronomists and Food: Contributions and Challenges*, American Society of Agronomy, Madison, Wis., 1977, p. 37.
2. M. Y. Nuttonson, *Wheat-Climate Relationships and the Use of Phenology in Ascertaining the Thermal and Photo-Thermal Requirements of Wheat*, American Institute of Crop Ecology, Washington, D.C., 1955, 388 pp.

3. J. Percival, *The Wheat Plant. A Monograph,* Dutton, New York, 1921, 463 pp.
4. R. F. Peterson, *Wheat. Botany, Cultivation, and Utilization,* Leonard Hill Books, London, and Wiley-Interscience, New York, 1965, 241 pp.
5. B. P. Pal, *Wheat, I.C.A.R. Cereal Crop Ser. No. 4,* Indian Council of Agricultural Research, New Delhi, 1966, 370 pp.
6. L. P. Reitz, in K. S. Quisenberry and L. P. Reitz, Eds., *Wheat and Wheat Improvement,* American Society of Agronomy, Madison, Wis., 1967, p. 1.
7. J. W. Schmidt, in M. D. Thorne, Ed., *Agronomists and Food: Contributions and Challenges,* American Society of Agronomy, Madison, Wis., 1977, p. 45.
8. P. E. Waggoner, *Agric. Meterology* **20,** 41 (1979).
9. R. H. Hageman, *Factors Affecting Yield of Cereal Grains via Physiological Processes, Speech Presented at Symposium on Plant Growth Regulators, Potential Use in World Food Production, Plant Growth Regulator Working Group (PGRWG) Meeting, Hot Springs, Arkansas, Aug. 9–11, 1977,* Dept. Agronomy, University of Illinois, Urbana, Ill., 28 pp.
10. L. T. Evans, in L. T. Evans, Ed., *Crop Physiology. Some Case Histories,* Cambridge University Press, London, 1975, p. 1.
11. G. Stanhill, *Agro-Ecosystems* **3,** 1 (1976).
12. C. T. de Wit, in J. J. Landsberg and C. V. Cutting, Eds., *Environmental Effects on Crop Physiology,* Academic Press, London, 1977, p. 365.
13. R. A. Fischer and R. Maurer, *Aust. J. Agric. Res.* **29,** 897 (1978).
14. H. Kutsch, *Der Tropenlandwirt* **80,** 120 (1979) (In Ger., Engl. sum.).
15. W. Feekes, *De Tarwe en Haar Milieu, Vers. XVII Tech. Tarwe Comm.,* Groningen, The Netherlands, 1941, p. 560. cited in (16).
16. E. C. Large, *Plant Path.* **3,** 128 (1954).
17. L. A. Jensen and H. R. Lund, *How Cereal Crops Grow,* N. Dak. State Univ. Coop. Ext. Bull. No. 3, 1971, 39 pp.
18. D. R. Tottman, R. J. Makepeace, and H. Broad, *Ann. Appl. Biol.* **93,** 221 (1979).
19. D. R. Tottman, *Ann. Appl. Biol.* **87,** 213 (1977).
20. R. P. Waldren and A. D. Flowerday, *Agron. J.* **71,** 391 (1979).
21. Centro International de Mejoramiento de Maiz Y Trigo, *Speeding the Breeding,* Mexico City, Mexico, 1979, 20 pp.
22. G. W. Robertson, *Int. J. Biometeorol.* **12,** 191 (1968).
23. A. M. Feyerherm, E. T. Kanemasu, and G. M. Paulsen, *Response of Winter and Spring Wheat Grain Yields to Meteorological Variation. Final Report: Contract NAS9-14282, Feb., 1977. Prepared for Natl. Aeronautics and Space Admin., Johnson Space Center, Houston, Texas,* Kansas State University, Manhattan, Kan., 1977, 64 pp., 5 appendices.
24. A. M. Feyerherm, *Planting Date and Wheat Yield Models. Final Report: Contract NAS9-14533, Sept., 1977, Prepared for Natl. Aeronautics and Space Admin., Johnson Space Center, Houston, Texas,* Kansas State University, Manhattan, Kan., 1977, 46 pp., 3 appendices.
25. A. M. Feyerherm, *Application of Wheat Yield Model to United States and India. Final Report: Contract NAS9-14533, Dec., 1977, Prepared for Natl. Aeronautics and Space Admin., Johnson Space Center, Houston, Texas,* Kansas State University, Manhattan, Kan., 1977, 34 pp.
26. T. Hodges and E. T. Kanemasu, *Agron. J.* **69,** 974 (1977).
27. F. L. Milthorpe, *Ann. Bot.* **14,** 79 (1950).
28. P. C. Owen, *J. Exp. Bot.* **3,** 188 (1952).

29. A. D. Hanson, *New Phytol.* **72**, 1063 (1973).
30. M. Idris and M. Aslam, *Can. J. Bot.* **53**, 1328 (1975).
31. I. I. Chaudhuri and H. H. Wiebe, *Plant Soil* **28**, 208 (1968).
32. A. Gul and R. E. Allan, *Crop Sci.* **16**, 611 (1976).
33. A. Gul and R. E. Allan, *Crop Sci.* **16**, 615 (1976).
34. M. J. Lindstrom, R. I. Papendick, and F. E. Koehler, *Agron. J.* **68**, 137 (1976).
35. R. de Jong and K. F. Best, *Can. J. Soil Sci.* **59**, 259 (1979).
36. R. E. Prebble, *Exp. Agric.* **6**, 303 (1970).
37. T. W. Hegarty, *Plant Cell Environ.* **1**, 101 (1978).
38. M. S. Carceller and A. Soriano, *Can. J. Bot.* **50**, 105 (1972).
39. R. H. Helmerick and R. P. Pfeifer, *Agron. J.* **46**, 560 (1954).
40. L. M. Powell and R. P. Pfeifer, *Agron. J., 48*, 555 (1956).
41. J. S. Boyer and H. G. McPherson, *Adv. Agron.* **27**, 1 (1975).
42. J. C. O'Toole and T. T. Chang, *Drought and Rice Improvement in Perspective, IRRI Res. Paper Ser. No. 14*, The Int. Rice Res. Inst., Manila, Philippines, 1978, 27 pp.
43. B. B. Bayles, J. W. Taylor, and A. T. Bartel, *J. Am. Soc. Agron.* **29**, 40 (1937).
44. C. B. Kenway, H. B. Peto, and K. W. Neatby, *Can. J. Res.* **20C**, 397 (1942).
45. A. W. Platt and J. G. Darroch, *Sci. Agric.* **22**, 521 (1942).
46. S. I. Dolgov and G. B. Vinogradova, *Sov. Soil Sci.* **9**, 613 (1977) (In Engl., translated from Russian).
47. P. J. Gregory, M. McGowan, P. V. Biscoe, and B. Hunter., *J. Agric. Sci.* **91**, 91 (1978).
48. E. L. Greacen, P. Ponsana, and K. P. Barley, in O. L. Lange, L. Kappen, and E.-D. Schulze, Eds., *Water and Plant Life. Problems and Modern Approaches*, Springer-Verlag, Berlin, 1976, p. 86.
49. M. J. Pinthus, *Crop Sci.* **9**, 267 (1969).
50. P. J. Gregory, M. McGowan, and P. V. Biscoe, *J. Agric. Sci.* **91**, 103 (1978).
51. D. J. Connor, *Aust. J. Plant Physiol.* **2**, 353 (1975).
52. J. B. Passioura, *Aust. J. Agric. Res.* **23**, 745 (1972).
53. W. S. Meyer and A. M. Alston, *Agron. J.* **70**, 981 (1978).
54. N. T. Singh and G. S. Brar, *Ind. J. Agric. Sci.* **49**, 266 (1979).
55. J. H. Monyo and W. J. Whittington, *J. Agric. Sci.* **74**, 329 (1970).
56. E. A. Hurd, in T. T. Kozlowski, Ed., *Water Deficits and Plant Growth*, Vol. IV, *Soil Water Management, Plant Responses, and Breeding for Drought Resistance*, Academic Press, New York, 1976, p. 317.
57. T. F. Townley-Smith and E. A. Hurd, in H. Mussell and R. C. Staples, Eds., *Stress Physiology in Crop Plants*, Wiley-Interscience, New York, 1977, p. 447.
58. B. M. Robertson, J. G. Waines, and B. S. Gill, *Crop Sci.* **19**, 843 (1979).
59. J. MacKey, in E. R. Sears and L. M. S. Sears, Eds., *Proceedings of the Fourth International Wheat Genetics Symposium*, University of Missouri, Columbia Mo., 1973, p. 827.
60. E. A. Hurd, in K. L. Larson and J. D. Eastin, Eds., *Drought Injury and Resistance in Crops, CSSA Spec. Pub. No. 2*, Crop Science Society of America, Madison, Wis., 1971, p. 77.
61. S. Tomasovic, *Arhiva Za Poljoprivredne Nauke* **31**, 127 (1978) (English sum.).
62. G. B. Adjei and M. B. Kirkham, *Euphytica* **29**, 155 (1980).
63. O. S. Aamodt and W. H. Johnston, *Can. J. Res.* **14C**, 122 (1936).
64. L. O'Brien, *Aust. J. Agric. Res.* **30**, 587 (1979).

65. J. Hudig, *Landbouwk. Tijdschr.* **51,** 372 (1939). cited in (64).
66. H. M. Taylor and B. Klepper, *Adv. Agron.* **30,** 99 (1978).
67. G. J. Burch, *Aust. J. Agric. Res.* **30,** 279 (1979).
68. J. J. Landsberg and N. D. Fowkes, *Ann. Bot.* **42,** 493 (1978).
69. J. B. Passioura, *J. Aust. Inst. Agric. Sci.* **43,** 117 (1977).
70. W. S. Meyer, E. L. Greacen, and A. M. Alston, *J. Exp. Bot.* **29,** 1451 (1978).
71. G. P. Wind, *Neth. J. Agric. Sci.* **3,** 259 (1955).
72. H. B. So, in J. L. Harley and R. S. Russell, Eds., *The Soil-Root Interface,* Academic Press, London, 1979, p. 99.
73. W. S. Meyer and J. T. Ritchie, *Plant Physiol.* **65,** 33 (1980).
74. K. A. Seaton and J. J. Landsberg, *Aust. J. Agric. Res.* **29,** 913 (1978).
75. W. N. Herkelrath, *Water Uptake by Plant Roots,* Ph.D. Thesis, University of Wisconsin, Madison, 1975, 111 pp.
76. W. N. Herkelrath, E. E. Miller, and W. R. Gardner, *Soil Sci. Soc. Am. J.* **41,** 1033 (1977).
77. W. N. Herkelrath, E. E. Miller, and W. R. Gardner, *Soil Sci. Soc. Am. J.* **41,** 1039 (1977).
78. B. Klepper and H. M. Taylor, in J. L. Harley and R. S. Russell, Eds., *The Soil-Root Interface,* Academic Press, London, 1979, p. 51.
79. W. J. Busscher and D. D. Fritton, *Soil Sci.* **125,** 1 (1978).
80. M. B. Kirkham, in *Proceedings of the International Soil Tillage Research Organization, ISTRO,* University of Hohenheim, Stuttgart, West Germany, 1979, p. 271.
81. D. L. Karlen and D. A. Whitney, *Agron. J.* **72,** 281 (1980).
82. D. C. Thill, R. E. Witters, and R. I. Papendick, *Agron. J.* **70,** 1041 (1978).
83. W. E. Ramsay and D. E. McCloud, *Soil and Crop Sci. Soc. Flor. Proc.* **37,** 114 (1977).
84. J. T. Musick and D. A. Dusek, *Agron. J.* **72,** 45 (1980).
85. J. Sippola, T. Yläranta, and H. Jansson, *Ann. Agr. Fenniae* **17,** 158 (1978).
86. C. A. Campbell and H. R. Davidson, *Can. J. Plant Sci.* **59,** 603 (1979).
87. A. B. Frank, J. F. Power, and W. O. Willis, *Agron. J.* **65,** 777 (1973).
88. R. R. Johnson, N. M. Frey, and D. N. Moss, *Crop Sci.* **14,** 728 (1974).
89. B. S. Sandhu and M. L. Horton, *Agron. J.* **69,** 357 (1977).
90. B. S. Sandhu and M. L. Horton, *Agron. J.* **69,** 361 (1977).
91. J. Bingham, in *Proc. Fourth FAO/Rockefeller Foundation Wheat Seminar, Tehran, Iran, 21 May–2 June 1973,* Food and Agric. Org. of the United Nations, Rome, 1974, p. 127.
92. M. A. Khan and S. Tsunoda, *Jpn. J. Breeding* **20,** 305 (1970).
93. M. A. Khan and S. Tsunoda, *Jpn. J. Breeding* **20,** 344 (1970).
94. R. Kaul, *Can. J. Plant Sci.* **54,** 811 (1974).
95. Centro Internacional de Mejoramiento de Maiz Y Trigo, *CIMMYT Report on Wheat Improvement 1975. A Report on the Wheat Improvement Program of the International Maize and Wheat Improvement Center, El Batan, Mexico,* Mexico City, Mexico, 1975, 207 p.
96. E. G. Krenzer, Jr., and D. N. Moss, *Crop Sci.* **15,** 17 (1975).
97. W. Dedio, D. W. Stewart, and D. G. Green, *Can. J. Plant Sci.* **56,** 243 (1976).
98. M. Aslam and L. A. Hunt, *Planta* **141,** 23 (1978).
99. M. K. Singh and S. Tsunoda, *Photosynthetica* **12,** 280 (1978).
100. C. Planchon, *Euphytica* **28,** 403 (1979).

101. A. H. Ryhiner and M. Matsuda, *Neth. J. Agric. Sci.* **26,** 200 (1978).
102. F. E. Clements, J. E. Weaver, and H. C. Hanson, *Plant Competition. An analysis of Community Functions, Carnegie Inst. Washington Pub. No. 398,* Carnegie Inst. Washington, Washington, D. C., 1929, 340 pp.
103. E. J. Russell and D. J. Watson, *The Rothamsted Field Experiments on the Growth of Wheat, Imp. Bur. Soil Sci. Tech. Commun. No. 40,* Imperial Bureau of Soil Science, Harpenden, England, 1940, 163 pp.
104. M. E. Jensen, Ed., *Consumptive Use of Water and Irrigation Water Requirements,* American Society of Civil Engineers, New York, New York, 1973, 215 pp.
105. I. D. Teare, C. J. Peterson, and A. G. Law, *Crop Sci.* **11,** 496 (1971).
106. J. F. Ledent and M. F. Jouret, *Biol. Plant.* **20,** 287 (1978).
107. D. W. Sheriff, *J. Exp. Bot.* **28,** 1399 (1977).
108. G. Frimmel, *Angew. Botanik.* **51,** 333 (1977) (In Ger., Engl. sum.).
109. K. E. Miskin and D. C. Rasmusson, *Crop Sci.* **10,** 575 (1970).
110. M. H. Salim and G. W. Todd, *Agron. J.* **57,** 593 (1965).
111. J. E. Birdsall and K. W. Neatby, *Can. J. Res.* **C22,** 38 (1944).
112. K. E. Miskin, D. C. Rasmusson, and D. N. Moss, *Crop Sci.* **12,** 780 (1972).
113. D. Shimshi and J. Ephrat, *Agron. J.* **67,** 326 (1975).
114. H. G. Jones, *J. Appl. Ecol.* **13,** 605 (1976).
115. H. G. Jones, *J. Exp. Bot.* **28,** 162 (1977).
116. H. G. Jones, in H. Mussell and R. C. Staples, Eds., *Stress Physiology in Crop Plants,* Wiley, New York, 1979, p. 407.
117. R. A. Fischer, M. Sanchez, and J. R. Syme, *Exp. Agr.* **13,** 341 (1977).
118. O. T. Denmead and B. D. Millar, *Agron. J.* **68,** 297 (1976).
119. D. W. Lawlor and J. V. Lake, *J. Appl. Ecol.* **13,** 639 (1976).
120. Å. Nordin, *Physiol. Plant.* **37,** 157 (1976).
121. H. M. Rawson, J. E. Begg, and R. G. Woodward, *Planta,* **134,** 5 (1977).
122. M. B. Kirkham and R. M. Ahring, *Agron. J.* **70,** 657 (1978).
123. M. B. Kirkham *New Phytol.* **82,** 89 (1979).
124. H. G. Jones, *J. Agric. Sci.* **92,** 83 (1979).
125. G. W. Burton, W. W. Hanna, J. C. Johnson, Jr., D. B. Leuck, W. G. Monson, J. B. Powell, H. D. Wells, and N. W. Widstrom, *Crop Sci.* **17,** 613 (1977).
126. T. B. Paltridge and M. P. C. de Vries, *Aust. J. Agric. Res.* **24,** 633 (1973).
127. C. Willemot and L. Pelletier, *Can. J. Plant Sci.* **59,** 639 (1979).
128. J. M. Morgan, *Aust. J. Plant Physiol.* **4,** 75 (1977).
129. H. G. Jones, *J. Agric. Sci.* **88,** 267 (1977).
130. A. G. Abdel-Samie and M. Talha, *The Desert Inst. Bull. (Egypt)* **22(2),** 1 (1972).
131. W. L. Ehrler, S. B. Idso, R. D. Jackson, and R. J. Reginato, *Agron. J.* **70,** 251 (1978).
132. W. L. Ehrler, S. B. Idso, R. D. Jackson, and R. J. Reginato, *Agron. J.* **70,** 999 (1978).
133. R. Kaul and W. L. Crowle, *Z. Pflanzenzüchtg.* **65,** 233 (1971).
134. J. F. Angus and M. W. Moncur, *Aust. J. Agric. Res.* **28,** 177 (1977).
135. E. L. Skidmore, L. J. Hagen, and I. D. Teare, *Crop Sci.* **15,** 443 (1975).
136. O. T. Denmead and B. D. Millar, *Agron. J.* **68,** 307 (1976).
137. B. D. Millar and O. T. Denmead, *Agron. J.* **68,** 303 (1976).
138. S. E. Simmelsgaard, *Physiol. Plant.* **37,** 167 (1976).
139. K. A. Seaton, J. J. Landsberg, and R. H. Sedgley, *Aust. J. Agric. Res.* **28,** 355 (1977).

140. P. V. Biscoe and J. N. Gallagher, in J. J. Landsberg and C. V. Cutting, Eds., *Environmental Effects on Crop Physiology*, Academic Press, London, 1977, p. 74.
141. H. G. Jones, *J. Appl. Ecol.* **15,** 613 (1978).
142. R. A. Fischer and G. D. Kohn, *Aust. J. Agric. Res.* **17,** 255 (1966).
143. P. J. Salter and J. E. Goode, *Crop Responses to Water at Different Stages of Growth, Chapt. 2, Cereals, Res. Rev. No. 2, Commonwealth Bur. Hort. and Plantation Crops, East Malling, Maidstone, Kent,* Commonwealth Agr. Bur. Farnham Royal, Bucks, England, 1967, p. 15.
144. A. G. O. Whiteside, *Sci. Agric.* **21,** 320 (1941).
145. A. T. Bartel, *J. Agric. Res.* **74,** 97 (1947).
146. T. C. Hsiao, E. Acevedo, E. Fereres, and D. W. Henderson, *Phil. Trans. R. Soc. London Ser. B.* **273,** 479 (1976).
147. J. M. Cutler and D. W. Rains, *Crop Sci.* **17,** 329 (1977).
148. J. M. Cutler, D. W. Rains, and R. S. Loomis, *Physiol. Plant.* **40,** 225 (1977).
149. A. D. Hanson, C. E. Nelsen, and E. H. Everson, *Crop Sci.* **17,** 720 (1977).
150. I. M. Vassiliev and M. G. Vassiliev, *Plant Physiol.* **11,** 115 (1938).
151. D. W. Lawlor, *Photosynthetica* **10,** 431 (1976).
152. B. I. Shcherbakov, *Soviet Plant Physiol.* **24(Part 2)**, 87 (1977).
153. A. S. Sandhu and H. H. Laude, *Agron. J.* **50,** 78 (1958).
154. J. M. Morgan, *Nature (London)* **270,** 234 (1977).
155. J. M. Morgan, *Aust. J. Exp. Agric. Anim. Husb.* **11,** 349 (1971).
156. R. A. Fischer and G. D. Kohn, *Aust. J. Agric. Res.* **17,** 269 (1966).
157. R. A. Fischer, in R. O. Slatyer, Ed., *Plant Response to Climatic Factors,* United Nations Educational, Scientific and Cultural Organization, Paris, 1973, p. 233.
158. J. Calvert, *Protoplasma* **24,** 505 (1935).
159. G. W. Todd, F. W. Ingram, and C. A. Stutte, *Proc. Okla. Acad. Sci.* **42,** 55 (1961).
160. W. Dedio, *Can. J. Plant Sci.* **55,** 369 (1975).
161. J. J. Lehane and W. J. Staple, *Can. J. Soil Sci.* **42,** 180 (1962).
162. R. D. Asana and R. N. Basu, *Ind. J. Plant Physiol.* **6,** 1 (1963).
163. D. Aspinall, P. B. Nicholls, and L. H. May, *Aust. J. Agric. Res.* **15,** 729 (1964).
164. D. Aspinall, *Aust. J. Agric. Res.* **16,** 265 (1965).
165. M. J. Brown and R. L. Vanderlip, *Agron. J.* **61,** 95 (1969).
166. C. A. Campbell and W. S. Ferguson, *Can. J. Plant Sci.* **49,** 129 (1969).
167. C. A. Campbell, D. S. McBean, and D. G. Green, *Can. J. Plant Sci.* **49,** 29 (1969).
168. A. D. Day and S. Intalap, *Agron. J.* **62,** 27 (1970).
169. G. M. Wright, *N.Z. J. Agric. Res.* **15,** 43 (1972).
170. E. Seif and D. G. Pederson, *Aust. J. Agric. Res.* **29,** 1107 (1978).
171. R. A. Peck and M. B. Kirkham, *Proc. Okla, Acad. Sci.* **59,** 53 (1979).
172. D. M. Greene and M. B. Kirkham, *Irrigation Sci.* **1,** 241 (1980).
173. O. T. Bonnett, *Inflorescences of Maize, Wheat, Rye, Barley, and Oats: Their Initiation and Development. Ill. Agr. Exp. Stn. Bull. 721,* 1966, 105 pp.
174. R. D. Asana and V. S. Mani, *Ind. J. Genet. Plant Breeding* **18,** 187 (1958).
175. R. D. Asana and A. D. Saini, *Physiol. Plant.* **11,** 666 (1958).
176. R. D. Asana, A. D. Saini, and D. Ray, *Physiol. Plant.* **11,** 655 (1958).
177. R. A. Fischer and G. D. Kohn, *Aust. J. Agric. Res.* **17,** 281 (1966).
178. W. Baier and G. W. Robertson, *Can. J. Plant Sci.* **47,** 617 (1967).

179. C. A. Campbell, *Can. J. Plant Sci.* **48**, 313 (1968).

180. E. L. Smith, *The Genetics of Wheat Architecture*, in *The Grasses and Grasslands of Oklahoma, Ann. Okla. Acad. Sci. Pub. No. 6*, The Samuel Roberts Noble Foundation, Ardmore, Okla., 1976, p. 117.

181. J. Bingham, *Ann. Appl. Biol.* **57**, 365 (1966).

182. E. L. Skidmore, L. J. Hagen, D. G. Naylor, and I. D. Teare, *Agron. J.* **66**, 501 (1974).

183. W. J. Staple and J. J. Lehane, *Can. J. Agric. Sci.* **34**, 552 (1954).

184. A. Hadjichristodoulou, *J. Agric. Sci.* **87**, 489 (1976).

185. R. D. Jackson, R. J. Reginato, and S. B. Idso, *Water Resources Res.* **13**, 651 (1977).

186. D. R. Laing and R. A. Fischer, *Euphytica* **26**, 129 (1977).

187. J. B. Bole and S. Dubetz, *Can. J. Soil Sci.* **58**, 321 (1978).

188. V. P. Rasmussen and R. J. Hanks, *Agron. J.* **70**, 940 (1978).

189. R. A. Fischer, *J. Aust. Inst. Agric. Sci.* **45**, 83 (1979).

190. R. M. Gifford, *Aust. J. Plant Physiol.* **6**, 367 (1979).

191. B. W. Greb, *Reducing Drought Effects on Croplands in the West-Central Great Plains, Agr. Information Bull. 420*, U. S. Dept. Agr., Sci. Educ. Admin., Washington, D.C., 1979, 31 pp.

192. R. A. Christ, *J. Exp. Bot.* **29**, 611 (1978).

193. I. Zelitch, in I. Zelitch, Ed., *Stomata and Water Relations in Plants, Conn. Agric. Exp. Stn. Bull. 664*, New Haven, Connecticut, 1963, p. 18.

194. J. Gale and R. M. Hagan, *Annu. Rev. Plant Physiol.* **17**, 269 (1966).

195. F. Bidinger, R. B. Musgrave, and R. A. Fischer, *Nature (London)* **270**, 431 (1977).

196. H. M. Rawson, A. K. Bagga, and P. M. Bremner, *Aust. J. Plant Physiol.* **4**, 389 (1977).

197. G. N. Thorne, *Physiology of Grain Yield of Wheat and Barley*, in *Rothamsted Report for 1973, Part 2*, Rothamsted Exp. Sta., Harpenden Herts., England, 1973, p. 5.

198. J. N. Gallagher, P. V. Biscoe, and B. Hunter, *Nature (London)* **264**, 541 (1976).

199. R. R. Johnson and D. N. Moss, *Crop Sci.* **16**, 697 (1976).

200. I. F. Wardlaw, *Aust. J. Biol. Sci.* **24**, 1047 (1971).

201. L. T. Evans, J. Bingham, P. Jackson, and J. Sutherland, *Ann. Appl. Biol.* **70**, 67 (1972).

202. L. T. Evans, in L. T. Evans, Ed., *Crop Physiology. Some Case Histories*, Cambridge University Press, London, 1975, p. 327.

203. L. T. Evans, I. F. Wardlaw, and R. A. Fischer, in L. T. Evans, Ed., *Crop Physiology. Some Case Histories*, Cambridge University Press, London, 1975, p. 101.

204. H. G. McPherson and J. S. Boyer, *Agron. J.* **69**, 714 (1977).

205. R. H. McNeal and M. A. Berg, *Euphytica* **28**, 89 (1979).

206. S. J. Yang and E. de Jong, *Can. J. Soil Sci.* **51**, 211 (1971).

207. A. J. Segovia and R. H. Brown, *Crop Sci.* **18**, 90 (1978).

208. L. T. Evans, R. L. Dunstone, H. M. Rawson, and R. F. Williams, *Aust. J. Biol. Sci.* **23**, 743 (1970).

209. J. W. Patrick, *Aust. J. Biol. Sci.* **25**, 455 (1972).

210. I. F. Wardlaw, in I. F. Wardlaw and J. B. Passioura, Eds., *Transport and Transfer Processes in Plants*, Academic Press, New York, 1976, p. 381

211. H. M. Neghassi, D. F. Heermann, and D. E. Smika, *Trans. Am. Soc. Agric. Eng.* **18**, 549 (1975).

212. A. Livnè and Y. Vaadia, in T. T. Kozlowski, Ed., *Water Deficits and Plant Growth*, Vol. 3, *Plant Responses and Control of Water Balance*, Academic Press, New York, 1972, p. 255.

213. A. W. Naylor, in T. T. Kozlowski, Ed., *Water Deficits and Plant Growth*, Vol. 3, *Plant Responses and Control of Water Balance*, Academic Press, New York, 1972, p. 241.
214. S. A. Quarrie and H. G. Jones, *Ann. Bot.* **44,** 323 (1979).
215. C. Bengston, S. O. Falk, and S. Larsson, *Physiol. Plant.* **41,** 149 (1977).
216. C. Bengtson, B. Klockare, R. Klockare, S. Larsson, and C. Sundqvist, *Physiol. Plant.* **43,** 205 (1978).
217. M. G. Pitman and D. Wellfare, *J. Exp. Bot.* **29,** 1125 (1978).
218. H. Göring and K. Zoglauer, *Biochem. Physiol. Pflanzen* **174,** 568 (1979).
219. S. T. C. Wright, *Planta* **144,** 179 (1979).
220. A. D. Hanson and C. E. Nelsen, *Plant Physiol.* **62,** 305 (1978).
221. R. G. Wyn Jones and R. Storey, *Aust. J. Plant Physiol.* **5,** 817 (1978).
222. W. D. Hitz and A. D. Hanson, *Phytochemistry* **19,** 2371 (1980).
223. R. van Brackle, Ed., *Western Hay and Grain Grower. Southern Edition,* **9(2),** 22 (1980).
224. W. R. Luellen, Ed., *Crops and Soils* **32(4),** 23 (1980).
225. D. Shimshi and U. Kafkafi, *Irrigation Sci.* **1,** 27 (1978).
226. H. R. B. Hack, *Ann. Bot.* **42,** 509 (1978).
227. W. Larcher, *Physiological Plant Ecology*, Springer-Verlag, Berlin, 1975, 252 pp.
228. H. W. Woolhouse, *Endeavor* (*N.S.*) **2,** 35 (1978).
229. C. C. Black, *Adv. Ecol. Res.* **7,** 87 (1971).
230. H. L. Shantz and L. N. Piemeisel, *J. Agric. Res.* **34,** 1093 (1927).

GLOSSARY

AERENCHYMA. Lysigenous gas spaces through which oxygen diffuses to supply submerged roots with oxygen.

ALBEDO. Portion of incoming solar radiation reflected from the surface of an object.

ANASTAMOSING. Interconnection of branches to form a net.

APOPLAST. Part of the plant tissue matrix outside the membranes enclosing the cell contents, i.e. outside the plasmalemma.

AVAILABLE SOIL MOISTURE. Water retained by a soil between field capacity and the wilting point.

AXIAL. In the direction of the main axis.

AXIAL RESISTANCE. Resistance to water flow along the lengthwise axis of plant roots or stems.

BOLL. Seed-containing fruit of cotton.

BOUNDARY LAYER. Layer of air near the earth's surface in which the fluid (air) is affected by shearing forces originating at the surface. In this book the term refers to the boundary layer developed within a few meters of a cropped surface (internal boundary layers) rather than the earth's boundary layer (external boundary layers).

BOWEN RATIO. Ratio of sensible heat to latent heat.

BULK DENSITY. Mass of oven-dry soil per unit bulk volume including the pore space.

CATENA. Closely linked series of components in a biological or physical system which together constitutes an identifiable association, e.g., the soil–plant–atmosphere continuum in the water cycle.

COLEORHIZA. Sheath investing the radicle in some plants through which the roots burst.

CONSUMPTIVE USE. Total amount of water lost from the surface of a crop. Calculations of consumptive use consider crop differences through use of crop consumptive use coefficients that apply to specific crops.

CUTICULAR RESISTANCE. Resistance to the diffusion of a gas (usually water vapor or carbon dioxide) through the cuticle. The values given in Chapter 3 are for water vapor.

CUTOUT. Cessation of rapid vegetative development in cotton often associated with fruiting.

DEVELOPMENTAL PLASTICITY. Ability of a plant to change its phenological development to suit the environmental conditions.

DORSIVENTRAL. Having distinct upper and lower sides, as for a cotton leaf.

DRIVING FORCE. Potential gradient between two reference points in the soil.

DRY ADIABATIC LAPSE RATE. Rate of decrease of temperature with height of a parcel of dry air lifted adiabatically (without the transfer of heat across the boundaries of the parcel) through the atmosphere. The rate is 9.8°C per km.

ELASTICITY. Ability of cell structures to resume their original size and shape after deformation.

EMISSIVITY. Ratio of the emittance of a given surface at a specified wavelength and temperature to the emittance of an ideal blackbody at the same wavelength and temperature.

ENDARCH (XYLEM). Arrangement of vessels within a xylem strand in which maturation of the cells progresses centrifugally, i.e., the oldest vessels are closest to the center of the axis.

EVAPORATION. Physical process for the change of liquid water to gaseous water vapor.

EVAPOTRANSPIRATION. Process of water vapor transfer into the atmosphere from vegetated land surfaces. It includes water evaporated from the soil surface and water transpired by plants.

FETCH. Upwind distance between the leading edge of a field and the location at which measurements are made.

FIELD CAPACITY. Near-saturated soil water condition where gravitational forces have removed as much water from a soil as possible, usually corresponding to a matrix potential of 0.1 to 0.2 bar.

FREE SPACE. Space existing in a plant tissue matrix outside the cell membranes and occurring between or around structural fibres.

FREE ENERGY. Force tending to return a system to equilibrium.

FULL TURGOR. Turgor of a fully hydrated cell, i.e., the turgor potential at a relative water content of 100% or a water potential of zero.

GOSSYPOL. Poisonous pigment ($C_{30}H_{30}O_8$) found in lysigenous glands in cotton.

GRAVITATIONAL POTENTIAL (ψ_g). Force required to maintain soil water suspended in the soil profile above some gravitational elevation.

HYDRAULIC CONDUCTIVITY. Expression of the ability for water flow per unit time through a unit area of soil in response to a given potential gradient. Also known by the following terms: permeability, transmissivity, or conductivity.

HYDROSTATIC HEAD. Positive force exerted on tension-free soil water due to the mass of tension-free soil water above a reference point.

HYSTERESIS. Phenomenon describing why water potential is not a unique function of soil water content, but also demonstrates dependency on the previous moisture history of the soil.

INCOMING LONG-WAVE RADIATION. Radiation from the sky, concentrated primarily in the 3.0–4.0 μm waveband of the electromagnetic spectrum, which is received at the earth's surface.

INTERSTICES. Small spaces between adjoining cells in a plant tissue matrix.

ISOTHERMAL. Of equal or constant temperature. In this book it refers to constant temperature with increasing height above the ground.

LATENT HEAT FLUX. Transfer of energy in the evaporation process. The amount of energy transferred is equal to the latent heat of vaporization times the quantity of water evaporated.

LEAF PLASTOCHRON. Number of days between successive leaf appearances.

LEAF RESISTANCE. Resistance to the diffusion of a gas (usually water vapor or carbon dioxide) through the epidermis, i.e., the cuticle and stomata.

LONG-WAVE RADIATION. Radiation that is concentrated primarily in the 3.0–40 μm wavelength range of the electromagnetic spectrum.

LYSIGENOUS (GLAND). Glands within the intercellular spaces formed by the dissolution of surrounding cell walls.

MATRIC POTENTIAL (ψ_m). Attractive force between soil water and soil solids that reduces the free energy of soil water.

MATRIX. Solid portion of soil comprising the volumes and surfaces of the bulk material.

MICRONAIRE. Measure of fiber fineness.

NECTARY. Multicellular, glandular structure that secretes a sugary liquid.

NET RADIATION. Net difference between the total upward and downward radiation fluxes. It is the sum of the net short-wave radiation balance plus the net long-wave radiation balance.

NODAL ROOTS. Roots initiated on the basal nodes of the plant stem.

ONTOGENY. Course of development of a cell, organ, or organism.

PAN EVAPORATION. Water evaporated from an exposed pan, generally a class A Weather Bureau pan. The amount of water evaporated from a pan is often interpreted to estimate potential evapotranspiration.

PED. Unit of soil structure such as a naturally formed aggregate.

PERMANENT WILTING COEFFICIENT. Minimum water content at which a plant is not capable of extracting sufficient water to prevent wilting, usually corresponds to a soil water potential of −15 bars.

PHENOLOGICAL DEVELOPMENT. Development of a plant through the stages of floral initiation, flowering, fruit set, and maturity.

PHYLLOTAXY. Arrangement of leaves on a stem.

PLANT-AVAILABLE WATER SUPPLY. Amount of available soil moisture.

PLASMALEMMA. Thin membrane enclosing the cell contents.

PLASMODESMATA. Small connections between adjacent cells bound by a continuous outer membrane, i.e., the plasmalemma, and strand of cytoplasm.

PSYCHROMETRIC CONSTANT. Defined as the ratio $C_P P/(LM_w/M_a)$ where C_P is the specific heat of air, P is the atmospheric pressure, L is the latent heat of vaporization, M_w is the molecular weight of water, and M_a is the molecular weight of dry air.

RADIATION. Process by which heat is transferred from one surface to another through space by electromagnetic waves. A surface will radiate energy at a rate proportional to the fourth power of its absolute temperature.

RHIZOSPHERE. Zone surrounding the root surface. It includes the space between the root and soil that is sometimes occupied by mucilaginous root exudates and the layer of soil adjacent to the root surface which is often occupied by root hairs.

ROOT PLASTOCHRON. Number of days between successive root appearances.

SATURATED FLOW. Movement of soil water under conditions where gravitational potential exceeds matrix potential.

SEMINAL ROOTS. Roots originating in the seed embryo.

SENSIBLE HEAT FLUX. Transfer of heat from the air to a surface if the air is warmer than the surface or from the surface to the air if the surface is warmer than the air.

SHORT-WAVE RADIATION. Radiation that is concentrated primarily in the 0.3–4.0 μm wavelength range of the electromagnetic spectrum.

SOIL HEAT FLUX. Transfer of heat into or out of the soil through the conduction process.

SOIL MOISTURE. Wetness, or relative water content, of the soil expressed in relation to soil mass or soil volume.

SOIL MOISTURE RETENTION CURVE. Graphical relationship between total soil water potential and soil water content.

SOIL–PLANT–ATMOSPHERE CONTINUUM. Association between the soil, plant, and atmosphere by virtue of a common process or component.

SOLUTE POTENTIAL (ψ_s). Reduction in the free energy of soil water due to electrostatic attraction/repulsion of dissolved solutes and soil water molecules.

SQUARE. Unopened floral bud of cotton enclosed in bracts.

STOMATAL RESISTANCE. Resistance to the diffusion of a gas (usually water vapor or carbon dioxide) through the stomatal pores. The values in Chapter 3 are for water vapor.

SYMPLAST. Component of plant tissue enclosed within the plasmalemma. It is continuous between cells by virtue of connecting plasmodesmata.

TENSION. Force of attraction between soil water and soil solids.

TETRARCH. Arrangement of protoxylem strands into four separate poles.

TONOPLAST. Inner cell membrane enclosing the vacuole.

TORTUOSITY. "Roundabout" flow path water actually takes while flowing through the series of pores in a soil, compared to an apparent straight flow path.

TOTAL SOIL WATER POTENTIAL (ψ_{soil}). Specific potential energy of soil water as compared to free water, representing a sum of all factors in the soil system capable of decreasing the potential energy of free water.

TRANSPIRATION. Evaporation of water that has been absorbed through plant roots and transported through the plant from the leaves.

TRICHOME. Hairlike projections originating from the epidermis.

UNSATURATED FLOW. Movement of soil water while under potentials less than -0.1 bar but less than -31 bars. Also known as capillary conductivity.

VAPOR PRESSURE. Pressure exerted by the molecules of a vapor. In a mixture of gases it is the vapor's contribution to the total pressure (i.e., its partial pressure). In this book vapor pressure refers to the partial pressure of water vapor in the atmosphere.

WATER CONTENT (WEIGHT, %). Ratio of the weight of water in a soil sample to the oven-dry weight of that sample, expressed as a percentage.

WATER POTENTIAL. Potential energy per unit mass of water with reference to pure water at zero potential. The gradient of water potential is the driving force for water movement into and through a plant.

WILTING POINT. Soil water potential where plants are unable to extract water.

ZERO TURGOR. Turgor potential of zero, i.e., when the leaf water potential just equals the osmotic potential of a cell.

INDEX